Sustainable Development and Biodiversity

Volume 10

Series editor

Kishan Gopal Ramawat
Botany Department, M.L. Sukhadia University, Udaipur, Rajasthan, India

This book series provides complete, comprehensive and broad subject based reviews about existing biodiversity of different habitats and conservation strategies in the framework of different technologies, ecosystem diversity, and genetic diversity. The ways by which these resources are used with sustainable management and replenishment are also dealt with. The topics of interest include but are not restricted only to sustainable development of various ecosystems and conservation of hotspots, traditional methods and role of local people, threatened and endangered species, global climate change and effect on biodiversity, invasive species, impact of various activities on biodiversity, biodiversity conservation in sustaining livelihoods and reducing poverty, and technologies available and required. The books in this series will be useful to botanists, environmentalists, marine biologists, policy makers, conservationists, and NGOs working for environment protection.

More information about this series at http://www.springer.com/series/11920

Vijay Rani Rajpal · S. Rama Rao
S.N. Raina
Editors

Gene Pool Diversity and Crop Improvement

Volume 1

Springer

Editors
Vijay Rani Rajpal
Delhi University
Delhi
India

S.N. Raina
Amity University of Biotechnology
Noida, Uttar Pradesh
India

S. Rama Rao
Department of Biotechnology
and Bioinformatics
North Eastern Hill University
Shillong, Megalaya
India

ISSN 2352-474X ISSN 2352-4758 (electronic)
Sustainable Development and Biodiversity
ISBN 978-3-319-27094-4 ISBN 978-3-319-27096-8 (eBook)
DOI 10.1007/978-3-319-27096-8

Library of Congress Control Number: 2015957776

Printed on acid-free paper

This Springer imprint is published by SpringerNature
The registered company is Springer International Publishing AG Switzerland

Preface

To respond to an ever-increasing need of food and fibre by the growing world's population, standing today at 7 billion and expected to reach 9 billion by 2050, there is a pressing need to increase crop productivity. This can be achieved by developing cultivars with better grain yield and high nutritive value through plant breeding in traditional crop plants and in supplementary crops identified by International Plant Genetic Resources Institute (IPGRI) and Consultative Group on International Agriculture (CGIAR). The genetic improvement can also be achieved by using biotechnological tools for alien gene transfer and engineering the traits in plants that are otherwise difficult using conventional plant breeding approaches, by newer methods to precisely and rapidly screen for traits of interest in the progeny, and by cytogenetic manipulations. Knowledge of genetic diversity, locked in the germplasm resources of the crop plants and wild relatives constituting primary, secondary, and tertiary gene pools, and the genome(s) characterization is essential for crop improvement and developing gene transfer strategies by multitude of tools that are now available.

This book addresses aforementioned issues in several crop species. Each chapter elucidates an authoritative account on the topic. We are sincerely grateful to all the authors for their valuable contributions. We would like to acknowledge cooperation, patience, and support of our contributors, who have put in their serious efforts to ensure a high scientific quality of this book with up-to-date information. We thank Dr. K.G. Ramawat for motivating us to take up this assignment. Sincere thanks are due to Khushboo Arora for her help during the editing process. This work could not be completed without the active support of Springer team who took pains in streamlining the production process. We particularly appreciate Dr. Valeria for her continued support. Vijay Rani Rajpal is sincerely grateful to her husband Susheel Rajpal and daughter Navya Rajpal for their patience and support during the entire period of this book project.

Plant breeders, taxonomists, geneticists, cytogeneticists, molecular biologists, and biotechnologists will greatly benefit from this book. We sincerely hope that this book will serve as a milestone towards achieving meaningful plant genetic improvement to meet the ever-increasing requirements of food and fibre of this world.

Vijay Rani Rajpal
S. Rama Rao
S.N. Raina

Contents

Contributors

Renuka Agrawal Department of Botany, Miranda House, University of Delhi, Delhi, India

Anila Badiyal Molecular Cytogenetics and Tissue Culture Laboratory, Department of Crop Improvement, CSK Himachal Pradesh Agricultural University, Palampur, India

S.S. Banga Department of Plant Breeding and Genetics, Punjab Agricultural University, Ludhiana, India

Shashi Banga Department of Plant Breeding and Genetics, Punjab Agricultural University, Ludhiana, India

Igor V. Bartish Department of Genetic Ecology, Institute of Botany, Academy of Sciences of Czech Republic, Pruhonice, Czech Republic

Niti Pathak Bhaduri Department of Botany, University of Delhi, Delhi, India

Vinay Bhardwaj Central Potato Institute Shimla, Shimla, HP, India

Atul Bhargava Amity University Uttar Pradesh (Lucknow Campus), Lucknow, Uttar Pradesh, India

Kangila Venkataramana Bhat National Bureau of Plant Genetic Resources, Pusa Campus, New Delhi, India

Harinder Kumar Chaudhary Molecular Cytogenetics and Tissue Culture Laboratory, Department of Crop Improvement, CSK Himachal Pradesh Agricultural University, Palampur, India

Upasna Chettry Plant Molecular Biology Laboratory, UGC-Centre for Advanced Studies in Botany, North-Eastern Hill University, Shillong, India

Nikhil K. Chrungoo Plant Molecular Biology Laboratory, UGC-Centre for Advanced Studies in Botany, North-Eastern Hill University, Shillong, India

Lashaihun Dohtdong Plant Molecular Biology Laboratory, UGC-Centre for Advanced Studies in Botany, North-Eastern Hill University, Shillong, India

R.N. Gohil Department of Botany, University of Jammu, Jammu, India

Yasir Serag Alnor Gorafi Arid Land Research Center, Tottori University, Tottori, Japan; Agricultural Research Corporation, Wad Medani, Sudan

Neha Goyal Department of Environmental Studies, University of Delhi, Delhi, India

Navdeep Singh Jamwal Molecular Cytogenetics and Tissue Culture Laboratory, Department of Crop Improvement, CSK Himachal Pradesh Agricultural University, Palampur, India

Vineeta Kaila Molecular Cytogenetics and Tissue Culture Laboratory, Department of Crop Improvement, CSK Himachal Pradesh Agricultural University, Palampur, India

Veenu Kaul Department of Botany, University of Jammu, Jammu, India

SM Paul Khurana Faculty of Science, Engineering & Technology, University Science Instrumentation Centre, Amity University, Gurgaon, Haryana, India

Sachin Kumar Department of Botany, University of Delhi, Delhi, India

Suman Lakhanpaul Department of Botany, University of Delhi, Delhi, India

Ankur Maheshwari Department of Zoology, Zakir Husian Delhi College, University of Delhi, New Delhi, India

Sonal Malik Department of Botany, University of Delhi, Delhi, India

S. Manickam ICAR-Central Institute for Cotton Research, Regional Station, Coimbatore, India

S.N. Nigam Hyderabad, Telangana, India

Deepak Ohri Amity University Uttar Pradesh (Lucknow Campus), Lucknow, UP, India

A.H. Prakash ICAR-Central Institute for Cotton Research, Regional Station, Coimbatore, India

Shoukat Ahmad Rather Molecular Cytogenetics and Tissue Culture Laboratory, Department of Crop Improvement, CSK Himachal Pradesh Agricultural University, Palampur, India

Gyan Prakash Sharma Department of Environmental Studies, University of Delhi, Delhi, India

Amrita Singh Department of Botany, University of Delhi, Delhi, India

Anurudh K. Singh Gurgaon, Haryana, India

BP Singh Central Potato Institute Shimla, Shimla, HP, India

Vibhuti Singh Department of Botany, University of Delhi, Delhi, India

Avinash Srivastava Central Potato Research Station, Shillong, Meghalaya, India

Hisashi Tsujimoto Arid Land Research Center, Tottori University, Tottori, Japan

Sooraj Umdale National Bureau of Plant Genetic Resources, Pusa Campus, New Delhi, India

Ruchi Vir Department of Botany, University of Delhi, Delhi, India

Chapter 1
Leymus racemosus: A Potential Species of Gene Pool Enrichment for Wheat Improvement

Yasir Serag Alnor Gorafi and Hisashi Tsujimoto

Abstract *Leymus racemosus* is a wild species belonging to tribe Triticeae (Poaceae), which includes important cereal crops such as bread wheat and barley. This perennial species grows along the coast and in dry lands, and is reportedly tolerant to various biotic and abiotic stresses. Although *L. racemosus* is evolutionarily distant from wheat (*Triticum* spp.) within the tribe, it has been successfully hybridized with wheat, and several wheat–*L. racemosus* chromosome introgression lines have been selected from among the backcrossed progenies of the hybrid. *L. racemosus* is, therefore, a promising wild species for wheat improvement. The production of wheat, as one of the world's most important staple cereals, must increase to provide food for growing population, but wheat is threatened by the effects of climate change, especially increased drought and heat. In addition, depletion of natural resources makes the likelihood of success for increasing future productivity unclear. Maximizing yield to help achieve food security under such challenging conditions will not be easy. Wheat productivity could be improved by enhancing tolerance to biotic and abiotic stress, and tolerance to soil macro- and micronutrient deficiencies or toxicities. However, the genetic base of variation available for most of these traits is very narrow in the available elite germplasm. Wild relatives of wheat, including *L. racemosus*, are an important source of wheat genetic variation and have contributed much to the improvement of wheat productivity. This review describes the potential of this wild wheat relative as a germplasm for wheat improvement, and the characteristics that affect its efficient use in breeding, including its chromosomes, traits, and some genes that have been identified and transferred from this species to common wheat.

Keywords *Leymus* · Wheat · Germplasm enhancement · Drought · Chromosome introgression line

---

Y.S.A. Gorafi · H. Tsujimoto (✉)
Arid Land Research Center, Tottori University, 1390 Hamasaka, Tottori 680-0001, Japan
e-mail: tsujim@alrc.tottori-u.ac.jp

Y.S.A. Gorafi
Agricultural Research Corporation, PO Box 126, Wad Medani, Sudan

V.R. Rajpal et al. (eds.), *Gene Pool Diversity and Crop Improvement*,
Sustainable Development and Biodiversity 10,
DOI 10.1007/978-3-319-27096-8_1

1.1 Introduction

Wheat (*Triticum* spp.) is the world's most widely grown grain and provides nearly 20 % of the calories humans require (FAOstat 2007). The world's population is growing exponentially and will exceed nine billion by 2050. According to recent FAO estimates, projected demand will require global agricultural production to increase by about 60 % relative to yields obtained between 2005 and 2007. Between 2010 and 2012, nearly one in eight people in the world (870 million) had less than the minimum amount of food required. Of these, 852 million comprised up to 15 % of the total population of developing countries (Ogbonnaya et al. 2013). Further, impact of climate change is expected to reduce wheat gain yield (Yang et al. 2013; Lobell et al. 2011), which will further complicate food production in the future. The requirement for higher productivity under more limiting conditions, whether by increasing yield or acreage, is an urgent objective that must be met. However, the enhancement of yield is limited by biotic and abiotic stresses and limited availability of arable land.

The modern bread wheat (*Triticum aestivum*) , which accounts for 95 % of the production of all wheat species, has evolved only 8500 years ago from a rare interspecific hybridization event between tetraploid and diploid wheat (Kihara 1944, 1966; McFadden and Sears 1944; Qi et al. 2007). Because this event is thought to have occurred only once (Curtis and Halford 2014), all modern wheat is thought to have descended from that single cross. Consequently, the genetic variability of hexaploid bread wheat germplasm has a rather narrow base, and after domestication and several decades of extensive selection and breeding, this already limited genetic base has become even narrower (Reif et al. 2005). With such a narrow base, the required increases in wheat yields are hard to imagine. Therefore, there is now growing interest in the great genetic diversity present in wild relatives of wheat. Certain traits in some of these wild relatives have already been characterized and successfully used to improve the tolerance of wheat to abiotic (such as heat, salinity, and drought) and biotic stresses (e.g., diseases, insects, and kit-nematodes), and to improve nutrient use efficiency, grain yield, and bread making quality (Eastwood et al. 1994; Jiang et al. 1994; Cox et al. 1995; Gatford et al. 2002; Marais et al. 1994; Martín-Sánchez et al. 2003; Dreccer et al. 2004; Wang et al. 2010; Garg et al. 2009; Liu et al. 2013), via molecular biology and cytogenetic techniques.

This review sheds light on the potential of an important wild relative of wheat, *L. racemosus* for wheat improvement, and describes several characteristics; including its chromosomes, useful traits, and genes identified to date that could be transferred to improve common wheat.

1.2 Botanical Information for *Leymus racemosus*

Among the wild relatives of wheat, the genus *Leymus* (wildrye) is a distantly related source of many potentially adaptive alien genes that could be useful for the breeding of common wheat. *Leymus* is an allopolyploid genus with chromosomes numbers ranging from $2n = 4\times = 28$ to $2n = 12\times = 84$ in the tribe Triticeae, which comprises about 50 species and subspecies worldwide (Löve 1984). The genome of tetraploid *L. racemosus* is comprised of two genomes of different origins ($2n = 28$, NsNsXmXm); the Ns genome originated from *Psathyrostachys*, while the origin of the Xm genome remains unknown (Zhang and Dvorak 1991; Wang and Jensen 1994; Wang et al. 1994, 2006).

L. racemosus (Lam.) Tzvelev (Fig. 1.1a), is a perennial grass that is also known by the common name 'mammoth wildrye' as well as other names including *Elymus arenarius* var. *giganteus*, *E. giganteus*, *E. racemosus*, and *L. giganteus* (PLANTS Database 2010; http://plants.usda.gov/). The species *L. racemosus* distributes in a range of dry coastal or inland sites from Central Asia to Eastern Europe (Kishii et al. 2004). It is also found growing in a variety of environments including saline, alkaline, dry, or semi-dry lands and forests (Fan et al. 2009). Different from the other species, in *Leymus* only *L. racemosus* and *L. mollis* have exceptionally large spikes, strong rhizomes, and vigorous growth (Fig. 1.1b).

L. racemosus has been used for ecological purposes such as stabilizing and protecting sand dunes from wind erosion, revegetating mine tailings, and making firebreaks and green strips. It is also used to create wildlife habitat for birds (St. John 2010). Monsen et al. (2004) reported that *L. racemosus* is generally unsuitable and unpalatable for grazing animal.

Fig. 1.1 Morphology of *Leymus racemosus* and its addition lines. **a** *L. racemosus* maintained in a pot of 50 cm diameter. This plant is perennial, stress-tolerant and forms long rhizomes as indicated by the *arrow*. **b** The long rhizome of *L. mollis*, the species most closely resembling *L. racemosus*. **c** Spikes of wheat–*L. racemosus* chromosome introgression lines (CILs). The morphology of the CILs is similar to common wheat

L. racemosus is quite salt- and drought-tolerant (McGuire and Dvorak 1981) and is also resistant to various diseases including scab (Mujeeb-Kazi et al. 1983). Our observations of *L. racemosus* plants growing at the Arid Land Research Center, Tottori, Japan (35°32′N, 134°13′E), confirmed the ability of this species to tolerate heat stress; plants grew vigorously and flowered from May to June, and grain-filling, which is particularly a heat-sensitive stage, succeeded from July through August during high summer temperatures and humidity.

1.3 Crossability of *Leymus racemosus* with Wheat and Chromosome Introgression Lines

Several *Leymus* species, including *L. racemosus,* have been successfully hybridized and backcrossed with wheat. The first pentaploid hybrid between wheat and *L. racemosus*, which included the genomes ABDNsXm was developed by Mujeeb-Kazi and Rodriguez (1981). The second attempt of hybridization was made in China, aimed at introducing the genes for scab resistance from *L. racemosus* into the common wheat. Wang et al. (1986, 1991) characterized the cytogenetics of the F_1 hybrid between wheat and *L. racemosus* and its BC_1 and BC_2 derivatives. Subsequently, seven disomic addition lines and two di-telosomic addition lines were developed and screened for scab resistance (Chen et al. 1993, 1995; Ren et al. 1996; Sun et al. 1997, 1998). Lu et al. (1995) developed one disomic addition line and one double-disomic addition–substitution line via anther culture. Another five disomic addition lines and several additional double-disomic addition, di-telosomic addition, and disomic substitution lines were characterized in 23 lines derived from the bread wheat cultivar 'Chinese Spring' (CS) (Qi et al. 1997). Kishii et al. (2004) produced ten chromosome addition lines using a *L. racemosus* accession collected along the Bulgarian Black Sea coast (Fig. 1.1c). Table 1.1 summarizes the wheat–*L. racemosus* chromosome introgression lines maintained in the Tottori Alien Chromosome Bank for Wheat (TACBOW) that is funded by the National Bio-resources project (NBRP-wheat), and the chromosome information for each.

1.4 Use of Genes from *Leymus racemosus* for Wheat Improvement

There are two possible strategies currently available to use genes from *L. racemosus*, or other wild relatives of wheat, for wheat improvement. In the first approach, the desired phenotype of wild species is identified by direct examination of the species. This is followed by crossing wheat and the wild species to produce a sterile F_1 hybrid by embryo rescue. Then, the F_1 plants are repeatedly backcrossed to wheat to generate CILs (including addition, substitution, and translocation lines) that harbor

Table 1.1 List of the wheat–*Leymus* chromosome introgression lines produced or maintained at the Tottori Alien Chromosome Bank of Wheat supported by NBRP-wheat

Strain ID	Strain name and chromosome name	Homoeologous group	Number of chromosomes	References
TACBOW0001[a]	*Leymus racemosus* A addition	2	44	Kishii et al. (2004)
TACBOW0003	*L. racemosus* E addition	ND[b]	44	Kishii et al. (2004)
TACBOW0004	*L. racemosus* F addition	4	44	Kishii et al. (2004)
TACBOW0005	*L. racemosus* H addition	3	44	Kishii et al. (2004)
TACBOW0006	*L. racemosus* I addition	5	44	Kishii et al. (2004)
TACBOW0008	*L. racemosus* k addition	6	44	Kishii et al. (2004)
TACBOW0009	*L. racemosus* l addition	2	44	Kishii et al. (2004)
TACBOW0010	*L. racemosus* n addition	3,7	44	Kishii et al. (2004)
TACBOW0011	*L. racemosus* H substitution	3	42	Kishii et al. (2004)
TACBOW0012 (TA#7643)[a]	*L. racemosus* 2Lr#1 addition	2	44	Qi et al. (1997)
TACBOW0013 (TA#7646)	*L. racemosus* 5Lr#1 addition	5	44	Qi et al. (1997)
TACBOW0014 (TA#7648-1)	*L. racemosus* 7Lr#1 addition	6	44	Qi et al. (1997)
TACBOW0015 (TA#7648-3)	*L. racemosus* 7Lr#1 addition	3, 7	44	Qi et al. (1997)
TACBOW0016 (TA#7652)	*L. racemosus* ? Lr#1 addition	NA[c]	44	Qi et al. (1997)
TACBOW0017 (TA#7644)	*L. racemosus* 2Lr#1 substitution	2	42	Qi et al. (1997)

[a]TA#7643: Number from the Kansas State University gene bank
[b]ND: not determined
[c]NA: not available

entire or partial chromosomes from the wild species. Finally, these CILs are screened to identify the line that has the desired phenotype. The advantage of this method is reliability; and because useful traits can be easily followed from the wild species to common wheat and it may be particularly efficient for qualitative traits. However, this strategy is time-consuming, and expected traits from the wild species may not exhibit sufficient expression in the genetic background of common wheat. This happened especially in case of quantitative characters. In addition, wild relatives such as *L. racemosus* are often perennials with substantial vegetative biomass, small seeds, and poor agronomic traits. This nature hinders proper evaluation of their traits and response to different environmental conditions. Thus, it is not an easy task to hunt for the useful genes or traits from wild species.

The second strategy for using genes from *L. racemosus* or other wild relatives of wheat depends on phenotyping the CILs and not the wild species itself. The unique property of these CILs is that they are phenotypically similar to wheat (Fig. 1.1c), which eases evaluation for any useful genes they might carry. The advantage of this method is that the presence of these genes in the wheat genetic background allows evaluation of their positive or negative impact results from interactions with the wheat genome, and the influence of the wheat genome on the expression of the alien genes. Once a CIL with desired phenotype is identified, translocation lines could be developed for a target chromosome that bears the desired phenotype. Translocation lines could then be screened to identify lines with the desired phenotype.

The utility of CILs is not limited to wheat breeding; they can also be used for other applications. For example, Kikuchi et al. (2009) and Cho et al. (2011a) used wheat–*L. racemosus* CILs to investigate the effects of heavy-ion beams and the DNA methylation inhibitor zebularine on chromosome structure and rearrangement. As those CILs harbor a pair of alien chromosomes that serve as markers that can be easily detected by genomic in situ hybridization (GISH), Cho et al. (2011b) used double monosomic addition lines carrying *L. racemosus* and *L. mollis* chromosomes to study meiotic chromosome pairing. The line having two homologous chromosome painted in different colors by GISH was quite useful for investigating factors affecting meiotic chromosome pairing and recombination.

1.5 Impact of *Leymus* Genes on Improvement of Wheat Traits

Within the last few decades several wheat–*L. racemosus* CILs have been developed and evaluated, and useful chromosomes and traits have been determined, and analyzed (Chen et al. 2005; Qi et al. 2008; Wang and Chen 2008; Subbarao et al. 2007; Mohammed et al. 2013, 2014). These lines include useful genes for disease resistance, biological nitrification inhibition, aluminum tolerance, and heat stress tolerance.

1.5.1 Scab Resistance

Mujeeb-Kazi et al. (1983) reported that *L. racemosus* is highly resistant to scab, also known as Fusarium head blight. A wheat breeding program was later initiated in China to transfer this scab tolerance to common wheat. Three scab-resistant wheat–*L. racemosus* disomic addition lines (DALr.2, DALr.7, and DALr.14) were identified (Chen et al. 1993, 1995). Qi et al. (1997) reported that the additional chromosomes in DALr.2 and DALr.14 belong to homoeologous group 7 and 5, respectively, so they were designated 7Lr#1 and 5Lr#1. From these disomic addition lines, new wheat–*L. racemosus* translocation lines were then developed by inducing chromosome breakage with irradiation and gametocidal genes (Chen et al. 2005). In that study, nine scab-resistant wheat–*L. racemosus* translocation lines were identified. In one of these lines, NAU614, the long arm of 5Lr#1 had been translocated to wheat chromosome 6B. In four other lines, NAU601, NAU615, NAU617, and NAU635, part of the short arm of 7Lr#1 had been transferred to various wheat chromosomes. Another four lines, NAU611, NAU634, NAU633, and NAU618 contained translocations involving *Leymus* chromosome Lr.7 and various wheat chromosomes. The translocation lines containing a single alien chromosome segment were more resistant to scab than was the susceptible wheat parent CS but were less resistant than the resistant parent *L. racemosus*. At least three resistance genes were identified in *L. racemosus*, including one located on chromosome Lr.7, another that could be assigned to the long arm of 5Lr#1, and a third that could be assigned to the short arm of 7Lr#1. One novel resistance gene designated *Fhb3* was also mapped to the distal region of the short arm of chromosome 7Lr#1, and specific PCR-based markers were also developed to facilitate marker-assisted selection in scab resistance breeding programs (Qi et al. 2008).

1.5.2 Biological Nitrification Inhibition

Biological nitrification is a serious problem in agriculture that results in costly nitrogen loss. Subbarao et al. (2007) reported the first example of biological nitrification inhibition (BNI) identified among the wild relatives of cereals in *L. racemosus*, then introduced and successfully expressed this BNI trait in common wheat. *L. racemosus* releases 20-fold higher amounts of BNI compounds than does cultivated wheat. In their study, root exudates of cultivated wheat were unable to inhibit biological nitrification in the soil, but the root exudates of *L. racemosus* could suppress NO_3^- formation and maintain over 90 % of the inorganic nitrogen in the soil as NH_4^+ for 60 days. Chromosomes Lr#n, Lr#J, and Lr#I were associated with the high BNI capacity of *L. racemosus* (Fig. 1.2; Subbaro et al. 2007). The BNI trait from the Lr#n chromosome was introduced into wheat by inducing a Robertsonian translocation between wheat and the Lr#n chromosome (Kishii et al. 2008). Recently, a program was established at CIMMYT to identify and map the genes responsible for BNI and introduce them into all commercial wheat cultivars in order to reduce production costs and minimize the nitrogen pollution generated by wheat production systems.

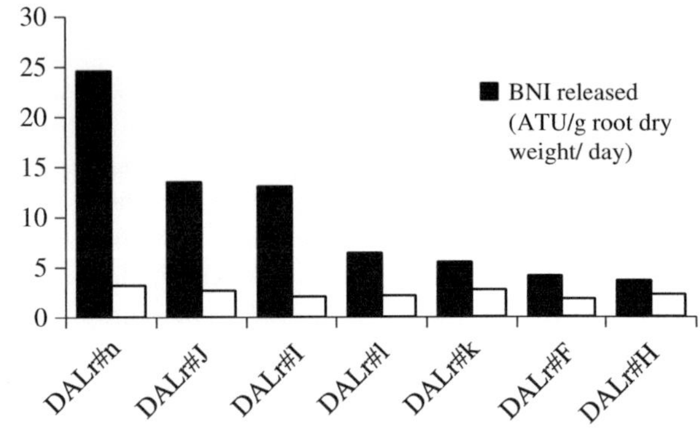

Fig. 1.2 BNI release and dry matter production of some wheat–*L. racemosus* disomic addition lines (Subbarao et al. 2007)

1.5.3 Aluminum Tolerance

Aluminum toxicity is a key factor that limits wheat production in the acidic soil with pH below 5.5 (Evans and Kamprath 1970). Mohammed et al. (2013) screened 15 wheat–*L. racemosus* CILs for aluminum (Al) toxicity tolerance. *L. racemosus* chromosomes Lr#A and Lr#E were found to significantly enhance relative root growth (Fig. 1.3a, b; Mohammed et al. 2013) under Al stress and thereby increase the Al tolerance of wheat. A line containing Lr#E chromosome was more aluminum-tolerant than a line containing Lr#A chromosome at the highest Al concentration tested (200 μM) (Mohammed et al. 2013). The introgressed chromosomes did not affect the aluminum uptake or expression of the wheat *ALMT1* gene that controls the exclusion of Al from root tips, but a gene or genes on chromosome Lr#E did improve wheat cell membrane integrity of wheat under Al stress conditions (Fig. 1.3c; Mohammed et al. 2013). The results suggest that there are Al tolerance mechanisms other than exclusion of Al from the root tips. Eventually, chromosome engineering using these two lines could result in wheat cultivars with enhanced Al tolerance.

1.5.4 Heat Stress

Mohammed et al. (2014) examined the response of 14 wheat–*L. racemosus* CILs to high temperature to determine whether they have potential for breeding improved wheat cultivars. These introgression lines and their parent CS were evaluated either in a growth chamber at the seedling stage and in the field at the reproductive stage in two heat-stressed environments in Sudan. To be sure that the plants would be exposed to heat stress at the reproductive stage; optimum and late planting strategies were used. For seedlings under growth chamber conditions, the presence of chromosomes Lr#A, Lr#I, Lr#l, and 2Lr#1 were found to improve heat stress tolerance as measured by improved photosynthesis and mitochondrial electron transport (Fig. 1.4a, b; Mohammed et al. 2014). Under field conditions, the

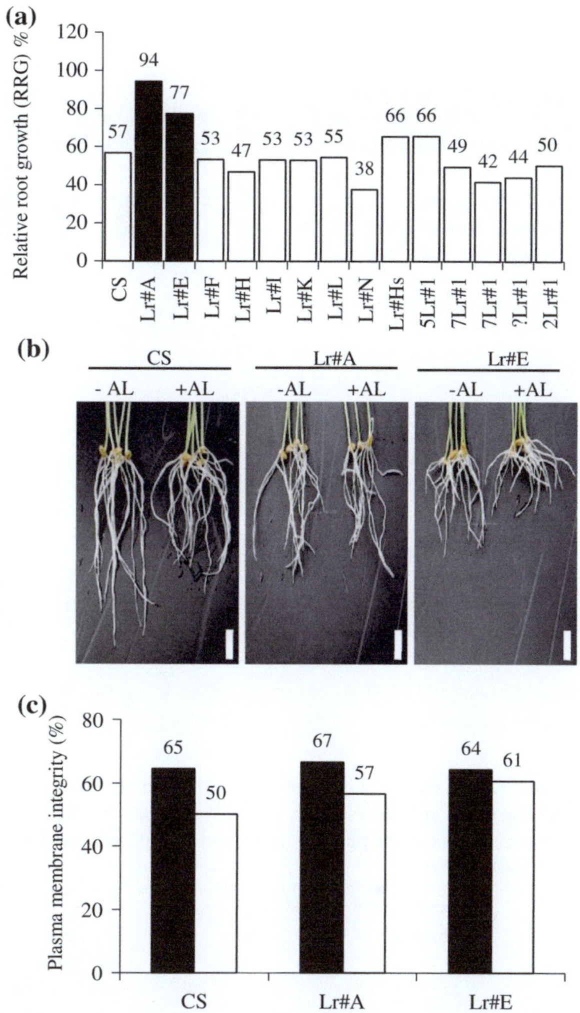

Fig. 1.3 Performance of common wheat cultivar Chinese Spring (CS) and *Leymus racemosus* chromosome addition lines under aluminum (Al) toxicity conditions. **a** Relative root growth (RRG) of CS and addition lines grown under Al toxicity conditions. *Dark bars* indicate the Al tolerant lines. **b** Sensitivity of CS and tolerant lines Lr#A and Lr# E to Al toxicity. **c** Plasma membrane integrity of CS and tolerant lines Lr#A and Lr#E under Al toxicity conditions measured as percentage of electrolyte leakage, dark bars indicate the control and the *white bars* indicate the Al treatment (Mohammed et al. 2013)

presence of chromosomes Lr#I and Lr#n enhanced grain number per spike and heat tolerance (Table 1.2; Mohammed et al. 2014). Their results also indicated that chromosome 7Lr#1 might improve yield potential as it has higher grain yield than the wheat parent CS under non-stressed conditions (Table 1.2, Mohammed et al. 2014). Based on a heat susceptibility index (HSI) calculated from grain yield obtained from optimum and late sowings, *L. racemosus* chromosomes Lr#I and 2Lr#1 were classified as highly heat tolerant, whereas Lr#A and 5Lr#1 were classified as moderately heat tolerant (Table 1.2; Mohammed et al. 2014). This study clearly indicated that several *Leymus* chromosomes have good potential for improving wheat adaptation and tolerance to heat stress (Mohammed et al. 2014).

In addition to scab resistance, BNI, aluminum tolerance, and heat stress tolerance, lines carrying the Lr#J and Lr#l chromosomes also expressed gene(s) for

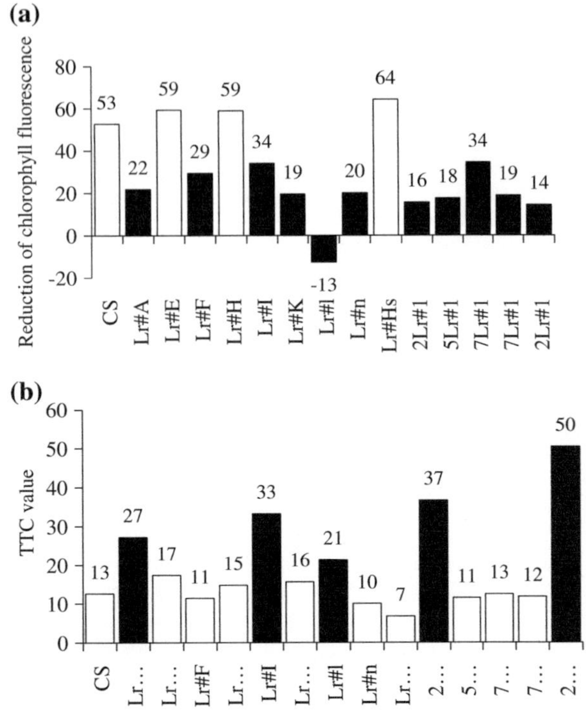

Fig. 1.4 Reduction in chlorophyll fluorescence (**a**) and triphenyl tetrazolium chloride (TTC) reduction assay values (**b**) in CS and *L. racemosus* addition lines grown in a growth chamber under normal and heat-stressed conditions; *dark bars* indicate lines with significant differences from CS (Mohammed et al. 2014)

Table 1.2 Kernel number per spike, grain yield, and heat susceptibility index (HSI) of some addition lines and their parent cultivar 'Chinese Spring' (CS) during evaluation for heat stress tolerance at optimum planting (OP) and late planting (LP) in Shambat (2011/2012) and Gezira (2012/2013), Sudan

	Kernel number/spike				Grain yield				HSI
	Shambat		Gezira		Shambat (g/plant)		Gezira (g/m^2)		
Line	OP	LP	OP	LP	OP	LP	OP	LP	
Lr#A	53^b	50^b	49^a	32^a	1.9^a	1.7^a	122^a	56^a	0.59
Lr#I	68^a	58^a	47^a	29^b	2.3^a	1.7^a	132^a	78^a	0.24
Lr#n	59^a	58^a	45^b	26^b	2.1^a	1.9^a	133^a	50^a	0.71
2Lr#1	54^b	49^b	36^b	27^b	2.1^a	1.7^a	121^a	74^a	0.25
5Lr#1	43^c	43^c	32^b	14^c	1.9^a	1.8^a	110^a	56^a	0.59
7Lr#1	49^c	57^b	47^a	36^a	2.1^a	1.9^a	265^a	29^a	1.12
CS	55^b	51^b	37^b	26^b	1.5^b	1.3^b	51^b	14^b	1.40

$^{a,\,b}$Significant difference from CS
^{c}Compatible to CS

resistance to leaf or stem rust, and the Lr#k line possessed a gene for a novel seed storage protein (Kishii 2011). Further, the Lr#H chromosome addition line exhibited early heading (Kishii et al. 2004).

1.6 Genetic Maps and Molecular Markers

Genetic maps and genomic information are essential for modern breeding programs as they facilitate the assessment of genetic diversity for germplasm enhancement, and ease the transfer of desired genetic traits by marker-assisted breeding (Pérez-de-Castro 2012). Linkage mapping has been used extensively to identify genes or QTLs in plant genomes that control important traits. It relies on random recombination at meiosis and segregation of the recombinant chromosomes among offsprings (Dear 2001). For genes introduced on alien chromosomes, it is difficult to perform genetic analysis and construct genetic maps using CILs because no recombination occurs at meiosis due to absence of homology between the alien chromosome and the wheat chromosomes. Therefore, deletion mapping is the best way to identify and map the genes located on the alien chromosome to allow their transfer into wheat.

To create an efficient and informative deletion map, specific molecular markers for the target chromosome are essential. No molecular markers specific for *L. racemosus* have been developed to date. However, expressed sequence tags (EST) markers from barley have been used to develop molecular markers that are polymorphic between wheat and some important wild species (Hagras et al. 2005). About 182 markers polymorphic between wheat and *L. racemosus* were identified in their study. Although genetic maps are not yet available for *L. racemosus*, several genetic maps have been constructed for other *Leymus* species. Molecular genetic maps were constructed for two full-sib families (TTC1 and TTC2) derived from *L. cinereus* and *L. triticoides* (Wu et al. 2003). The genetic maps of the TTC1 and TTC2 families included a combined total of 1583 AFLP markers and 67 heterologous anchor markers (from other grass species) in 14 linkage groups. Two sets of seven homoeologous linkage groups were tentatively identified based on synteny of the heterologous anchor markers between wheat (*Triticum* spp.), barley (*Hordeum vulgare*), and cereal rye (*Secale cereale*) (Wu et al. 2003; Larson et al. 2006). These populations were found to segregate for genes and QTLs controlling growth habit (Larson et al. 2006), forage quality (Larson and Mayland 2007), and seed shattering (Larson and Kellog 2009).

From a collection of 28786 ESTs derived from subterranean rhizome and tiller buds of *L. triticoides* × *L. cinereus* hybrids, 11281 unigene contigs were assembled (Bushman et al. 2008). Of these, 9389 had at least one match with the rice genome. In that study, 1798 SSR primers were designed of which 1575 were polymorphic (Bushman et al. 2008). Further, *Leymus* EST-SSR primers that aligned to rice chromosome 2 were tested for synteny in *Leymus* (Bushman et al. 2008; Kaur et al. 2008).

Using the same mapping populations, Larson et al. (2012) developed an integrated consensus map including 375 AFLP, 350 SSR, and 48 heterologous markers that cover 2381 cM. A total of 146 of these markers were tested in 17 wheat–*L. racemosus* CILs and were found to be polymorphic between wheat and *L. racemosus*. All CILs tested positively for between 8 and 24 *Leymus* EST-SSR marker loci representing at least one homoeologous group; most of these matched the previous chromosome identification (Larson et al. 2012).

Using suppressive-subtractive hybridization, Habora et al. (2012) identified 112 osmotic-stress-responsive genes (ESTs) from *L. mollis* that are differentially expressed under osmotic stress. All these types of genetic information that are available could be used to generate molecular markers for *L. racemosus* to facilitate the introduction of its useful adaptive genes in wheat breeding.

1.7 Conclusions and Future Prospects

The objectives of wheat breeding programs are linked with the demands of cumulative population growth and the necessity to increase food production. Due to the narrow genetic base available in modern bread wheat germplasm, achieving this objective will be the challenge of the century. Therefore, finding broader sources of adaptive genes in the wheat gene pool will be essential for introducing traits that could help increase yields in the warmer, drier climate predicted for most of the world's breadbasket. *L. racemosus* represents a potentially valuable source of genetic diversity for the improvement of stress tolerance specifically, and wheat production generally. In this review, we have described *L. racemosus* genes that will be potentially useful for improving the tolerance of wheat to biotic and abiotic stresses and improving its nitrogen use efficiency. However, for the most efficient exploitation of *L. racemosus* genes in wheat breeding, the targeted screening of wheat–*L. racemosus* cytogenetic stocks for resistance to biotic and abiotic stresses should be continued. The wealth of molecular genetic information that is becoming available should be efficiently exploited for breeding purposes, and the shortage should be replenished through genome wide approaches.

Acknowledgments We thank Dr. M. Kishii, CIMMYT, who gave us valuable information about *L. racemosus* addition lines.

References

Bushman BS, Larson SR, Mott IW, Cliften PF, Wang RRC, Chatterton NJ, Hernandez AG, Ali S, Kim RW, Thimmapuram J, Gong G, Liu L, Mikel MA (2008) Development and annotation of perennial Triticeae ESTs and SSR markers. Genome 51:779–788

Chen PD, Wang ZT, Wang SL, Huang L, Wang YZ, Liu DJ (1993) Transfer of scab resistance from *Elymus giganteus* into common wheat. In: Li ZS, Xin ZY (eds) Proceedings of the 8th International wheat genetic symposium. Agricultural Science Tech Press, China, pp 153–157

Chen PD, Wang ZT, Wang SL, Huang L, Wang YZ, Liu DJ (1995) Transfer of useful germplasm from *Leymus racemosus* Lam, to common wheat. III. Development of addition lines with wheat scab resistance. Chin J Genet 22:119–124

Chen P, Liu W, Yuan J, Wang X, Zhou B, Wang S, Zhang S, Feng Y, Yang B, Liu G et al (2005) Development and characterization of wheat–*Leymus racemosus* translocation lines with resistance to *Fusarium* head blight. Theor Appl Genet 111:941–948

Cho S-W, Ishii T, Matsumoto N, Tanaka H, Eltayeb AE, Tsujimoto H (2011a) Effects of the cytidine analogue zebularine on wheat mitotic chromosomes. Chromosome Sci 14:23–28

Cho S-W, Moritama Y, Ishii T, Kishii M, Tanaka H, Eltayeb AE, Tsujimoto H (2011b) Homology of two alien chromosomes during meiosis in wheat. Chromosome Sci 14:45–52

Cox TS, Sears RG, Bequette RK, Martin TJ (1995) Germplasm enhancement in winter wheat x *Triticum tauschii* backcross populations. Crop Sci 35:913–919

Curtis T, Halford NG (2014) Food security: the challenge of increasing wheat yield and the importance of not compromising food safety. Ann Appl Biol 164:354–372

Dear PH (2001) Genome mapping. Encyclopedia of life science. Macmillan Publishers Ltd, Nature Publishing Group www.els.net

Dreccer F, Ogbonnaya FC, Borgognone G (2004) Sodium exclusion in primary synthetic wheats. Proc XI Wheat Breed Assembly:118–121

Eastwood RF, Lagudah ES, Appels R (1994) A direct search for DNA sequences tightly linked to cereal cyst nematode resistance genes in *Triticum tauschii*. Genome 37:311–319

Fan X, Sha LN, Yang RW, Zhang HQ, Kang HY, Ding CB, Zhang L, Zheng YL, Zhou YH (2009) Phylogeny and evolutionary history of *Leymus* (Triticeae; Poaceae) based on a single copy nuclear gene encoding plastid acetyl-CoA carboxylase. BMC Evol Biol 9:247

FAOstat (2007) The statistics division, food and agriculture organization of the United Nations, Rome. http://faostat.fao.org

Garg M, Tanaka H, Ishikawa N, Tanaka K, Yanaka M, Tsujimoto H (2009) *Agropyron elongatum* HMW-glutenins have a potential to improve wheat end-product quality through targeted chromosome introgression. Breed Sci 50:358–363

Gatford KT, Hearnden P, Ogbonnaya F, Eastwood RF, Halloran GM (2002) Novel resistance to pre-harvest sprouting in Australian wheat from the wild relative *Triticum tauschii*. Euphytica 126:67–76

Habora MEE, Eltayeb AE, Tsujimoto H, Tanaka K (2012) Identification of drought stress responsive genes from *Leymus mollis*, a wild relative of wheat (*Triticum aestivum* L.) Breed Sci 62:78–86

Hagras AA, Kishii M, Sato K, Tanaka H, Tsujimoto H (2005) Extended application of barley EST markers for analysis of alien chromosomes added to wheat genetic background. Breed Sci 55:335–341

Jiang J, Friebe B, Gill BS (1994) Recent advances in alien gene transfer in wheat. Euphytica 73:199–212

St. John L, Ogle DG, Stannard M, Pavek P (2010) Plant guide for mammoth wild rye (*Leymus racemosus*). USDA-Natural Resources Conservation Service, Aberdeen, ID Plant Materials Center

Kaur P, Larson SR, Bushman BS, Wang RRC, Mott IW, Hole D, Thimmapuram J, Gong G, Liu L (2008) Genes controlling plant growth habit in *Leymus* (Triticeae): maize *barren stalk1 (ba1)*, rice *lax* panicle, and wheat *tiller inhibition (tin3)* genes as possible candidates. Funct Integr Genomics 8:375–386

Kihara H (1944) Discovery of the DD-analyser, one of the ancestors of *Triticum vulgare* (abstr). Agric Hortic 19:889–890 (in Japanese)

Kihara H (1966) Factors affecting the evolution of common wheat. Indian J Genet 26A:14–28

Kikuchi S, Saito Y, Ryuto H, Fukunishi N, Abe T, Tanaka H, Tsujimoto H (2009) Effects of heavy-ion beams on chromosomes of common wheat, *Triticum aestivum*. Mutat Res/Fund Mol Mech Mutagen 669:63–66

Kishii M (2011) Production of wheat-*Leymus racemosus* translocation lines. Wheat Inf Serv 111:11–13

Kishii M, Yamada T, Sasakuma T, Tsujimoto H (2004) Production of wheat-*Leymus racemosus* chromosome addition lines. Theor Appl Genet 109:255–260

Kishii M, Ban T, Subbarao GV, Ortiz-Monasterio I (2008) Transferring of the biological nitrification inhibition (BNI) character from *Leymus racemosus* to wheat. In: Appels R,

Eastwook R, Laguday E, Langridge P, Mackay M, McIntyre L, Sharp P (eds) Proceedings of the 11th Int Wheat Genetics Symposium. Brisbane, Australia, pp 226–227

Larson SR, Kellog EA (2009) Genetic dissection of seed production traits and identification of a major-effect seed retention QTL in hybrid *Leymus* (Triticeae) wild ryes. Crop Sci 49:29–40

Larson SR, Mayland HF (2007) Comparative mapping of fiber, protein, and mineral content QTLs in two interspecific *Leymus* wildrye full-sib families. Mol Breed 20:331–347

Larson SR, Wu XL, Jones TA, Jensen KB, Chatterton NJ, Waldron BL, Robins JG, Bushman BS, Palazzo AJ (2006) Comparative mapping of growth habit, plant height, and flowering QTLs in two interspecific families of *Leymus*. Crop Sci 46:2526–2539

Larson SR, Kishii M, Tsujimoto H, Qi LL, Chen PD, Lazo GR, Jensen KB, Wang RR-C (2012) *Leymus* EST linkage maps identify 4NsL-5NsL reciprocal translocation, wheat–*Leymus* chromosome introgressions, and functionally important gene loci. Theor Appl Genet 124:189–206

Liu J, Chang Z, Zhang X, Yang Z, Li X, Jia J, Zhan H, Guo H, Wang J (2013) Putative *Thinopyrum intermedium*-derived stripe rust resistance gene *Yr50* maps on wheat chromosome arm 4BL. Theor Appl Genet 126:265–274

Lobell DB, Schlenker W, Cost-Roberts J (2011) Climate trends and global crop production since 1980. Science 333:616–620

Löve A (1984) Conspectus of the Triticeae. Feddes Repert 95:425–521

Lu RJ, Chen PD, Liu DJ (1995) Transfer of useful germplasm from *Leymus racemosus* Lam to common wheat. IV. Development of double alien disomic addition and substitution lines by anther culture. J Nanjing Agric Univ 18:1–6

Marais GF, Marais AS (1994) The derivation of compensating translocations involving homoeologous group 3 chromosomes of wheat and rye. Euphytica 79:75–80

Martín-Sánchez JA, Gómez-Colmenarejo M, Del Moral J, Sin E, Montes MJ, González-Belinchón C, López-Braña I, Delibes A (2003) A new Hessian fly resistance gene (*H*30) transferred from the wild grass *Aegilops triuncialis* to hexaploid wheat. Theor Appl Genet 106:1248–1255

McFadden ES, Sears ER (1944) The artificial synthesis of *Triticum spelta*. Rec Genet Soc Am 13:26–27

McGuire PE, Dvorak J (1981) High salt-tolerance potential in wheat grasses. Crop Sci 21:702–705

Mohammed YSA, Eltayeb AE, Tsujimoto H (2013) Enhancement of aluminum tolerance in wheat by addition of chromosomes from the wild relative *Leymus racemosus*. Breed Sci 63:407–416

Mohammed YSA, Tahir ISA, Kamal NM, Eltayeb AE, Ali AM, Tsujimoto H (2014) Impact of wheat-*Leymus racemosus* added chromosomes on wheat adaptation and tolerance to heat stress. Breed Sci 63:450–460

Monsen S, Stevens R, Shaw N (2004) Grasses. USDA forest service Gen. Technical report. RMRS-GTR-136, pp 199–294

Mujeeb-Kazi A, Rodriguez R (1981) An intergeneric hybrid of *Triticum aestivum* L. X *Elymus giganteus*. J Hered 72:253–256

Mujeeb-Kazi A, Bernard M, Bekele GT, Minard JL (1983) Incorporation of alien genetic information from *Elymus giganteus* into *Triticum aestivum*. In: Sakamoto S (ed) Proceedings of the 6th International wheat genetic symposium, Kyoto, pp 223–231

Ogbonnaya FC, Abdalla O, Mujeeb-Kazi A, Kazi AG, Xu SS, Gosman N, Lagudah ES, Bonnett D, Sorrells ME, Tsujimoto Hisashi (2013) Synthetic hexaploids: harnessing species of the primary gene pool for wheat improvement. Plant Breed Rev 37:35–122

Pérez-de-Castro AM, Vilanova S, Cañizares J, Pascual L, Blanca JM, Díez MJ, Prohens J Picó B (2012) Application of genomic tools in plant breeding. Curr Genomics 13:179–195

PLANTS Database. Conservation plant characteristics for *Leymus racemosus*. http://plants.usda.gov. Accessed 25 Aug 2010

Qi LL, Wang S, Chen PD, Liu DJ, Friebe B, Gill BS (1997) Molecular cytogenetic analysis of *Leymus racemosus* chromosomes added to wheat. Theor Appl Genet 95:1084–1109

Qi LL, Friebe B, Zhang P, Gill BS (2007) Homoeologous recombination, chromosome engineering and crop improvement. Chromosome Res 15:3–19

Qi LL, Pumphrey MO, Friebe B, Chen PD, Gill BS (2008) Molecular cytogenetic characterization of alien introgressions with gene *Fhb3* for resistance to Fusarium head blight disease of wheat. Theor Appl Genet 117:1155–1166
Reif JC, Zhang P, Dreisigacker S, Warburton ML, van Ginkel M, Hoisington D, Bohn M, Melchinger AE (2005) Wheat genetic diversity trends during domestication and breeding. Theor Appl Genet 110:859–864
Ren XQ, Sun WX, Chen PD, Liu DJ (1996) Development and identification of *T. aestivum-L. racemosus* addition lines. J Nanjing Agric Univ 19:1–5
Subbarao GV, Ban T, Kishii M, Ito O, Samejima H, Wang HY, Pearse SJ, Gopalakrishnan S, Nakahara K, Tsujimoto H, Berry HW (2007) Can biological nitrification inhibition (BNI) genes from perennial *Leymus racemosus* (Triticeae) combat nitrification in wheat farming? Plant Soil 299:55–64
Sun WX, Chen PD, Liu DJ (1997) Transfer of useful germplasm from, *Leymus racemosus* Lam. to common wheat. V. Development and identification of three *T. aestivum-L. racemosus* disomic addition lines. J Nanjing Agric Univ 20:6–11 (in Chinese)
Sun WX, Chen PD, Liu DJ (1998) Transfer of useful germplasm from *Leymus racemosus* Lam. to common wheat. VI. Development and identification of *T. aestivum-L. racemosus* telosomic addition lines. Acta Genet Sinica 25:259–264 (in Chinese)
Wang LS, Chen PD (2008) Development of *Triticum aestivum–Leymus racemosus* ditelosomic substitution line 7Lr#1S (7A) with resistance to wheat scab and its meiotic behavior analysis. Chin Sci Bull 53:3522–3529
Wang RR-C, Jensen KB (1994) Absence of the J genome in *Leymus* species (Poaceae: Triticeae): evidence from DNA hybridization and meiotic pairing. Genome 37:231–235
Wang YN, Chen PD, Liu DJ (1986) Transfer of useful germplasm from *E. giganteus* to common wheat. I. Production of wheat *E. giganteus* hybrids. J Nanjing Agric Univ 1:10–14
Wang YN, Chen PD, Wang ZT, Liu DA (1991) Transfer of useful germplasm from *E. giganteus* to common wheat. II. Cytogenetics and scab resistance of backcross derivatives. J Nanjing Agric Univ 14:1–5
Wang RRC, von Bothmer R, Dvorak J, Linde-Laursen I, Muramatsu M (1994) Genome symbols in the Triticeae (Poaceae). In: Wang et al (eds) Proceedings of the second international Triticeae symposium. Utah State University Press, Logan, pp 29–34
Wang RRC, Zhang JY, Lee B, Jensen KB, Kishii M, Tsujimoto H (2006) Variations in abundance of two repetitive sequences in *Leymus* and *Psathyrostachys* species. Genome 49:511–519
Wang S, Yin L, Tanaka H, Tanaka K, Tsujimoto H (2010) Identification of wheat alien chromosome addition lines for breeding wheat with high phosphorus efficiency. Breed Sci 60:372–379
Wu XL, Larson SR, Hu ZM, Palazzo AJ, Jones TA, Wang RRC, Jensen KB, Chatterton NJ (2003) Molecular genetic linkage maps for allotetraploid *Leymus* (Triticeae). Genome 46:627–646
Yang Y, Liu DL, Anwar MR, Zuo H, Yang Y (2014) Impact of future climate change on wheat production in relation to plant-available water capacity in a semiarid environment. Theor Appl Climatol 115:391–410
Zhang HB, Dvorak J (1991) The genome origin of tetraploid species of *Leymus* (Poaceae: Triticeae) inferred from variation in repeated nucleotide sequences. Am J Bot 78:871–884

Chapter 2
Arachis Gene Pools and Genetic Improvement in Groundnut

Anurudh K. Singh and S.N. Nigam

Abstract Groundnut (*Arachis hypogaea* L.) is an important oilseed and food crop in the world. The crop is predominantly grown in low input production systems in developing countries in Asia and Africa. There are several production constraints, both biotic and abiotic, to groundnut. Some of these are global in nature and the others are either regional or local. Four *Arachis* gene pools contain 80 species, distributed among nine sections are native to five countries of South America. Section *Arachis* contains tetraploid cultivated groundnut, divided into two subspecies and six botanical varieties and a number of cross-compatible diploid species with rich genetic diversity. International efforts have made significant progress in collection and conservation of these genetic resources, facilitating genetic improvement. Groundnut is an autogamous crop. The pedigree and bulk selection methods are more commonly used by the groundnut breeders. Conventional breeding, including cytogenetic manipulations introgressing genes from cross-compatible wild diploid species has been effective in some areas, while in others it has been tardy due to lack of proper and effective phenotyping tools and limited understanding of the genomics, genetics/inheritance, and underlying mechanisms influencing targeted traits. A greater diversification of parental resources (both cultivated and wild *Arachis* species) in breeding programs is required to develop new cultivars with diversified genetic backgrounds, which will enable them to perform better under the changing climatic/adverse conditions. Molecular breeding is in infancy. Infrequent and low polymorphisms have restricted the progress in the development and application of genetic maps, except in cases where polymorphic chromosomal regions have been introgressed from diploid wild *Arachis* species into *A. hypogaea*. Both conventional and nonconventional crop improvement efforts in groundnut need to concentrate on bridging

---

A.K. Singh (✉)
Plot No. 2924, Sector -23, Gurgaon 122017, Haryana, India
e-mail: anurudhksingh@gmail.com

S.N. Nigam
Plot No. 125, Road No. 74 Jubilee Hills, Hyderabad 500 033, Telangana, India
e-mail: snnigam1947@gmail.com

V.R. Rajpal et al. (eds.), *Gene Pool Diversity and Crop Improvement*,
Sustainable Development and Biodiversity 10,
DOI 10.1007/978-3-319-27096-8_2

the yield gap between the potential yield and the realized yield by alleviating major production constraints particularly in rainfed environment.

Keywords *Arachis hypogaea* · *Arachis* gene pool · Center of origin/diversity · Core collection · Genetic improvement · Molecular breeding · Genetic transformation

2.1 Introduction

Groundnut or peanut (*Arachis hypogaea* L.), an annual legume of indeterminate growth habit, is primarily grown for its high-quality edible oil (44–56 %) and easily digestible protein (22–30 %) in its seeds. Groundnut seeds also contain carbohydrates (10–25 %) and are a rich source of vitamins (E, K, and B complex), minerals (Ca, P, Mg, Zn, and Fe), and fiber. It ranks sixth in edible oil production among the oilseed crops and thirteenth among the food crops (production utilized directly as food or in confections) in the world. It is also the third most important source of vegetable protein in the world. It is grown predominantly for food use in North America, Southern Africa, West Africa, Southeast Asia, and Europe and predominantly for edible oil use in South America and Southwest Asia. In East Africa and East Asia, both food and edible oil uses are important. On a global basis, 41 % of the production goes for food use and 49 % for extraction of edible oil. The remaining 10 % is used as feed and seed or goes waste. The commercial production of groundnut occurs principally between 40°N and 40°S and about 114 countries grow groundnut. The crop is predominantly grown in low-input production system in Asia and Africa with yield ranging between 700 and 1000 kg/ha and in high input system in USA, Australia, Argentina, Brazil, China, and South Africa where yields of 2000–4000 kg/ha are obtained.

2.2 *Arachis* Gene Pool and Taxonomy

Genus *Arachis* gene pool contains 80 species (Krapovickas and Gregory 1994; Valls and Simpson 2005). Based on distribution, character clustering, and cross-compatibilities, Krapovickas and Gregory (1994) classified the genus into nine sections. Of these, section *Trierectoides* contains 2, *Erectoides* 14, *Extranervosae* 10, *Triseminatae* 1, *Heteranthae* 6, *Caulorhizae* 2, *Procumbentes* 10, *Rhizomatosae* 4, and *Arachis* 31 species. Section *Arachis* contains cultivated groundnut (*A. hypogaea*), another tetraploid species *A. monticola* and 29 diploid wild species. Based on high degree of genetic isolation and the comparative cytology (Fernandez and Krapovickas 1994), Krapovickas and Gregory (1994) inferred that five sections, *Trierectoides*, *Erectotoides*, *Triseminatae*, *Extranervosae,* and *Heteranthae*, are primitive compared to sections, *Procumbentes, Caulorhizae, Rhizomatosae,* and *Arachis,* except for section *Erectoides*. Sections *Rhizomatosae* and *Arachis* have evolved feature of shorter peg length, while sections *Caulorrhizae* and *Rhizomatosae* have new methods of vegetative propagation: stolons and rhizomes, respectively, and appear to be advanced. Besides, the annual character represents an adaptive advantage that permits the species to avoid droughts in the northeast Brazil (*Heteranthae*) and in the foot hills of the Andes, as well as the flooding of the Paraguay River watershed (*Arachis*). Additionally, section *Arachis* appears to be spreading to new territories invading the areas of other sections. Its species grow intermixed with populations of *Extranervosae* in the upper Paraguay basin and occupy common ground with section *Procumbentes* in the Gran Pantanal. They have reached the shores of La Plata and the southeastern coast of Brazil and grow from Yala in northwest Argentina to the Tocantins in northeast Brazil, besides worldwide adaptation of *A. hypogaea*. Cross-compatibility recorded between *A. paraguariensis* ssp. *paraguariensis,* the southernmost taxon in the section *Erectoides, A. rigonii,* the westernmost species in the section *Procumbentes,* and *A. duranensis*, one of the most western species of section *Arachis* (Gregory and Gregory 1979) and recently between some more species of these sections (Singh 1998; Mallikarjuna 2005; Mallikarjuna and Hoisington 2009) and *A. glabrata* of *Rhizomatoseae* (Mallikarjuna and Sastri 2002) corroborate the closeness and advance nature of these sections. Figure 2.1 illustrates the overall phylogenetic relationships between various sections.

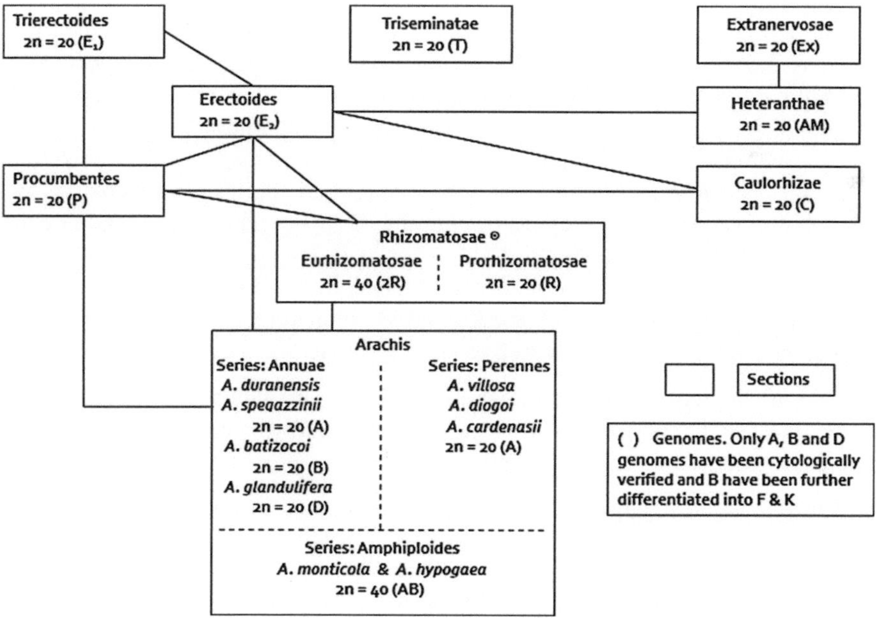

Fig. 2.1 The sections of genus *Arachis* with connecting lines displaying intersectional crossability. All species within a section are crossable, except *Eurhizomatosae* x *Prorhizomatosae*

Harlen and de Wet (1971) proposed the *gene pool* concept in order to provide a genetic perspective to relationship of cultivated plant species to other components of genetic diversity, the wild relatives, based on cross-compatibility into (1) primary gene pool (GP-1), (2) secondary gene pool (GP-2), and tertiary gene pool (GP-3). Application of this principle facilitates clearer understanding of phylogenetic relationships between the wild and cultivated species and helps to identify appropriate breeding strategies for incorporating desired genes into conventionally usable form of cultivated species for designing new cultivars. This helps to facilitate conventional cytogenetic manipulations to establish fertile hybrids, improve genetic recombination for incorporation of desirable genes into a usable form and hybridization using pre- or post-fertilization manipulations to establish hybrids. Alternatively, biotechnological approaches may be applied to access genes through sexual or parasexual means of genetic transformation. This approach has been used in groundnut classifying the genetic diversity into four gene pools (Smartt 1990; Singh and Simpson 1994):

1. The primary gene pool consists of landraces and traditional cultivars of groundnut from primary and secondary centers of genetic diversity in South America and other groundnut-growing countries and wild *A. monticola* found in northwest Argentina having free crossability with *A. hypogaea* producing normal segregants.
2. The secondary gene pool consists of diploid species from section *Arachis*, cross-compatible with *A. hypogaea*, despite ploidy differences, producing sterile to partially fertile hybrids.
3. The tertiary gene pool includes species of section *Procumbentes*, which have crossed with diploid species of section *Arachis* (Gregory and Gregory 1979; Mallikarjuna 2005; Mallikarjuna and Hoisington 2009) and probably coevolved with series *perennes* of section *Arachis*; *Erectoides,* whose species are weakly cross-compatible with diploid species of section *Arachis* and *A. hypogaea* (Singh 1998); and *Rhizomatosae*, whose tetraploid species can be crossed both with diploid species of section *Arachis* and *A. hypogaea* (Gregory and Gregory 1979; Mallikarjuna and Sastri 2002).
4. The quaternary gene pool of the remaining *Arachis* species that are cross-incompatible or very weakly cross-compatible to species of section *Arachis* and are classified into five other sections.

Based on heritable genetic variation observed in cultivated *A. hypogaea*, Krapovikas and Gregory (1994) divided it into the following two subspecies and six botanical varieties.

1. Subsp. *hypogaea*: Characterized by absence of flowers on main axis and regular alternation of vegetative and reproductive branches on the laterals and long life cycle.
2. Subsp. *fastigiata*: Characterized by presence of flowers on main axis and no specific order of vegetative and reproductive branches on the laterals and shorter life cycle.

Subsp. *hypogaea* is divided into two botanical varieties as follows:

i. Var. *hypogaea*: Characterized by leaflets with glabrous dorsal surface, short central or main axis, prostrate to erect growth habit, simple inflorescence, and 2–3 seeded pods.
ii. Var. *hirsuta*: Characterized by leaflets with entire dorsal surface hairy (1–2 mm), long central or main axis, prostrate growth habit, and 2–4 seeded pods.

Subsp. *fastigiata* is divided into four botanical varieties as follows:

i. Var. *fastigiata*: Characterized by leaflets with glabrous dorsal surface and hair only on the midrib, 3–5 seeded pods with smooth or lightly marked reticulation without highlighting the longitudinal ribs, reproductive branches mostly short and thin, little branched, curved branches, erect growth habit, and simple inflorescence.
ii. Var. *peruviana*: Characterized by leaflets with glabrous dorsal surface and hair only on the midrib, pods with very marked reticulation and with prominent longitudinal ribs and long and strong reproductive branches (5–10 cm) with strong central axis and lateral branches.
iii. Var. *aequatoriana*: Characterized by leaflets with entire dorsal surface hairy (1–2 mm), long reproductive branches, mainly the lateral branches, central axis mostly with short inflorescence and reproductive branches, deep pod reticulation, purple stems, more branched and erect growth habit.
iv. Var. *vulgaris*: Characterized by pods mostly 2-seeded, bunched fruits pointing to the base of the plant, erect growth habit, more branched and compound inflorescence.

Commercially, var. *hypogaea* is also known as Virginia type (large-seeded) or Runner type (small-seeded), var. *hirsuta* as Peruvian runner, var. *fastigiata* as Valencia type, and var. *vulgaris* as Spanish type.

Besides the variability of primary gene pool of cultivated *A. hypogaea*, the wild *Arachis* species have attracted groundnut workers because of their resistance to diseases and insect pests for which the genetic variation in primary gene pool is limited. The most accessible variability of primary and secondary gene pools have been successfully utilized in groundnut improvement and their potential value is now much more predictable and productive. The exploitation of tertiary and quaternary gene pools waits for advancement in the biotechnological techniques and policy decision with regard to release of transgenic varieties at global level.

2.3 Extent of Distribution, Center of Origin, and Genetic Diversity

The genus *Arachis* is naturally distributed in east of the Andes, south of Amazon, north of La Plata from northwest Argentina to northeast Brazil, including Argentina, Bolivia, Brazil, Paraguay, and Uruguay, i.e., from the mouth of the Amazon (0°) to south across the Sao Francisco and the Jequitinhonha, and into the mild temperate zone to 34° S on the shores of the South Atlantic in the southern Uruguay (Fig. 2.2). *Arachis* species grow from sea level to 650 m above mean sea level (amsl) on the Planalto, from southern Mato Grosso to southern Goias, to

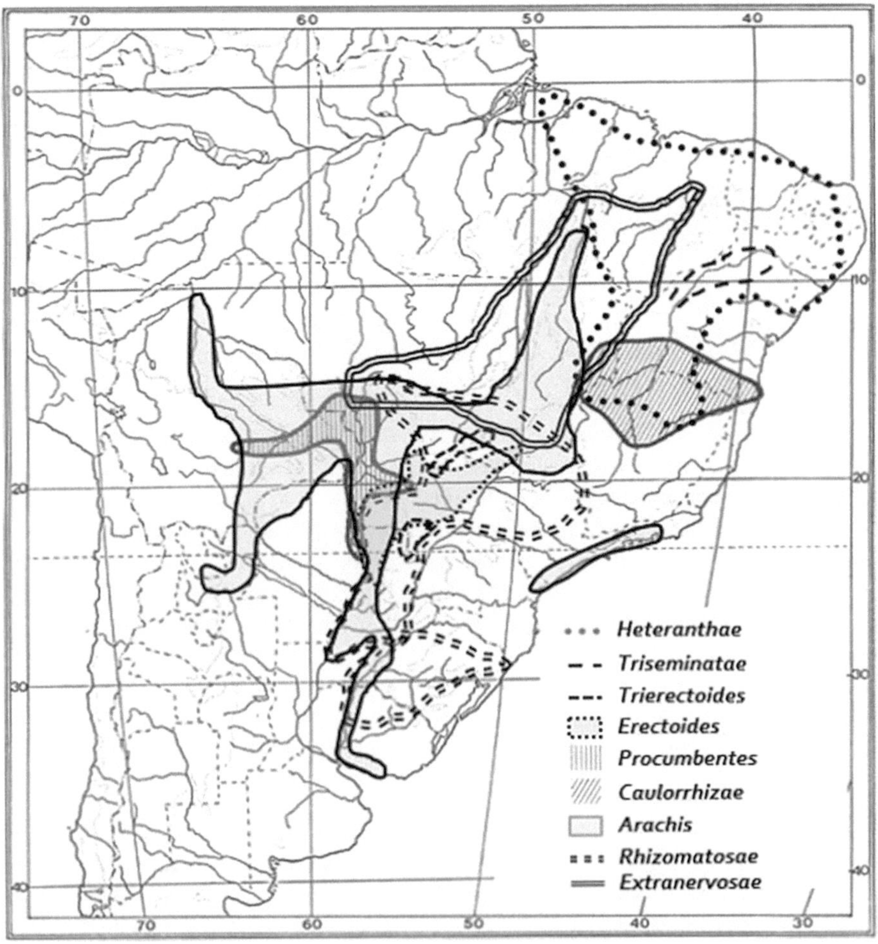

Fig. 2.2 Distribution and extent of various sections of genus *Arachis* in South America (based on Krapovickas and Gregory 1994/2007)

1450 m near Jujuy. They are found mixed in the vegetation of mixed forest to open grassland. The species may grow submerged, among stones bathed with water, in dry gravel and in flood plain alluvium. For these reasons *Arachis* species are found from semiarid region to the tropical locations receiving an average rainfall more than 2000 mm and subjected either to intense drought or flooding. Adaptation of wild *Arachis* species to such diverse conditions has resulted in generation of great genetic variability and resilience to grow under diverse and adverse conditions. This probably led to the development of geocarpy and tuberiform roots to overcome the harshness of dry conditions and to escape the seasonal fires. However, in cultivated groundnut (*A. hypogaea*), selection pressure against the tubiriform roots led to elimination of this trait, but the geocarpy providing protection to fruits from adverse external environment, ensuring regeneration was retained.

Krapovickas and Gregory (1994) considered that the genus *Arachis* originated in the Sierra de Amambay, on the border between Mato Grosso do Sul (Brazil) and Paraguay, where grew, possibly, the oldest species of the genus, *A. guaranitica* (Gregory et al. 1980). It has been difficult to understand how the genus could have dispersed to some 4000 km, both toward northeast up to Amazon as well as to the west, up to the Andes. Fluvial system associated with rivers, streams and the deposits, and landforms created by them must have played an important role in dispersal, as many species have distribution associated with the watershed of the great Paraguay, Uruguay, and Parana or Sao Francisco rivers. The species generally live near watercourse, in places where the water evidently reaches only during the higher floods. The geocarpic habit also indicates possible support to long distance dispersal of species through the rivers and streams. For this reason, Gregory et al. (1973, 1980) postulated that most ancient species were found in higher elevations, their immediate descendants occupied the next lower eroded surfaces, while the distantly evolved species occupied still lower and more recently eroded surfaces. Further, as seeds moved to lower elevations, they became isolated in major river valleys; thus probably different sections of the genus evolved independently in parallel fashion. This perception, however, is changing with record of overlapped distribution of species belonging to some sections. Dispersion of species has also occurred by animal and human movement (Singh et al. 2004).

Based on Krapovickas and Gregory (1994), Fig. 2.2 presents the extent and distribution of genus *Arachis* and its various sections in South America. Section *Trierectoides* lives in the highest places of the divide between the watersheds of the Paraguay and Parana rivers, 400–700 m amsl. The northern limits are found in Jatai, in Goias at 700 m amsl between the Araguaia and Paranaiba rivers. Section *Erectoides* is characteristic of the ‘cerrado’ with red soil, which surrounds the Mato Grosso Pantanal and is nearly exclusive to Mato Grosso do Sul, with some species going beyond and others extending into Paraguay. The other group of this section lives in the southwest extreme of the section’s range. All species of section *Extranervosae* live in state of Goias, Tocantins, the central part of Mato Grosso, and the northern part of the Mining Triangle in Minas Gerais. A few extend beyond these limits. *A. villosulicarpa*, the other cultivated species of genus, is grown by the indigenous people of west central Mato grosso. The majority of the species in this section grow on a very special soil type, frequently encountered in the "cerrado", constituted by a thin layer of soil over a stony substrate. The lone member of *Triseminatae* grows in the state of Bahia, in the south of Pernambuco, in the north of Minas Gerais, and in the vicinity of the São Francisco River. Section *Heteranthae* is a typical of northeastern region of Brazil. Section *Caulorrhizae* grows in the border area between the Brazilian states of Goiás, Bahia, and Minas Gerais, reaching as far as the Atlantic coast, where the type specimen of *A. pintoi* was collected. Section *Procumentes* extends itself along the Paraguay River from Concepción on the Tropic of Capricorn, toward the north, flanking the Pantanal in Mato Grosso on the south and the north, and then expands westward as far as Santa Cruz de la Sierra, in Bolivia, living in soils that are periodically flooded. Section *Rhizomatosae's* tetraploid species occupy a central position within the overall range of the genus *Arachis* and the diploid *A. burkartii* growing more to the south. Section *Arachis* is distributed along an axis that coincides more or less with the 57th and 58th meridians, that encompasses the watersheds of the Paraguay and Uruguay rivers and ends at the La Plata River. It is the most widely distributed section invading/overlapping the areas of other sections. The perennial species of section are found along water house and some are adapted to flooding. Further, two arms of section extend toward the north and correspond to the basin of the Tocantins River to the east and to the Mamoré and Guaporé River system to the west, between Trinidad and Guayaramerín, in Bolivia (Fig. 2.2). In these two expansions, the species encountered are annuals, adapted to prolonged inundation. *A. stenosperma*, an annual growing on the sands of the Atlantic coast, isolated at the eastern extreme, evidently was carried by the humans (Singh et al. 2004). Expansion of section toward the southwest is constituted by annual species, adapted to conditions of periodic drought. They extend from the dry "chaco" up to the first foothills of the Andes: *A. batizocoi* (300–950 m amsl) and *A. duranensis* (250–1250 m amsl) together with *A. monticola* (1350–1560 m amsl) grow at the highest elevations.

Highest numbers of species representing eight sections of *Arachis* are reported from Brazil, of which four are nearly endemic. Bolivia has the second largest number of species followed by Paraguay, Argentina, and Uruguay. Most species occurring in Brazil are confined to the west central region, with a group of species endemic to the semiarid region of northeast. Further differentiation in patterns of genetic variability in different sections occurred as a result of their adaptation to different ecological niche, where they were caught with a series of land uplifts during their movement downstream in the associated drainage systems. Genetic isolation among the species of section *Arachis* is not that strongly marked as among the species of other sections.

Regarding the origin of cultivated groundnut, *A. hypogaea*, Krapovickas (1969) proposed southern Bolivia and northwestern Argentina, which is the range of the diploid species considered to be involved in its origin. As per Hammons (1994), it probably first occurred in the valleys of the Parana and Paraguay River systems in the Gran Chaco area. Krapovickas (1969) suggested the eastern foothills of the Andes for domestication, based on wide range of ecological diversity and uses of groundnut. This area is also an important center of diversity of primitive subsp. *hypogaea*. Archeological evidence suggests that groundnut has been in cultivation for over 3500 years. Early European explorers found local Indians cultivating groundnut in many islands in the Antilles, on the northeast coasts of Brazil, in all the warm regions of the Rio de la Plata basin, extensively in Peru and sparsely in Mexico.

Of the two subspecies of *A. hypogaea*, the primitive subsp. *hypogaea* has its most important center of variation in Bolivia. In southeast Bolivia, on the first foothills of the Andes in the departments of Tarija and Chuquisaca, samples of cultivated groundnut with the greatest amount of primitive characters have been collected. *A. hypogaea* subsp. *hypogaea* var. *hirsuta* Köhler was found in the archeological deposits from the coast of Peru. *A. hypogaea* subsp. *fastigiata* var. *fastigiata* (Valencia types) has its most important center of variation in Paraguay and is the most widespread variety in all of South America. *A. hypogaea* subsp. *fastigiata* var. *peruviana* is grown in almost all of Peru, especially in the basin of the Marañón River, and is common in Ecuador. Its southern limit is found in northern Bolivia, where a few samples were found in Rurrenabaque on the Beni River and in the department of Pando. A few samples were also obtained in Acre state of Brazil. *A. hypogaea* subsp. *fastigiata* var. *aequatoriana* is nearly confined to Ecuador, primarily cultivated in the provinces of El Oro and Loja. It is also cultivated sporadically in northern Peru. *A. hypogaea* subsp. *fastigiata* var. *vulgaris* (Spanish types) is grown in South America in Uruguay, in Argentina (Santa Fe, Entre Ríos and Corrientes), in southern Brazil and to some extent, in Paraguay. Based on above occurrence, Krapovickas (1969) recognized five and Gregory and Gregory (1976) six centers of genetic diversity, while recent explorations added Ecuador as the seventh center with distinct group of landraces referred as var. *aequatoriana* (Singh and Nigam 1997). These centers (Fig. 2.3) are given below:

1. The eastern foothills of the Andes in Bolivia
2. The Guarani region
3. Goias and Minas Gerais (Brazil)
4. Rondonia and northwest Mato Grosso (Brazil)
5. Peru
6. Northeastern Brazil
7. Ecuador

These centers of diversity present a very high level of genetic variation due to natural introgressive hybridization between divergent types, followed by human selection resulting in production of typical hybrid swarms. The first center (eastern foothills of the Andes in Bolivia) is a center of diversity of subsp. *hypogaea* var. *hypogaea* with few landraces of var. *fastigiata*. In Bolivia, there are indications of introgression between the two subspecies, and 'Overo' and 'Cruceno' types of groundnut are probably the product of such introgression (Krapovickas 1969; Gregory and Gregory 1976). The second center of diversity is the Guarani region, dominated by the erect subsp. *fastigiata* (Valencia types) with rare occurrence of subsp. *hypogaea* and with little evidence of introgression between the two

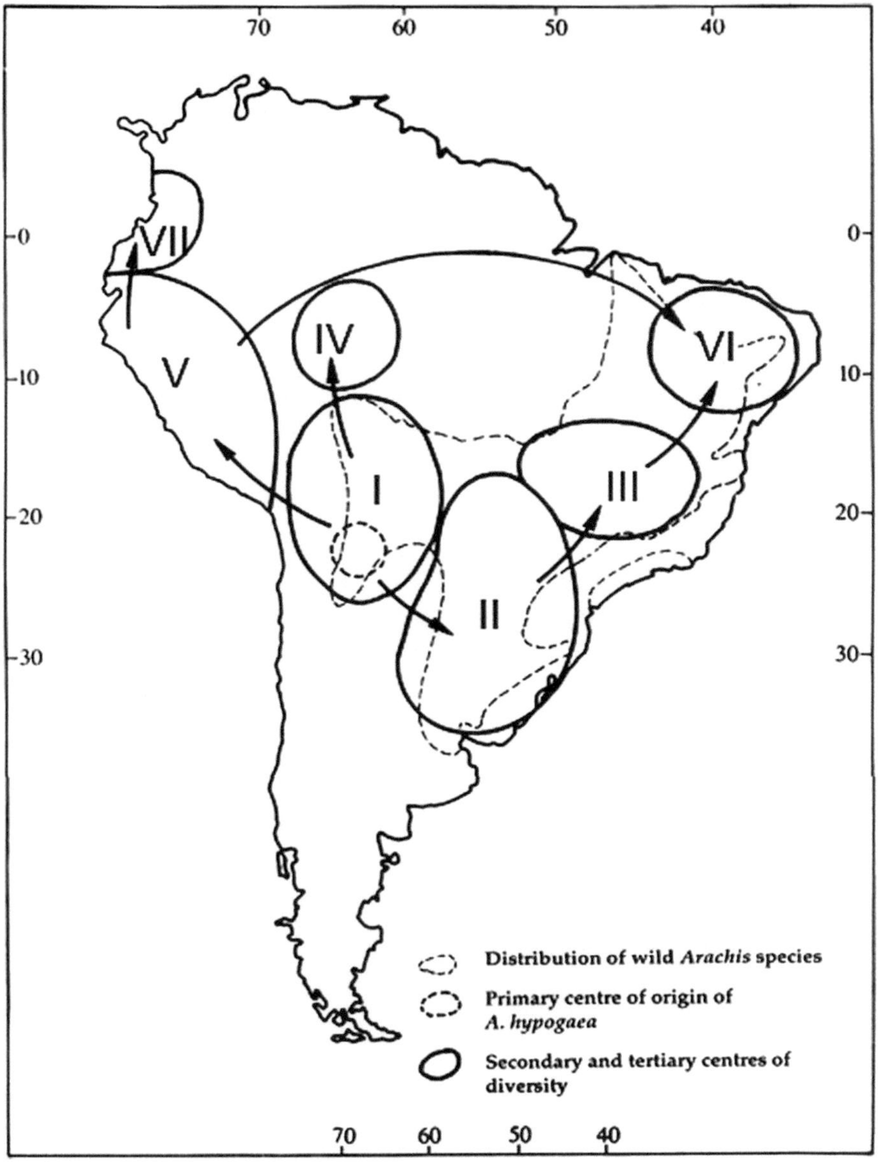

Fig. 2.3 Centers of origin and diversity of *Arachis hypogaea* in South America, I. The eastern foothills of Andes in Bolivia; II. The Guarani region; III. Goias and Minas Gerais; IV. Rondonia and northwest Mato Grosso; V. Peru; VI. Northeastern Brazil and VII. Ecuador

subspecies. However, hybrid swarms of intermediate landraces between the two botanical varieties of subsp. *fastigiata, fastigiata* and *vulgaris*, do exist, consequently both var. *fastigiata* (Valencia), Porto Alegre and var. *vulgaris* (Spanish)

Negrito are identified. It is possible that spread of Valencia types to other parts of the world might have occurred from Paruguay or central Brazil, but more likely point of embarkation is from the northeast coasts. The Guarani region is also center of diversity for var. *vulgaris*, the Spanish types, which probably were disseminated from this region (Krapovickas 1969; Gregory and Gregory 1976). The third center of genetic diversity, Goias and Minas Gerais, has distinct varietal pattern from the Guarani, but is still dominated by erect subsp. *fastigiata* (Valencia types), with very few representative of subsp. *hypogaea*. Landraces of botanical varieties, *fastigiata* and *vulgaris*, are found without much indication of introgression. Rondonia, in the fourth center of diversity represents the nambyquarae types of subsp. *hypogaea*.

The fifth center of diversity, Peru, is represented by the collection of three distinct types. One type includes like the one found in pre-Columbian tombs, with fruits having prominent constriction, veins, and beak and belongs to subsp. *hypogaea*, referred as Chinese type in USA and another with similar fruit characteristics, but belonging to subsp. *fastigiata* var. *fastigiata*. The two together have been referred as 'peruvian' type by Dubard (1906). And a third type with smooth pods, three to five seeds per pod and almost no beak, belonging to subsp. *fastigiata* var. *fastigiata* has also been collected. The sixth center, the northeast of Brazil, is regarded as the tertiary center of diversity (Krapovickas 1969; Gregory and Gregory 1976) with almost all botanical types. Seventh center, Ecuador, represents types similar to var. *fastigiata* from Peru, but morphologically distinct and might even be considered as intermediate between vars. *fastigiata* and *hypogaea*. Williams (1991) studied the region of the north Beni of Bolivia/Peru and collected some more extraordinary types, appearing to be intermediate between subsp. *fastigiata* and subsp. *hypogaea*.

The global dispersal of cultivated groundnut occurred in early 1500, at least in two distinct forms—a two-seeded Brazilian and three-seeded Peruvian types dispersed soon after the discovery of New World (Dubard 1906). Many authorities believe that the Portuguese carried two-seeded varieties from Brazil to Africa, to the Malabar Coast of southwestern India and possibly to the Far East. The Peruvian type (*A. hypogaea* var. *hirsuta*) went to the Western Pacific, to China, to Indonesia (Java), and to Madagascar. Their most plausible path was up the west coast from Peru to Mexico, thence across the Pacific as an item of trade between Acapulco and Manila. Gibbons et al. (1972) recorded cultivar clusters of subsp. *fastigiata* var. *vulgaris* representing both the Guariani region and the region of the eastern slopes of Andes in Bolivia and parts of western Brazil in Africa. In Africa and Asia, groundnut readapted to environment and specialized agricultural production systems. Africa received groundnut from Brazil in West Africa and probably from west coast of South America in the east coast through Philippines, China, and India and became important center of diversity. The Spanish type of groundnut was introduced into Europe from South America (Krapovickas 1969). Higgins (1951) speculated that variety *hypogaea* was introduced to the southeastern United States from Europe, while Simpson et al. (2001) suggested that cultivated peanut traveled in slave ships from Africa into the southeastern United States, Central America, and northeast South America, thus returning as modified germplasm. By the nineteenth century, groundnut became an important food crop in West Africa, Southeast and South Asia, and USA, generating rich genetic diversity.

2.4 Collection, Conservation, and Ensuring Availability of Genetic Resources

Exploration for collecting seeds and living plants of cultivated groundnut varieties and wild *Arachis* species started in mid-twentieth century by USDA and CSIRO scientists. The first exploration dedicated to collection of germplasm was conducted in Argentina in 1945 with the initiation of plant breeding program at the Manfredi Agricultural Experiment Station (Cordoba) and with the organization of the Department of Plant Exploration and Introduction (DEIP) of the Ministerio de Agricultura de la Nación, under the direction of E. C. Clos. Since then to early 70s, extensive explorations were made by Krapovickas (CONICET) and Gregory (USDA) collecting live specimens of wild species and samples of cultivated groundnuts. It was followed with introductions of these collections to other parts of the world. Banks (1976) emphasized the need to make additional collections of the cultivated groundnut and the wild species before destruction of their habitats. Consequent to the support of the International Board for Plant Genetic Resources (IBPGR), FAO, and the International Crops Research Institute for the Semi-Arid Tropics (ICRISAT), 17 expeditions were undertaken between 1976 and 1983 to the centers of origin and diversity of *Arachis* in South America, surveying almost entire area of distribution of the genus (Valls et al. 1985). These efforts continued, enriching the available genetic diversity/collections, till the time of enforcement of Convention on Biological Diversity (CBD), which provided ownership to the nations of their biological resources and made them responsible for their maintenance and conservation, to facilitate their use. Each expedition provided additional locations of both wild and cultivated groundnut, as a result the map of natural resources for different sections was greatly modified from the earlier one presented by Gregory et al. (1980). Additionally, cultivated groundnut accessions were collected from groundnut-growing areas of various countries, included landraces, farmers'/traditional varieties, material developed by the breeders and/or released varieties, and the genetic stocks identified with special features or sources of resistance to biotic and abiotic stresses, representing different botanical varieties and cultivar groups. Groundnut germplasm is conserved as pods or seeds, except for wild *Arachis* species belonging to section *Rhizomatosae*, which rarely produce seed and if produced, progenies are highly heterogeneous and therefore are conserved as live plants under controlled conditions providing an environment close to their habitat. Globally, several repositories have facilities for processing and ex situ conservation/storage of seeds, facilitating prolonged shelf life.

Short-term chambers: These are maintained at 20 ± 5 °C and relative humidity (RH) 35–45 %. Pods/seeds in these chambers remain viable for several years without much loss of viability.

Medium-term chambers: These are prefabricated modules maintaining temperature between 4 to 10 °C and RH between 30 to 40 %. The seeds are dried to 8–10 % moisture level and kept in moisture-proof containers. The seeds remain viable for 25–35 years without much loss of viability.

Long-term chambers: These are also prefabricated modules maintaining temperature around minus 18 °C and host around 1000–1500 seeds of each accession. The seeds are dried to a moisture level of 4–5 %, hermetically sealed in vacuum sealed aluminum pouches, and transferred to the chambers.

The major repositories of the world groundnut germplasm collections are at ICRISAT, India; in the USA at Southern Regional Plant Introduction in Georgia, at North Carolina State University, Raleigh, and at Texas Agricultural Experiment Station, Stephenville; in Brazil at Empressa Brasileria De Pesquisa Agropecuaria (EMBRAPA)/Centro Nacional de Recursos Geneticos (CENARGEN), Brasilia and Instituto Agronomico, Campinas; and in Argentina at Instituto Botanica del Nordeste (IBONE), Corrientes and Instituto Nacional de Technologia Agropecuaria (INTA), Manfredi. The germplasm collections include both cultivated and wild *Arachis* species accessions. ICRISAT maintains over 14,966 from 91 countries, while USDA and NBPGR (India), 9225 and 13,337 accessions, respectively, and provide basic genetic stocks to the international and national scientific community. Table 2.1 presents the germplasm status at some of the major repositories. The low seed multiplication rate and large seed size of groundnut are the major concerns in germpasm multiplication and storage.

Despite assembly of large collections, the use of germplasm accessions in groundnut improvement has remained limited, keeping the genetic base of most groundnut cultivars narrow and vulnerable to biotic and abiotic stresses and natural vagaries. Initially, this was circumvented by ensuring the availability and supply of global genetic resources. For example, ICRISAT since 1980 distributed 38,362 accessions to Indian researchers, leading to increased use of exotic germplasm in national breeding programs. Consequently, during 1985–1995, 46 % of new groundnut variety proposals in All India Coordinated Research Project on Oilseeds had exotic germplasm in their parentage, accruing an increase in national production from 794 kg in 1980 to 988 kg in 1994, about 1.4 % per annum (Singh and Nigam 1996). At the same time to improve information on agronomic and economic traits of each accession that requires extensive evaluation and for manageable quantification of genetic diversity, core collection concept advocated by Brown (1989) was applied, leading to creation of a set of 10 % of total collections retaining most variability of the entire collections. It was hoped that it would facilitate easier access to genetic resources, enhance their use in crop improvement and also simplify their management in genebank. Initially, core collections were developed by stratifying germplasm accessions by country of origin and botanical varieties, followed by the use of data on quantitative morphological traits for principal component/multivariate analysis and clustering, and randomly selecting 10 % of

Table 2.1 Groundnut germplasm holding at some important repositories

Repository	Number of accessions conserved		Additional information
	Primary gene pool	Other gene pools	
ICRISAT, India[a]	Var. *hypogaea* 6838 + *vulgaris* 5493 + *fastigiata* 2351 + *aequitoriana* 14 + *peruviana* 251 + *hirsuta* 19 = **14966**	Representing Section, *Arachis, Rhizomatosae, Extranervosae, Erectoides, Procumbentes Caulorhizae, Triseminatae* = **453**	Representing 91 countries
GRIN and USA[b]	*A. hypogaea* 6804 + ssp. *fastigiata* 361 + ssp. *hypogaea* 141 + var. *aequatoriana* 62 + var. *fastigiata* 1149 + var. *peruviana* 24 + var. *vulgaris* 128 + var. *hirsuta* 29 + var. *hypogaea* 527 = **9225**	Representing Section, *Arachis, Rhizomatosae, Procumbentes, Heteranthae, Extranervosae, Erectoides, Trierectoides, Triseminatae* = **641**	Texas A&M, Experiment Station, Stephenville, maintains 1200 acc. of 70 wild species, 400 hybrids and two mapping populations, while North Carolina State University, USA 740 acc. of primary and 406 of other gene pools
NBPGR, Delhi and DGR, Junagadh, India (*personal communication*)	**13337** at NBPGR, and var. *hypogaea* 2386 + ssp. *fastigiata* 4458 + others 2280 = **9129** at DGR Junagadh, India	Representing *Arachis, Caulorhizae, Heteranthae, Rhizomatosae, Procumbentes, Erectoides* = **112**, **105** respectively	Represent 90 and 84 countries respectively, NBPGR maintains duplicates of ICRISAT accession of Indian origin
Oil Crops Research Institute, Chinese Academy of Agricultural Sciences, and Crops Research Institute, Guaeglong Academy of Agricultural Sciences, China[c]	**7837** + **4210**	**246**	Core of 576 and mini-core of 298
CENARGEN, Brazil and Instituto Agronomicas Campinas, Brazil[c]	**1200** + **2140**	Representing Section, *Arachis, Caulorhizae, Rhizomatosae, Extranervosae, Erectoides, Triseminatae* = **1220**	
INTA, Manfredi and IBONE, Argentina[c]	**3534**	**106** + **472**	

Source
[a]//ICRISAT// Groundnut Crop: www.icrisat.org/crop-groundnut-genebank.htm
[b]www.ars-grin.gov/npgs/cgc_reports/Status11.pdf
[c]Pandey et al. (2012)

collections to constitute a core collection. Using this methodology, first core collection of 831 accessions was developed on US germplasm collection of 7,432 by Holbrook et al. (1993); followed by the development of a core collection of 1,704 accessions on 14,310 world collections assembled at ICRISAT (Upadhyaya et al. 2003); a core collection of 576 from a collection of 6,390 accessions in China (Jiang et al. 2007) and another core collection of 576 accessions, and a mini-core collection of 298 accessions from a collection of 6,839 conserved at the Oil Crops Research Institute of Chinese Academy of Agricultural Sciences at Wuhan (Jiang et al. 2013). Using statistical methods it was ensured that these core collections retained the diversity index and phenotypic correlation of different traits to that of the entire collections so that they represented most spectrum of variability and were effective in the genetic improvement of groundnut. These core collections were further evaluated in multilocations for identification of regional core (Upadhyaya et al. 2005), development of mini-cores (Upadhyaya et al. 2002a), for identification of variability for specific traits (Upadhyaya et al. 2006; Jiang et al. 2013); characterizing the core collections using specific molecular markers to enable better quantification of genetic variability (Kottapalli et al. 2007), and identifying accessions with specific trait/resistance associated with molecular markers (Chamberlin et al. 2010). These efforts have been extended to characterization of diversity using association mapping for exploring the molecular basis of phenotypic variations, demonstrating a great potential of integrating the association analysis and marker-assisted breeding by utilizing the mini-core collection (Ren et al. 2014). Attempts are also made to purify the accessions of mini-core and register them on the basis of morphological, biochemical, and resistance traits (Chen et al. 2013a). Comparison of core collections developed in different parts of the world showed different traits contributing to variability in different set of collections, associated to the dominance of subspecies and botanical varieties in a collection and selection pressure (Jiang et al. 2008), indicating want of a universal core collection for groundnut improvement meeting everyone needs. Most cores are proportionally limited in variability from vars. *hirsuta, peruviana,* and *aequatoriana.*

2.5 Major Constraints to Groundnut Production

Groundnut suffers from several biotic and abiotic production constraints. Some of them are global in nature; and others are either regional or local.

Biotic constraints: Among the foliar fungal diseases, early leaf spot (ELS; *Cercospora arachidicola* Hori.), late leaf spot [LLS; *Phaeoisariopsis personata* (Berk.&M.A. Curtis) Van Arx], and rust (*Puccinia arachidis* Spegazzini) are wide spread and are prevalent wherever groundnut is grown. Other foliar fungal diseases, which could be important in certain regions/countries, include web blotch (*Phoma arachidicola* Marasas, Pauer & Boerema) and pepper spot and leaf scorch [*Leptosphaerulina crassiasca* (sechet) Jackson &Bell]. Among the seed and seedling fungal diseases, preemergence seed and seedling rots [*Aspergillus flavus*

Link ex Fries, *A. niger* van Tieghem, *Macrophomina phaseolina* (Tassi) Goidanich, *Sclerotium rolfsii* Saccardo, *Rhizoctonia solani* Kühn, *Lasiodiplodia theobromae* (Pat.) Griffon & Maubl., *Rhizopus* spp., *Penicillium* spp., *Phythium* spp., and *Fusarium* spp.], Aspergillus crown rot/ collar rot (*A. niger* van Tieghem), yellow mold (*A. flavus* Link ex Fries), and Rhizoctonia damping off (*Rhizoctonia solani* Kühn) are wide spread. Important stem, root, and pod diseases caused by fungi include stem and pod rots (*Sclerotium rolfsii* Saccardo), Sclerotinia blight [*Sclerotinia sclerotiorum* (Lib.) de Bary], Cylindrocladium black rot [CBR; *Cylindrocladium crotalariae* (Loos) Bell & Sobers], Botrytis blight (*Botrytis cinerea* Pers. Ex Fries), Fusarium wilt (*Fusarium oxysporum* Schlechtend. Emend Snyder & Hans.), charcoal rot [*Macrophomina phaseolina* (Tassi) Goidanich], and pod rot [*Pythium myriotylum* Dreschler, *Rhizoctonia solani* Kühn, *Fusarium solani* (Mart.) Saccardo f. sp phaseoli (Burkholder) Snyder & Hans., *Fusarium oxysporum* Schlechtend. Emend Snyder & Hans., and *Macrophomina phaseolina* (Tassi) Goidanich] are important. The groundnut pod and kernels can also get infected while developing with *A. flavus* Link ex Fries/*A. parasiticus* Speare leading to their contamination by aflatoxin. The only bacterial disease of significance is bacterial wilt [BW; *Ralstonia solanacearum* (E.F. Smith)], which is wide spread in East and Southeast Asia.

Significant virus diseases include peanut stripe (PStV; *Peanut stripe virus*) in East and Southeast Asia, peanut clump (PCV; *Peanut clump virus*) in West Africa, peanut bud necrosis (PBND; *Peanut bud necrosis virus*) in South Asia, tomato spotted wilt (TSWV; *Tomato spotted wilt virus*) in North America, peanut stem necrosis (PSND; *Tobacco streak virus*) in South India, and groundnut rosette disease (GRD; a complex of *groundnut rosette virus*, *groundnut rosette assistor virus,* and a satellite RNA) in Africa. Important diseases caused by nematodes are root knot [*Meloidogyne arenaria* (Neal) Chitwood, *M. hapla* Chitwood, *M. javanica,* and *M. incognita*; the first two are wide spread], root lesion [*Pratylenchus brachyurus* (Godfrey) Filipjev & Sch. Stekh.], and Kalahasti malady (*Tylenchorhynchus brevilineatus* Williams) in Andhra Pradesh in India.

Defoliators, tobacco caterpillar/tobacco armyworm (*Spodoptera litura* Fab.), hairy caterpillars (*Amsacta albistriga* Walk., *A. moori* Butler), Bihar hairy caterpillar [*Spilosoma* (*Diacrisia*) *oblique* (Walk.)], gram pod borer (*Helicoverpa armigera* Hübner) and groundnut leaf miner (*Aproaerema modicella* Deventer), sucking pests, aphids (*Aphis craccivora* Koch.), thrips (*Scirtothrips dorsalis* Hood., *Thrips palmi* Karny., *Frankliniella schultzei* Trybom) and *Caliothrips indicus* Bagnall and jassids/leaf hopper (*Empoasca kerri* Pruthi), root and pod feeders, white grub [*Lachnosterna* (= *Holotrichia*) *consanguinea* Blanch.] and *L. serrata* (Fab.), termites/white ants (*Microtermes* spp. and *Odontotermes* spp.) and earwig (*Anisolabis stali* Dohrn) and storage pests groundnut, bruchid (*Caryedon serratus* Oliver), and red flour beetle (*Tribolium castaneum* Herbst) are the major insect pests of groundnut.

Abiotic constraints: As a majority of groundnut is grown under rainfed conditions, drought is the most significant abiotic stress affecting groundnut production. Drought can occur at any stage—early-season, mid-season, end-of-season, and intermittent. Drought also predisposes groundnut pods to aflatoxin contamination by *A. flavus*. Other abiotic constraints include low soil fertility, salinity, iron chlorosis, aluminum toxicity, cold temperature at germination, and high temperature at podding and harvest.

2.6 Searching Potential Genetic Resources and Managing Constraints with Genetic Improvement

For assessment of genetic variability and identification of genetic resources with desired features, the groundnut germplasm assembled at various places have been characterized and evaluated based on the common groundnut descriptors developed by IBPGR/ICRISAT (1992). Variability analysis has shown greater variation in landraces and in the accessions collected from the primary and secondary centers of diversity in South America, particularly for resistance to biotic and abiotic stresses and agronomic features like seed mass (Singh and Simpson 1994). An assessment of genetic diversity on world collection at ICRISAT for 16 morphological and 10 agronomic traits has shown vast diversity in size and shape of pods and seeds. Principal Component analysis using 38 traits and clustering on first seven PC scores produced three clusters; consisting North America, middle East, and East Asia in the first cluster, South America in the second cluster, and West Africa, Europe, Central Africa, South Asia, Oceania, Southern Africa, Eastern Africa, Southeast Asia, Central America, and Caribbean in the third cluster. The means for different agronomic traits differed significantly among regions, while the variances for all the traits among regions were heterogeneous. South America cluster showed 100 percent range variation for 12 of the 16 morphological traits and revealed highest range of variation. Assessment of phenotypic diversity in core collection revealed significant variation. The average phenotypic diversity index was higher in the *fastigiata* group (0.146) than the *hypogaea* group (0.141). The *hypogaea* group showed significantly greater mean pod length, pod width, seed length, seed width, yield per plant, and 100-seed weight than the *fastigiata* group in both rainy and postrainy seasons whereas it was opposite for plant height, leaflet length, leaflet width, and shelling percentage with *fastigiata* group showing significantly greater means. Principal coordinate and principal component analyses showed that 12 morphological descriptors and 15 agronomic traits were important in explaining multivariate polymorphism. Leaflet shape and surface, color of standard petal markings, seed color pattern, seed width, and protein content did not significantly account for variation in the first five principal coordinates or components of *fastigiata* and *hypogaea* types, indicating their relatively low importance. The average phenotypic diversity index was similar in both subspecies. The Shannon–Weaver diversity index varied among traits between the two and the diversity within a subspecies/group depended upon the season and traits recorded. Molecular profiling of joint composite collection developed by ICRISAT and EMBRAPA, using 21 SSRs showed rich allelic diversity, group-specific unique alleles, and common alleles sharing between subspecies and geographical groups. Gene diversity ranged from 0.559–0.926, with an average of 0.819. Group-specific unique alleles were 101 in wild *Arachis*, 50 in subsp. *fastigiata*, and only 11 in subsp. *hypogaea*. Accessions from Americas revealed the highest number of unique alleles (109), while Africa and Asia had only six and nine, respectively. The two subsp. *hypogaea* and *fastigiata* shared 70 alleles. In contrast, the wild *Arachis* shared only 15 alleles with *hypogaea* and 32 alleles with *fastigiata* (//ICRISAT// Groundnut Crop: www.icrisat.org/crop-groundnut-genebank.htm). Greater genetic diversity among the landraces originating from primary and secondary centers of diversity in South America is corroborated by the molecular characterization using various markers, for example, in Bolivian landraces (Husain and Mallikarjuna 2012).

For many traits, the primary gene pool has been found limited, but the wild *Arachis* species have been found with desired variability; for example, for PStV no resistant line was found in cultivated groundnut despite screening of 9,000 accessions, but several accessions of wild *Arachis* showed negative reaction (Culver et al. 1987; Prasada Rao et al. 1991). Often wild *Arachis* species have shown higher level of resistance than primary gene pool. Variability observed among the accessions of wild species for their reaction against specific constraints (Singh et al. 1996) demands thorough investigation for useful exploitation. Table 2.2 presents the number of accessions identified with useful diversity and used in breeding programs, and Table 2.3 lists representative wild species with multiple resistances.

2.6.1 Genetic Improvement Using Resources of Primary Gene Pool Through Conventional Breeding

Groundnut is a highly self-pollinated crop, though cross-pollination can reach as high as 10 % at locations and in seasons, where bee activity is high. Standard breeding methods, followed in groundnut for developing a cultivar, can be placed into two groups—(a) methods without hybridization, and (b) methods after hybridization. The former includes introduction, pure line/ mass selection and mutation breeding and the latter bulk selection, pedigree selection, bulk pedigree selection, single seed descent method, backcross method, and recurrent selection. Among these, pedigree and bulk selection methods are more commonly used by the breeders. Some breeders use a combination of breeding methods or make modification to conventional methods.

Examples of release of cultivars following introduction and selection include, among others, release of JL 24 in India, release of Indian cultivars, TMV 2 and JL 24, under different names in many countries in Southeast Asia and Africa, release of Makulu Red, Apollo, Egret, Chalimbana, Mani Pintar, and Malimba in Africa, and New Mexico Valencia C in the USA. Mutation breeding uses X-rays, gamma rays, and various chemicals to create mutations breaking specific linkages and enhancing variation for specific character in a genotype. Bhabha Atomic Research Centre

Table 2.2 Primary Gene Pool Genetic diversity for useful traits in world collections at ICRISAT

Stress/trait	Acc. screened	Acc. with desirable variability	Additional information
Fungal and bacterial diseases			
Early leaf spot	7000	37 (2)	15 (India) + 4 (Malawi) + 18 (West Africa)
Late leaf spot	13000	69 (26)	
Rust	13000	169 (35)	
Aflatoxin production	582	4	
Aflatoxin seed invasion	580	39 (4)	
Pod and stem rot	3222	24/9 (6)	4 (India) + 5 (USA)
Bacterial wilt		24	Screening in Indonesia and China
Viral diseases			
PBNV	7400	23	
PMV	6944	2	Promise in wild *Arachis* spp.
Insect pest			
Thrips	5345	15 (7)	Promise in wild *Arachis* spp.
Jassids	136	30/6 (7)	
Termites	520	20/9 (6)	
Leaf minor	600	18/4 (6)	
Aphids	300	4 (1)	
Abiotic stress			
Drought	820	46 (8)	
N fixation	342	4 (2)	
Nutritional traits			
High oil content	8868	20/44 (10)	
High protein content	8868	117/51	

Source Singh and Nigam (1997), Singh et al. (1997) Figure in parenthesis indicate number of commonly used source in breeding program

(BARC), Mumbai, India used gamma rays to create desired variation for further use in conjunction with other breeding methods and released 15 groundnut cultivars till date. Some of these are TG #s 19, 37A, 38B and 51, TAG 24, TGB 39, TPG 41, TLG 45.

Hybridization provides opportunity to combine genes from different parents and recombine them in a single genotype via single cross, three-way cross, four-way cross, convergent cross, diallel mating, and diallel selective mating. Selection is practiced in segregating generations following the method of selection of choice. Generally, this process takes 12–15 years, but can be expedited by taking multiple crops in a year under controlled greenhouse conditions or raising off-season nursery at other locations where environmental conditions are favorable to raise a crop.

Table 2.3 Representative desirable genetic diversity in secondary, tertiary, and quaternary gene pool of *Arachis*

Species/gene pool[a]	Early leafspot	Late leafspot	Rust	PStV[b]	GRD[c]	Thrips	Leaf hopper	Lepidoptera
Secondary gene pool (Sect. *Arachis*)								
A. diogoi	R	R			R			
A. duranensis	MR	–	I	I	–	S	I	HR
A. spegazzini	MR	_	I	R	–	R	HR	HR
A stenosperma	HR	HR	HR	–	R	HR	HR	HR
A. villosa	R	R	I	S	R	–	–	–
A. correntina	–	–	I	R	–	HR	HR	HR
A. cardenasii	HR	HR	I	R	R	HR	HR	HR
A. chacoense	HR	HR	I	S	–	HR	HR	HR
A. kempff-mercadoi	R	R	–	–	–	–	R	R
Tertiary gene pool								
A. appressipila (P)	R	–	R	–	R	–	–	–
A. rigonii (P)	–	–	–	S	–	HR	I	HR
A. benthamii (E)	MR	MR	–	I or R	–	–	–	–
A. paraguariensis (E)	R	MR	R	R	–	–	R	–
A. glabrata (R_2)	S or MR	S or MR	I	R or I	–	I	HR	HR
Quaternary gene pool								
A. repens (C)	R	R	R	I	–	I	HR	HR
A. lutescens (Ex)	R	HR			–			
B. macedoi (Ex)	R	–	–	–	–	I	I	–
A. villosulicarpa (Ex)	HR	HR	I	–	–	–	–	–
A. pusilla (Tri)	R	R	I	–	–	I	I	HR
A. triseminata (Tri)	R	R	R		R		R	R

Source Stalker and Moss (1987), Upadhyaya et al. (2011)
[a]Symbol in parenthesis for section; Reaction- *MR* Moderately resistant, *R* Resistance, *HR* Highly resistant, *I* Immune, *S* Susceptible, it may vary between accessions of same species
[b]PStV = Peanut stunt virus
[c]GRD = Groundnut rosette disease

In spite of limited DNA polymorphism, there is abundant morphological variability present for most of the traits among germplasm accessions of the cultivated species. However, only an insignificant portion of this large variability has been utilized for crop improvement for reasons described earlier. To promote intensified and diversified use of genetic resources in crop improvement, recently, core and mini-core collections in groundnut have been developed (Holbrook et al. 1993;

Upadhyaya et al. 2002a, 2003, 2010), which capture representative variability in cultivated groundnut germplasm collection. However, in any applied breeding program where breeders have to maintain physical, chemical, and esthetic quality parameters as per the market requirements, breeders are often reluctant to use primitive germplasm because of linkage drag, which takes a long time to get rid off. Any variation in these qualities discourages processors to accept new genotypes as these variations affect the quality of their products and efficiency of their processing operations. Having used primitive germplasm in the beginning of the program, the breeders prefer to use second- or third-generation breeding lines with desired genes for use in breeding programs. In a recent publication, Janila and Nigam (2013) have reviewed the phenotyping protocols for various biotic and abiotic stresses, which are being followed in groundnut improvement programs. Murthy and Reddy (1993) and Reddy and Murthy (1996) have summarized the results of various genetic and inheritance studies covering most of the traits in groundnut. For details, readers are advised to refer to these publications. The status of up-to-date efforts, made for genetic improvement of groundnut in relation to various traits of significance for an overall engineering of cultivars incorporating both desirable agronomic features and resilience to stress factors, is described below.

Yield and yield-related traits: Yield is a complex trait with quantitative inheritance. In addition to yield, pod and seed characters are also important for esthetic and commercial considerations. Most of the pod and seed characters with few exceptions are governed by a few genes. Selection either for higher pod yield or for greater harvest index is essential for improvement of yield potential in future cultivars. Remobilization of reserves from vegetative biomass to pods under conditions of source limitations (falling temperature, defoliation by pathogens or water stress) is likely to be significant in maintaining yields but may limit response to improved conditions specially with high partitioning. Newer high-yielding cultivars in the USA allocate a greater proportion of biomass to reproductive tissue early in the growth cycle with greater reproductive efficiency and have more spreading growth habit and greater seed and pod weight than older cultivars (Wells et al. 1991; Seaton et al. 1992). However, the crop duration in the USA is much longer (140–160 days) than the one available to the crop in South and Southeast Asia and West Africa (<100 days).

Between 1944 and 1987, the average yearly genetic gain for yield in Virginia market type cultivars in the USA was 14.7 kg per ha. However, when the emphasis in breeding programs shifted to pest resistance, earliness, and quality, the new cultivars improved upon these traits but failed to combine them with increased yield potential (Mozingo et al. 1987). During 1980s and mid-1990s, the groundnut yield in India increased by 1.4 % per annum (Nigam et al. 1991). New Spanish-type cultivars in India have greater seed size, seed weight, and pod numbers per plant than the older cultivars (Rathnakumar et al. 2012). Increase of 0.43 % per year in seed yield, of 0.29 % per year in seed weight, and of 0.52 % per year in pod growth rate during 1948–2004 were obtained in Argentina (Haro et al. 2013).

Resistance to foliar fungal diseases: ELS, LLS, and rust are the most widely distributed and economically important diseases of groundnut in the world. Only one leaf spot dominates in a region, however, both pathogens can be observed in the same field. LLS and rust often occur together. Breeding for resistance to foliar diseases in groundnut got a real boost in the late 1970s and early 1980s when a massive field screening program for resistance to foliar fungal diseases (rust and LLS) of more than 13,000 germplasm accessions from 89 countries was launched at ICRISAT, Patancheru, India. Subsequently, many sources of resistance were identified and used in breeding programs (Singh et al. 1997). Most of these resistant sources are landraces from South America and have undesirable agronomic characters (low yield, poor pod and seed traits, and longer crop duration). The components of resistance include longer incubation period and latent period, reduced sporulation, smaller lesion diameter, lower infection frequency, and less defoliation in resistance sources. Combining high levels of resistance to ELS and LLS into high-yielding cultivars with acceptable market traits continues to be difficult.

The first-generation cultivars emanating from foliar diseases resistance programs in India, ICG (FDRS) 10 and Girnar 1, did not find acceptance among the farmers and traders in spite of their higher yields under heavy disease pressure due to unattractive pod and seed characteristics. However, when these cultivars were recycled again in breeding programs, the resultant second-generation genotypes had better pod and seed characteristics and more acceptability among farmers and traders in spite of some dilution in their levels of resistance.

ELS, is more serious in Southern Africa and the USA. Resistant/tolerant *A. hypogaea* genotypes have been identified in Malawi, West Africa, India, and the USA with a disease score ranging between 3.6 and 6.3 on a 1–9 scale, where 1 = no disease and 9 = more than 81 % foliage destroyed. However, resistant sources reported from the USA (NC 3033 and PI #s 270806, 259747, and 350680) were found susceptible in India and Malawi. Excessive use of chemicals to control the disease in the USA is suspected to have led to variation in pathogen. Genotypes ICG #s 6284, 6902, 7878, 10000, 10948, and 13917 show some resistance at more than one location. Rate-reducing resistance is quantitative in nature and controlled by both additive and nonadditive gene effects including maternal effects. Duplicate recessive inheritance is also observed. Narrow sense heritability varies from low to high. Some of the ELS tolerant cultivars released in India are ICGS 44, ICGS 76, M 335, BG 3, Somnath, CSMG 84-1, M 522, Prutha, and GG 7 and in the USA are VA 81B, DP 1, Georganic, C-99R, Bailey, Florida 07, and Sugg.

LLS is predominant in warmer areas. Sixty-nine *A. hypogaea* genotypes tolerant to LLS with disease score ranging between 3 and 5 on a 1–9 scale (described earlier) have been identified. Forty-nine of these resistant sources are landraces belonging to var. *peruviana* with low yield and shelling outturn and other undesirable pod and seed characters. Resistance to LLS is quantitative in nature and governed by both additive and nonadditive gene effects including maternal effects. Duplicate recessive inheritance is also reported. Tolerant cultivars released in India include RG 141, ICG(FDRS) 10, ICGV #s 86590 and 86325, K 134, Girnar 1, GBPD 4, R 8808, ALR #s 1, 2, and 3, BSR 1, VRI 5, and CSMG 84-1 and in the

USA, Southern Runner, Florida MDR 98, TUFRunner TM '727', Florida 07, and C-99R, among others.

For rust, of the 169 *A. hypogaea* genotypes reported resistant (a score of five or less on a 1–9 scale), 135 are landraces belonging to var. *peruviana*. Many of these (ICG #s 7896, 7897, 7899, 10014, 10030, 10052, 10053, 10067, 10933, 10939, 10940, and 10943) have a disease score of <3 but are agronomically poor (low shelling outturn, thick pod shell, strong pod reticulation, and unacceptable seed coat color). New sources of resistance—ICG #s 10056, 10567, 10925, 10932, 11108, 12059, 12112, and 12113 and the interspecific derivatives involving *A. batizocoi* and *A. duranensis*—have high levels of resistance with good agronomic potential and resistance to other biotic stresses. Resistance to rust is reported to be recessive, partial dominant, or dominant with duplicate recessive, digenic, trigenic, or multigenic inheritance. Resistant cultivars released in India include ICG (FDRS) 10, ICGV 86590, and GBPD 4, among others.

Resistance to rust and LLS is correlated ($r = 0.48–0.60$). Forty-two LLS resistant genotypes are also resistant to rust. Of these, ICG #s 1703, 4995, and 10920 and interspecific derivative ICG 13917 [259-2 (red)] are useful in multiple resistance breeding, the last one being resistant to all the three pathogens. Other useful sources of resistance to both LLS and rust with agronomic potential are *A. hypogaea* genotypes ICG #s 6330, 7884, 10023, 10035, and 11182 and interspecific derivatives ICG #s 11312, 11317 (also resistant to ELS), 11321, 11325, 11337, 13916, 13917, 13919, 13920, and 13922. Cultivars reported with multiple resistances to foliar diseases, among others, are ICG (FDRS) 4, ICG (FDRS) 10, Girnar 1, ICGV 86590, Somnath, GBPD 4, VRI 2, VRIGn 5, and ALR #s 1, 2, and 3 in India and Azivivi, Nkosour, Adepa, and Jenkaa in Ghana.

Resistance to soil-borne fungal diseases: Breeding for resistance to soil-borne fungal diseases continues to be difficult as creating uniform disease pressure in the disease screening nursery remains challenging. Breeding for resistance to *A. flavus/A. parasiticus* and aflatoxin contamination has received the most attention among the soil-borne fungal diseases. Other diseases, where breeding efforts are in progress, include pod and stem rots, cylindrocladium black rot (CBR), and sclerotinia blight.

Efforts on breeding for resistance to *A. flavus/A. parasiticus* invasion and aflatoxin production in the USA, where *A. parasiticus* is dominant, and in other countries in Asia and Africa, where *A. flavus* predominates, are in progress. However, they have not yet succeeded in ensuring complete freedom from *A. flavus/A. parasiticus* infection and aflatoxin contamination in groundnut cultivars. Liang et al. (2009) and Nigam et al. (2009) have reviewed the progress in breeding for resistance to *A. flavus* invasion and aflatoxin contamination at ICRISAT, India/West Africa and Guangdong Academy of Agricultural Sciences in China, respectively.

There are three barriers to *A. flavus/A. parasiticus* infection and aflatoxin production in groundnut seed—pod wall, seed coat, and cotyledons. Resistance to pod infection is attributed to shell wall structure and that of seed coat to thickness and density of palisade layers, absence of fissures and cavities, and presence of wax and cutin layers on the seed coat. Cotyledons are where the toxin is produced. Three resistance mechanisms—preharvest resistance, seed coat resistance [in vitro seed colonization (IVSC)], and cotyledon resistance (aflatoxin production) are independently inherited and provide opportunity for gene pyramiding (Upadhyaya et al. 2002b). The genetics of resistance is not clearly understood. There are a few published reports on broad sense heritability of three mechanisms of resistance (low to moderate) and combining ability of resistance sources (Rao et al. 1989; Upadhyaya et al. 1997a). A pair of major genes with additive value of 0.38 and a pair of minor genes with additive value of 0.12 were reported to be responsible for resistance to seed infection by *A. flavus* (Zhou et al. 1999; Zhou and Liang 2002). Predominantly, nonadditive genetic variance for aflatoxin production is also reported. Sources of resistant to preharvest infection [≤2 % infection; 21 genotypes —ICG #s 1122, 1173, 1323, 1326 (J 11)*, 1859, 1994, 3263 (U 4-47-7)*, 3267, 3336*, 3700*, 4589, 4749 (PI 337394 F)*, 4888, 7633 (UF 71513)*, etc.; *consistent across locations], IVSC (≤15 % seed colonized; PI #s 337394F* and 337409*, UF 71513, Ah 78223, J 11*, US-47-7, Var 27, Faizpur, Monir 240-30; * consistent across locations and pathogen pressure), and aflatoxin production (<0.7 μg per kg: ICG #s 10609, 11682, 10615, 6760, 9610) are available, but none of these is completely free from infection or aflatoxin production. Containment of preharvest infection is essential as once infected, the seed cannot be disinfected and the infection is carried forward. Seed coat resistance provides postharvest protection in storage. Recommended genotypes for use in breeding because of their multiple resistances include ICG #s 1326, 1859, 3263, 3336, 3700, 4749, 7633, 9407, 9610, 10094, etc.

Drought predisposes groundnut to aflatoxin contamination. Some drought-tolerant lines also show low preharvest seed infection and aflatoxin production (Holbrook et al. 2000a). Fatty acid composition is also reported to influence directly or indirectly aflatoxin contamination (Holbrook et al. 2000c; Xue et al. 2005). N_2 fixation and related traits have negative and significant effects on kernel infection and aflatoxin contamination especially under drought conditions (Arunyanark et al. 2012). Girdthai et al. (2010) suggested inclusion of SLA [specific leaf area; positively correlated with aflatoxin traits (seed infection and aflatoxin contamination)] and SCMR (SPAD chlorophyll meter reading; negatively correlated with aflatoxin traits), among other traits, in selection for resistance to aflatoxin contamination. It is advisable to have more number of replications in field screening nursery as plot to plot and plant to plant variations within a plot for preharvest seed infection, despite sufficient fungal propagules being present in the soil, which is often large. Screening for resistance to in vitro seed colonization and aflatoxin production is done in laboratory following protocols prescribed by various researchers. During field and laboratory screening, it is not unusual to find nil preharvest infection but presence of aflatoxin in the same genotype and the reverse is also observed. Conventional breeding alone does not ensure complete freedom from aflatoxin contamination; at best it is able to combine the level of resistance available in resistant parents with high yield and other agronomic characters. Attempts to pyramid resistance genes of different resistance mechanisms have also not changed the situation much. Elite breeding lines giving good performance in India and Mali/Niger include ICGV #s 88145, 89104, 91278, 91283, 91284, 87084, 87094, and 87110 and in China include ICGV #s 95440, 95422, 94435, and 95435 and UF 71315. Yueyou #s 9 and 20 are released cultivars in China which are resistant to *A. flavus* invasion (Liang et al. 2009).

Stem and pod rots (stem rot also known as white mold, southern blight, sclerotium rot or white mold), caused by *Sclerotium rolfsii*, is wide spread in major groundnut-growing areas in the world. Some screening techniques have been described by Mehan et al. (1995). Field screening is more consistent than screening in the greenhouse. Uniformity and level of inoculum in the sick plot can be enhanced by adding sterilized oat seed inoculum of *S. rolfsii*, but individual plants may still escape the infection. The 'agar disk technique' is used to screen individual plants. Sources of moderate resistance include NC 2, NC Ac #s 18016 and 18416, ICG #s 15233, 15234, 15235, and 15236, and ICGV #s 86590 and 87160, among others. Bera et al. (2014) screened 286 interspecific derivatives in a sick plot in the field and in concrete blocks and pots filled with sick soil and found NRCGCS #s 47, 99, 131, and 319 highly promising. Cultivars with moderate resistance released in the USA are Southern Runner, Toalson, Pronto, Georgia Browne, Sunbelt Runner, and Tamrun 96.

Cylindrocladium black rot (CBR), caused by *Cylindrocladium parasiticum*, is largely reported from the USA, particularly from North Carolina. The screening for resistance to CBR is done at naturally infested hot spot locations. In general, the Spanish cultivars are most resistant, the Valencia cultivars most susceptible, and the Virginia cultivars moderately susceptible. The inheritance of resistance is complex (Green et al. 1983). The resistance delays the onset of epidemics rather than the rate of disease progress (Culbreath et al. 1991). NC #s 8 C, 10 C, and 12 C are the partial resistant cultivars released in the USA.

Sclerotinia blight (SB), caused by *Sclerotinia miner* Jagger, is important in Virginia and Oklahoma in the USA. 'Detached shoot technique', which relies on rate of lesion growth and development and disease infected fields (hot spot) are used for screening. In general, cultivars with Spanish ancestry are more resistant than those with Valencia and Virginia ancestries (Akem et al. 1992). Resistance to the disease is quantitative (Wildman et al. 1992). Interspecific lines derived from *A. hypogaea* × *A. cardenasii* cross are highly resistant to the disease. Spanish cultivars Toalson and Tamspan 90 have good resistance to sclerotinia blight.

Resistance to virus diseases: The status of genetic management of virus diseases in groundnut has recently been reviewed by Nigam et al. (2012). The conventional breeding efforts have concentrated only on three virus diseases—peanut bud necrosis disease (PBND) in India, tomato spotted wilt virus (TSWV) disease in the USA, and groundnut rosette disease (GRD) in Africa.

PBND, caused by peanut bud necrosis virus (PBNV), is economically important in South and Southeast Asia. It is transmitted by thrips species *Thrips palmi*; the virus is acquired by the larvae but transmission is done exclusively by the adults in a persistent manner. The virus is not seed transmitted. Several germplasm accessions with consistently low disease incidence in the field (ICG #s 848, 851, 852,

862, 869, 885, 2271, 2306, 2307, 2323, 2741, 3042, 3806, 3873, 5030, 5024, 5043, 5044, 6135, 6317, 6323, 7676, 7892, and others) and breeding lines/cultivars DRG 18, ICG 7812, ICG (FDRS) 10, JSSP 3, KNG 22, PI 393516, and ICGV #s 80325, 86031, and 86388, among others, have been identified at ICRISAT. The last two breeding lines (ICGV #s 86031 and 86388) are resistant to both, the vector and the virus (Dwivedi et al. 1995). Sources of resistance to the vector include NC Ac #s 2242, 2214, 2243, 2240, 2232, 2230, and others. The resistance is stable across environments.

Three factors with additive gene effects are reported to be responsible for low disease incidence (Buiel 1996). Significant gca, sca, and reciprocal effects are also observed for disease incidence with the gca effects being predominant (Pensuk et al. 2002). Because of significant reciprocal effects, the resistant source should be used as female parent in hybridization. Nonadditive gene effects are also reported for low PBND incidence (Pensuk et al. 2004). In another study, additive gene effects were found to be major contributors to PBND resistance besides additive × additive and dominance gene effects (Poledate et al. 2007). Resistant cultivars released are CO 3, ICGS #s 11, 44 (ICGV 87128), and 37 (ICGV 87187), R 8808 (KRG 2), R 9251, K 134, DRG 12, RSHY 1, and Kadiri 4 in India and Khon Kaen 6 in Thailand.

Culbreath et al. (2003) have done an extensive review of epidemiology and management of TSWV disease of groundnut in the USA, where it is a major production constraint. TSWV is transmitted by thrips species, *Frankliniella fusca* (Hinds) (tobacco thrips), *F. occidentalis* (Pergande) (western flower thrips), *F. intonsa*, *F. schultzei*, *S. dorsalis*, *Thrips tabaci*, *T. palmi,* and *T. setosus*, in a persistent manner; the first two are the primary vectors. The virus is not seed or pollen borne. Sources of resistance in cultivated groundnut include PI #s 203396 (also resistant to LLS), 196621, 339967, and 341267. Significant gca and sca (Anderson et al. 1990) and transgressive segregation for TSWV resistance (Holbrook et al. 2003) are reported. However, genetic mechanism of resistance is not elucidated. Breeding lines derived from var. *hypogaea* and var. *hirsuta* have higher resistance to TSWV (Culbreath et al. 2005). TSWV resistant/tolerant cultivars released in the USA are Southern Runner, Georgia Browne, Georgia Green, Tamrun 96, Georgia Bold, Georgia Hi-O/L, Georgia 01R, C-99R, Florida MDR 98, Tifguard, Georganic (highest level of field tolerance among cultivars), and others. However, they may suffer significant damage during extreme epidemics.

Waliyar et al. (2007) have summarized a century of research on GRD and its management. GRD is confined to the African continent and its surrounding islands. It has a complex of three causal agents—groundnut rosette assistor virus (GRAV), groundnut rosette virus (GRV), and a satellite RNA (SatRNA). These three agents synergistically act with each other for survival and spread. GRV is dependent on GRAV for transmission by aphid vector *Aphis craccivora* and SatRNA, which is responsible for rosette symptoms, is itself dependent on GRV for replication. GRV and SatRNA alone do not produce GRD symptoms. GRAV on its own can cause mild yellowing/chlorosis of leaves and can cause reduction in plant growth and yield. GRV and SatRNA must be packaged within GRAV coat protein to be aphid transmissible. GRV is dependent on its SatRNA for encapsidation in coat protein.

GRV on its own produces transient symptoms only. GRV and SatRNA are always found together in nature. These three causal agents are not seed borne. There are two variants of GRD symptoms—chlorotic rosette and green rosette. Chlorotic rosette occurs throughout sub-Saharan Africa and green rosette, which earlier was largely confined to West Africa, is now also reported from Southern and Eastern Africa. SatRNA is responsible for variation in symptoms.

Using viruliferous aphids and grafting, genotypes can be evaluated for resistance to all the three causal agents (Olorunju et al. 1992) in greenhouse. Mechanical sap inoculation can be used only for screening for resistance to GRV and SatRNA. Resistance to GRAV can be evaluated using vector aphids fed on GRAV-infected groundnut plants or by grafting scions on to plants under test from these plants. Resistance to GRD was first found in groundnut germplasm originating from Burkina Faso and Côte d'Ivoire in mid-1950s. Several pure line selections made in late maturing Virginia landraces in Burkina Faso, such as 48-7, 48-14, 48-15A, 48-21, 48-34, 48-35, 48-36, 48-37, 48-44, 48-45, and 48-70A, were resistant to GRD. Subsequently, evaluation of 12,500 germplasm accessions form ICRISAT's gene bank resulted in identification of 150 resistant sources (130 long-duration Virginia types and 20 short-duration Spanish types) (Subrahmanyam et al. 1998; Olorunju et al. 2001). Sources of resistant to aphid vector, EC 36892 (ICG 5240), and ICG 12991, show less GRD but they are susceptible to all the three agents of GRD.

All the sources of resistance to GRD are resistant to GRV and its SatRNA but they are susceptible to GRAV. The resistance to GRAV in cultivated groundnut is yet to be found. Resistance to GRD (effective against GRV and its SatRNA) in cultivated types is governed by two independent recessive genes which are effective against both chlorotic and green rosette (De Berchoux 1960; Nigam and Bock 1990; Olorunju et al. 1992). Resistant cultivars released in Africa include RMP #s 12 and 91, 69-101, KH #s 241D and 149A, RG 1, Nyanda (ICGV 93437), ICG 12991, ICGV-SM #s 90704, 99568, 99555, 99557, 01711, and 01721, and Samnut #s 23 (ICGV-IS 96894), 21 (UGA 2), and 22 (M572.801).

Resistance to bacterial wilt disease: Bacterial wilt, caused by *Ralstonia solanacearum*, is one of the major production constraints in groundnut in Southeast and East Asia. Hot spot locations with naturally infested soils are used to screen germplasm and breeding populations. Breeding for resistance to bacterial wilt was probably the first disease resistance breeding activity in groundnut which was initiated in Indonesia. It led to release in 1925 of Schwarz 21, a bacterial wilt resistant variety selected from a local population, in Indonesia. Since then, several sources of resistance, mostly belonging to subspecies *fastigiata*, have been reported from Indonesia and China (Mehan et al. 1994). These resistance sources belonging to the subspecies *fastigiata* are also early maturing and tolerant of acid soils and poor soil fertility. Nature of the resistance is dependent on the genetic background. In the Chinese dragon type groundnut, the resistance is reported to be partially dominant with cytoplasmic effect (Shan et al. 1998). In *fastigiata* types, the resistance is reported to be partially dominant involving three pairs of major genes and some minor genes (Liao et al. 1986) and partially dominant in some crosses and partially recessive in some crosses (Shan et al. 1998). Several bacterial wilt resistant cultivars have been released—such as Xiekongchung, Teishansanliyue, Yue You #s 13, 589, 92, 256, 200, 256, and 79, Wu You 4, Gui You 28, E Hua 5, Zhong Hua 2, Lu Hua 3, Yuanza 9307, and others in China and Schwarz 21, Gajah, Matjan, Kidang, Banteng, Macan, and others in Indonesia.

Insect pest resistance: Breeding for resistance to insect pests has received limited attention due to difficulty in screening large number of germplasm lines and segregating breeding populations under sporadic and variable natural insect pressure. In most cases, limited screening under field and laboratory or controlled conditions has been carried out leading to identification of sources of tolerance/resistance to major insect pests and characterization of reaction of advanced breeding lines (Amin et al. 1985; Lynch 1990; Wightman and Ranga Rao 1994). Many genotypes with resistance to multiple insect pests are also reported (Nigam et al. 1991). To enhance the natural pressure of insect pests, rows of susceptible genotypes (infester rows) are planted at regular intervals with test materials. The cultured population of the insect pests is also released to raise the levels of insect pressure in the screening nursery. The resistance mechanisms may involve any one of the following or their combinations: repellence, antibiosis, tolerance, physical structures, and avoidance.

Sucking pests such as thrips, jassids, and aphids not only cause direct yield losses but some of them (thrips and aphids) also act as vectors of the virus diseases. Several genotypes resistant to thrips and jassids are reported. Some of the thrips resistant/ tolerant genotypes are listed in sections of PBND and TSWV virus diseases. High density, distribution and length of trichomes (NC Ac #s 2214, 2230, and 2240), and thick leaf cuticle (NC Ac #s 2242 and 2243) are important factors associated with resistance to thrips and jassids. In aphid resistant genotypes, NC Ac 343, EC 36892, and ICGV 86030, antibiosis operates by reducing growth and fecundity (Padgham et al. 1990). Nonadditive genetic variance was predominant for all trichome characters; for trichome length and jassid damage additive genetic variance was also important (Dwivedi et al. 1986). For resistance to complex of pests (thrips, jassids, and *Helicoverpa*) in North Carolina, USA, additive genetic variance was predominant (Holley et al. 1985). Breeding lines tolerant to jassid are ICGV #s 86388, 86462, 86252, 86393, and 86455, among others.

Among defoliators, leaf miner (*Aproaerema modicella* (Deventer)) and tobacco caterpillar (*Spodoptera litura* F.) are important. Screening for resistance to defoliators under natural field conditions is difficult because of variation in infestation in space and time. No-choice cage technique is used to screen for resistance to *S. litura*. Nuclear insect culture is maintained on artificial diet. A known number of first- or third-instar larvae are released for varying period of time on 15-day old greenhouse grown plants which are kept inside a plastic jar cage with wire mesh screen windows. Observations on insect survival (number of surviving larvae and larval weight) and leaf area damage are recorded. For leaf miner, natural infestation is relied upon, which can be enhanced by planting soybean as an infestor crop and creating prolonged drought. It is difficult to devise no-choice cage screening for leaf miner. Breeding lines, ICGV #s 86031, 87154, and 87160, and germplasm accessions, ICG #s 2271 and 1697, showed resistance to both tobacco caterpillar and leaf miner. Other genotypes showing promise against leaf miner are NC Ac #s 343 and 17090 and ICG (FDRS) 4.

Nematode resistance: Breeding for resistance to nematodes has received little attention elsewhere except in the USA, China and to some extent in India. Screening for resistance to *Meloidogyne arenaria* (root knot nematode) in the USA resulted in identification of several genotypes that supported less egg production per gram of fresh root weight (Holbrook and Noe 1992; Holbrook et al. 2000b). COAN was the first groundnut cultivar resistant to *M. arenaria* in the USA. The resistance in COAN was conditioned by a single dominant gene from TxAG 7, which is a backcross derivative of TxAG-6, a complex interspecific derivative involving *A. cardenasii*, *A. batizocoi*, and *A. diogoi*. But it was susceptible to TSWV. In the USA, root knot nematode resistant variety Nema TAM was the first variety developed using marker-assisted selection (MAS), but it was also susceptible to TSWV. Subsequently Tifguard, resistant to both root knot nematode and TSWV, was developed following conventional breeding. 'Kalahasti disease' caused by stunt nematode (*Tylenchorhynchus brevilineatus*) was first noticed in 1975/1976 in Kalahasti area of Andhra Pradesh, India. From replicated screening of 1599 genotypes in a hot spot location in a farmer's field in Kalahasti during 1985/86–1986/87, 14 resistant genotypes were identified. Most of these genotypes had undesirable pod/seed characteristics with the exception of TCG 1518, an advanced Virginia bunch breeding line, which was later released as Tirupati 3 for cultivation in disease-affected areas (Mehan et al. 1993). In another screening exercise of 39 genotypes during 1992–1994, TCGS #s 307, 313, and 320 (released as Kalahasti) were also identified as resistant to the disease with the last two having pod yield exceeding 3 t/ha (Naidu and Moses 2000).

Resistance to abiotic stresses: Drought is the overriding stress factor in rainfed groundnut. Other emerging issues are salinity and heat tolerance. Although considered a drought-tolerant legume, it can still suffer early-season, mid-season, end-of-season, or intermittent droughts impacting adversely on yield and yield-related traits including quality of the produce due to reduced photosynthesis, N_2 fixation, and calcium uptake by developing pods. The impact will depend on the timing of occurrence, duration, and intensity of drought. A 20/25-day moisture stress soon after crop emergence is beneficial to the crop as it forces roots to go deeper into the soil in search of moisture and when the moisture stress is released, it induces profuse flowering resulting in synchronized and uniform maturity and increased yield. The adverse effect of end-of-season drought can be overcome by developing short-duration varieties with their life cycle matching the period of soil moisture availability. It is the mid-season drought that is a cause of worry as insufficient water at the time of flowering and fruiting reduces the yield significantly. Direct selection for yield under drought is effective but it is resource consuming and lacks repeatability across different environments. Drought tolerance can be enhanced by improvements in soil water extraction ability (T) or improvements in water-use efficiency (TE). Genetic variation for root system (Songsri et al. 2008b) and transpiration efficiency (g dry matter per kg of water transpired) is reported, but these traits are difficult to measure. Easily measurable surrogates for these traits are needed for use in a large-scale breeding programs. Transpiration efficiency is negatively correlated with $\Delta\ ^{13}C$ (carbon isotope discrimination) in leaves, which is rapid but expensive to measure. $\Delta\ ^{13}C$ is highly positively correlated with specific leaf area (SLA, ratio of leaf area to leaf dry weight), which is easy and inexpensive to measure (Wright et al. 1994; Nageswar Rao and Wright 1994). SLA has inverse relationship with relative leaf water content (RWC) and the low SLA types are drought tolerant as they are able to maintain higher RWC. However, SLA is influenced by the time of sampling and age of the leaf (Wright et al. 1996). SLA is inversely correlated with SPAD chlorophyll meter reading (SCMR) (Nageswar Rao et al. 2001), which, in turn, is positively correlated with TE. SCMR is measured by a hand held device which is easy to operate and can rapidly record observations. Thus, for fast screening, SCMR can be used in a large-scale breeding programs aiming to improve drought tolerance in groundnut. SLA and SCMR can be recorded any time after 60 days of crop growth, preferably under moisture deficit conditions (Nigam and Aruna 2008a). However, the utility of SLA and SCMR in screening for drought tolerance has been questioned in some studies (Devi et al. 2011). Sufficient variation for physiological traits such as SLA, T, TE, and HI (Nageswar Rao and Wright 1994; Wright et al. 1994, 1996; Nageswar Rao and Nigam 2001) and in tolerance to mid-season and/or terminal droughts is reported (Nageswar Rao et al. 1989; Nigam et al. 2003). High heritability for HI, SCMR, and $\Delta\ ^{13}C$ and medium to high heritability for SLA are reported (Songsri et al. 2008a; Chen et al. 2013b). Both additive and additive $\times$ additive gene effects for SLA and HI and additive gene effects for $\Delta\ ^{13}C$ are reported (Jayalakshmi et al. 1999; Nigam et al. 2001). The segregating populations are screened in the field under imposed drought conditions and selections are made based on pod yield, pod number, and pod filling. In selected populations, surrogates SCMR or SLA can also be used along with pod yield and other characters. Both empirical (yield-based) and trait-based approaches are effective in selecting for drought tolerance (Nigam et al. 2005). In the case of trait-based approach, TE is the major contributor to pod yield, which indicates more efficient utilization of available water. However, in the case of empirical approach, it is T which is a major contributor to pod yield, which indicates better mining of water from soil layers. The better mining does not necessarily mean better utilization of water. In case of limited water availability, enough T may not occur thus impacting on pod yield. It is advisable to integrate surrogates of TE in the selection scheme for drought tolerance. Some of the drought-tolerant breeding lines/cultivars released are ICGS #s 44 and 76, ICG(FDRS) 10, ICGV #s 91114 and 00351, R 8808, GPBD 4, Dh 86, and Kadiri 5 in India, 796, 55-437, and TS 32-1 in West Africa, and BARI 2011 in Pakistan.

When drought occurs, temperature also rises. Drought and heat tolerance appear to be correlated. Besides, breeding for tolerance to high temperatures has also become essential to meet the challenges of changing climate. In vitro pollen germination, pollen tube growth and membrane thermostability, growth rates, fruit set, and partitioning have been used to measure response of groundnut genotypes to high temperature (Craufurd et al. 2003; Ntare et al. 2001; Hamidou et al. 2013). The heat tolerant genotypes are 796, 55-437, ICG 1236, TMV 2, ICGS 11 and ICGV #s 86021, 87281, and 92121, among others.

Breeding for adaptation traits: Growth and development in groundnut is largely driven by temperature. The optimum temperature (T_0) for growth and development in groundnut ranges between 27 and 32 °C. The base temperature (T_b) in groundnut ranges between 9 and 13 °C below which the growth ceases (Williams and Boote 1995). There is variation in T_b for different phenological stages and among genotypes. At lower temperatures, growth is slowed down and it takes longer for crop to mature. The reverse is observed at relatively higher temperatures. But, the growth stops at temperatures exceeding 45 °C as protein gets denatured. At temperatures above the optimum, significant reduction in dry matter production and partitioning of dry matter to pods are observed but flower production is not affected. Photoperiod does not affect flowering in groundnut but it affects partitioning (Nigam et al. 1994), however, these effects are genotypic specific. Irradiance also plays a role together with temperature in determining the crop duration.

Short duration: Breeding for early maturity in groundnut has been reviewed by Nigam and Aruna (2008b). Selection based on days to first flower alone is ineffective in identifying early maturing lines as there are other processes also involved in reaching to maturity. Instead of calendar days, use of cumulative thermal time (CTT) measured in day-degrees (°Cd), is recommended for selecting for early maturity at a given location (Rao et al. 1991). The CTT is measured in day-degrees (°Cd) above the base temperature and is calculated on successive days by subtracting the base temperature from the mean daily temperature and adding each value to the subtotal accumulated since the seed was sown ($\mathrm{CTT}(^{\circ}\mathrm{Cd}) = \sum((T_{\max} + T_{\min})/2 - T_b)$). In photoperiod-insensitive genotypes the CTT for maturity does not differ across environments barring the influence of environmental factors other than photoperiod. For photoperiod-sensitive genotypes, the CTT will vary with photoperiod over the photoperiod-sensitive range. It must be remembered that early maturity is a relative term; in India early maturing varieties are less than 100-day duration whereas in China and USA a variety of 120-day duration will qualify as early maturing variety.

Incorporating large seed size in short-duration cultivars is unlikely to succeed as large seeds take more time to emerge on sowing, and to develop and mature. Similarly, combining higher levels of resistance to foliar diseases and short duration will be difficult to achieve through conventional breeding. On the other hand, a moderate level of resistance will have only limited influence on crop duration and would also stabilize productivity in a cropping system. In breeding for early maturity, it is helpful to partition crop duration into different segments/stages and examine the possibility of shortening their duration individually and collectively with an overall aim to reduce crop duration. These segments/stages include days to germination and emergence, days to first flower after emergence, days from opening of first flower to opening of a given number of flowers per plant, and days from opening a flower to maturation into seeds. Based on the botanical characteristics and physiological behavior of the crop, the following characteristics could be visualized for attaining short duration of the crop: short plant stature (plant height in case of subspecies *fastigiata* and plant spread in case of subspecies *hypogaea*) with smaller internodal length, faster germination and emergence, fewer days to first flowering, and accumulation of a maximum number of early flowers, more flowers per node, absence of late flowers, fewer days after fertilization for a peg to enter soil, faster pod and seed growth, high seed partitioning, and high shelling turnover. To capitalize on the full potential of the genotypes with aforementioned traits, it would be essential to modify crop husbandry to accommodate larger numbers of plants per unit area to provide quick ground cover and to provide plant with required nutrients and other inputs. The following considerations in breeding strategy will help to achieve the objective of early maturity along with high yield: (i) Selection for low T_b and CTT for various phenological stages, (ii) Selection for tolerance to high temperature, (iii) Selection for photoperiod-insensitive genotypes, (iv) Selection for high crop growth rate and partitioning, (v) Selection for high water-use efficiency, and (vi) Evaluation in target environments/cropping systems. Inheritance of earliness and its components has been reviewed by Nigam and Aruna (2008b). From a single gene to 4–5 genes, from complete dominance to incomplete dominance of late maturity over early maturity, epistatic gene effects, absence of reciprocal differences, higher gca variance than sca variance, additive genetic variance, additive and dominance gene effects, and generally high heritability are reported in the literature. Various sources of early maturity identified are Chico, Gangapuri, JL 24, and ICG #s 3540, 3631, 4558, 4729, 4890, 9427, 9930, and 11605, 91776, 91176, Dh 40, ALG (E) 57, TG #s 1E, 2E, and 3E, etc. Of these, Chico, which matures in 75–80 days, has been very extensively used in the breeding programs. Some of the early maturing cultivars released in different countries are Pronto and Spanco in the USA, Dh 40, TNAU 97, ALG (E) 57, GG #s 3, 5, 7, and 12, TG 26, R 9251, M 522, RS 138, K 134, JL 220, VRI 3, and C0 4 in India and 55-437, TS 32-1, 73-30, KH #s 149A and 241D, Te 3, and Fleur 11 in West Africa in different habit groups.

Seed dormancy: A majority of the groundnut in developing countries is grown under rainfed conditions characterized by uncertain and irregular rainfall. The groundnut crop is very often caught in rains at the time of harvest, which results in *in situ* germination in Spanish and Valencia cultivars, thus causing significant losses in yield and quality of the produce. Incorporation of 2–3 weeks fresh seed dormancy in Spanish and Valencia cultivars will help to avoid these losses, which could reach up to 40 % (Reddy et al. 1985). Depending on their genetic constitution, different seed parts—seed coat, cotyledons, and embryo—have been reported to have a role in imparting dormancy (Bandyopadhyay et al. 1999; Nautiyal et al. 2001). Fresh seed dormancy is more under control of testa than cotyledons. Complexity arises in studying the inheritance of seed dormancy when both maternal (testa) and zygotic (cotyledons) tissues are involved in its control. From monogenic control with seed dormancy dominant over nondormancy (Upadhyaya and Nigam 1999; Yaw et al. 2008) to quantitative inheritance with additive, dominance, and digenic epistatic effects (Khalfaoui 1991; Nautiyal et al. 1994) are reported. Several Spanish breeding lines/cultivars with fresh seed dormancy are available now (Upadhyaya et al. 1997b). Most of these originate from

Virginia × Spanish/Valencia crosses. Instead of screening for seed dormancy in early generations, the advanced generation Spanish/Valencia breeding lines are screened for fresh seed dormancy in laboratory and under field conditions.

Salinity: There is no targeted program in progress to breed groundnut tolerant to soil salinity. In limited studies, genotypes have been screened and tolerant genotypes based on plant survival and seed yield per plant have been identified. The tolerant genotypes include germplasm, breeding lines, and cultivars such as NRCG #s 2588, 4659, 5513, 6131, 6450, 6820, 6919, and 7206, TMV 2 NLM, TG 33, JNDS-2004-15, VRI 3, UF 70-103, TKG 19 A, S 206, Tirupati 4, M 522, Punjab 1, BG 3, Somnath, and ICGV 86590 (Singh et al. 2008, 2010).

Quality traits: Oxidative stability and shelf life of groundnut and its products can be enhanced by improving oleic-to-linoleic fatty acid ratio, which normally ranges between 0.8 and 2.5 in old commercial cultivars. These two fatty acids constitute about 80 % [55 % oleic acid (18:1) and 25 % linoleic acid (18:2)] of the oil content of groundnut (Knauft et al. 1993). Of these two, linoleic fatty acid is less saturated and less stable than oleic acid. In peanut breeding program at the University of Florida in 1987, two breeding lines originating from F 435, a high oleic acid spontaneous mutant, with 80 % oleic and 2 % linoleic acid composition were identified (Norden et al. 1987). With simple inheritance (single recessive or two recessive genes and some possible modifiers depending upon the parents involved in the crosses), it is easy to transfer high-oleate trait to other genotypes through backcross breeding program (Moore and Knauft 1989; Knauft et al. 1993; Lopez et al. 2001). Cultivars developed with high O/L ratio in the USA through conventional breeding are SunOleic 95R, SunOleic 97R, Tamrun OL01, Georgia 04S, Andru II, Florida-07, and Hull, through chemical mutagenesis are Mycogen-Flavorunner and M 2-225 and through Gamma radiation are Georgia-02 C and Georgia Hi-high. Varieties with high levels of oleic acid, when consumed, have beneficial effect on human and animal health.

Improved flavor: Since 1980, the flavor of roasted groundnut has become an important consideration in breeding programs engaged in developing Virginia varieties for direct consumption as it influences consumers' acceptance. Several roasted groundnut quality sensory attributes are heritable (Pattee et al. 1998). Thus, choice of parents becomes critical in ensuring good flavor of roasted groundnut in breeding lines. Jenkins Jumbo, one of the ancestors of USA-bred Virginia varieties, was found responsible for their poor roasted flavor. The parents selected for hybridization should have at least acceptable roasted flavor to ensure consumers' acceptability for new cultivars. During the selection process all plants with off-type flavor should be rejected.

2.6.2 Genetic Improvement Using Resources of Secondary Gene Pool

2.6.2.1 Phylogeny of *A. hypogaea*

Information on phylogenetic relationship between cultivated and the species of other gene pool is an essential prerequisite to facilitate gene transfer. A series of initial cytogenetic investigations in section *Arachis* that includes tetraploid ($2n = 40$) cultivated groundnut, *A. hypogaea*, wild tetraploid *A. monticola,* and 29 diploid ($2n = 20$) species crossable with *A. hypogaea,* falling in secondary gene pool revealed that basically there are two genomes, A and B distributed among the diploid species of section *Arachis*, which together contribute to the evolution of cultivated *A. hypogaea* with genomic constitution AABB (Smartt and Gregory 1967; Smartt et al. 1978; Singh and Moss 1982, 1984b; Gardner and Stalker 1983; Singh 1988). These and further studies indicated that most diploid species of section *Arachis* contain A genome, while *A batizocoi*, *A. ipaënsis* and four more, including *A. hoehnei* contain B genome (Mallikarjuna et al. 2006), and K30091, 30098, 30099, and 30100 (*A. glandulifera*), probably another genome D (Stalker 1991). Recently, based on FISH mapping of rDNA loci and heterochromatin detection, two new genome types (F and K) have been described for some of the species formerly considered in the B genome group (Robledo and Seijo 2010). *Arachis benensis* and *A. trinitensis* are now classified as having an F genome and *A. batizocoi*, *A. cruziana,* and *A. krapovickasii*, a K genome. These two genomes have centromeric bands on most of the chromosomes, differing from each other in the amount and distribution of heterochromatin. However, the exact phylogenetic relationships of the F, K, and D genomes with the A and the B genomes need further study. Cross-compatibility, chromosome pairing, and hybrid fertility suggest that A and B genomes are homoeologus and they together evolved *A. hypogaea*, a segmental allopolyploid, with A. *batizocoi* or *A. hoehnei* contributing B genome and *A. duranensis*, *A. villosa,* or *A. cardenasii* A genome (Smartt et al. 1978; Singh and Moss 1984b; Singh 1986a, 1988; Mallikarjuna et al. 2006). Molecular markers affinity suggested *A. ipaënsis* as contributor of B genome, supported by genomic in situ hybridization (Raina and Mukai 1999; Seijo et al. 2004) and *A. duranensis* of A genome (Kochert et al. 1991, 1996; Burow et al. 2009; Moretzsohn et al. 2013), though needing response to some basic questions raised by Singh and Smartt (1998). All studies, including biochemical profile (Singh et al. 1991) supported broad genomic grouping of section *Arachis*. Thus based on cross-compatibility, chromosome pairing, and hybrid fertility, direct introgression of genes from section *Arachis* diploid wild species is possible through direct hybridization with tetraploid *A. hypogaea,* with or without ploidy and cytogenetic manipulations. The other sections species are genomically distant for direct introgression of gene(s).

2.6.2.2 *Accessing Secondary Gene Pool with or Without Ploidy Manipulations of Hybrids*

The cytogenetic information generated on genomic constitution and relationship between cultivated tetraploid *A. hypogaea* and the diploid species of section *Arachis* helped to visualize the implications of direct hybridization (*A. hypogaea* 4x × *A.* sp. 2x) and with autotetraploid and amphidiploid of diploid species with different genomic combinations, on the hybrid fertility, levels of meiotic recombination and on gene transfer, illustrating merits of various options (Singh 1985; Singh and Gibbons 1985; Singh et al. 1990; Simpson 1991). Direct hybridization is the first logical option for transfer of genes into cultivated groundnut from diploid species. Krapovickas and Rigoni (1951) were the first to report hybrid between *A. hypogaea* and *A. villosa*. Subsequently, triploid were produced involving a number diploid *Arachis* species (Smartt and Gregory 1967; Singh et al. 1980; Singh 1985) with the objective of gene transfer. Triploids produced some seeds and seedlings (Smartt and Gregory 1967, Singh and Moss 1984a) consequent to production of haploid, hyperdiploid, and unreduced gametes. Eighty-two percent F_2 of these seedlings were hexaploid, while other had chromosome ranging from 40 to 58 (Singh and Moss 1984a). However, fertility in triploid hybrids was restored by doubling of chromosomes to produce hexaploids (Smartt and Gregory 1967; Singh et al. 1980; Singh 1985). At ICRISAT, triploids were initially produced with the objective of incorporating genes conferring resistance to foliar diseases, involving cultivars of tetraploid *A. hypogaea* and eight diploid species such as *A. cardenasii* (resistant to LLS, rust and groundnut rosette), *A. diogoi* (called *A. chacoense* earlier) (resistant to ELS, rust, and groundnut rosette), *A. stenosperma* (field resistant to both leaf spots), *A. batizocoi* and *A. duranensis* (resistant to rust). Both synthetic hexaploids and the partial fertile triploids were backcrossed to recurrent *A. hypogaea* parents to effect gene transfer, which produced progenies with chromosomes ranging from 40 to 60. Backcrossed progenies were further backcrossed with recurrent *A. hypogaea* parents and intermittently selfed to regain agronomic traits of cultivated groundnut, and the produced backcross and selfed progenies were screened to select and progress with progenies incorporating desired resistance without dilution (Singh and Gibbons 1985; Singh et al. 1990; Ouedraogo et al. 1994; Simpson 2001).

2.6.2.3 Accessing Gene Pool with Ploidy Manipulations of Diploid Species or Hybrids

Autotetraploids and amphidiploids of various diploid species of section *Arachis* were produced in intra- and intergenomic combinations and crossed with tetraploid *A. hypogaea* (Gardner and Stalker 1983; Singh 1985, 1986a, b). It was expected that the resultant hybrids would have improved fertility with removal of ploidy difference and more so in complementary genomic combination. Further, as visualized, use of autotetraploid helped in increasing the dosage of desired trait exploiting homologous intergenomic pairing, while hybridization with amphidiploids from crosses between "A" and "B" genome species produced relatively more fertile hybrids with greater recombination through preferential autosyndetic pairing between wild and cultivated species chromosomes, effecting gene transfer (Singh 1986a, b; Srikanth et al. 2012). Autotetraploids were initially established in eight diploid species, of which six were crossed with *A. hypogaea* as male parents. Similarly, amphidiploids were produced involving eight diploid species of section *Arachis* in 31 combinations, both in intra- (AAAA) and intergenomic (AABB), and 23 were successfully crossed with *A. hypogaea*. As expected, the intergenomic amphidiploids produced greater number of seeds than intragenomic (Singh 1985, 1986b; Mallikarjuna et al. 2011). A number of resultant hybrids from these crosses were backcrossed with groundnut cultivars and intermittently selfed resulting in production of *A. hypogaea*-like stable tetraploid derivatives, which were screened against various pathogens (Singh et al. 1990). These approaches were effective in incorporating resistance to rust and LLS and ELS from several of wild *Arachis* species (Gardner and Stalker 1983; Singh and Gibbons 1985), giving encouragement for full exploitation of the secondary gene pool with concerted efforts on target genes in future.

Using these breeding options by 1989, 209 *A. hypogaea*-like interspecific derivatives incorporating genes conferring resistance to various groundnut diseases were produced at ICRISAT, in addition to hybrid populations received North Carolina State University, produced by Smartt and Gregory (1967). Screening and multilocation yield trials of these interspecific derivatives were conducted in collaboration of national agricultural research systems (NARS) in India and abroad, identifying genotypes well adapted, high yielding, and resistant to prevailing stresses (Singh and Gibbons 1985). Many of these were dual-purpose types with potential for both haulm (fodder) and kernel yield. Interspecific derivatives, 83/372-2-2-22-B1 with resistance to groundnut rosette virus (Moss et al. 1993), ICGV 86699 with multiple disease and insect resistance (Reddy et al. 1996), ICGV 87165 with multiple disease resistance (Moss et al. 1997), ICGV 86715 with foliar disease resistance (Moss et al. 1998), and ICGV #s 99001, 99003, 99004, and 99005 with LLS and rust resistance (Singh et al. 2003) were registered and form the basis of foliar disease resistance breeding programs worldwide. Some, like ICGV 86775 was released as variety in Mauritius. The resulting increase in harvest due to this work is estimated to be some US$ 500 million (Sasson 1996).

2.6.3 Accessing Tertiary Gene Pool and Beyond with Alternative Manipulations

2.6.3.1 Bridge Crosses

The successful hybridization between diploid species of section *Arachis* and those belonging to section *Erectoides* and *Procumbentes* (Gregory and Gregory 1979), but without development of normal seed (Singh 1998), suggests that such cross-ability can be exploited in establishing hybrids with or without pre-fertilization manipulations and/or embryo rescue, that can provide access to significant portions of tertiary gene pool. Such hybrids between diploid species of section *Arachis* and those of *Erectoides* and *Procumbentes* have potential to work as bridge to carry genetic information to *A. hypogaea* from other cross-incompatible species of *Trierectoides*, *Heteranthae,* and *Caulorhizae* (Fig. 2.1). However, the usefulness of such manipulations in genetic improvement of groundnut is yet to be tried and established.

2.6.3.2 *Nonconventional Manipulations and Embryo Rescue*

Several methods, such as mentor pollen, in vitro fertilization, hormone treatments after pollination to overcome prezygotic incompatibility and embryo rescue are possible for direct access the gene conferring resistance to various biotic and abiotic stresses from diploid species section *Procumbentes, Erectoides,* and tetraploid species of *Rhizomatosae*. Interspecific hybrids were produced between *A. hypogaea* and *A. chiquitana* and *A. kretschmeri* of *Procumbentes* by applying growth regulators to pollinated pistils, and hybrid plants were obtained by germinating embryos in vitro (Mallikarjuna 2005; Mallikarjuna and Hoisington 2009). The possibility of establishing hybrids between diploid *Erectoides* and diploid and tetraploid species of section *Arachis* has been corroborated using such manipulations (Singh 1998). In vitro embryo rescue overcoming postzygotic incompatibility has helped establish hybrids between *A. hypogaea* cv. MK 374 and *A. glabrata* (Mallikarjuna and Sastri 2002). These approaches has also helped improve the success rate between some difficult crosses within section *Arachis*, like the success rate in cross, *A. hypogaea* ($2n = 40$) × *A. kempff-mercadoi* ($2n = 20$) increased significantly by culturing immature seeds in vitro (Mallikarjuna et al. 2004). Thus approaches to access the genetic resources (diversity) from tertiary gene pool are under initial stages of hybrid establishment and need further efforts for incorporation of desired genes into stable tetraploid *A. hypogaea*-like interspecific derivatives for use in conventional breeding efforts.

2.6.4 Molecular Breeding

Molecular markers and dense genetic linkage map are necessary for the application of marker-assisted breeding in crop improvement. Infrequent and low polymorphisms have restricted the progress in the development and application of genetic maps in groundnut breeding except in cases where polymorphic chromosomal regions have been introgressed into *A. hypogaea* from diploid relatives. Pandey et al. (2012) reviewed the advances in *Arachis* genomics. Their publication lists *Arachis* markers in public domain, main populations used in *Arachis* genomics research, details of some major genetic maps constructed in *Arachis* species and QTLs identified for some economically important traits in groundnut.

2.6.4.1 Genetic Maps of Cultivated Groundnut

A few maps constructed earlier were based on diploid or interspecific tetraploid populations (Halward et al. 1993; Moretzsohn et al. 2005, 2009). Varshney et al. (2009) were the first to report the construction of a genetic map for cultivated groundnut by screening 1145 SSR markers on two genotypes (TAG 24 and ICGV 86031), which are the parents of a recombinant inbred line (RIL) population. A total of 135 SSR loci were mapped into 22 linkage groups. Hong et al. (2010) constructed a composite linkage map from three individual linkage maps constructed from each of the three RIL populations which had common female parent Yueyou 13. The composite linkage map consisted of 22 composite linkage groups with 175 SSR markers covering a composite map length of 885.4 cM with an average marker density of 5.8 cM. Based on segregation data from RIL population of cross TAG 24 × GBPD 4, Khedikar et al. (2010) developed a partial linkage map with 56 SSR loci over 14 linkage groups. In an integrated map derived from two cultivated × cultivated RIL populations, Qin et al. (2012) anchored 324 SSR markers covering 1352.1 cM map distance with 21 linkage groups. Gautami et al. (2012b) constructed a consensus genetic map for drought tolerance traits based on three genetic maps developed from three RIL populations—reference map based on TAG 24 × ICGV 86031RIL population with 191 SSR loci (Varshney et al. 2009) and two other new maps based on—ICGS 76 × CSMG 84-1 RIL population with 119 SSR loci and ICGS 44 × ICGS 76 RIL population with 82 SSR loci. The consensus map spanned 2840.8 cM map distance with 293 SSR loci distributed over 20 linkage groups. Based on 11 populations, Gautami et al. (2012a) constructed an international reference consensus map for tetraploid groundnut with 897 marker loci (895 SSR loci and two cleaved amplified polymorphic sequence (CAPS)) distributed on 20 linkage groups and spanning a map distance of 3863.6 cM with an average map density of 4.4 cM.

2.6.4.2 Marker-Assisted Breeding

Identification of molecular markers associated with traits of interest and detection of QTLs through linkage mapping are the two prerequisites for application of marker-assisted breeding in crop improvement program. Stalker and Mozingo (2001) and Dwivedi et al. (2003) have comprehensively reviewed the history of marker development in groundnut. Initially, RAPD and RFLP markers were used to screen groundnut germplasm and/ or tetraploid interspecific breeding lines. However, they were not ideal for marker-assisted breeding for various reasons. Now simple sequence repeat (SSR, also known as microsatellites) are the markers of choice for molecular breeding in most crops. The SSR markers are preferred because of their abundance and uniformity of distribution throughout most of the genome, their multiallelic, codominance inheritance, and their highly polymorphic and reproducible nature where analysis is simple and readily transferable (Weber 1990). Different types of markers have been reported for almost all biotic stresses [root knot nematode (Burow et al. 1996), ELS and LLS (Stalker and Mozingo 2001; Khedikar et al. 2010), rust (Varma et al. 2005; Mondal et al. 2008, 2012), aflatoxin contamination (Milla et al. 2005), sclerotinia blight (Chenault and Maas 2005), aphids (Herselman et al. 2004), bruchid (Mondal et al. 2014)] and other traits [high oleic trait (Patel et al. 2004)] in groundnut. The number of polymorphic SSR markers for different traits are increasing fast.

Marker-assisted backcrossing (MABC) is extensively used to introgress transgene or major loci or a major QTL into a cultivar. Depending on the population size and considering one or two target loci, two to three backcrosses are generally sufficient to recover most of the recipient genome. The marker-assisted recurrent selection (MARS) and genomic selection (GS) approaches are practiced to accumulate favorable alleles with small effects in a genotype under improvement. The latter rather than relying on mapped loci uses breeding values, which are calculated based on high density genotypic data and historical phenotypic data from a ‘training population’ usually made up of breeding lines. The status of marker-assisted breeding for different traits in groundnut is summarized below.

Foliar fungal diseases: From composite interval mapping based on genotypic and phenotypic data from RIL population of TAG 24 × GPBD 4 cross, Khedikar et al. (2010) identified 11 QTLs for LLS in three environments explaining only 1.7–6.5 % phenotypic variation. Employing bulk segregant analysis, Shoba et al. (2012) identified three primers among the polymorphic SSR markers, which were able to distinguish between LLS resistant and susceptible bulks and individuals in $F_{2:3}$ progenies of TMV 2 (susceptible parent) × COG 0437 (resistant parent) cross. In single marker analysis, they found seven markers linked to LLS severity score, which explained 32–59 % phenotypic variation. They recommended use of PM 384 marker in marker-assisted breeding over a wide range of genetic backgrounds. Shoba et al. (2013) identified one QTL each for 100-kernel weight and LLS severity score. The former explained 6.1 % variation and the latter 37.9 % phenotypic variation in respective characters. The QTL for LLS can be effectively utilized in marker-assisted breeding for resistance to LLS.

Varma et al. (2005) studied F_2 populations of ICGV 99003 × TMV 2 and ICGV 99005 × TMV 2 crosses and identified two SSR alleles in the former and seven in the latter associated with rust resistance. Mondal et al. (2008) studied F_2 population of 117 individuals of VG 9514 × TAG 24 cross. Contrary to the earlier published reports of rust resistance being recessive and governed by a few genes, they reported it to be dominant and governed by a single gene in this cross between cultivated types. In their study, only 11 RAPD markers out of 160 showed polymorphism between two parents. Using bulk segregant analysis, they identified J 7 (1300) as a suitable marker for marker-assisted selection. From another study, Mondal et al. (2012) identified two EST-SSR markers (SSR_GO340445 and SSR_HO115759) closely linked to rust resistance, which were suitable candidates for marker-assisted selection.

Aflatoxin contamination: Using microarray analysis in A 13 cultivar, which is resistant to drought and preharvest aflatoxin contamination by *A. parasiticus*, Lu et al. (2005) identified 25 upregulated, commonly expressed genes when the cultivar was challenged by both drought and preharvest *A. parasiticus* infection. Of these, 20 were validated by real-time PCR. After characterization of each of these genes, appropriate gene probes can be developed for application in breeding programs. Liang et al. (2009) reported six QTLs, each located on a different linkage group, for resistance to *A. flavus* invasion, which could explain phenotypic variation ranging from 6.2 to 22.7 %.

Tomato spotted wilt virus (TSWV): Qin et al. (2012) identified two major QTLs for resistance to TSWV disease.

Insect pests: Mondal et al. (2014) identified two main QTLs for component traits associated with bruchid resistance. The QTL for total developmental period (TDP) explained 57–82 % phenotypic variation and that for adult emergence (AE) explained 13–21 % phenotypic variation. Additionally, three QTLs for TDP, AE and number of holes and one QTL for pod weight loss, which were also identified, explained 14–39 % phenotypic variation.

Nematode resistance: Marker-assisted selection in groundnut was first used in breeding for resistance to root knot nematode (*M. arenaria*). Burow et al. (1996) identified three RAPD markers linked to root knot nematode resistance, which was due to a single dominant gene. Subsequent studies by Choi et al. (1999), while confirming the single dominant gene nature of the resistance in some populations, also indicated the possibility of a second gene for resistance and evaluated the utility of these markers as selectable markers. Using marker-assisted selection, Chu et al. (2011) combined the root knot nematode resistance and high oleic trait leading to the development of Tifguard High O/L genotype. In addition to saving time, they also reported a significant reduction in the amount of breeding material carried through the breeding program by following marker-assisted selection.

Drought: Varshney et al. (2009) reported 2–5 QTLs for T, TE, SLA, and SCMR, which explained only 3.5–14.1 % phenotypic variation for these traits. Based on identification of few major and many minor QTLs and QTL × QTL interactions, Ravi et al. (2011) confirmed the complex and quantitative nature of drought tolerance in groundnut. Gautami et al. (2012b) identified 153 main effect QTLs and 25 epistatic QTLs with drought-tolerance-related traits. As no major QTL for drought adaptation was identified, Ravi et al. (2011) and Gautami et al. (2012b) recommended adoption of MARS and GS approaches to introgress a large number of QTLs to breed drought-resistant groundnut genotypes.

High oil/oleic acid content: Huang et al. (2012) reported that three SSR alleles associated with high oil content in wild *Arachis* species, are absent in cultivated groundnut. Using wild *Arachis* species, the oil content of cultivated groundnut can be increased. From the study of Yuanza 9102 × Zhonghua 5 RIL population, Huang et al. (2011) found 2A5-250/240 SSR marker tightly linked to oil content trait (2A5-250 with low oil content; 2A5-240 with high oil content). High O/L trait is reported to be dependent on two homeologous oleoyl-PC desaturase genes, *ahFAD2A* and *ahFAD2B* (Chu et al. 2009). The 4th backcross progenies, developed following marker-assisted selection, had all combinations of the two genes except ol_2ol_2 homologous mutant. The highest oleic acid content was found in progeny with all four mutant alleles (ol_1ol_1 ol_2ol_2) (Mienie and Pretorius 2013). Chu et al. (2011) developed Tifguard High O/L cultivar after three accelerated backcrossing and following marker-assisted selection among progenies to combine nematode resistance from Tifguard and high O/L trait from Georgia-02C and Florida-07.

2.6.5 *Genetic Transformation*

In a recent publication, Sunkara et al. (2013) discussed the progress and prospects of transgenic interventions in the improvement of groundnut. They have also listed responses of various explants and hormones on in vitro shoot regeneration and an update on genetic transformation of groundnut. The commonly used methods for DNA delivery or gene transfer into organogenic or embryogenic cultures of groundnut are either biological using *Agrobacterium tumefaciens* or by direct gene transfer using microprojectile/particle bombardment or by electroporation. The choice of method depends on several factors including laboratory facilities and technical skills available and the cultivar and regeneration system used. The first successful transformation and accompanying plant generation using microbombardment technique in groundnut was reported by Ozias-Akins et al. (1993). However, the efficiency of genetic transformation was low and the process took many months for plants to mature. Sharma and Anjaiah (2000) developed a different protocol for genetic transformation with *Agrobacterium tumefaciens*, which works with a wider range of groundnut genotypes. The direct regeneration system adopted in the above protocol favors genetic transformation because of advantages of de novo production of shoot primordial synchronous with the period of cellular differentiation, rapidity of morphogenesis and lack of requirement for frequent subcultures.

Ozias-Akins (2007) compiled a list of stable transformation up to 2005 in cultivated groundnut. Genetic transformation efforts in groundnut cover a wide range of abiotic and biotic stresses. These include drought and salinity in abiotic stresses and LLS, ELS, rust, *Aspergillus flavus* and sclerotinia blight in fungal diseases, GRD, PSND, PBND, TSWV, PStV, and PCV in virus diseases, bacterial wilt, *Spodoptera litura*, *Helicoverpa armigera* and lesser corn stalk borer in insect pests and vitamin A biofortification, oil quality, and herbicide tolerance. The genetic transformation research in groundnut is at a slow pace because of the restriction in testing and ban on commercializing transgenics in many countries.

Biotic stresses: In most cases, the level of resistance achieved through transgenic is more or less similar to that achieved through conventional breeding. However, transgenics do provide opportunity to combine conventional resistance with that of nonconventional resistance to improve the level of protection against pathogen or stress factors.

Several genes have been used to develop transgenic events with resistance to fungal diseases in groundnut. These are listed in the paper by Sunkara et al. (2013) and include glucanase, chitinase, SniOLP, and Rs-AFP2 for LLS and ELS, chitinase for rust, oxalate, glucanase, and chitinase for sclerotinia blight and Stilbene synthase, glucanase, chitinase, mod1, anionicperoxidase, synthetic peptide D4E1, LOX 1, Nonheme chloroperoxidase (cpo), and Pn LOX 3 for *A. flavus* infection and aflatoxin biosynthesis. These genes suppressed the disease, delayed the onset of disease, enhanced resistance, and decreased disease incidence. In the case of sclerotinia blight, reduced lesion area and in the case of *A. flavus,* reduced aflatoxin contamination was also observed. Transgenic lines of Okrun cultivar with rice chitinase and an alfalfa glucanase gene showed up to 43–100 % reduction in incidence of sclerotinia blight compared to the parent variety in the field (Chenault et al. 2005).

Virus disease: Compared to fungal diseases, virus diseases have received greater attention in transgenic research. The protein-mediated resistance, in general, offers only moderate protection against a broad range of related viruses, while RNA-mediated resistance offers high levels of protection, but only against closely related strains of a virus (Dawson 1996). RNAi technology (RNA silencing or cosuppression of homologous genes) provides a significant tool for developing virus resistant groundnut genotypes (Wang et al. 2000).

In case of PBND, both *A. tumefaciens*- and microprojectile-mediated genetic transformation approaches using PBNV nucleocapsid gene encoding for viral coat protein are being pursued. Transgenic events with $PBNV_{np}$ gene, developed at ICRISAT, showed lower incidence and delayed onset of disease and also recovery from disease suggesting only a modest tolerance to PBNV. Currently, RNAi-mediated approach is being followed to counter the effect of nonstructural silencing suppressor gene (NSs gene) in the PBNV genome.

For TSWV, the protection of transgenic plants is under both RNA- and protein-mediated control (Yang et al. 1998). These approaches include using both sense and antisense TSWV nucleocapsid protein gene (N gene) expression. Nucleocapsid protein gene (NP) has been introduced via microprojectile bombardment into New Mexico Valencia A cultivar (Li et al. 1997) and a runner cultivar (Chenault and Payton 2003). *A. tumefaciens*-mediated transformation is also followed. AT 120 (with antisense nucleocapsid gene) (Magbanua et al. 2000) and Marc 1 (with coat protein gene) (Ozias-Akins et al. 2002) cultivars were also transformed. Expression of sense or antisense NP gene from TSWV delayed expression of symptoms and prevented systemic virus infection but did not provide complete resistance to the disease. This single gene resistance may be short-lived because of highly heterogeneous population of the virus. Use of stable pathogen-derived resistance based on homology dependent RNA silencing for durable TSWV resistance has been suggested by Bucher et al. (2003).

For PStV, transgenic plants of Gajah and NC 7 cultivars containing one of the two forms of PStV coat protein gene (*cp* 2 or *cp* 4) exhibited high levels of RNA-mediated resistance (Higgins et al. 2004). The PStV resistance in transgenic groundnut cv. Gajah was stable up to seven generations of selfing (Hapsoro et al. 2005, 2007).

PSND transgenics produced following *A. tumefaciens*-mediated transformation with TSV coat protein gene (TSV_{cp} gene) showed three symptoms—blockage of systemic movement of the virus within the plants, recovery from an initial infection and subsequent new growth devoid of disease symptoms and susceptible reaction. Transgenic lines cv. JL 24 containing sense and antisense coat protein gene of TSV were developed using *A. tumefaciens*-mediated transformation (Bag et al. 2007). However, these lines are yet to be tested for disease reaction at hotspot locations under field conditions.

Transgenic lines having $IPCV_{cp}$ and $IPCV_{rep}$ genes of Indian peanut clump were produced following *A. tumefaciens*-mediated transformation and tested under containment facilities at ICRISAT. Some events showed resistant phenotype where the virus titer declined with maturity.

For GRD, pathogen-derived resistance (introduction of GRAV or GRV genomic sequences or genes or *Sat*RNA-derived sequences that down regulate GRV replication) is a potential strategy for controlling the disease through generation of transgenic plants (Taliansky et al. 1996). Groundnut transgenics having $GRAV_{cp}$ gene were developed at ICRISAT and are currently being tested in South Africa.

Insect pests: Synthetic genes, *cry1 EC* against *Spodoptera litura* (Tiwari et al. 2008), *cry1 X* against *Helicoverpa armigera* and *S. litura* (Entoori et al. 2008), and *cry1 Ac* against lesser cornstalk borer (Singsit et al. 1997) have shown good promise against respective insect pests.

Abiotic stresses: In drought, the DREB group of transcription factors has received greater attention in developing drought-tolerant varieties in various crops through transgenic research. Selected transgenic events of JL 24 cultivar containing DREB 1A transcription factor driven by rd29A promoter showed higher TE under both well-watered and water-limiting conditions with one event recording as high as 40 % more TE over untransformed control (Bhatnagar-Mathur et al. 2007). All DREB 1A transgenic events had significantly higher seed filling under drought and displayed 20–30 % lower pod yield reduction than their untransformed counterpart under drought stress (Bhatnagar-Mathur et al. 2014). Water stress promotes rooting growth more strongly in DREB 1A transgenic events than in the wild type especially in deeper soil layers leading to increased water extraction. Qin et al. (2011) reported that regulated expression of isopentenyltransferase gene (IPT) significantly improved drought tolerance in groundnut. Transgenic plants maintained higher photosynthetic rates, higher stomatal conductance, higher transpiration and recorded higher yields than wild types under reduced irrigation conditions.

For salinity, transgenic events with AtNHX1 gene have been studied in a limited way. Asif et al. (2011) reported that over expression of AtNHX1 gene, isolated from Arabidopsis and driven by 35S promoter, in groundnut not only improved salt tolerance but also drought tolerance in transgenic events. Banjara et al. (2012) also reported increased tolerance of salt in transgenic events carrying AtNHX1 gene in groundnut. AhNHX1 gene from groundnut has been isolated and its important role in salt tolerance in groundnut has been confirmed.

Nutritional quality: Zmpsy 1 gene from maize and *β*-lycopene cyclase gene from tomato are being used to enrich groundnut seeds with *β-carotene* (pro- vitamin A). Second-generation transgenic events showed many fold increase in vitamin A content (Bhatnagar-Panwar et al. 2013). For Oleic/Linoleic fatty acid ratio (O/L ratio), an FAD2 gene RNAi construct was transformed into groundnut to reduce content of linoleic acid and increase the stability of groundnut oil (Zhang et al. 2007; Huang et al. 2008; Yin et al. 2009). Seeds from the transgenic plants showed an increased O/L ratio (Huang et al. 2008). Endogenous *allergens*, Ara h 2 and Ara h 6, were silenced by introducing RNAi construct targeting homologous coding sequence and human IgE binding to these proteins was greatly reduced (Dodo et al. 2008; Chu et al. 2008)

Herbicide tolerance: *Agrobacterium*-mediated transgenic groundnuts over expressing pEGAD-EPSPS with altered kinetics of enzyme showed improved tolerance to glyphosate (Manjunatha et al. 2008).

2.7 Conclusions

Arachis gene pool consisting of cultivated groundnut from 95 countries and 80 wild relatives naturally distributed in five major countries of South America offers a reservoir of genetic diversity for genetic improvement of groundnut crop. Significant progress has been made in collection and conservation of the available natural genetic variation in the repositories located in major groundnut-growing regions/countries. However, the assembled genetic variation is skewed and is limited in case of vars. *aequatoriana, peruviana* and *hirsuta.* Similarly, wild species gene pool that offers significant variability, particularly for biotic and abiotic stresses, has been understood very little in terms of genetic diversity within species. Collection of large amount of genetic diversity/collections has created a problem of plenty and use of genetic variability has been very limited, producing cultivars with very narrow genetic base. To improve understanding and quantification of genetic diversity, core collection approach has been vigorously advocated in the last two decades. However, there is no hard data to suggest that it has improved use of genetic diversity and needs further look to make these core collections more dynamic and true representative of total genetic diversity, particularly useful diversity of breeding value to facilitate management of yield constraints, and to meet the needs of diverse agroclimatic and production conditions. The successful exploitation of several wild species of secondary gene pool, incorporating gene(s) conferring resistance to major biotic stresses once again highlight the importance of wild relatives in genetic enhancement/improvement of crop species. However, access to genes from tertiary gene pool is still limited, confined to initial hybrids, while the quaternary gene pool is yet to be tapped.

Conventional breeding has been effective in some areas, while in others it has been tardy due to lack of proper and effective phenotyping tools and limited understanding of the underlying mechanisms influencing targeted traits. Being largely a rainfed crop, the genetic gains in yield potential are likely to be low and slow to come by. In such situations, resistance breeding efforts are going to be more rewarding in improving realized yield. Resistance to soil-borne diseases including *A. flavus* infection and aflatoxin contamination and insect pests requires greater attention in groundnut improvement programs. However, for effective genetic enhancement in these areas, better and effective screening methods/tools and a clear understanding of underlying mechanisms of resistance are required. Information on inheritance/genetics of several traits is either lacking or limited. This knowledge gap needs to be filled in to devise better strategies for crop improvement. A greater diversification of parental resources in breeding programs is required to develop new cultivars with diversified genetic backgrounds, which will enable them to perform better under adverse conditions. Along with crop improvement research, the crop management research also needs to be pursued vigorously to harness the full potential of improved cultivars in a synergistic manner.

2.8 Future Perspective

More exploration is required in areas of distribution of *aequatoriana, peruviana,* and *hirsuta* varieties to obtain their comparative representation in collections and in core collection of global genetic diversity. Recognizing the presence of higher genetic variation, allelic diversity and presence of greater unique alleles in wild *Arachis,* greater efforts are needed for searching new genes/alleles in wild *Arachis* with intensive evaluation and proper characterization.

To improve the use of genetic diversity in groundnut improvement, the core collection needs to be made more dynamic. To achieve this the gene pool concept can be extended to total collections, stratifying them for specific diversity of breeding value, such as early maturity, accessions with genes conferring resistance to various biotic and biotic stresses, nutritional characters, etc. This is followed by principal component/multivariate analysis on quantitative agronomic traits and clustering, and selective picking of accessions representing taxonomic and geographic affinities, facilitating encompassing of total spectrum of useful variability to formulate an *active core collection* that can meet most requirements for genetic improvement of groundnut with precise breeding.

Crop improvement efforts, both conventional and nonconventional in groundnut needs to concentrate on bridging the yield gap between the potential yield and the realized yield, by alleviating major production constraints particularly in rainfed environment. The specific issues that require attention are listed below.

- Most of the foliar diseases resistant cultivars have a high level of resistance to rust. However, levels of resistance to both leaf spots in cultivars need further improvement without compromising on agronomic characters including crop duration as it often gets enhanced with higher levels of resistance. It may be desirable to intermate foliar diseases resistant second- or third-generation advanced breeding lines originating from different parents including inter-specific derivatives to improve the level of resistance without bringing in linkage drag. Further, resistance to foliar diseases needs to be incorporated in short-duration cultivars without affecting their duration.
- To enhance effectiveness of aflatoxin resistance breeding, sampling procedures and screening methods need major refinement to improve characterization and precise estimation of infection and aflatoxin production. Ascertaining allelic relationship among resistant sources would help gene pyramiding.
- Breeding for resistance to soil-borne diseases needs impetus as these have increased over time resulting in significant plant mortality in the field. For this to happen, sources with higher levels of resistance and effective screening techniques are required. Resistance of wild *Arachis* species, where available, should be effectively exploited.
- Combining virus resistance with that of the vectors shall help reinforce the resistance against virus diseases. This may include application of newer approaches such as RNAi technology. In the case of GRD, the off-season survival of the disease causing agents is still a mystery and needs to be investigated for better disease management. There is a need to look for diversified sources of resistance to GRD and identify sources of resistance to GRAV in primary gene pool. Efforts for incorporation of GRAV resistance from wild *Arachis* species needs to be initiated. Additionally, allelic relationship between *A. hypogaea* sources and wild *Arachis* species should be studied to identify new resistance genes. For further reinforcement of resistance, aphid resistance should be incorporated. In Africa, developing GRD-resistant, short-duration, and high-yielding varieties with traits acceptable to farmers, traders and consumers should remain a high priority.
- More studies are needed to understand genetics and mechanisms of resistance to bacterial wilt disease. To obtain stable resistance against the disease, harnessing genes from diverse sources including wild *Arachis* species is required, besides combining it with resistance to rust and leaf spots.
- To breed resistance to insect pest, wild *Arachis* species, which show high levels of resistance, should be accessed with refined field screening techniques ensuring uniform desired pressure of insect pests. The resistance to nematodes may be combined with resistance to other stresses to derive larger benefits.

- Drought-tolerant cultivars are needed in different maturity groups along with resistance/tolerance to aflatoxin contamination. Adoption of marker-assisted recurrent selection (MARS) to accumulate several QTLs with small effects on drought tolerance in a single genotype will be helpful.
- Diversification of sources of earliness and studies on their genetic constitution and allelic relationships is needed to identify different genes for earliness which could be accumulated in a desired genotype.
- More efforts are needed in genomics research to saturate the linkage map of groundnut so that effective use of marker-assisted selection could be made in groundnut improvement.

The success achieved in genetic improvement of groundnut using wild species of secondary gene pool emphasizes the need for utilization of more wild species in genetic enhancement of *A. hypogaea* to produce better genetic resource of genes/alleles that can help in widening the genetic base of crop with sustainable resilience against biotic stresses and thereby yield. To access genes from tertiary and quaternary gene pools, efforts need to be extended to recombinant DNA technology using cis-transgenic approach, which shall partially dispel the negative apprehensions of environmentalists.

Emerging molecular tools provide an opportunity to enhance efficiency and effectiveness of the conventional breeding particularly for complex traits, which are multigenic. A holistic approach integrating conventional breeding, molecular breeding and transgenics will provide solutions to complex problems being currently faced in groundnut improvement. However, in the case of transgenic research, issues related to biosafety need to be dealt appropriately. In future, breeding programs will have to focus on developing customized cultivars to meet the requirements of the food industry. The new cultivars will have to be climate resilient to face the looming challenges of the climate change felt across the world.

Acknowledgments The authors are grateful to Dr. C. E. Simpson, Professor Emeritus, Texas A & M University, Experimental Agricultural Station, Stephenville, USA for sharing literature and information on germplasm holdings and staff of JS Kanwar Library, ICRISAT, India for providing access to literature as and when requested.

References

Akem CN, Melouk HA, Smith OD (1992) Field evaluation of peanut genotypes for resistance to sclerotinia blight. Crop Prot 11:345–348

Amin PW, Singh KN, Dwivedi SL, Rao VR (1985) Sources of resistance to jassid (*Empoasca kerri* Purthi), thrips (*Frankliniella shultzei* Trybom) and termites (*Odontotermes* sp.) in groundnut (*Arachis hypogaea* L.). Peanut Sci 12:58–60

Anderson WF, Patanothai A, Wynne JC, Gibbons RW (1990) Assessment of a diallel cross for multiple foliar pest resistance in peanut. Olèagineux 45:373–378

Arunyanark A, Pimratch S, Jogloy S, Wongkaew S, Vorasoot N, Akkasaeng C, Kesmala T, Patanothai A, Holbrook CC (2012) Association between aflatoxin contamination and N_2 fixation in peanut under drought conditions. Int J Plant Prod 6:161–172

Asif MA, Zafar Y, Iqbal J, Iqbal MM, Rashid U, Ali GM, Arif A, Nazir F (2011) Enhanced expression of AtNHX1 in transgenic groundnut (*Arachis hypogaea* L.) improves salt and drought tolerance. Mol Biotechnol 49:250–256

Bag S, Singh RS, Jain RK (2007) *Agrobacterium*-mediated transformation of groundnut with coat protein gene of Tobacco streak virus. Ind J Virol 18:65–69

Bandyopadhyay A, Nautiyal PC, Radhakrishnan T, Gor HK (1999) Role of testa, cotyledon and embryonic axis in seed dormancy of groundnut (*Arachis hypogaea* L). J Agron Crop Sci 182:37–41

Banjara M, Zhu LF, Shen GX, Payton P, Zhang H (2012) Expression of an Arabidopsis sodium/proton antiporter gene (AtNHX1) in peanut to improve salt tolerance. Plant Biotechnol Rep 6:59–67

Banks DJ (1976) Peanuts: germplasm resources. Crop Sci 16:499–502

Bera SK, Kasundra SV, Kamdar JH, Ajay BC, Lal C, Thirumalasamy PP, Dash P, Maurya K (2014) Variable response of interspecific breeding lines of groundnut to *Sclerotium rolfsii* infection under field and laboratory conditions. Electron J Plant Breed 5:22–29

Bhatnagar-Mathur P, Devi MJ, Reddy DS, Lavanya M, Vadez V, Serraj R, Yamaguchi-Shinozaki K, Sharma KK (2007) Stress-inducible expression of At DREB 1A in transgenic peanut (*Arachis hypogaea* L.) increases transpiration efficiency under water-limiting conditions. Plant Cell Rep 26:2071–2082

Bhatnagar-Mathur P, Rao JS, Vadez V, Dumbala SR, Rathore A, Yamaguchi-Shinozaki K, Sharma KK (2014) Transgenic peanut over expressing the *DREB1A* transcription factor has higher yields under drought stress. Mol Breed 33:327–340

Bhatnagar-Panwar M, Bhatnagar-Mathur P, Bhaaskarla VV, Dumbala SR, Sharma KK (2013) Rapid, accurate and routine HPLC method for large-scale screening of pro-vitamin A carotenoids in oilseeds. J Biochem Biotechnol. doi:10.1007/s13562-013-0239-1

Brown AHD (1989) The case of core collection. In: Brown AHD et al. (eds) The use of plant genetic resources. Cambridge University Press, Cambridge, pp 135–156

Bucher E, Sijen T, De Haan P, Goldbach R, Prins M (2003) Negative-strand tospoviruses and tenuiviruses carry a gene for a suppressor of gene silencing at analogous genomic positions. J Virol 77:1329–1336

Buiel AAM (1996) Quantitative resistance to peanut bud necrosis tospovirus in groundnut. Ph D dissertation, Wageningen Agricultural University, The Netherlands

Burow MD, Simpson CE, Faries MW, Starr JL, Paterson AH (2009) Molecular biogeographic study of recently described A- and B-genome *Arachis* species, also providing new insights into the origin of cultivated peanut. Genome 52:107–119

Burow MD, Simpson CE, Paterson AH, Starr JL (1996) Identification of peanut (*Arachis hypogaea* L.) RAPD markers diagnostic of root-knot nematode [*Meloidogyne arenaria* (Neal) Chitwood] resistance. Mol Breed 2:369–379

Chamberlin Chenault Kelly D, Melouk Hassan A, Payton Mark E (2010) Evaluation of the U.S. peanut mini core collection using a molecular marker for resistance to *Sclerotinia* minor Jagger. Euphytica 172:109–115

Chen CY, Barkley Noelle A, Wang Ming L, Holbrook Corley C, Dang Phat M (2013a) Registration of purified accessions for the US peanut mini-core germplasm collection. J Plant Regis 8(1):77–85

Chen CY, Nuti RC, Rowland DL, Faircloth WH, Lamb MC, Harvey E (2013b) Heritability and genetic relationships for drought-related traits in peanut. Crop Sci 53:1392–1402

Chenault KD, Maas A (2005) Identification of a simple sequence repeat (SSR) marker in cultivated peanut (*Arachis hypogaea* L.) potentially associated with sclerotinia blight resistance. Proc Am Peanut Res Educ Soc Inc 37:24-25

Chenault KD, Melouk HA, Payton ME (2005) Field reaction to sclerotinia blight among transgenic peanut lines containing antifungal genes. Crop Sci 45:511–515

Chenault KD, Payton ME (2003) Genetic transformation of runner-type peanut with the nucleocapsid gene of tomato spotted wilt virus. Peanut Sci 30:112–115

Choi K, Burow MD, Church G, Paterson AH, Simpson CE, Starr JL (1999) Genetics and mechanisms of resistance to *Meloidogyne arenaria* in peanut germplasm. J Nematol 31:283–290

Chu Y, Faustinelli P, Ramos ML, Hajduch M, Thelen JJ, Maleki SJ, Ozias-Akins P (2008) Reduction of IgE binding and non-Procmotion of *Aspergillus flavus* fungal growth by simultaneously silencing Ara h2 and Ara h6 in peanut. J Agric Food Chem 56:11225–11233

Chu Y, Holbrook CC, Ozias-Akins P (2009) Two alleles of *ahFAD2B* control the high oleic acid trait in cultivated peanut. Crop Sci 49:2029–2036

Chu Y, Wu CL, Holbrook CC, Tillman BL, Person G, Ozias-Akins (2011) Marker-assisted selection to pyramid nematode resistance and high oleic trait in peanut. Plant Genome 4: 110–117

Craufurd PQ, Prasad PVV, Kakani VG, Wheeler TR, Nigam SN (2003) Heat tolerance in groundnut. Field Crops Res 80:63–77

Culbreath AK, Beute MK, Campbell CL (1991) Spatial and temporal aspects of epidemics of *Cylindrocladium* black rot resistant and susceptible peanut genotypes. Phytopathology 81: 144–150

Culbreath AK, Gorbet DW, Martinez-Ochoa N, Holbrook CC, Todd JW, Isleib TG, Tillman BL (2005) High levels of field resistance to tomato spotted wilt virus in peanut breeding lines derived from *hypogaea* and *hirsuta* botanical varieties. Peanut Sci 32:20–24

Culbreath AK, Todd JW, Brown SL (2003) Epidemiology and management of tomato spotted wilt in peanut. Annu Rev Phytopathol 41:53–75

Culver JN, Sherwood JL, Melouk HA (1987) Resistance to Peanut Stripe Virus in *Arachis* germplasm. Plant Dis 71:1080–1082

Dawson WO (1996) Gene silencing and virus resistance: a common mechanism. Trends Plant Sci 1:107–108

De Berchoux C (1960) La rosette l'arachide en Haute-Volta. Comportement des lignes resistantes. Olèagineux 15:229–233

Devi MJ, Bhatnagar-Mathur P, Sharma KK, Serraj R, Anwar SY, Vadez V (2011) Relationships between transpiration efficiency and its surrogate traits in rd29A: DREB 1Atransgenic lines of groundnut. J Agron Crop Sci 197:271–283

Dodo HW, Konan KN, Chen FC, Egnin M, Viquez OM (2008) Alleviating peanut allergy using genetic engineering: The silencing of the immunodominant allergen Ara h 2 leads to its significant reduction and a decrease in peanut allergencity. Plant Biotechnol J 6:135–145

Dubard M (1906) De l'origine de l'Arachide. Mus. Nat. d'His. Naturelle, Paris Bull 5:340–344

Dwivedi SL, Amin PW, Rasheedunisa Nigam SN, Nagabhushanam GVS, Rao VR, Gibbons RW (1986) Genetic analysis of trichome characters associated with resistance to jassid (*Empoasca kerri* Pruthi) in peanut. Peanut Sci 13:15–18

Dwivedi SL, Crouch JH, Nigam SN, Ferguson ME, Paterson AH (2003) Molecular breeding of groundnut for enhanced productivity and food security in the semi-arid tropics: opportunities and challenges. Adv Agron 80:153–221

Dwivedi SL, Nigam SN, Reddy DVVR, Reddy AS, Ranga Rao GV (1995) Progress in breeding groundnut varieties resistant to peanut bud necrosis virus and its vector. In: Lenne JM (ed) Buiel AAM, Parlevliet JE., Recent studies on peanut bud necrosis disease: Proceedings of a meeting 20 March 1995, ICRISAT Asia Center, Patancheru, AP, India. International Crops Research Institute for the Semi-Arid Tropics; Dept Plant Breed, Wageningen, pp 35–40

Entoori K, Sreevathsa R, Manoj KA, Kumar PA, Kumar ARV, Madhusudhan B, Udayakumar (2008) A chimeric *cry1X* gene imparts resistance to *Spodoptera litura* and *Helicoverpa armigera* in the transgenic groundnut. EurAsian J BioSci 2:53–65

Fernández A, Krapovickas A (1994) Cromosomas y evolución en *Arachis* (*Leguminosae*). Bonplandia 8:187–220

Gardner MEB, Stalker HT (1983) Cytology and leafspot resistance of section *Arachis* amphiploids and their hybrids with *Arachis hypogaea*. Crop Sci 23:1069–1074

Gautami B, Foncěka D, Pandey MK, Moretzsohn MC, Sujay V, Qin H, Hong Y, Faye I, Chen X, Bhanu Prakash A, Shah TM, Gowda MVC, Nigam SN, Liang X, Hoisington DA, Guo A, Bertioli DJ, Jean-Francois Rami, Varshney RK (2012a) An international reference consensus genetic map with 897 marker loci based on 11 mapping populations for tetraploid groundnut (*Arachis hypogaea* L.). PLoS ONE 7:1–11

Gautami B, Pandey MK, Vadez V, Nigam SN, Ratnakumar P, Krishnamurthy L, Radhakrishnan T, Gowda MVC, Narasu ML, Hoisington DA, Knapp SJ, Varshney RK (2012b) Quantitative trait locus analysis and construction of consensus genetic map for drought tolerance traits based on three recombinant inbred line populations in cultivated groundnut (*Arachis hypogaea* L.). Mol Breed 30:757–772

Gibbons RW, Bunting AH, Smartt JS (1972) The classification of varieties of groundnut (*Arachis hypogaea* L.). Euphytica 21:78–85

Girdthai T, Jogloy S, Vorasoot N, Akkasaeng C, Wongkaew S, Holbrook Patanothai A (2010) Heritability of and genotypic correlations between aflatoxin traits and physiological traits for drought tolerance under end of season drought in peanut (*Arachis hypogaea* L.). Field Crops Res 118:69–176

Green CC, Wynne JC, Beute MK (1983) Genetic variability and heritability estimates based on the F_2 generation from crosses of large seeded Virginia type peanuts with lines resistant to *Cylindrocladium* black rot. Peanut Sci 10:47–51

Gregory MP, Gregory WC (1979) Exotic germplasm of *Arachis* L.: interspecific hybrids. J Hered 70:185–193

Gregory WC, Gregory MP (1976) Groundnuts. In: Simmonds NW (ed) Evolution of crop plants. Longman, London, pp 151–154

Gregory WC, Gregory MP, Krapovickas A, Smith BW, Yarbrough JA (1973) Structures and genetic resources of peanuts. In: Wilson CT (ed) Peanuts—Culture and uses. Am Peanut Res Educ Assoc Stillwater, OK 3:47–134

Gregory WC, Krapovickas A, Gregory MP (1980) Structure, variation, evolution and classification. In: Summerfield RJ, Bunting AH (eds) *Arachis*. Advances in legume science, vol 2. Kew, London, pp 469–481

Halward T, Stalker HT, Kochert G (1993) Development of an RFLP linkage map in diploid peanut species. Theor Appl Genet 87:379–384

Hamidou F, Halilou O, Vadez V (2013) Assessment of groundnut under combined heat and drought stress. J Agron Crop Sci 199:1–11

Hammons RO (1994) The origin and history of Groundnut. In: Smartt J (ed) The groundnut crop. A scientific basis for improvement. Chapman & Hall, London, pp 24–42

Hapsoro D, Aswidinnoor H, Jumanto Suseno R, Sudarsono J (2007) Resistance to peanut stripe virus (PStV) in transgenic peanuts (*Arachis hypogaea* L.) carrying PStV *cp* gene was stable up to seven generations of selfing. Biota 12:83–91

Hapsoro D, Aswidinnoor H, Suseno R, Sudarsono J (2005) *Agrobacterium*-mediated transformation of peanuts (*Arachis hypogaea* L.) with PStV *cp* gene. J Trop Agric 10:85–93

Harlan JR, de Wet JMJ (1971) Towards a rational classification of cultivated plants. Taxon 20:509–517

Haro RJ, Baldessari J, Otegui ME (2013) Genetic improvement of peanut in Argentina between 1948 and 2004: seed yield and its components. Field Crops Res 149:76–86

Herselman L, Thwaites R, Kimmins FM, Courtois B, van der Merwe PJA, Seal SE (2004) Identification and mapping of AFLP markers linked to peanut (*Arachis hypogaea* L.) resistance to the aphid vector of groundnut rosette disease. Theor Appl Genet 109:1426–1433

Higgins BB (1951) Origin and early history of the peanut. The Peanut- The Unpredictable Legume. Nat Fert Assoc, Washington DC, pp 118–127

Higgins CM, Hall RM, Mitter N, Cruikshank A, Dietzen RG (2004) Peanut stripe potyvirus resistance in peanut (*Arachis hypogaea* L.) plants carrying viral coat protein gene sequences. Transgenic Res 13:59–67

Holbrook CC, Anderson WF, Pittman RN (1993) Selection of a core collection from the U.S. germplasm collection of peanut. Crop Sci 33:859–861

Holbrook CC, Kvien CK, Ruker KS, Wilson DM, Hook JE, Matheron ME (2000a) Preharvest aflatoxin contamination in drought-tolerant and drought-intolerant peanut genotypes. Peanut Sci 27:45–48

Holbrook CC, Noe JP (1992) Resistance to the peanut root-knot nematode (*Meloidogyne arenaria*) in *Arachis hypogaea*. Peanut Sci 19:35–37

Holbrook CC, Timper P, Culbreath AK (2003) Resistance to tomato spotted wilt virus and root-knot nematode in peanut interspecific breeding lines. Crop Sci 43:1100–1113

Holbrook CC, Timper P, Xue HQ (2000b) Evaluation of the core collection approach for identifying resistance to *Meloidogyne arenaria* in peanut. Crop Sci 40:1172–1175

Holbrook CC, Wilson DM, Matheron ME, Hunter JE, Knauft DA, Gorbet DW (2000c) *Aspergillus* colonization and aflatoxin contamination in peanut genotypes with reduced linoleic acid composition. Plant Dis 84:148–150

Holley RH, Wynne JC, Campbell WV, Isleib TG (1985) Combining ability for insect resistance in peanuts. Olèagineux 40:203–207

Hong Y, Chen Y, Liang X, Liu H, Zhou G, Li S, Wen S, Holbrook CC, Guo B (2010) A SSR-based composite genetic linkage map for the cultivated peanut (*Arachis hypogaea* L.) genome. BMC Plant Biol 10:10–17

Huang BY, Zhang XY, Miao LJ, Yan Z, Hai Y, Yi ML, Xu J, Chen ZK (2008) RNAi transformation of Ah *FAD2* gene and fatty acid analysis of transgenic seeds. Chin J Oil Crop Sci 30:290–293

Huang L, Jiang H, Ren X, Chen Y, Xiao Y, Zhao X Tang M, Huang J, Upadhyaya HD, Liao B (2012) Abundant microsatellite diversity and oil content in wild *Arachis* species. PLoS ONE 7 (11):e50002

Huang L, Zhao X, Zhang W, Fan Z, Ren X, Liao B, Jiang H, Chen Y (2011) Identification of SSR markers linked to oil content in peanut (*Arachis hypogaea* L.) through RIL population and natural population. Acta Agron Sinica 37:1967–1974

Husain F, Mallikarjuna N (2012) Genetic diversity in Bolivian landrace lines of groundnut (*Arachis hypogaea* L.). Ind J Genet. Plant Breed 72(3):384–389

IBPGR/ICRISAT (1992) Groundnut descriptors. IBPGR Descriptors Series, IBPGR, Rome and ICRISAT, Patancheru, 125 pp

Janila P, Nigam SN (2013) Phenotyping for groundnut (*Arachis hypogaea* L.) improvement. In: Panguluri SK, Kumar AA (eds) Phenotyping for plant breeding: applications of phenotyping methods for crop improvement. Springer Science + Business Media, New York, pp 129–167

Jayalakshmi V, Rajareddy C, Reddy PV, Nageswar Rao RC (1999) Genetic analysis of carbon isotope discrimination and specific leaf area in groundnut (*Arachis hypogaea* L.). J Oilseeds Res 16:1–5

Jiang H, Ren X, Chen Y, Huang L, Zhou X, Huang J, Froenicke L, Yu J, Guo B, Liao B (2013) Phenotypic evaluation of the Chinese mini-mini core collection of peanut (*Arachis hypogaea* L.) and assessment for resistance to bacterial wilt disease caused by *Ralstonia solanacearum*. Plant Genet Resour 11(01):77–83

Jiang H, Ren X, Shou Liao B, Huang J, Chen Y (2007) Establishment of peanut core collection in China. Wuhan Zhiwuxue Yanjiu 25(3):289–293

Jiang H, Ren X, Liao B, Huang J, Lei Y, Chen B, Guo B, Holbrook CC, Upadhyaya HD (2008) Peanut core collection established in China and compared with ICRISAT mini core collection. Acta Agron Sin 34:25–30

Khalfaoui JLB (1991) Inheritance of seed dormancy in a cross between two Spanish peanut cultivars. Peanut Sci 18:65–67

Khedikar YP, Gowda MVC, Sarvamangala C, Patgar KV, Upadhyaya HD, Varshney RK (2010) A QTL study on late leaf spot and rust revealed one major QTL for molecular breeding for rust resistance in groundnut (*Arachis hypogaea* L.). Theor Appl Genet 121:971–984

Knauft DA, Moore KM, Gorbet DW (1993) Further studies on the inheritance of fatty acid composition in peanut. Peanut Sci 20:74–76

Kochert G, Halward T, Branch WD, Simpson CE (1991) RFLP variability in peanut (*Arachis hypogaea* L.) cultivars and wild species. Theor Appl Genet 81:565–570

Kochert G, Stalker HT, Gimenes M, Galgaro L, Lopez CR, Moore K (1996) RFLP and cytogenetic evidence on the origin and evolution of allotetraploid domesticated peanut, *Arachis hypogaea* (Leguminosae). Am J Bot 83:1282–1291

Kottapalli KR, Burow MD, Burow G, Burke J, Puppala N (2007) Molecular characterization of the US peanut mini core collection using microsatellite markers. Crop Sci 47(4):1718–1727

Krapovickas A (1969) The origin, variability and spread of the groundnut (*Arachis hypogaea*) (English translation by Smartt J) In: Ucko RJ, Dimbledy CW (eds) The domestication and exploitation of plants and animals. Duckworth, London, pp 427–441

Krapovickas A, Rigoni VA (1951) Estudios citológicos en el género *Arachis*. Revista Invest Agríc 5(3):289–293

Krapovikas A, Gregory WC (1994) Taxonomia del genero *Arachis* (*Leguminosae*). *Bonplandia* VIII:1-187. (In Spanish). (English translation by Williams DE, Simpson CE 2007). Taxonomy of the genus *Arachis* (Leguminosae). Bonplandia 16 (suppl):1–205

Li ZJ, Jarret RL, Demski JW (1997) Engineered resistance to tomato spotted wilt virus in transgenic peanut expressing the viral nulceocapsid gene. Transgenic Res 6:297–305

Liang X, Zhou G, Hong Y, Chen X, Liu H, Li S (2009) Overview of research progress on peanut (*Arachis hypogaea* L.) host resistance to aflatoxin contamination and genomics at the Guangdong Academy of Agricultural Sciences. Peanut Sci 36:29–34

Liao BS, Li WR, Sun DR (1986) A study on inheritance of resistance to *Pseudomonas solanacearum* E.F. Smith in *Arachis hypogaea* L. Oil Crops China 3:1–8

Lopez Y, Smith OD, Senseman SA, Rooney WL (2001) Genetic factors influencing high oleic acid content in Spanish market-type peanut cultivars. Crop Sci 41:51–56

Lu M, Liang XQ, Dang P, Holbrook CC, Bausher MG, Lee RD, Guo BZ (2005) Microarray-based screening of differentially expressed genes in peanut in response to *Aspergillus parasiticus* infection and drought stress. Plant Sci 169:695–703

Lynch RE (1990) Resistance in peanut to major arthropod pests. Florida Entomologist 73:422–445

Magbanua ZV, Wilde HD, Roberts JK, Chowdhury K, Abad J, Moyer JW, Wetzstein HY, Parrott WA (2000) Field resistance to tomato spotted wilt virus in transgenic peanut (*Arachis hypogaea* L.) expressing an antisense nucleocapsid gene sequence. Mol Breed 6:227–236

Mallikarjuna N (2005) Production of hybrids between *Arachis hypogaea* and *A. chiquitana* (section *Proccumbentes).* Peanut Sci 32(2):148–152

Mallikarjuna N, Hoisington DA (2009) Peanut improvement: production of fertile hybrids and backcross progeny between *Arachis hypogaea* and *A. kretschmeri*. Food Secur 1(4):457–462

Mallikarjuna N, Pande S, Jadhav DR, Sastri DC, Rao JN (2004) Introgression of disease resistance genes from *Arachis kempff-mercadoi* into cultivated groundnut. Plant Breed 123(6):573–576

Mallikarjuna N, Sastri DC (2002) Morphological, cytological and disease resistance studies of the intersectional hybrid between *Arachis hypogaea* L. and *A. glabrata* Benth. Euphytica 126 (2):161–167

Mallikarjuna N, Senthilvel S, Hoisington DA (2011) Development of new sources of tetraploid *Arachis* to broaden the genetic base of cultivated groundnut (*Arachis hypogaea* L.). Genet Res Crop Evol 58(6):889–907

Mallikarjuna N, Tandra SK, Jadhav DR (2006) *Arachis hoehnei* the probable B genome donor of *Arachis hypogaea* based on crossability, cytogenetical and molecular studies. J SAT Agri Res 2(1):1–2

Manjunatha SB, Suma TC, Sreevathsa R, Devendra R, Udayakumar M, Prasad TG (2008) Evaluation of advanced generation transgenic groundnut lines resistant to herbicide-glyphosate. Ind J Weed Sci 40:162–165

Mehan VK, Liao BS, Tan YJ, Robinson-Smith A, McDonald D, Hayward AC (1994) Bacterial wilt of groundnut. Information Bulletin no. 35. International Crops Research Institute for the Semi-Arid Tropics, Patancheru, AP, India, 28 pp

Mehan VK, Mayee CD, Brenneman TB, McDonald D (1995) Stem and pod rots of groundnut. Information Bulletin no. 44. International Crops Research Institute for the Semi-Arid Tropics, Patancheru, AP, India; Peanut Collaborative Research Support Program, Griffin, GA, USA, 28 pp

Mehan VK, Reddy DDR, McDonald D (1993) Resistance in groundnut genotypes to Kalahasti malady caused by the stunt nematode, *Tylenchorhynchus brevilineatus*. Int J Pest Manag 39:201–203

Mienie CMS, Pretorius AE (2013) Application of marker-assisted selection for *ahFAD2A* and *ah FAD2B* genes governing high oleic acid trait in South African groundnut cultivars (*Arachis hypogaea* L.). Afr J Biotechnol 12:283–289

Milla SR, Isleib TG, Tallury SP (2005) Identification of AFLP markers linked to reduced aflatoxin accumulation in *A. cardenasii*-derived germplasm lines of peanut. Proc Am Peanut Res Educ Soc Inc 37: 90

Mondal S, Badigannavar AM, D'Souza SF (2012) Development of genic molecular marker linked to a rust resistance gene in cultivated groundnut (*Arachis hypogaea* L.). Euphytica 188: 163–173

Mondal S, Badigannavar AM, Murty SSS (2008) RAPD markers linked to a rust resistance gene in cultivated groundnut (*Arachis hypogaea* L.). Euphytica 159:233–239

Mondal S, Hadapad AB, Hande PA, Badignnavar AM (2014) Identification of quantitative trait loci for bruchid (*Caryedon serratus* L.) resistance components in cultivated groundnut (*Arachis hypogaea* L.). Mol Breed 33:961–973

Moore KM, Knauft DA (1989) The inheritance of high oleic acid in peanut. J Heredity 80: 252–253

Moretzsohn MC, Barbosa AVG, Alves-Freitas DMT, Teixeira C, Leal-Bertioli SCM, Guimarães PM, Pereira R, Lopez CR, Cavallari MM, Valls JFM, Bertioli DJ, Gimenes MA (2009) A linkage map for the B genome of *Arachis* (Fabaceae) and its synteny to the A-genome. BMC Plant Biol 9:40

Moretzsohn MC, Leoi L, Procite K, Guimarães PM, Leal-Bertioli SCM, Gimenes MA, Martins WS, Valls JFM, Grattapaglia D, Bertioli DJ (2005) A microsatellite-based, gene-rich linkage map for the AA genome of *Arachis* (*Fabaceae*). Theor Appl Genet 111:1060–1071

Moretzsohn MC, Gouvea EG, Inglis PW, Leal-Bertioli SC, Valls JFM, Bertioli DJ (2013) A study of the relationships of cultivated peanut (*Arachis hypogaea*) and its most closely related wild species using intron sequences and microsatellite markers. Ann Bot 111(1):113–126

Moss JP, Singh AK, Nigam SN, Hildebrand GL, Govinden N, Ismael FM, Subrahmanyam P, Reddy LJ (1998) Registration of ICGV-SM 86715 Peanut Germplasm. Crop Sci 38:572

Moss JP, Singh AK, Reddy LJ, Nigam SN, Subrahmanyam P, McDonald D, Reddy AGS (1997) Registration of ICGV 87165 peanut germplasm line with multiple resistances. Crop Sci 37:1028

Moss JP, Singh AK, Subrahmanyam P, Hildebrand GL, Murant AF (1993) Transfer of resistance to groundnut rosette disease from a wild *Arachis* species into cultivated groundnut. Int Arachis Newslett 13:22–23

Mozingo RW, Coffelt TA, Wynne JC (1987) Genetic improvement in large-seeded Virginia-type peanut cultivars since 1944. Crop Sci 27:228–231

Murthy TGK, Reddy PS (eds) (1993) Genetics of groundnuts. In: Cytogenetics and genetics of groundnuts. Antercept, Andover, UK, pp 144–268

Nageswar Rao RC, Nigam SN (2001) Genetic options for drought management in groundnut. In: Saxena NP (ed) Management of agricultural drought: agronomic and genetic options. Oxford and IBH, New Delhi, pp 123–141

Nageswar Rao RC, Talwar HS, Wright GC (2001) Rapid assessment of specific leaf area and leaf N in peanut (*Arachis hypogaea* L.) using chlorophyll meter. J Agron Crop Sci 189:175–182

Nageswar Rao RC, Williams JH, Singh M (1989) Genotypic sensitivity to drought and yield potential of peanut. Agron J 81:887–893

Nageswar Rao RC, Wright GC (1994) Stability of the relationship between specific leaf area and carbon isotope discrimination across environments in peanut. Crop Sci 34:98–103

Naidu PH, Moses GJ (2000) Identification and evaluation of sources of resistance to Kalahasti malady in groundnut. Ind Phytopath 53:395–398

Nautiyal PC, Bandyopadhyay A, Zala PV (2001) In situ sprouting and regulation of fresh seed dormancy in Spanish type groundnut (*Arachis hypogaea* L.). Field Crops Res 70:233–241

Nautiyal PC, Ravindra V, Bandyopadhyay A (1994) Peanut seed dormancy. ACIAR- Food Legumes Newslett 22:2

Nigam SN, Aruna R (2008a) Stability of Soil Plant Analytical Development (SPAD) Chlorophyll Meter Reading (SCMR) and specific leaf area (SLA) and their association across varying soil moisture stress conditions in groundnut (*Arachis hypogaea* L.). Euphytica 160:111–117

Nigam SN, Aruna R (2008b) Improving breeding efficiency for early maturity in peanut. Plant Breed Rev 30:295–322

Nigam SN, Bock KR (1990) Inheritance of resistance to groundnut rosette virus in groundnut (*Arachis hypogaea* L.). Ann Appl Biol 117:553–560

Nigam SN, Chandra S, Sridevi KR, Bhukta M, Reddy AGS, Rao RCN, Wright GC, Reddy PV, Deshmukh MP, Mathur RK, Basu MS, Vasundhara S, Varman PV, Nagda AK (2005) Efficiency of physiology trait-based and empirical selection approaches for drought tolerance in groundnut. Ann Appl Biol 146:433–439

Nigam SN, Dwivedi SL, Gibbons RW (1991) Groundnut breeding: constraints, achievements and future possibilities. Plant Breed Abst 61:1127–1136

Nigam SN, Nageswar Rao RC, Wright GC (2003) Breeding for increased water-use efficiency in groundnut. In: Rai M, Singh H, Hegde DM (eds) National seminar on stress management in oilseeds for attaining self-reliance in vegetable oils: thematic papers. Directorate of Oilseed Research, Rajendranagar, Hyderabad, AP, India, pp 305–318

Nigam SN, Nageswara Rao RC, Wynne JC, Williams JH, Fitzner M, Nagabhushanam GVS (1994) Effect and interaction of temperature and photoperiod on growth and partitioning in three groundnut (*Arachis hypogaea* L.) genotypes. Ann Appl Biol 125:541–552

Nigam SN, Prasada Rao RDVJ, Bhatnagar-Mathur P, Sharma KK (2012) Genetic management of virus diseases in peanut. Plant Breed Rev 36:293–356

Nigam SN, Upadhyaya HD, Chandra S, Nageswar Rao RC, Wright GC, Reddy AGS (2001) Gene effects for specific leaf area and harvest index in three crosses of groundnut (*Arachis hypogaea* L.). Ann Appl Biol 139:301–306

Nigam SN, Waliyar F, Aruna R, Reddy SV, Lava Kumar P, Craufurd PQ, Diallo AT, Ntare BR, Upadhyaya HD (2009) Breeding peanut for resistance to aflatoxin contamination at ICRISAT. Peanut Sci 36:42–49

Norden AJ, Gorbet DW, Knauft DA, Young CT (1987) Variability in oil quality among peanut genotypes in the Florida breeding program. Peanut Sci 14:7–11

Ntare BR, Williams JH, Dougbedji F (2001) Evaluation of groundnut genotypes for heat tolerance under field conditions in a Sahelian environment using a simple physiological model for yield. J Agric Sci 136:81–88

Olorunju PE, Kuhn CW, Demski JW, Misari SM, Ansa OA (1992) Inheritance of resistance in peanut to mixed infections of groundnut rosette virus (GRV) and groundnut rosette assistor virus and a single infection of GRV. Plant Dis 76:95–100

Olorunju PE, Ntare BR, Pande S, Reddy SV (2001) Additional sources of resistance to groundnut rosette disease in groundnut germplasm and breeding lines. Ann Appl Biol 159:259–268

Ouedraogo M, Smith OD, Simpson CE, Smith DH (1994) Early and late leaf spot resistance and agronomic performance of nineteen interspecific derived peanut lines. Peanut Sci 21:99–104

Ozias-Akins P (2007) Peanut. In: Pua EC, Davey MR (eds) Biotechnology in agriculture and forestry, vol 61. Springer, Berlin, pp 81–105

Ozias-Akins P, Schnall JA, Anderson WF, Singsit C, Clemente TE, Adang MJ, Weissinger AK (1993) Regeneration of transgenic peanut plants from stably transformed embryogenic callus. Plant Sci 93:185–194

Ozias-Akins P, Yang P, Culbreath AK, Gorbet DW, Weeks JR (2002) Field resistance to *Tomato spotted wilt virus* in a transgenic peanut (*Arachis hypogaea* L.). Proc Am Res Educ Soc Inc 34:70

Padgham DE, Kimmins FM, Ranga Rao GV (1990) Resistance in groundnut (*Arachis hypogaea* L.) to *Aphis craccivora* (Koch.). Ann Appl Biol 117:285–294

Pandey MK, Monyo E, Ozias-Akins P, Liang X, Guimarães P, Nigam SN, Upadhyaya HD, Janila P, Zhang X, Guo B, Cook DR, Bertioli DJ. Michelmore R, Varshney RK (2012) Advances in *Arachis* genomics for peanut improvement. Biotechnol Adv 30:639–651

Patel M, Jung S, Moore K, Powell G, Ainsworth C, Abbott A (2004) High-oleate peanut mutants result from a MITE insertion into the FAD2 gene. Theor Appl Genet 108:1492–1502

Pattee HE, Isleib TG, Giesbrecht FG (1998) Variation in intensity of sweet and bitter sensory attributes across peanut genotypes. Peanut Sci 25:63–69

Pensuk K, Jogloy S, Wongkaew S, Patanothai A (2004) Generation means analysis of resistance to peanut bud necrosis caused by peanut bud necrosis tospovirus in peanut. Plant Breed 123: 90–92

Pensuk K, Wongkaew S, Jogloy S, Patanothai A (2002) Combining ability for resistance in peanut (*Arachis hypogaea*) to peanut bud necrosis tospovirus (PBNV). Ann Appl Biol 141:143–146

Poledate A, Laohasiriwong S, Jaisil P, Vorasoot N, Jogloy S, Kesmala T, Patanothai A (2007) Gene effects for parameters of peanut bud necrosis virus (PBNV) resistance in peanut. Pakistan J Biol Sci 10:1501–1506

Prasada Rao RDVJ, Reddy AS, Chakarbarty SK, Reddy DVR, Rao VR, Moss JP (1991) Identification of Peanut Stripe Virus resistance in wild *Arachis* germplasm. Peanut Sci 18:1–2

Qin H, Gu Q, Zhang JL, Sun L, Kuppu S, Zhang YZ, Burow M, Payton P, Blumwald E, Zhang H (2011) Regulated expression of an isopentenyltransferase gene (IPT) in peanut significantly improves drought tolerance and increases yield under field conditions. Plant Cell Physiol 52:1904–1914

Qin HD, Feng SP, Chen C, Guo YF, Knapp S, Culbreath A, He GH, Wang ML, Zhang XY, Holbrook CC, Ozias-Akins P, Guo BZ (2012) An integrated genetic linkage map of cultivated peanut (*Arachis hypogaea* L.) constructed from two RIL populations. Theor Appl Genet 124:653–664

Raina SN, Mukai Y (1999) Genomic in situ hybridization in *Arachis* (Fabaceae) identifies the diploid wild progenitors of cultivated (*A. hypogaea*) and related wild (*A. monticola*) peanut species. Plant Syst Evol 214:251–262

Rao MJV, Nigam SN, Huda AKS (1991) Use of thermal time concept in selection for earliness in peanut. Peanut Sci 19:7–10

Rao MJV, Nigam SN, Mehan VK, McDonald D (1989) *Aspergillus flavus* resistance breeding in groundnut: progress made at ICRISAT Center. In: McDonald D, Mehan VK (eds) Aflatoxin contamination of groundnut. Proc Int Workshop, 6–9 Oct 1987, ICRISAT Center. International Crops Research Institute for the Semi-Arid Tropics, Patancheru, AP, India, pp 345–355

Rathnakumar AL, Hariprasanna K, Lalwani HB (2012) Genetic improvement in Spanish type groundnut. J Oilseeds Res 27:1–7

Ravi K, Vadez V, Isobe S, Mir RR, Guo Y, Nigam SN, Gowda MVC, Radhakrishnan T, Bertioli DJ, Knapp SJ, Varshney RK (2011) Identification of several small main-effect QTLs and a large number of epistatic QTLs for drought tolerance related traits in groundnut (*Arachis hypogaea* L.). Theor Appl Genet 122:1119–1132

Reddy LJ, Nigam SN, Singh AK, Moss JP, Subrahmanyam P, McDonald D, Reddy AGS (1996) Registration of ICGV 86699 peanut germplasm line with multiple disease and insect resistance. Crop Sci 36:821

Reddy PS, Murthy TGK (1996) Current genetic research on groundnut in India. Genetica 97:263–277

Reddy PS, Zade VR, Desmukh SN (1985) A new Spanish bunch groundnut cultivar with fresh seed dormancy. J Oilseed Res 2:103–106

Ren X, Chen Y, Zhou X, Xia Y, Huang J, Lei Y, Yan L, Wan L, Liao B (2014) Diversity Characterization and association analysis of agronomic traits in a Chinese peanut (*Arachis hypogaea* L.) mini-core collection. J Integr Plant Biol 56(2):159–169

Robledo G, Seijo G (2010) Species relationships among the wild B genome of *Arachis* species (section *Arachis*) based on FISH mapping of rDNA loci and heterochromatin detection: a new proposal for genome arrangement. Theor Appl Genet 121:1033–1046

Sasson A (1996) Biotechnologies and the use of plant genetic resources for industrial purpose: benefits and constraints for developing countries. In: Castri F di, Younes J (eds) Biodiversity, science and development: towards a new partnership. CAB International, pp 469-487
Seaton ML, Coffelt TA, Van Scoyoc SW (1992) Comparison of vegetative and reproductive traits of 14 peanut cultivars. Olèagineux 47:471–474
Seijo JG, Lavia GI, Fernandez A, Krapovikas A, Ducasse D, Moscone EA (2004) Physical mapping of the 5S and 18S-25S $_R$RNA gene by FISH as evidence that *Arachis duranensis* and *A. ipaensis* are the wild diploid progenitors of *A. hypogaea* (Leguminosae). Am J Bot 91: 1294–1303
Shan ZH, Duan NX, Jiang HF, Tan YJ, Li D, Liao BS (1998) Inheritance to bacterial wilt in Chinese dragon groundnuts. In: Prior P, Allen C, Elphinstone J (eds) Bacterial wilt disease. Springer, Berlin, pp 300–305
Sharma KK, Anjaiah VV (2000) An efficient method for the Production of transgenic plants of peanut (*Arachis hypogaea* L.) through *Agrobacterium tumefaciens*-mediated genetic transformation. Plant Sci 159:7–19
Shoba D, Manivannan N, Vindhiyavarman P, Nigam SN (2012) SSR markers associated for late leaf spot disease resistance by bulk segregant analysis in groundnut (*Arachis hypogaea* L.). Euphytica 188:265–272
Shoba D, Manivannan N, Vindhiyavarman P, Nigam SN (2013) Identification of quantitative trait loci (QTL) for late leaf spot disease resistance in groundnut (*Arachis hypogaea* L.). Legumes Res 36:467–472
Simpson CE (1991) Pathways for introgression of pest resistance into *Arachis hypogaea* L. Peanut Sci 18:22–26
Simpson CE (2001) Use of Wild *Arachis* species/Introgression of Genes into *A. hypogaea*. Peanut Sci 28:114–116
Simpson CE, Krapovickas A, Valls JFM (2001) History of *Arachis* including evidence of *A. hypogaea* L. progenitors. Peanut Sci 28:78–80
Singh AK (1985) Genetic introgression from compatible *Arachis* species into Groundnut. In: Proceedings of international workshop on cytogenetics of *Arachis*, Oct 31–Nov 2 1983. International Crops Research Institute for the Semi-Arid Tropics, Patancheru, AP, India, pp 107–117
Singh AK (1986a) Utilization of wild relatives in the genetic improvement of *Arachis hypogaea* L. Part 8. Synthetic amphidiploids and their importance in interspecific breeding. Theor Appl Genet 72:433–439
Singh AK (1986b) Utilization of wild relatives in the genetic improvement of *Arachis hypogaea* L. Part 7. Autotetraploid production and prospects in interspecific breeding. Theor Appl Genet 72:164–169
Singh AK (1988) Putative genome donors of *A. hypogaea* L. Evidence from cross with synthetic amphidiploids. Plant Syst Evol 160:143–153
Singh AK (1998) Hybridization barriers among species of *Arachis* L., namely of the section *Arachis* (including the groundnut) and *Erectoides*. Genet Res Crop Evol 45:41–45
Singh AK, Dwivedi SL, Pande S, Moss JP, Nigam SN, Sastri DC (2003) Registration of rust and late leaf spot resistant peanut germplasm lines. Crop Sci 43(1):440–441
Singh AK, Gibbons RW (1985) Wild species in crop improvement: groundnut—A case study. In: Gupta PK, Bahl JR (eds) Advances in genetics and crop improvement. Rastogi Publication, Meerut, pp 297–308
Singh AK, Mehan VK, Nigam SN (1997) Sources of resistance to groundnut fungal and bacterial diseases: an update and appraisal. Information Bulletin no. 50. International Crops Research Institute for the Semi-Arid Tropics, Patancheru, AP, India, 48 pp
Singh AK, Moss JP (1982) Utilization of wild relatives in genetic improvement of *Arachis hypogaea* L. Part 2. Chromosome complement of species in section *Arachis*. Theor Appl Genet 61:305–314

Singh AK, Moss JP (1984a) Utilization of wild relatives in genetic improvement of *Arachis hypogaea* L. VI. Fertility in triploids: cytological basis and breeding implications. Peanut Sci 11:17–21

Singh AK, Moss JP (1984b) Utilization of wild relatives in the genetic improvement of *Arachis hypogaea* L. Part 5. Genome analysis in section Arachis and its implications in gene transfer. Theor Appl Genet 68:355–364

Singh AK, Nigam SN (1996) Gains in groundnut productivity illustrates success of resistance breeding based on exotic germplasm. Diversity 12:59–60

Singh AK, Nigam SN (1997) Groundnut In: Fuccillo D, Sears L, Stapleton P (eds) Biodiversity in trust—conservation and use of plant genetic resources in CGIAR center. Cambridge University Press Cambridge UK, pp 113–124

Singh AK, Sastri DC, Moss JP (1980) Utilization of wild *Arachis* species at ICRISAT. In: ICRISAT (International Crops Research Institute for the Semi-Arid Tropics) Proceedings of the international groundnut workshop (13–17 Oct 1980), Patancheru, AP, India, pp 82–90

Singh AK, Simpson CE (1994) Biosystematic and Genetic Resources. In: Smartt J (ed) The groundnut crop. A scientific basis for improvement. Chapman and Hall, London, pp 96–138

Singh AK, Sivaramakrishnan S, Mengesha MH, Ramaiah CD (1991) Phylogenetic relationships in section *Arachis* based on seed protein profile. Theor Appl Genet 82:593–597

Singh AK, Smartt J (1998) The genome donors of the groundnut/peanut (*Arachis hypogaea* L.) revisited. Genet Res Crop Evol 45:113–118

Singh AK, Stalker HT, Moss JP (1990) Cytogenetics and use of alien genetic variation in groundnut (*Arachis*) improvement. In: Tsuchiya T, Gupta PK (eds) Chromosome engineering in plants: genetics, breeding & evolution. Elsevier, Amsterdam, pp 65–77

Singh AK, Subrahmanyam P, Gurtu S (1996) Variation in a wild groundnut species, *Arachis duranensis* Krapov. & W.C. Gregory. Genet Res Crop Evol 43:135–142

Singh AL, Hariprasanna K, Chaudhary V, Gor HK, Chikani BM (2010) Identification of groundnut (*Arachis hypogaea* L.) cultivars tolerant of soil salinity. J Plant Nutr 33:1761–1776

Singh AL, Hariprassana K, Solanki RM (2008) Screening and selection of groundnut genotypes for tolerance of soil salinity. Aust J Crop Sci 1:69–77

Singh Anurudh K, Smartt J, Rakesh Singh (2004) Variation studies in a wild groundnut species, *Arachis stenosperma* Krapov. & W.C. Gregory nov. sp. Plant Genet Resour 2:99–106

Singsit C, Adang MJ, Lynch RE, Anderson WF, Wang A, Cardineau C, Ozias-Akins P (1997) Expression of a *Bacillus thuringiensis cry1A(c)* gene in transgenetic peanut plants and its efficacy against lesser cornstalk borer. Transgenic Res 6:169–176

Smartt J (1990) Grain legumes: evolution and genetic resources, 2nd edn. Cambridge University, Cambridge, 531 pp

Smartt JS, Gregory WC (1967) Interspecific crosscompatibility between the cultivated peanut *Arachis hypogaea* L. and other members of the genus *Arachis*. Oléagineux 22(7):455–459

Smartt JS, Gregory WC, Gregory MP (1978) The genomes of *Arachis hypogaea*. 1. Cytogenetic studies of putative genome donors. Euphytica 27:665–675

Songsri P, Jogloy S, Kesmala T, Vorasoot N, Akkasaeng C, Patonothai A, Holbrook CC (2008a) Heritability of drought resistance traits and correlation of drought resistance and agronomic traits in peanut. Crop Sci 48:2245–2253

Songsri P, Jogloy S, Vorasoot N, Akkasaeng C, Patonothai A, Holbrook CC (2008b) Root distribution of drought-resistant peanut genotypes in response to drought. J Agron Crop Sci 194:92–103

Srikanth S, Shilpa K, Jadhav D, Satyanarayana T, Mallikarjuna N (2012) Meiotic study of three synthesized tetraploid groundnut. Ind J Genet Plant Breed 72(3):332–335

Stalker HT (1991) A new species in section *Arachis* of peanuts with a D genome. Am J Bot 78 (5):630–637

Stalker HT, Moss JP (1987) Speciation, cytogenetics and utilization of *Arachis* species. Adv Agron 41:1–40

Stalker HT, Mozingo LG (2001) Molecular markers of *Arachis* and marker-assisted selection. Peanut Sci 28:117–123

Subrahmanyam P, Hildebrand GL, Naidu RA, Reddy LR, Singh AK (1998) Sources of resistance to groundnut rosette disease in global groundnut germplasm. Ann Appl Biol 132:473–485

Sunkara S, Bhatnagar-Mathur P, Sharma KK (2013) Transgenic interventions in peanut crop improvement: progress and prospects. In: Mallikarjuna N, Varshney RK (eds) Genetics, Genomics and Breeding of peanuts. CRC Press, Taylor and Francis, Boca Raton, Florida, pp 179–216

Taliansky ME, Robinson DJ, Murant AF (1996) Complete nucleotide sequence and organization of the RNA genome of groundnut rosette umbravirus. J Gen Virol 77:2335–2345

Tiwari S, Mishra DK, Singh A, Singh PK, Tuli R (2008) Expression of a synthetic *cry1EC* gene for resistance against *Spodoptera litura* in transgenic peanut (*Arachis hypogaea* L.). Plant Cell Rep 27:1017–1025

Upadhyaya HD, Bramel PJ, Ortiz R, Singh S (2002a) Developing a minicore of peanut for utilization of genetic resources. Crop Sci 42:2150–2156

Upadhyaya HD, Mallikarjuna Swamy BP, Kenchana Goudar PV, Kullaiswamy BY, Singh S (2005) Identification of diverse groundnut germplasm through multienvironment evaluation of a core collection for Asia. Field Crops Res 93:293–299

Upadhyaya HD, Nigam SN (1999) Inheritance of fresh seed dormancy in peanut. Crop Sci 39: 98–101

Upadhyaya HD, Nigam SN, Mehan VK, Lenne JM (1997a) Aflatoxin contamination of groundnut- prospects of a genetic solution through conventional breeding. In: Mehan VK, Gowda CLL (eds) Aflatoxin contamination problems in groundnut in Asia: proceedings of the first working group meeting, 27–29 May 1996. Ministry of Agriculture and Rural Development, Hanoi, Vietnam, ICRISAT, Patancheru, India, pp 81–85

Upadhyaya HD, Nigam SN, Rao MJV, Reddy AGS, Yellaiah N, Reddy NS (1997b) Registration of five Spanish peanut germplasm with fresh seed dormancy. Crop Sci 37:1027

Upadhyaya HD, Nigam SN, Thakur RP (2002b) Genetic enhancement for resistance to aflatoxin contamination in groundnut In: Summary Proc of the seventh ICRISAT regional groundnut meeting for Western and Central Africa, 6-8 December 2000, Cotonou, Benin. International Crops Research Institute for the Semi-Arid Tropics, Patancheru, AP, India, pp 29–36

Upadhyaya HD, Ortiz R, Bramel PJ, Singh S (2003) Development of a groundnut core collection using taxonomical, geographical and morphological descriptors. Genet Res Crop Evol 50: 139–148

Upadhyaya HD, Reddy LJ, Gowda CLL, Singh S (2006) Identification of diverse groundnut germplasm: sources of early maturity in a core collection. Field Crops Res 97(2–3):261–271

Upadhyaya HD Sharma S and Dwivedi SL (2011) *Arachis*. In: Kole C (ed) Wild crop relatives: genomics and breeding resources, Legume Crops and Forages. Springer, Berlin, pp 1–19

Upadhyaya HD, Yadav D, Dronavalli N, Gowda CLLL, Singh S (2010) Minicore germplasm collection for infusing genetic diversity in plant breeding program. Electr J Plant Breed 1 (4):1294–1309

Valls JFM, Ramanatha Rao V, Simpson CE and Krapovickas A (1985) Current status of collection and conservation of South American groundnut germplasm with emphasis on wild species of *Arachis*. In: Proceedings of international workshop on cytogenetics of *Arachis*, Oct 31–Nov 2 1983. International Crops Research Institute for the Semi-Arid Tropics, Patancheru, AP, India, pp 15–35

Valls JFM, Simpson CE (2005) New species of *Arachis* (Leguminosae) from Brazil, Paraguay and Bolivia. Bonplandia 14(1–2):35–63

Varma TSN, Dwivedi SL, Pande S, Gowda MVC (2005) SSR markers associated with resistance to rust (*Puccinia arachidis* Speg.) in groundnut (*Arachis hypogaea* L.). SABRAO J Breed Genet 37:107–119

Varshney RK, Bertioli DJ, Moretzsohn MC, Vadez V, Krishnamurthy L, Aruna R, Nigam SN, Moss BJ, Seetha K, Ravi K, He G, Knapp SJ, Hoisington DA (2009) The first SSR-based genetic linkage map for cultivated groundnut (*Arachis hypogaea* L.). Theor Appl Genet 118:729–739

Waliyar F, Kumar PL, Ntare BR, Monyo E, Nigam SN, Reddy AS, Osiru M, Diallo AT (2007) A century of research on groundnut rosette disease and its management. Information Bulletin no. 75. International Crops Research Institute for the Semi-Arid Tropics, Patancheru, AP, India, 40 pp

Wang M, Abbott D, Waterhouse PM (2000) A single copy of a virus derived transgene encoding hairpin RNA gives immunity to barley yellow dwarf virus. Mol Plant Path 1:401–410

Weber JL (1990) Informativeness of huan (dC-dA) (dG-dT) in polymorphism. Genomics 7: 524–530

Wells R, Bi T, Anderson WF, Wynne JC (1991) Peanut yield as a result of fifty years of breeding. Agron J 83:957–961

Wightman JA, Ranga Rao GV (1994) Groundnut pests. In: Smartt J (ed) The groundnut crop: a scientific basis for improvement. Chapman and Hall, London, pp 395–479

Wildman LG, Smith OD, Simpson CE, Taber RA (1992) Inheritance of resistance to *Sclerotinia minor* in selected Spanish peanut crosses. Peanut Sci 19:31–35

Williams ED (1991) Exploration of Amozonian Bolivia yields rare peanut landraces. Diversity 5:12–13

Williams JH, Boote KJ (1995) Physiology and modeling-predicting the unpredictable legume. In: Pattee HE, Stalker HT (eds) Advances in peanut science. Am Peanut Res Educ Soc Stillwater 301–353

Wright GC, Nageswar Rao RC, Basu MS (1996) A physiological approach to the understanding of genotype by environment interactions—A case study on improvement of drought adaptation in peanut. In: Cooper M, Hammer GL (eds) Plant adaptation and crop improvement. CAB International, Wallingford, pp 365–381

Wright GC, Nageswar Rao RC, Farquhar GD (1994) Water-use efficiency and carbon isotope discrimination in peanut under water deficit conditions. Crop Sci 34:92–97

Xue HQ, Isleib TG, Payne GA, Novitzky WF, OBrian G (2005) Aflatoxin production in peanut lines selected to represent a range of linoleic acid concentrations. J Food Protect 68:126–132

Yang H, Singcit C, Wang A, Gonsalves D, Ozias-Akins P (1998) Transgenic peanut plants containing a nucleocapsid protein gene of tomato spotted wilt virus show divergence levels of gene expression. Plant Cell Rep 17:693–699

Yaw AJ, Richard A, Safo-Kantanka O, Adu-Dapaah HK, Ohemeng-Dapaah S, Agyeman A (2008) Inheritance of fresh seed dormancy in groundnut. Afr J Biotechnol 7:421–424

Yin DM, Tang HT, Tai GQ, Yang QY, Cui DQ (2009) Expression analysis and establishment of regeneration system of oleate desaturase gene in peanut. Scientia Agricultura Sinica 42: 1827–1832

Zhang XQ, Shan L, Tang GY, Teng N, Bi YP (2007) Transformation of RNAi suppressed expression vector containing of Δ^{12} fatty acid desaturase gene via *Agrobacterium* infection in peanut (*Arachis hypogaea* L.). Chin J Oil Crop Sci 29:409–415

Zhou GY, Liang XQ (2002) Analysis of major-minor genes related to resistance to infection by *Aspergillus flavus* in peanut. J Peanut Sci 31:11–14

Zhou GY, Liang XQ, Li YC, Li XC, Li SL (1999) Evaluation and application of introduced peanut cultivars for resistance to *Aspergillus flavus* invasion. J Peanut Sci 32:14–17

Chapter 3
Genetic Resources of Chickpea (*Cicer arietinum* L.) and Their Utilization

Deepak Ohri

Abstract Chickpea is a prominent grain legume crop providing cheap source of protein to the humankind. It originated in the Near East from the progenitor species *Cicer reticulatum* having a narrow distribution and genetic base. Moreover, during the course of domestication chickpea experienced various bottlenecks resulting in still narrow genetic variation in its two major forms 'Kabuli' and 'Desi'. Further genetic improvement would therefore depend on the exploration and introduction of useful genes from its wild relatives. The genus *Cicer* has 49 taxa including nine annual species. The genetic relationships among these and with the cultigen have been analyzed and elaborated by diverse methods including morphology, seed proteins, isozymes, karyotypes, FISH and various DNA markers. All these studies have resulted in demarcating primary, secondary and tertiary gene pools and show a very close relationship of the cultigen with two annual species *C. reticulatum* and *C. echinospermum* besides some perennial species. However, direct transfer of genes by hybridization has proved to be nearly impossible as the cultigen shows very poor or no crossability with any of the wild species except the progenitor species. This problem is being addressed by QTL mapping of mostly disease resistance loci from the RIL's produced from intra as well as interspecific crosses. Further efforts are being made to integrate genetic maps with physical maps. These methods provide a strong basis for genetic and genomic analysis of chickpea genome and facilitate further the use of molecular methods in breeding.

Keywords *Cicer arietinum* · Origin · Domestication · Interspecific relationships · Molecular maps

---

D. Ohri (✉)
Amity University Uttar Pradesh (Lucknow Campus), Malhaur (Near Railway Station), Gomti Nagar Extension, Lucknow 226028, UP, India
e-mail: dohri@lko.amity.edu; ohri_deepak@rediffmail.com

V.R. Rajpal et al. (eds.), *Gene Pool Diversity and Crop Improvement*, Sustainable Development and Biodiversity 10,
DOI 10.1007/978-3-319-27096-8_3

3.1 Introduction

Chickpea (*Cicer arietinum* L.) is the only cultivated species of the genus *Cicer*, and as an important cool season grain legume it ranks second most important pulse crop being grown in about 50 countries on an area of 12 m ha with a total production of 11 m tons and productivity of 910 kg ha^{-1} (FAOSTAT 2012). With two third of the total world production occurring in India the other major producing countries are Pakistan, Myanmar, Iran, Turkey, Mexico, Canada and USA (FAOSTAT 2012). Chickpea provides a relatively cheap source of dietary protein and its seeds contain 20.3 % protein, approximately 40 % carbohydrates and 3–6 % oil (Gil et al. 1996). It is also rich in minerals and is a good source of calcium, magnesium, potassium, phosphorus, iron, zinc and manganese besides a number of vitamins (Ibricki et al. 2003; Wood and Grusak 2007). It also contains higher amount of carotenoids such as β-carotene than genetically engineered rice (Abbo et al. 2005). Moreover, in comparison to other legumes anti-nutritional factors are nearly absent (Muzquiz and Wood 2007). With all these nutritional components chickpea very well serves the purpose of a nutraceutical (Agharkar 1991; McIntosh and Topping 2000; Charles et al. 2002; Jukanti et al. 2012).

Two distinct types of chickpea are classified into, microsperma and macrosperma referring to the seed size (Cubero 1987). Commercially two distinct types are available 'Desi' with small angular dark brown seeds with rough surface, pink or purple flowers, anthocyanin pigments on the stems semi-erect or semi-spreading habit and 'Kabuli' with large ram-shaped seeds with smooth surface, white coloured flowers, lack of anthocyanin pigmentation and semi-spreading habit. These two types also have different centres of diversity as 'Kabuli' types with narrow genetic diversity mainly grow in Mediterranean region, central and West Asia, while 'Desi' with much wider genetic diversity in the Indian subcontinent and Ethiopia (van der Maesen 1972; Berger and Turner 2007). The 'Kabuli' types are generally considered to have evolved from 'Desi' types (Moreno and Cubero 1978; Hawtin and Singh 1980; Salimath et al. 1984; Gil and Cubero 1993) which is also supported by close similarity of seed coat texture between 'Desi' type of *C. arietinum* and *C. reticulatum* therefore implying a recent divergence of 'Kabuli' type from 'Desi' (Javadi and Yamaguchi 2004a). However, a white flower coloured mutation was isolated in M2 generation of some accession of *C. reticulatum*. This mutation also had cream coloured seeds as compared with dark coloured seeds in the parent *C. reticulatum* suggesting that this 'Kabuli' chickpea might have originated as a mutation of *C. reticulatum* (Toker 2009). Recently, the study on transcriptome sequencing of 'Kabuli' chickpea shows a higher similarity of transcripts between 'Kabuli' and 'Desi' as compared with these and the wild progenitor. It has been deduced that first cultivated chickpeas diverged from the wild progenitor and two cultivated types 'Kabuli' and 'Desi' diverged soon after that (Agarwal et al. 2012). However, the authors have not ruled out the possibility of both the cultivated types originating directly from the wild progenitor (Agarwal et al. 2012). The distinct genetic backgrounds of these two types has been shown by RAPD and

ISSR markers where they form two separate groups which, however, do not correlate with their geographical origin (Iruela et al. 2002). However, the analysis by STMS markers does not clearly demarcate 'Kabuli' and 'Desi' types (Rizvi et al. 2014). Now the differences between these two types are slowly merging due to plant breeding programmes requiring to combine the large seed size with local adaptation and vigour of 'Desi' types (Yadav et al. 2004). A third type named as intermediate type or 'pea shaped' has been identified with small to medium round pea-shaped seeds (Sharma et al. 2013).

3.2 Origin and Domestication

Chickpea is one of the founder grain crops having originated when the humans started domesticating the various plant species at the beginning of agriculture in the Fertile Crescent (Near East) 12,000–10,000 years ago with the cultivation of seven grain crops (*Triticum monococcum* L., *Triticum turgidum* L., *Hordeum vulgare* L., *Pisum sativum* L. *Lens culinaris*, *Cicer arietinum* L., *Vicia ervilia* (L.) Willd. and *Linum usitatissimum* L., a fiber crop) called founder crop package (Zohary and Hopf 2000; Lev-Yadun et al. 2000). Although many evidences have been recorded about chickpea cultivation, the earliest most authentic record is of 7260 B.C at Tell el-Kerkh, Syria where seed samples of both chickpea and progenitor *C. reticulatum* were clearly distinguishable (Tanno and Willcox 2006).

Chickpea has undergone many physiological and morphological changes since its evolution from the nearest parental species *C. reticulatum*. *C. reticulatum* has a very restricted distribution, currently reported from 18 locations in southeastern Turkey (Berger et al. 2003). The modern chickpea, therefore, has a narrow genetic base because of genetic bottlenecks it has experienced at various stages of domestication. According to Abbo et al. (2003) four such bottlenecks are the restricted distribution of the progenitor *C. reticulatum*, founder effects because of narrow genetic sampling in the initial stages, shift in the growing season from winter spring sowing to escape ascochyta blight, and finally replacement of locally existing land races by elite modern cultivars evolved by the breeders. However, despite the problems like disease susceptibility and manipulation of the growing season experienced by the early farmers the chickpea cultivation was necessitated by the advantages it provided in terms of nutritional superiority as some chickpea stocks analyzed were found to have high tryptophane levels (Karem et al. 2007; Abbo et al. 2005).

3.3 Taxonomy

The genus *Cicer* belongs to the family Fabaceae in the monogeneric tribe Cicerae Alef (Kupicha 1981). It consists of 49 taxa including nine annual species (van der Maesen et al. 2007; Donmez 2011; Ozturk et al. 2013). Traditionally, the genus

Cicer has been divided into two subgenera (*Pseudononis* and *Viciastrum*) and four sections *Cicer, Chamaecicer, Polycicer* and *Acanthocicer* based mainly on morphological characters and geographical distribution (Popov 1929; van der Maesen 1972, 1987). Recently, Davies et al. (2007), analyzed 104 characters in 35 recognized taxa of *Cicer* by multivariate statistics and proposed a new subgeneric classification with three subgenera, five sections and two series.

Genus Cicer
Subgenus Cicer
Section Cicer
Series Cicer

C. arietinum L.
C. echinospermum P.H. Davies
C. reticulatum Ladiz.

Series Pinnatifida

C. bijugum Rech.f
C. judaicum Boiss.
C. pinnatifidum Jaub. Sapach

Section Chamaecicer

C. atlanticum Coss. Ex Maire
C. incisum (Willd.) K. Maly

Sugenus Viciastrum
Section Annua

C. chorassanicum (Bunge) Popov
C. yamashitae Kitam

Section Polycicer

C. floribundum Fenzl.
C. graceum Orph.
C. heterophyllum Contandr. Pamuk C & Quezel
C. isauricum P.H. Davies
C. montbretii Jaub. & Spach.

Section Vicioides

C. acanthophyllum Borris
C. anatolicum Alef.
C. balcaricum Galushko
C. baldshuanicum (Popov) Lincz.
C. fedtschenkoi Lincz
C. flexuosum Lipsky
C. grande (Popov) Korotkova
C. incanum Korotkova

C. korshinsky Lincz.
C. laetum Rassulova & Sharipova
C. luteum Rassulova & Sharipova
C. macracanthum Popov
C. microphyllum Benth.
C. multijugum Maesen
C. nuristanicum Kitam
C. paucijugum (Popov) Nevski
C. pungens Boiss
C. rassulovie Lincz.
C. rechingeri Podlech
C. songaricum Steph. Ex DC
C. stapfianum Rech.f
C. subaphyllum Boiss
C. tragacanthoides Jaub & Spach

Subgenus Stenophylloma

C. canariense A.G. Guerra & G.P. Lewis
C. cuneatum Hochst. Ex A. Rich

Furthermore, a classification based on nuclear ITS and chloroplast trnK/matK and trns-trnG regions has been proposed (van der Maesen et al. 2007). This grouping of species on the basis of molecular data clearly brought out inadequacies of the earlier systems as section *Cicer* (subg. *Pseudononis*) and section *Acanthocicer* (subg. *Viciastrum*) have been shown to be polyphyletic and only section *Polycicer* (subg. *Viciastrum*) forms a well-supported monophyletic group. Furthermore, two African species *C. canariense* (section Polycicer) and *C. cuneatum* (section *Cicer*) form a highly supported basal clade in the phylogenetic tree (van der Maesen et al. 2007).

3.4 Phylogenetic Relationships Between Species

A proper assessment of the genetic variation present in various wild taxa and their phylogenetic relationship with each other and with the cultigen is of utmost importance to introduce wild characters of agronomic importance.

3.4.1 Morphological Characters

Morphological characters have been used to define relationships between different species of *Cicer*. A study on 228 accessions belonging to eight annual species and 20 'Kabuli' chickpea lines shows that the cultigen is more variable than wild species and it also differs from the latter in terms of leaf area, growth habit, plant height, first pod height, pod dehiscence and 100 seed weight, the characters which changed during domestication. Further, it was found that *C. reticulatum, C. echinospermum, C. bijugum* were closest to the cultigen (Robertson 1997). Javadi and Yamaguchi (2004b) divided 17 species belonging to all the four sections into six plumule types. The type PI with spiral form of compound leaf and two adnate or separated stipular parts is characterized by *C. arietinum* and its two closest relatives *C. reticulatum* and *C. echinospermum*. The type PII with narrowly spiral form of compound leaf and two relatively close stipular parts included *C. yamashitae* and *C. chorassanicum* and the PIII included *C. pinnatifidum* and *C. judaicum* while the PIV included *C. cuneatum* and *C. canariense* and PV included mostly perennial species and PVI *C. bijugum* (Javadi and Yamaguchi 2004b).

Recently, Ozturk et al. (2013) investigated 17 *Cicer* taxa growing in Turkey by generating a data matrix prepared from 143 morphological, palynological and seed characters. The *Cicer* species were divided into two major groups with first group including *C. arietinum, C. reticulatum, C. echinospermum, C. bijugum, C. pinnatifidum* (section *Cicer*), *C. incisum* subsp. *incisum* and *C. incisum* subsp. *serpentinicola* (section Chamaecicer) and the second group comprised *C. heterophyllum* var. *heterophyllum, C. heterophyllum* var. *kassianum, C. uludereensis, C .isauricum, C. montbretii, C. floribundum* var. *floribundum, C. floribundum* var. *amanicola* (section Polycicer) and *C. anatolicum* (section *Vicioides*).

3.4.2 Karyotypes and Physical Mapping

Karyotypic comparison provides a basis for comparative study of gross structural changes taking place in the genome of different species within a genus. A number of studies have described minor variability in the karyotype of various accessions of both 'Kabuli' and 'Desi' *C. arietinum* (Kutarekar and Wanjari 1983; Mukherjee and Sharma 1987; Ohri and Pal 1991; Galasso and Pignone 1992; Venora et al. 1995; Akter and Alam 2005; Kordi et al. 2006). The karyotypes of the eight annual species have been described both by feulgen staining (Ohri and Pal 1991; Ocampo et al. 1992; Ahmad 2000) and by banding techniques (Tayyar et al. 1994; Galasso et al. 1996) while only four perennial species have been studied, e.g. C. *anatolicum* (Ahmad 1989; Hejazi 2011), *C. songaricum* (Ohri 1999), *C. oxyodon* (Hejazi 2011) and *C. canariense* (Pundir et al. 1993). All the species studied are diploid having $2n = 16$.

Karyotypes of five accessions of *C. arietinum* belonging to both 'Kabuli' and 'Desi' types studied by Ohri and Pal (1991) are more or less similar as the first and the longest chromosome which is median has a satellite on the longer arm and the rest of the chromosomes are median point, median, median submedian or submedian. The karyotypes of *C. reticulatum* and *C. echinospermum* are similar to that of *C. arietinum* but in case of *C. reticulatum* first two pairs which are median also have a satellite each on the longer arms. The karyotypes of these three species fall in

1b class of Stebbins (1958). The complement of *C. bijugum* has a satellite on the second pair the secondary constriction lies very near the centromere, and it also has a subterminal and a submedian pair. *C. pinnatifidum* has a satellite on the smallest pair and three submedian pairs, and both these species fall in 2a class of Stebbins (1958). In case of *C. cuneatum* the first pair has a satellite on the longer arm and it is peculiar in having three submedian pairs and therefore comes under 2b class. *C. judaicum* which has smallest complement of all the species, is also the most asymmetrical in having two subterminal pairs and a secondary constriction on the second pair, it falls in 3b class. In addition to these *C. yamashitae* studied by Ocampo et al. (1992) has mostly median and submedian chromosomes and the third pair is satellited. Tayyar et al. (1994) studied C-banding in all the nine species of *Cicer*. Mainly, centromeric C-bands were observed in addition to some intercalary bands which facilitated proper identification and pairing. The smallest haploid genome length was observed in *C. judaicum* and the longest in *C. arietinum*. There was no correlation between the amount of heterochromatin and the total haploid genome length as *C. chorassanicum* had the lowest (38.4 %) and *C. cuneatum* the highest (63.1 %) heterochromatin content. However, the C-banded karyotypes of *C. arietinum, C. reticulatum* and *C. echinospermum* were found to be quite similar. Tayyar et al. (1994) have classified two groups on the basis of heterochromatin content, i.e. *C. cuneatum* and *C. bijugum* with high heterochromatin content of 61.3 and 57.7 %, respectively and *C. pinnatifidum, C. judaicum, C. arietinum, C. reticulatum, C. yamashitae, C. echinospermum* and *C. chorassanicum* range from 38.4–46.0 %. Galasso et al. (1996) also studied C-banding in *C. arietinum, C. reticulatum* and *C. echinospermum* which showed similarity in the presence of mostly centromeric heterochromatic bands. The differences were found with regard to some intercalary bands as in case of chromosome B of *C. reticulatum* and the presence of satellites on the first two pairs in this species in contrast to other species which have a satellite only on the first pair. Fluorescent banding showed two pairs of chromosomes with CMA positive heterochromatin in all the three species. One site of this heterochromatin is located at the secondary constriction of a chromosome resembling chromosome A of *C. arietinum* and the other site in *C. echinospersmum* is present in subterminal position on the chromosome homoeologous to the chromosome B of *C. arietinum*, while in *C. reticulatum* this site is observed on the secondary constriction of second satellite chromosome. Karyotypes of three perennial species have been studied in some detail. In *C. anatolicum* secondary constriction is present on the longest chromosome and the rest of the chromosomes are either median or submedian, one smallest pair is median point. Though more asymmetrical, the kayotype of *C. anatolicum* resembles those of *C. arietinum, C reticulatum* and *C echinospermum* (Ahmad 1989). *C. songaricum* shows more symmetrical karyotype with three median point chromosomes while the others are either median or submedian and the third longest chromosome has a secondary constriction (Ohri 1999). In *C. oxyodon* satellite is present on the short arm of seventh pair and the remaining chromosomes are either median or submedian (Hejazi 2011).

Abbo et al. (1994) determined rDNA sites by fluorescent in situ hybridization (FISH). The cultigen which shows only one secondary constriction by feulgen staining, produces rDNA signals on three pairs of chromosomes. However, out of these only two pairs are regularly detected and the third is rarely observed perhaps due to low copy number. *C. reticulatum* as expected regularly has two pairs of hybridization sites corresponding to two pairs of satellite chromosomes as observed earlier (Ohri and Pal 1991; Ocampo et al. 1992; Tayyar et al. 1994). To account for three pairs of rDNA sites in the cultigens Abbo et al. (1994) have suggested a major translocation which removed one pair of satellite to another chromosome pair thus creating a major site, one of intermediate intensity and the remaining one with low intensity. Galasso et al. (1996), however, observed two pairs of hybridization sites each for the clone pTa71 (containing 18S-5.8S-25S rRNA genes) and clone pTa794 (containing 5S rRNA genes) in the three species, i.e. *C arietinum, C. reticulatum* and *C. echinospermum*. The presence of transcriptional activity by AgNOR staining revealed a major and a minor NOR pair in *C. reticulatum* but only one major active NOR pair in case of *C. arietinum* and *C. echinospermum* implying, therefore that the chromosome B showing CMA and pTa71 signaled an inactive remnant region which is active in *C. reticulatum*. This is further corroborated by the presence of four nucleoli (two large and two small) in interphase nuclei of *C. reticulatum* and only two large ones in *C. arietinum* and *C. echinospermum* (Galasso et al. 1996). However, the similarity in the size of 5S and 18-25S rRNA units of *C. arietinum* and *C. reticulatum* confirm their close relationship as between *C. bijugum, C. chorassanicum* and *C. echinospermum* with a smaller unit while *C. cuneatum* has the smallest 18-25S rRNA unit of all the *Cicer* species because of smallest intergenic spacer (Patil et al. 1995). It may be mentioned here that FISH has also been used on super stretched (extended 100 times) chromosomes of *C. arietinum* to increase the spatial resolution of neighbouring loci up to 70 kbp as compared to 5–10 mbp in case of metaphase chromosomes (Valarik et al. 2004).

Because of small size, the proper identification and pairing of the chromosomes of chickpea may be problematic even with banding techniques. This difficulty has been addressed by the physical mapping of molecular makers for specific chromosomes or arms. Gortner et al. (1998) used five simple sequence repeat oligonucleotides all of which produced hybridization signals with varying intensity and position, depending on the motif, on all the chromosomes. The metaphase chromosomes showed CA and GATA repeats mainly in the centromeric region while TA, A and AAC repeats occurring in dispersed manner. An *Arabidopsis* type of telomeric repeat (TTTAGGG)n produced a cluster of repeats on the short arm of chromosome B and a weaker signal on the short arm of chromosome A and very weak and inconsistent signals at the termini of other chromosomes (Gortner et al. 1998).

Similarly Staginnus et al. (1999) studied the physical mapping of four major repetitive families CaSat1, CaSat2, CaRep1 and CaRep2 on *Cicer arietinum* complement and their abundance and organization among eight other annual species. Major hybridization signals were observed with CaSat1 in the heterochromatin adjacent to secondary constriction of chromosome A and pericentric heterochromatin block of chromosome B and in both cases the repetitive family hybridizes near rDNA site. On the other hand CaSat2 hybridizes to pericentric heterochromatin blocks of all 16 chromosomes. The retrotransposon like sequence CaRep1 and CaRep2 hybridize mostly on the DAPI-positive pericentric heterochromatic region of all chromosomes. The presence and organization of two satellite (CaSat1 and CaSat2) probes was observed by Southern hybridization on RsaI digested genomic DNA which form a ladder-like sequence on all annual species except *C. cuneatum* where no hybridization was seen. However, the variation in intensity of banding was observed in case of other species. For example, CaSat1 produced the strongest signals in *C. reticulatum, C. arietinum, C. echinospermum* and *C. chorassanicum*, somewhat weak signals in *C. yamashitae*, and very faint in *C. bijugum, C. pinnatifidum* and *C. judaicum*. It is interesting to note that CaSat2 produced similar pattern on DNA from the perennial *C. anatolicum* and annual species (Staginnus et al. 1999). Another family CaRep3 belonging to highly repetitive Ty3-gypsy like retrotransposon was also mapped and shown to be present in the intercalary heterochromatin of all the chromosome and on the distal parts of satellite chromosome A (Staginnus et al. 2010). The hybridization signals were particularly absent from centromeric region and secondary constriction of chromosome A. The restriction pattern of CaRep3 sequence and their relative abundance was similar in *C. reticulatum, C. echinospermum* and the cultigen, very different in *C. bijugum*, and either absent or divergent in *C. chorassanicum* and *C. yamashitae*. Staginnus et al. (2010) also detected two other, an LTR (CaTy) and a non-LTR (CaLin) retrotransposon family. Their physical location showed the presence of CaTy in the distal parts of intercalary heterochromatin and adjacent euchromatic regions, a pattern observed in all chromosomes. However, CaLin has a low presence and reveals considerable heterogeneity and signals were present only on the chromosome pairs A, B and D (Staginnus et al. 2010).

Successful flow sorting of individual chickpea ('Kabuli') chromosomes was done for the first time by Vlacilova et al. (2002) and subsequently, of the 'Desi' type by Zatloukalova et al. (2011). While in the former case ('Kabuli') 5 peaks A, B, C, G and H could be assigned to individual chromosomes leaving three tightly spaced peaks represented by chromosomes D, E and F, the 'Desi' types depicted four peaks represented by chromosomes A, B, E and H and two composite peaks representing chromosomes C and D, and F and G. This corroborates minor chromosomal variation in 'Desi' and 'Kabuli' types as observed by Ohri and Pal (1991). Out of the eight chromosomes of 'Kabuli' the largest chromosome A showing a nucleolar organizing region (NOR) with 45S rDNA locus, the second large chromosome B showing a hybridization signal of 5S rDNA locus and a large interstitial band of *Arabidopsis* type telomeric repeat and the second smallest chromosome G with a 5S rDNA locus could be definitely identified. However, one smallest chromosome H could be assigned to linkage group LG8 of Winter et al. (1999, 2000) by the sequence tagged microsatellite site (STMS) markers (Vlacilova et al. 2002). Zatloukalova et al. (2011) used some probes for major DNA repeats such as CaSat1, CaSat2, CaRep1 and CaRep2 which produced similar results as obtained by Staginnus et al. (1999). In addition to these 57 BAC clones carrying inserts of 20–100 kb were used as probes for FISH on flow sorted chromosomes. This resulted in the identification of two clones localizing specifically to the chromosomes E and H which earlier lacked any cytological markers (Zatloukolova et al. 2011). Moreover, in this study STMS markers have been used to confirm LG8 to chromosome H, LG5 to chromosome A, LG4 to medium sized chromosome E and LG3 to the second largest chromosome B. However, the Chromosomes C and D were not flow sorted separately and jointly ascribed to LG6 and LG7 and likewise Chromosomes F and G to LG1 and LG2 (Zatloukalova et al. 2011).

3.4.3 Meiotic Associations in the Species

All the nine annual species including *C. arietinum* show normal formation of 8 bivalents at metaphase I and the chiasma frequency has been shown to be negatively correlated to genome length (Ahmad and Chen 2000). Likewise, a perennial species *C. canariense* also shows the normal meiosis with 8 bivalents (Pundir et al. 1993). In all the accessions of annual species studied by Ahmad and Chen (2000) only one chromosome pair was found to be associated with the nucleolus at pachytene and/or at diakinesis stage. This is an interesting observation especially in *C. reticulatum* where rRNA gene cluster has been mapped to two pairs of chromosomes (Abbo et al. 1994; Galasso et al. 1996). Additionally both rRNA gene sites are transcriptionally active albeit differentially and are thus capable of forming nuclei (Galasso et al. 1996). The association of only one pair with the nucleolus would thus indicate that perhaps the chromosome pair containing the lesser active of the rRNA site is not capable of associating with the nucleolus at pachytene/diakinesis since not a single PMC showed the expected association (Ahmad and Chen 2000).

3.4.4 Genome Size

Genome size has been determined for seven annual and one perennial species (Ohri and Pal 1991; Ohri 1999). The 2C DNA amounts range from 1.83 pg (*C. judaicum*) to 3.57 pg (*C. arietinum* ICC 5003). These seven species form three DNA groups whose means are separated by an interval of 0.8 pg. *C. judaicum* (1.83 pg) constitutes group I, *C. cuneatum* (2.50 pg), *C. bijugum* (2.54 pg), *C. pinnatifidum* (2.56 pg), *C. reticulatum* (2.65 pg) and *C. echinospermum* (2.56 pg) group II, while group III contains five cultivars of *C. arietinum* (3.30–3.57 pg) (Ohri and Pal 1991). *C. songaricum*, the only perennial species studied, shows a 2C DNA amount of 2.71 pg which is similar to the annual species included in group II (Ohri 1999). It is interesting to note that *C. reticulatum* has 22.3 % less DNA ($P < 0.01$) than *C. arietinum*. This has also been confirmed by Galasso et al. (1996) who reported 4C DNA amount of 5.30 and 5.22 pg for C. *reticulatum* and *C. echinospermum*, respectively and 6.57 pg for *C. arietinum*. These two studies while corroborating each other, however, do not agree with 2C DNA amount of 1.90 pg for *C. arietinum*

(Bennett and Smith 1976). Moreover, the genome size estimates by flow cytometry of 738 mb/1C (Arumuganathan and Earl 1991) and 2C of 1.74–1.80 pg (Ruperao et al. 2014) reported for different cultivars of *C. arietinum* and a kmer based estimate ($\sim$738 mb) of a Kabuli chickpea variety (Varshney et al. 2013) are significantly lower than reported by Ohri and Pal (1991) and Galasso et al. (1996).

3.4.5 Protein, Enzyme and DNA Markers

The electrophoretic data on seed proteins and allozyme/isozymes have been used by various studies to describe interspecific relationships. Ahmad and Slinkard (1992) analyzed both albumin and globulin fractions and found the profiles of *C. reticulatum* and *C. echinospermum* very similar to that of *C. arietinum,* and *C. judaicum* and *C. pinnatifidum* were established as different species to form a cluster with *C.bijugum* and *C. chorassanicum*, while *C. cuneatum* and *C. yamashitae* were placed in two separate groups.

Kazan and Muehlbauer (1991) studied isozyme variation at 30 loci and determined relationships between nine annual and one perennial species. A monophyletic origin of all annual species has been suggested because of the common presence of isozyme gene duplications. As expected a close allozyme similarity is observed in *C. reticulatum, C. echinospermum* and *C. arietinum* with *C. anatolicum* showing a close resemblance to this group. Similarly *C. bijugum, C. pinnatifidum* and *C. judaicum* form a second cluster and *C. yamashitae* and *C. chorassanicum* form the third group while *C. cuneatum* is distinctly separate from all other species. Other analyses by isozymes (Ahmad et al. 1992; Labdi et al. 1996; Tayyar and Waines 1996; Gargav and Gaur 2001) agree with this grouping except in case of some perennial species such as *C. microphyllum* (Gargav and Gaur 2001), and *C. anatolicum* and *C. songaricum* (Tayyar and Waines 1996) which cluster with *C. yamashitae* and *C. chorassanicum*. However, Sudupak and Kence (2004) placed *C. anatolicum* with two other perennial species *C. isauricum* and *C. montbretii* (section Polycicer) in a group separate from that containing six annual species including the cultigen and the progenitor species and also a perennial species *C. incisum* (section Chamaecicer) forming a cluster together.

The phylogenetic relationships of nine annual and some perennial species have also been studied using DNA-based molecular markers such as RAPD (Ahmad 1999; Sudupak et al. 2002; Iruela et al. 2002; Javadi and Yamaguchi 2004a; Talebi et al. 2009), ISSR (Rajesh et al. 2002; Iruela et al. 2002; Sudupak 2004; Amirmoradi et al. 2012; Ozturk et al. 2013), sequence tagged microsatellite sites (STMS) (Choumane et al. 2000; Sethy et al. 2006), AFLP (Sudupak et al. 2004; Nguyen et al. 2004; Shan et al. 2005), chloroplast sequence analysis (Javadi and Yamaguchi 2004c; Javadi et al. 2007), rDNA, RFLP and ITS sequences (Frediani and Caputo 2005; Singh et al. 2008; Javadi et al. 2007), start codon targeted (Scot) polymorphism and DAMD-PCR (Amirmoradi et al. 2012), EST markers (Buhariwala et al. 2005), iPBS retrotransposon markers (Andeden et al. 2013).

Some detailed studies have been done with larger taxon sampling comprising of different accessions of all the annual and some selected perennial species. Iruela et al. (2002) studied 75 accessions of 14 species including 8 annuals by RAPD. The dendrogram showed 4 groups and the first included all the perennial species of Asian origin i.e. *C. anatolicum, C. multijugum, C. macracanthum, C. microphyllum* and *C. oxyodon*, the second only *C. yamashitae*, the third *C. arietinum, C. reticulatum, C. echinospermum, C. pinnatifidum, C. judaicum* and *C. bijugum* while the fourth had the African species *C. cuneatum and C. canariense*. Javadi and Yamaguchi (2004a) did RAPD analysis of 35 accessions of 6 perennial and 5 annual species. The different accessions of 'Kabuli' and 'Desi' types clustered together and showed close relationship with those of *C. reticulatum*. This group is related to *C. echinospermum* while *C. bijugum* shows a distant relationship with other annual species in the first cluster and in the other cluster comprising of mostly perennial species, with the exception of *C. chorassanicum* which forms a subgroup, the other subgroup consists of *C. spiroceras, C. macracanthum, C. oxyodon, C. anatolicum, C. tragacanthoides* while *C. canariense* is distantly related to all other species. Likewise, Sudupak et al. (2002) studied 43 accessions of wild and cultivated species and found two main clusters and in the first cluster one subcluster is formed by the accessions of *C. reticulatum and C. arietinum* in keeping with their close relationship, while the accessions of *C. echinospermum* form another subcluster. In the other main cluster *C. bijugum*, and *C. pinnatifidum* form separate clusters while *C. judaicum* is grouped outside these subclusters which are joined by *C. incisum* and the other perennial species *C. isauricum, C. anatolicum* and *C. montbretii* form the other main cluster. This study showed that *C. incisum* is closest to the annual species.

In a comprehensive AFLP analysis of 95 accessions of 17 species all the perennial species, i.e. *C. multijugum, C. nuristanicum, C. microphyllum, C. songaricum, C. flexuosum, C. macracanthum, C. anatolicum* and *C. oxyodon* grouped together along with one annual *C. yamashitae,*while *C. pinnatifidum, C. bijugum and C. judaicum* formed a group nearer to the perennial species and *C. arietinum, C. reticulatum, C. echinospermum* formed a distinct group with very low genetic distances while *C. cuneatum* and *C. canariense* were most distantly placed with respect to all other species (Nguyen et al. 2004). Similarly, an AFLP study of 47 accessions of four perennial and six annual species grouped all the perennial species together i.e. *C. montbretii, C. isauricum, C. anatolicum* while the other cluster had two subclusters one of which included one perennial *C. incisum* along with *C. pinnatifidum, C. judaicum and C. bijugum*, the other had *C. arietinum, C. reticulatum* and *C. echinospermum. C. incisum* was found to be closest to the annual species (Sudupak et al. 2004). Another AFLP study of 146 accessions of 8 annual and one perennial species showed similar results except that *C. yamashitae* grouped with the perennial *C. anatolicum* and *C. cuneatum* as expected was the most distant to all the other species (Shan et al. 2005). This study also brought out geographical patterns of variation as maximum genetic variation of *C. reticulatum, C. echinospermum, C. pinnatifidum, C. bijugum* occurs in southeastern Turkey while *C. judaicum* shows maximum variation in Palestine region (Shan et al. 2005).

The relationship of 30 species based on combined consensus tree based on two plastid sequences and ITS revealed three well-supported clades. The species clearly segregated into four geographical groups in three clades as *C. arietinum, C. reticulatum, C. echinospermum, C. pinnatifidum, C. judaicum, C. bijugum* and *C. incisum* form a monophyletic group in clade III as they all belong to Middle East. The African group forms a monophyletic clade I comprising of *C. cuneatum and C. canariense*. Clade II is divided into two subgroups and consists of west central Asian *C. anatolicum, C. macracanthum, C. flexuosum, C. rechingeri, C. spiroceras, C. stapfianum, C. subaphyllum, C. kermanense, C. tragacanthoides, C. multijugum* with two annuals *C. chorassanicum* and *C. yamashitae* forming a sister group to this subgroup. The other subgroup is formed by the species of Aegean-Mediterranean distribution i.e. *C. floribundum, C. graecum, C. isauricum, C. montbretii* (Javadi et al. 2007). Earlier Javadi and Yamaguchi (2004c) obtained similar results on 25 species based on trn T-F region of chloroplast DNA. This also shows *C. anatolicum* forming a monophyletic group with other perennial species rather than with annuals. Similarly, Frediani and Caputo (2005) did cladistic analysis of ITS1 and ITS2 of 20 species of *Cicer* and noted two clades, one of which included two African species *C. canariense* and *C. cuneatum* and in the other clade *C. arietinum, C. reticulatum* and *C. echinospermum* form a closed group while *C. bijugum, C. judaicum, C. pinnatifidum* form a separate group and annual species *C. yamashitae and C. chorassanicum* belong to perennial species, such as *C. pungens, C. flexuosum, C. multijugum, C. macracanthum, C. songaricum, C. anatolicum, C. oxyodon, C. graecum, C. montbretii* and *C. microphyllum* (Frediani and Caputo 2005).

The ISSR polymorphism was used to study six annual and seven perennial species (Rajesh et al. 2002). Out of the three main clusters formed the first was comprised of *C. acanthophyllum, C. macracanthum, C. pungens, C. nuristanicum, C. arietinum, C. reticulatum* and *C. echinospermum* in which as expected the latter three species form a closed group. The second cluster had *C. yamashitae, C. bijugum* and *C. judaicum* where the latter two species showed higher similarity and the third cluster had *C. anatolicum, C. microphyllum* and *C. oxyodon*. The clustering of the species shows that annual species are polyphyletic as the perennial species do not form a single cluster. In a similar ISSR study, Sudupak (2004) showed that the perennial *C. incisum* is closest to the annual species, i.e. *C. judaicum, C. pinnatifidum* and *C. bijugum* and the accessions of *C. arietinum* and *C. reticulatum* form a single subgroup which is joined by *C. echinospermum*. Remarkably, *C. anatolicum* is most distantly place in relation to all other species.

Singh et al. (2008) made a phylogenetic analysis of 76 accessions of 10 species using RFLP and ITS sequences of nuclear ribosomal DNA. The tree generated from RFLP of rDNA formed 5 clades with all the accessions of *C. arietinum, C. reticulatum* and *C. echinospermum* in clade I, *C. bijugum*, four accessions of *C. judaicum* and one accession of *C. yamashitae* in clade II, rest of the *C. judaicum* and *C. pinnatifidum* accessions and *C. chorassanicum* form parts of clade III and IV, *C. cuneatum* and *C. yamashitae* form clade V. *C. microphyllum* the only perennial species studied forms a separate branch in the tree. This study shows that *C. bijugum* is completely separate from *C. pinnatifidum* and *C. judaicum* and two accessions of *C. yamashitae* are included in two different clades. Two clades were formed by ITS1 and ITS2 sequence analysis. One clade was constituted by *C. arietinum, C. reticulatum* and *C. echinospermum* the other clade consisted of *C. judaicum, C. chorassanicum, C. bijugum, C. cuneatum* and *C. microphyllum* where the latter two species are close together, while *C. pinnatifidum* and *C. yamashitae* constituted different branches in the tree. This study distinctly shows that *C. pinnatifidum* is distantly placed with respect to *C. bijugum* and *C. judaicum*.

Some other studies on smaller samples consisting of mostly annual species agree with the above reports. All the studies based on RAPD markers show a close relationship between *C. arietinum* and *C. reticulatum* with both of these along with *C. echinospermum* forming a group while *C. chorassanicum, C. yamashitae, C. pinnatifidum, C. judaicum, C. bijugum* cluster together and *C. cuneatum* shows a distant relationship with all other annual species (Ahmad 1999; Talebi et al. 2009). Choumane et al. (2000) and Sethy et al. (2006) analysed species relationships by STMSs and both the studies showed close genetic similarity between *C. arietinum, C. reticulatum* and *C. echinospermum*. This group was shown to be most closely related to the perennial species *C. anatolicum* by Choumane et al. (2000). Among the other species, *C. bijugum* and *C. pinnatifidum* are closely related as compared to *C. judaicum*, while *C. cuneatum* forms a distant group (Sethy et al. 2006). Similarly, Buhariwala et al. (2005) analysed EST-based markers to divide the species in three clusters, one cluster comprises of *C. arietinum, C. reticulatum* and *C. echinospermum*, the second *C. pinnatifidum, C. bijugum* and *C. judaicum*, and *C. yamashitae, C. chorassanicum* and *C. cuneatum* form the third cluster. Amirmoradi et al. (2012) obtained somewhat different grouping of 8 annual species, by using 3 marker types, i.e. start codon targeted (SCot) polymorphism, directed amplification of minisatellite DNA (DAMD-PCR) and ISSR. Five clusters were formed by ISSR where *C. arietinum* and *C. reticulatum* came together with *C. yamashitae* in the first, the second included *C. echinospermum*, the third *C. pinnatifidum*, fourth *C. cuneatum* and *C. bijugum* and the fifth *C. judaicum*. In Scot analysis, four clusters were formed. *C. arietinum* and *C. reticulatum* clustered with *C. yamashitae* and *C. pinnatifidum*, the second, third and fourth clusters were formed by *C. echinospermum, C. judaicum* and *C. bijugum* and *C. cuneatum*, respectively. Three clusters were observed in DAMD-PCR analysis and the first cluster had *C. arietinum* and *C. echinospermum*, the second *C. judaicum, C. bijugum* and *C. cuneatum* and the third *C. yamashitae* and *C. pinnatifidum*. In a recent study, Andeden et al. (2013) studied genetic diversity and relationships by iPBS-retrotransposons and ISSR markers, of 71 accessions of five annual species and the cultigen from its core area of origin and domestication. The combined ISSR and iPBS analysis divided the accessions in five groups in which *C. arietinum* and *C. reticulatum* form a single group. Another closely associated group belongs to *C. echinospermum* and the rest of the 3 groups are formed by *C. judaicum, C. pinnatifidum* and *C. bijugum*, respectively where *C. judaicum* and *C. bijugum* show greatest dissimilarity.

Genetic variation among 94 genotypes of eight annual species, including the cultigen, and one perennial species *C. microphyllum* has also been studied by single nucleotide polymorphism (SNP) and diversity array technology (DArT) by Roorkiwal et al. (2014). The UPGMA based on SNP markers formed two major groups, one consisting of cultivated genotypes and those of *C. reticulatum* and *C. echinospermum*, while in the other major group the genotypes of secondary gene pool and those of tertiary gene pools form different clusters. *C. reticulatum* shows particularly close relationship with its genotypes interspersed in the cultivated types. The other analysis (STRUCTURE) based on DArT data forms four clusters with a strong difference between cultivated and wild types, and the wild species form three clusters belonging to primary, secondary and tertiary gene pools. This study also brought out higher level of polymorphism among wild as compared to the cultivated genotypes. Moreover, *C. reticulatum* was found to be less diverse as compared to other wild species (Roorkiwal et al. 2014).

3.4.6 Interspecific Hybridization

It has already been mentioned that *C. arietinum* has a narrow genetic base, which crept in during its origin, as compared to its wild relatives (Abbo et al. 2003). This has been later confirmed by many studies using different DNA markers (Udupa et al. 1993; Choumane et al. 2000; Iruela et al. 2002; Nguyen et al. 2004; Choudhary et al. 2012a, b). This kind of situation makes it imperative to use the genetic variation present in wild relatives for further improvement with respect to the yield, nutritional quality and other characters providing resistance against various abiotic and biotic stresses. Many studies describe the extent of crossability of the cultigens with wild annual and some perennial species (Table 3.1).

Ladizinsky and Adler (1976a, b) studied crossability relationships between seven annual species (except *C. chorassanicum* and *C. yamashitae*) and meiotic behaviour of their hybrids. The cross between *C. arietinum* and *C. reticulatum* was most successful with fully viable F_1, regular meiosis and complete fertility. The F_1 was intermediate with respect to growth habit and seed structure and showed segregation in F_2 generation. The hybrid with one line of *C. arietinum*, however, showed a complex of four chromosomes and a bridge and a fragment at meiosis therefore indicating that the two parents differed by a translocation and an inversion (Ladizinsky and Adler 1976a). This again supports *C. reticulatum* as the wild progenitor of chickpea. *C. arietinum* and *C. echinospermum* show a low success rate, however, the F_1 which developed normally was highly sterile. Meiotic analysis showed 6II and a complex of four chromosomes as the two species differ by a translocation. Few seeds obtained from F_1 produced completely sterile F_2 progeny. *C. reticulatum* and *C. echinospermum* were very difficult to cross and only one F_1 showed normal development but was completely sterile. These two species also differed by a translocation as 6II and a complex of four chromosomes in seen at MI. Reciprocal crosses between *C. bijugum, C. pinnatifidum* and *C. judaicum* also

Table 3.1 Results of interspecific crosses involving *Cicer* species

	Author/s	Cross	F_1 status
1.	Ladizinsky and Adler (1976a, b)	*C. arietinum* × *C. reticulatum*	F_1 and F_2 fertile
		C. arietinum × *C. echinospermum*	F_1 Partially fertile
		C. reticulatum × *C. echinospermum*	F_1 fully sterile
		C. judaicum × *C pinnatifidum*	F_1 and F_2 partially fertile
		C. judaicum × *C. bijugum*	F_1 and F_2 partially fertile
		C. pinnatifidum × *C. bijugum*	F_1 and F_2 partially fertile
		C. arietinum × *C. cuneatum*	Failed
		C. judaicum × *C. cuneatum*	Failed
		C. pinnatifidum × *C. cuneatum*	Failed
2.	Mercy and Kakar (1975)	*C. arietinum* × *C. songaricum*	Failure of pollen germination and penetration of pollen tubes in style
3.	Pundir and Mangesha (1995)	*C. arietinum* × *C. echinospermum*	F_1 partially fertile
4.	Singh and Ocampo (1997)	*C. arietinum* × *C. reticulatum*	F_1 and F_2 Fertile
		C. arietinum × *C. echinospermum*	F_1 and F_2 partially fertile
5.	Badami et al. (1997)	*C. arietinum* × *C. pinnatifidum*	Albino plants obtained after embryo rescue
6.	Mallikarjuna (1999)	*C. arietinum* × *C. pinnatifidum*	Sterile F_1s after embryo rescue
7.	Stamigna et al. (2000)	*C. arietinum* × *C. judaicum*	Embryo abortion
		C. arietinum × *C. bijugum*	Embryo abortion
		C. arietinum × *C. pinnatifidum*	Embryo abortion
8.	Ahmad and Slinkard (2004)	*C. arietinum* × *C. echinospermum*	Viable embryo and seed formation
		C. echinospermum × *C. arietinum*	Embryo abortion
		C. arietinum × *C. pinnatfidum*	Embryo abortion
		C. pinnatifidum × *C. arietinum*	Embryo abortion
		C. arietinum × *C. judaicum*	Embryo abortion
		C. judaicum × *C. arietinum*	Embryo abortion
		C. arietinum × *C. chorassanicum*	Embryo abortion
		C. chorassanicum × *C. arietinum*	Embryo abortion
		C. arietinum × *C. yamashitae*	Embryo abortion
		C. yamashitae × *C. arietinum*	Embryo abortion
		C. arietinum × *C. cuneatum*	Embryo abortion
		C. cuneatum × *C. arietinum*	Embryo abortion
		C. arietinum × *C. bijugum*	Embryo abortion
9.	Clarke et al. (2006)	*C. arietinum* × *C. bijugum*	F_1 breakdown during embryogenesis
		C. arietinum × C. pinnatifidum	

(continued)

Table 3.1 (continued)

	Author/s	Cross	F_1 status
10.	Mallikarjuna et al. (2007)	C. arietinum × C. bijugum	Green F_1 plants selected after embryo rescue
11.	Kumari et al. (2011)	*C. arietinum* × *C. pinnatifidum*	Albino, partially green and green plantlets obtained after embryo rescue
		C. arietinum × *C. judaicum*	
12.	Clarke et al. (2011)	*C. arietinum* × *C. judaicum*	Albino, pale, green F_1 plants after embryo rescue
		C. arietinum × C. pinnatifidum	
13.	Abbo et al. (2011)	*C. judaicum* × *C. bijugum*	F_1 pollen stainability 50 %, F_2 breakdown
		C. judaicum × *C. pinnatifidum*	F_1 with protruding pistils & 30 % pollen stainability, F_2 breakdown
		C. cuneatum × *C. canariense*	50 % pollen stainability, F_2 breakdown
14.	Singh and Singh (2012)	*C. arietinum* × *C. judaicum*	F_1 partially sterile, 54 % pollen stainability
		C. judaicum × *C. arietinum*	

resulted in F_1s with intermediate morphology. Meiosis showed bivalents with variable univalents on the basis of which *C. pinnatifidum* was shown to be closer to *C. bijugum* than to *C. judaicum*. The hybrids though showed 30–50 % pollen fertility, did not result in any seed formation due to elongation of style (prezygotic barrier) at anthesis. However, hand pollination resulted in reasonably good seed production and F_2 progeny showing a close relationship between these three species. *C. cuneatum* which was crossed with *C. arietinum, C. judaicum* and *C. pinnatifidum* did not result in any viable seed though some empty pods were developed showing post zygotic barriers (Ladizinsky and Adler 1976a, b). Similar results were obtained by Abbo et al. (2011) in crosses involving *C. judaicum, C. bijugum* and *C. pinnatifidum*. The cross between *C. judaicum* and *C. bijugum* resulted in partially fertile F_1s which showed further breakdown in the F_2 and between *C. judaicum* and *C. pinnatifidum* resulted in F_1s with protruding styles which were backcrossed with *C. pinnatifidum* producing highly sterile BC_1F_1 plants. Interestingly, *C. cuneatum* × *C. canariense* cross succeeded resulting in F_1s with normal meiotic pairing and more than 50 % pollen stainability therefore supporting close relationship between these two species (Abbo et al. 2011; van der Maesen et al. 2007). Embryo rescue was used to obtain F_1 plants between *C. arietinum* and *C. judaicum* which showed intermediate characters and normal meiotic behaviour with 54 % pollen stainability (Singh and Singh 2012). There is no report of a successful cross between *C. arietinum* and any perennial species as with *C. songaricum* no hybrid seed was obtained despite large number of crosses (Mercy and Kakkar 1975). Hybridization of *C. arietinum* with *C. canariense* was possible as the pollen tubes germinated and the embryos grew up to globular stage and no plants were obtained (Mallikarjuna 2001).

3.4.7 *Barriers to Hybridization*

Barriers to interspecific crossability among the *Cicer* species occur at post zygotic level. The hybrid breakdown can occur due to various reasons such as embryo abortion (Ahmad et al. 1988; Bassiri et al. 1987; Badami et al. 1997; Mallikarjuna 1999; Stamigna et al. 2000; Ahmad and Slinkard 2004; Clarke et al. 2006; Mallikarjurna et al. 2011), albinism (Mallikarjuna and Jadhav 2008; Kumari et al. 2011; Clarke et al. 2011) or pollen sterility due to reduced chromosome pairing (Ladizinsky and Adler 1976a, b; Abbo et al. 2011). Another mechanism which may lead to failure of fertilization due to abnormal flower development causing protrusion of stigma in F_1 or F_2 progeny of certain crosses (Ladizinsky and Adler 1976b; Abbo et al. 2011). Mallikarjuna et al. (2011) have also described crosses between some annual and perennial species which resulted in various percentages of pod set but in no case a viable seedling was obtained.

3.5 Gene Pools of Chickpea

Redden and Berger (2007) included *C. reticulatum* along with various landraces and cultivars of *C. arietinum* in the primary gene pool, *C. echinospermum* in the secondary genepool, and rest of the annual and perennial species which are genetically highly differentiated from the cultigens comprised the tertiary gene pool. This demarcation has been altered a little by placing *C. reticulatum* in the secondary gene pool (Table 3.2, Mallikarjuna et al. 2011). This is quite appropriate considering the differential crossability success of *C. reticulatum* with various cultivars of

Table 3.2 Gene pools of *Cicer arietinum*

Primary	Secondary	Tertiary
Land races and cultivars of *C. arietinum*	*C. reticulatum*, *C. echinospermum*	*C. bijugum, C. judaicum, C. pinnatifidum, C. chorassanicum, C. yamashitae, C. cuneatum, C. atlanticum, C. incisum, C. incisum ssp. serpentinica, C. floribundum, C. floribundum var. amanicola, C. graecum, C. heterophyllum, C. heterophyllum* var. *kassianum, C. uludereensis, C. isauricum, C. montbretii, C. acanthophyllum, C. anatolicum, C. balcaricum, C. baldshuanicum, C. fedtschenkoi, C. flexuosum, C. grande, C. incanum, C. korshinskyi, C. laetum, C. luteum, C. macracanthum, C. microphyllum, C. multijugum, C. nuristanicum, C. paucijugum, C. pungens, C. rassuloviae, C. rechingeri, C. songaricum, C. stapfianum, C. subaphyllum, C. tragacanthoides, C. kermanense, C. mogoltavicum, C. oxyodon, C. spiroceras, C. canariense*

C. arietinum (used as female parent) and occurrence of inversions in crosses between *C. reticulatum* and some of the cultivars of *C. arietinum* (Ladizinsky and Adler 1976a) as well as by the differences in karyotypes and genome size (Ohri and Pal 1991; Galasso et al. 1996). Similar differences in crossability success have been shown between *C. echinospermum* and different lines of *C. arietinum* (Singh and Ocampo 1997; Collard et al. 2003; Mallikarjuna et al. 2011).

Many of the accessions of wild species belonging to secondary and tertiary gene pools have been identified for showing resistance to various abiotic and biotic stresses such as drought, suboptimal temperature, nutrient imbalance, salinity, ascochyta blight, fusarium wilt, botrytis grey mould, collar rot, leaf blight, pod borer, leaf minor, seed beetles, nematods, etc. (Toker et al. 2014). Nevertheless, it has already been mentioned that except in case of species belonging to the secondary gene pool the crosses of the cultigens with the species in tertiary gene pool invariably fail due to strong post zygotic barriers. In some cases even the plants obtained as a result of embryo rescue result in complete sterility (Mallikarjuna et al. 2011).

3.6 Molecular Maps

It has already been pointed out that the productivity of chickpea is adversely affected by some fungal diseases such as ascochyta blight and fusarium wilt in addition to some agronomic traits like flowering time, time to maturity, podding habit, etc. Efforts have been going on to map the genes of interest which may facilitate marker assisted selection and map based cloning of useful genes. However, the genetic variation within chickpea is minimal because of the bottlenecks it experienced during the course of domestication. Therefore, interspecific crosses have been attempted to maximize polymorphism for linkage analysis, though intraspecific crosses have also been used in some cases. To achieve this objective two types of mapping populations have been utilized to generate linkage maps, the F_2 population and recombinant inbred lines (RILs).

An integrated map has been prepared using 130 RILs from a wide cross between a *C. arietinum* cultivar resistant to fusarium wilt and *C. reticulatum*. A total of 354 markers including 118 STMSs, 96 DAFs, 70 AFLPs, 37 ISSRs, 17 RAPDs, 2 SCARs, 3 loci conferring resistance to various races of *Fusarium*, 8 isozymes and 3 cDNAs covered a distance of 2077.9 cM. Eight large and eight small linkage groups were identified with the average distance of 6.8 cM between the markers (Winter et al. 2000).

Another consensus map has been prepared by merging linkage maps from 10 different populations derived from five wide cross *C. arietinum* × *C. reticulatum* and five narrow 'Desi' × 'Kabuli' cross using STMS markers. The integrated map from wide crosses comprised of 555 loci including 135 STMSs and 33 cross genome markers distributed on eight linkage groups covering 652.67 cM. The map from narrow crosses involved 99 STMSs, 3 SCARs, 1 ASAP, *Fusarium* resistance gene, five morphological markers and RAPD and ISSR markers distributed on eight linkage groups covering 426.99 cM (Millan et al. 2010).

Similarly a high density map has been developed, based on RIL population between *C. arietinum* and *C. reticulatum*, with the help of SSR markers from bacterial artificial chromosome (BAC)-end sequences (BESs) and diversity array technology (DArT) markers. The map comprised of 1291 markers on eight linkage groups spanning 845.56 cM. The number of markers per linkage group ranged from 68 (LG8) to 218 (LG3) with an average inter marker distance of 0.65 cM (Thudi et al. 2011).

Choudhary et al. (2012a, b) developed different types of 487 novel EST-derived functional markers such as EST-SSRs, ITP, ESTPs, and SNPs to maximize the detection of polymorphisms in a mapping population of 129 RILs derived from *C. arietinum* (*Fusarium* resistant drought tolerant) × *C. reticulatum* (*Fusarium* wilt susceptible) cross. These markers were integrated with previously published STM markers to produce an advanced linkage map containing 406 loci distributed on eight linkage groups covering 1497.7 cM with the average marker density of 3.68 cM.

Santra et al. (2000) produced a map of nine linkage groups from an RIL population of *C. arietinum* and *C. reticulatum* cross. A total of 116 markers (isozymes, RAPDs, ISSRs) covered a map distance of 981.6 cM with an average distance of 8.4 cM between markers. Two quantitative trait loci (QTL-1 and QTL-2) conferring resistance to ascochyta blight have been tagged with different markers. Same RIL population was used by Takeoglu et al. (2002) to integrate 50 sequence tagged microsatellite (STMS) markers and a resistant gene analogue (RGA) locus to prepare a map covering 1174.5 cM with an average distance of 7.0 cM between markers on nine linkage groups. Six STMS markers were integrated into map region where 2 QTLs reported by Santra et al. (2000) were located. Also 2 DAFs were shown to be tightly linked to QTL-1 in the same RIL population (Rakshit et al. 2003). Cobos et al. (2006) used RILs from a cross of *C. arietinum* (resistant parent) and *C. reticulatum* (susceptible parent) to prepare a linkage map covering a distance of 601.2 cM in 10 linkage groups. However, the QTL for resistance to ascochyta blight was shown to be different as compared to previous studies as it was located on linkage group2 (LG2). Aryamanesh et al. (2010) studied interspecific F_2 population to identify 3 QTLs explaining 49 % of variation for ascochyta blight resistance on LG3 and LG4.

A composite linkage map was prepared using two RIL populations from *C. arietinum* and *C. reticulatum* cross showing segregation for resistance to ascochyta blight, fusarium and rust diseases. It was possible to map loci conferring resistance to ascochyta blight and fusarium wilt by RGA markers. Association was detected between RGAs and genes that controlled resistance to fusarium wilt caused by races 0 and 5 (Palomino et al. 2009).

Collard et al. (2003) prepared a linkage map from F_2 population from *C. arietinum* (susceptible to ascochyta blight) and *C. echinospermum* (resistant to ascochyta blight). Map covered a distance of 570 cM and at least two QTLs for seedling resistance were located on LG4.

With an objective of studying nutritional characters the F_2 population from a cross between *C. arietinum* and *C. reticulatum* was studied with 91 STMS and 2 CytP450 markers to generate a linkage map consisting of nine linkage groups and covering 344.6 cM. Four QTLs for beta-carotene concentration, 1 QTL for lutein concentration and 3 QTLs for seed weight were identified (Abbo et al. 2005).

Cho et al. (2004) used F7 derived RILs from intraspecific cross of susceptible and a resistant accession to prepare a linkage map and identified regions associated with blight resistance, a major QTL for resistance to pathotype II of *Ascochyta rabiei* and two QTLs for resistance to pathotype I. Flandez-Galvez et al. (2003) prepared a linkage map from F_2 population of chickpea cultivars showing contrasting disease reaction to *A. rabiei.* Fifty one STMS, 3 ISSR, and 12 RGA markers mapped on eight linkage groups. The map covered a distance of 534.5 cM with an average of 8.1 cM between markers. Chickpea derived STMS markers were distributed throughout the genome, but RGA markers clustered with ISSR markers on the linkage groups LGI, II and III. With an objective to map genetic loci associated with QTLs for ascochyta blight resistance, Taran et al. (2007) developed an F_2 population of 186 plants derived from a cross between a 'Kabuli' and a 'Desi' cultivars. A total of 144 SSR markers and I morphological marker were assigned to eight linkage groups in a map spanning 1285 cM. One QTL each for ascochyta blight resistance was found on linkage groups LG3, LG4, LG6. Madrid et al. (2008) analyzed an RIL population from *C. arietinum* and *C. reticulatum* cross and identified a QTL for chickpea rust resistance on LG7. Two STMS markers were identified flanking this resistance gene.

Radhika et al. (2007) developed a composite intraspecific map from two RIL populations with one common parent. Three yield related traits were analyzed with different markers to prepare a map covering a 739.6 cM. The characters of double podding and seeds per pod were tagged by different markers and 8 QTLs were found to influence seed weight.

In order to analyze the complex drought related traits two intraspecific mapping populations were studied for segregation of drought tolerance related root traits. This resulted in a consensus map consisting of 352 loci and identification of 9 QTL clusters containing QTLs for drought tolerance traits which can be targeted for molecular breeding (Varshney et al. 2014).

3.7 Conclusions

Chickpea holds a prominent position among grain legumes providing relatively cheap source of protein to the humankind. It originated in Near East from its progenitor species *C. reticulatum* which has a very restricted distribution. During the process of origin various bottlenecks have resulted in a very narrow genetic base in the cultigens. Today chickpea is available as two main types 'Kabuli' and 'Desi' which are considered to have diverged after originating from *C. reticulatum.* The genus *Cicer* comprises 49 taxa including nine annual species. The phylogenetic relationships of these species have been discussed on the basis of morphology, cytology, hybridization and molecular studies. These studies have resulted in the demarcation of primary secondary and tertiary gene pools. While crosses between taxa belonging to primary and secondary gene pools are feasible and result in hybrid progeny which is vegetatively and sexually viable, those involving tertiary gene pool are completely unsuccessful. Even the plants obtained by embryo rescue do not survive beyond a certain stage and are highly sterile. This produces a big constraint on the introduction of genes conferring resistance to various biotic and abiotic stresses and nutritional and yield components from the wild species to the cultigens. This problem is being addressed by QTL mapping of mostly disease resistance loci from the RIL's produced from intra as well as interspecific crosses. Further efforts are being made to integrate genetic maps with physical maps. These methods provide a strong basis for genetic and genomic analysis of chickpea genome and facilitate further the use of molecular methods in breeding.

References

Abbo S, Berger J, Turner NC (2003) Evolution of cultivated chickpea: four bottlenecks limit diversity and constraint adaptation. Funct Plant Biol 30:1081–1087

Abbo S, Mesghenne YT, Van Oss H (2011) Interspecific hybridization in wild *Cicer* species. Plant Breed 130:150–155

Abbo S, Miller TE, Reader SM, Dunford RP, King IP (1994) Detection of ribosomal DNA sites in lentil and chickpea by fluorescent in situ hybridization. Genome 37:713–716

Abbo S, Molina C, Jungmann R, Grusak A, Berkovitch Z, Reifen R, Kahl G, Winter P (2005) Quantitative trait loci governing carotenoid concentration and weight in seeds of chickpea (*Cicer arietinum* L.). Theor Appl Genet 111:185–195

Agarkar SP (1991) Medicinal plants of Bombay presidency. L Pbl Scientific Publishers, Jodhpur

Agarwal G, Jhanwar S, Priya P, Singh VK, Saxena MS, Parida SK, Rohini Garg, Tyagi AK, Jain M (2012) Comparative analysis of kabuli chickpea transcriptome with desi and wild chickpea provides a rich resource for development of functional markers. PLoS ONE 7:e52443

Ahmad F (1989) The chromosomal architecture of *Cicer anatolicum* Alef. a wild perennial relative of chickpea. Cytologia 54:753–757

Ahmad F (1999) Random amplified polymorphic DNA (RAPD) analysis reveals genetic relationships among the annual *Cicer* species. Theor Appl Genet 98:657–663

Ahmad F (2000) A comparative study of chromosome morphology among the nine annual species of *Cicer* L. Cytobios 101:37–53

Ahmad F, Chen Q (2000) Meiosis in *Cicer* L. species: the relationship between chiasma frequency and genome length. Cytologia 65:161–166

Ahmad F, Slinkard AE (1992) Genetic relationships in the genus *Cicer* L. as revealed by polyacrylamide gel electrophoresis of seed storage proteins. Theor Appl Genet 84:688–692

Ahmad F, Slinkard AE (2004) The extent of embryo and endosperm growth following interspecific hybridization between *Cicer arietinum* L. and related annual wild species. Genet Res Crop Evol 51:765–772

Ahmad F, Gaur PM, Slinkard AE (1992) Isozyme polymorphism and phylogenetic interpretations in the genus *Cicer* L. Theor Appl Genet 83:620–627

Ahmad F, Slinkard AE, Scoles GJ (1988) Investigations into the barriers to interspecific hybridization between *Cicer arietinum* L. and eight other annual *Cicer* species. Plant Breed 100:193–198

Akter S, Alam SS (2005) Differential fluorescent banding pattern in three varieties of *Cicer arietinum* L. (Fabaceae). Cytologia 70:441–445

Amirmoradi B, Talebi R, Karami E (2012) Comparison of genetic variation and differentiation among annual *Cicer* species using start codon targeted (SCoT) polymorphism, DAMD- PCR, and ISSR markers. Plant Syst Evol 298:1679–1688

Andeden EE, Baloch FS, Derya M, Kilian B, Ozkan H (2013) iPBS-retrotransposons-based genetic diversity and relationship among wild annual *Cicer species*. J Plant Biochem Biotechnol 22:453–466

Arumuganathan K, Earl ED (1991) Nuclear DNA content of some important plant species. Plant Mol Biol Rep 9:208–218

Aryamanesh N, Nelson MN, Yan G, Clarke HJ, Siddique KHM (2010) Mapping a major gene for ascochyta blight resistance and flowering time in a population between chickpea and *Cicer reticulatum*. Euphytica 173:307–319

Badami PS, Mallikarjuna N, Moss JP (1997) Interspecific hybridization between *Cicer arietinum* and *C. pinnatifidum*. Plant Breed 116:393–395

Bassiri A, Ahmad F, Slinkard AE (1987) Pollen grain germination and pollen tube growth following in vivo and in vitro self and interspecific pollinations in annual *Cicer* species. Euphytica 36:667–675

Bennett MD, Smith JB (1976) Nuclear DNA amounts in angiosperms. Phil Trans R Soc Bio Sci 274:227–274

Berger JD, Turner NC (2007) The ecology of chickpea. In: Yadav SS, Redden RJ, Chen W, Sharma B (eds) Chickpea breeding and management. CABI, Wallingford, pp 47–71

Berger JD, Abbo S, Turner NC (2003) Ecogeography of annual wild *Cicer* species: the poor state of the world collection. Crop Sci 43:1076–1090

Buhariwala HK, Jayashree B, Eshwar K (2005) ESTs from chickpea roots with putative roles for drought tolerance. BMC Plant Biol 5:16. doi:10.1.186/1471-2229-5-16

Charles MT, Dominique R, Kumar J, Gangi OP (2002) A preliminary study of the functional properties of chickpea leaves. In: Annual meeting of the Canadian society of food and nutrition, Edmenton, Alberta, Canada, pp 89–96

Cho S, Chen W, Muehlbauer FL (2004) Pathotype-specific factors in chickpea (*Cicer arietinum* L.) for quantitative resistance to ascochyta blight. Theor Appl Genet 109:733–739

Choudhary P, Khanna SM, Jain PK, Bharadwaj C, Kumar J, Lakhera PC, Srinivasan R (2012a) Genetic structure and diversity analysis of the primary gene pool of chickpea using SSR markers. Genet Mol Res 11:891–905

Choudhary S, Gaur R, Gupta S, Bhatia S (2012b) EST-derived genic molecular markers: development and utilization for generating an advanced transcript map of chickpea. Theor Appl Genet 124:1449–1462

Choumane W, Wnter P, Weigand F, Kahl G (2000) Conservation and variability of sequence tagged microsatellite sites from chickpea (*Cicer arietinum* L.) within the genus *Cicer*. Theor Appl Genet 101:269–278

Clarke HJ, Wilson JG, Kuo I, Lulsdorf MM, Mallikarjuna N, Kuo J, Siddique KHM (2006) Embryo rescue and plant regeneration in vitro of selfed chickpea (*Cicer arietinum* L.) and its wild annual relatives. Plant Cell Tiss Organ Cult 85:197–204

Clarke HJ, Kumari M, Khan TN, Siddique KHM (2011) Poorly formed chloroplasts are barriers to successful interspecific hybridization in chickpea following in vitro embryo rescue. Plant Cell Tiss Organ Cult 106:465–473

Cobos MJ, Rubio J, Strange RN, Moreno MT, Gil J, Millan T (2006) A new QTL for ascochyta blight resistance in an RIL population derived from an interspecific cross in chickpea. Euphytica 149:105–111

Collard BCY, Pang ECK, Ades PK, Taylor PWJ (2003) Preliminary investigation of QTLs associated with seedling resistance to ascochyta blight from *Cicer echinospermum*, a wild relative of chickpea. Theor Appl Genet 107:719–729

Cubero JI (1987) Morphology of chickpea. In: Saxena MC, Singh KB (eds) The chickpea. CAB International, Wallingford, pp 35–66

Davies AMR, Maxted N, van der Maesen LJG (2007) A natural infrageneric classification for *Cicer* (Leguminosae, Cicerae). Blumea 52:379–400

Donmez AA (2011) *Cicer uludereensis* Donmez: a new species of *Cicer* (chickpea) (Fabaceae) from around the Fertile Crescent, SE Turkey. Tuk J Bot 35:71–76

FAOSTAT (2012) Statistical database. http://www.fao.org

Flandez-Galvez H, Ford R, Pang ECK, Taylor PWJ (2003) An intra specific linkage map of the chickpea (*Cicer arietinum* L.) genome based on sequence tagged microsatellite site and resistant gene analog markers. Theor Appl Genet 106:1447–1456

Frediani M, Caputo P (2005) Phylogenetic relationships among annual and perennial species of the genus *Cicer* as inferred from ITS sequences of nuclear ribosomal DNA. Biol Plantarum 49:47–52

Galasso I, Pignone D (1992) Characterization of chickpea chromosomes by banding techniques. Genet Res Crop Evol 39:115–119

Galasso I, Frediani M, Maggiani M, Creminini R, Pignone D (1996) Chromatin characterization by banding techniques, in situ hybridization, and nuclear DNA content in *Cicer* L. (Leguminosae). Genome 39:258–265

Gargav S, Gaur PM (2001) Phylogenetic relationships of perennial chickpea species *Cicer microphyllum* Benth. with its annual relatives as revealed by allozyme polymorphism. Ind J Genet 61:218–222

Gil J, Cubero JI (1993) Inheritance of seed coat thickness in chickpea (*Cicer arietinum* L.) and its evolutionary implications. Plant Breed 111:257–260

Gil J, Nadal S, Luna D, Moreno MT, De Haro A (1996) Variability of some physicochemical characteristics in desi and kabuli chickpea types. J Sci Food Agric 71:179–184

Gortner G, Nenno M, Weising K, Zink D, Nagl W, Kahl G (1998) Chromosomal localization and distribution of simple sequence repeats and the *Arabidopsis*-type telomere sequence in the genome of *Cicer arietinum* L. Chromosome Res 6:97–104

Hawtin GC, Singh KB (1980) Kabuli-desi introgression problems and prospects. In: Workshop on chickpea improvement. ICRISAT, Hyderabad, India, pp 51–80 (In Proc Intern)

Hejazi SMH (2011) Karyological study on three *Cicer* L. species (Fabaceae) in Iran. Asian J Cell Biol 6:97–104

Ibricki H, Knewston SJB, Grusak MA (2003) Chickpea leaves as a vegetable green for humans: evaluation of mineral composition. J Sci Food Agric 83:945–950

Iruela M, Rubio J, Cubero JI, Gil J, Millan T (2002) Phylogenetic analysis in the genus *Cicer* and cultivated chickpea using RAPD and ISSR markers. Theor Appl Genet 104:643–651

Javadi F, Yamaguchi H (2004a) RAPD and seed coat morphology variation in annual and perennial species of the genus *Cicer* L. Genet Res Crop Evol 51:783–794

Javadi F, Yamaguchi H (2004b) A note on seed coat and plumule morphological variation in the genus *Cicer* L. (Fabaceae). Sci Rep Sch Agric Biol Sci Osaka Pref Univ 56:7–16

Javadi F, Yamaguchi H (2004c) Interspecific relationships of the genus *Cicer* L. (Fabaceae) based on trnT-F sequences. Theor Appl Genet 109:317–322

Javadi F, Wojciechowski MF, Yamaguchi H (2007) Geographical diversification of the genus *Cicer* (Leguminosae:Papilionoideae) inferred from molecular phylogenetic analysis of chloroplast and nuclear DNA sequences. Bot J Linn Soc 154:175–186

Jukanti AK, Gaur PM, Gowda CL, Chibbar RN (2012) Nutritional quality and health benefits of chickpea (*Cicer arietinum* L.): a review. Br J Nutr 108:11–26

Kazan K, Muehlbauer FJ (1991) Allozyme variation and phylogeny in annual species of *Cicer* (Leguminosae). Plant Syst Evol 175:11–21

Kerem Z, Lev-Yadun S, Gopher A, Weinberg P, Abbo S (2007) Chickpea domestication in the Neolithic Levant through the nutritional perspective. J Archeol Sci 34:1289–1293

Kordi FM, Majd A, Valizadeh M, Sheidai M, Sabaghpour H (2006) A comparative study of chromosome morphology among some genotypes of *Cicer arietinum* L. Pak J Biol Sci 9:1225–1230

Kumari M, Clarke HJ, des Francs-Small CC, Small I, Khan TN, Siddique KHM (2011) Albinism does not correlate with biparental inheritance of plastid DNA in interspecific hybrids in *Cicer* species. Plant Sci 180:628–633

Kupicha FK (1981) Vicieae. In: Polhill RM, Raven PH (eds) Advances in legume systematics. Royal Botanic Gardens, Kew, pp 377–381

Kutarekar DR, Wanjari KB (1983) Karyomorphological studies in some of the varieties of Bengal gram (*Cicer arietinum* Linn.). Cytologia 48:699–705

Labdi M, Robertson LD, Singh KB, Charrier A (1996) Genetic diversity and phylogenetic relationships among the annual *Cicer* species as revealed by isozyme polymorphism. Euphytica 88:181–188

Ladizinsky G, Adler A (1976a) The origin of chickpea (*Cicer arietinum* L.). Euphytica 25: 211–217

Ladizinsky G, Adler A (1976b) Genetic relationships among annual species of *Cicer* L. Theoret Appl Genet 48:197–203

Lev-Yadun S, Gopher A, Abbo S (2000) The cradle of agriculture. Science 288:1602–1603

Madrid E, Rubiales D, Moral A, Moreno MT, Millan T, Gil J, Rubio J (2008) Mechanism and molecular markers associated with rust resistance in a chickpea interspecific cross (*Cicer arietinum* × *C. reticulatum*). Eur J Plant Pathol 12:43–53

Mallikarjuna N (1999) Ovule and embryo culture to obtain hybrids from interspecific incompatible pollinations in chickpea. Euphytica 110:1–6

Mallikarjuna N (2001) Prospects of using *Cicer canariense* for chickpea improvement. Int Chickpea Pigeonpea Newsl 8:23–24

Mallikarjuna N, Jadhav D, Nagamani V, Amudhavalli C, Hoisington DA (2007) Progress in interspecific hybridization between *Cicer arietinum* and wild species *C. bijugum.* 5(1). SAT eJournal/http://ejournal.icrisat.org

Mallikarjuna N, Jadhav DR (2008) Techniques to produce hybrid between *Cicer arietinum* L. × *C. pinnatifidum* Jaub. Indian J Genet 68:1–8

Mallikarjuna N, Coyne C, Cho S, Rynearson S, Rajesh PN (2011) Cicer. In: Kole C (ed) Wild crop relatives: genomic and breeding resources, legume crops and forages. Springer, Heidelberg, pp 63–80

McIntosh GH, Topping DL (2000) Food legumes in human nutrition. In: Knight R (ed) Linking research and marketing opportunities for pulses in the 21st century. Kluwer Acedemic Publishers, Dordrecht, pp 655–666

Mercy ST, Kakkar SK (1975) Barriers to interspecific crossing in *Cicer*. Proc. Indian Natl Sci Acad Sect B 41:78–82

Millan T, Winter P, Jungling R, Gil J, Rubio J, Cho S, Cobos MJ, Iruela M, Rajesh PN, Tekeoglu M, Kahl G, Muehlbauer FJ (2010) A consensus genetic map of chickpea (*Cicer arietinum* L.) based on 10 mapping populations. Euphytica 175:175–189

Moreno M, Cubero JI (1978) Variation in *Cicer arietinum* L. Euphytica 27:465–485

Mukharjee S, Sharma AK (1987) Structure and behavior of chromosomes in the strains of *Cicer arietinum* L. Cytologia 52:707–713

Muzquiz M, Wood JA (2007) Antinutritional factors. In: Yadav SS, Redden RJ, Chen W, Sharma B (eds) Chickpea breeding and management. CABI, Wallingford, pp 143–166

Nguyen TT, Taylor PWJ, Redden RJ Ford R (2004) Genetic diversity estimates in *Cicer* using AFLP analysis. Plant Breed 123:173–179

Ocampo B, Venora G, Errico A, Singh KB (1992) Karyotype analysis in the genus *Cicer*. J Genet Breed 46:229–240

Ohri D (1999) Cytology of *Cicer songaricum* Steph. Ex DC, a wild relative of chickpea. Genet Res Crop Evol 46:111–113

Ohri D, Pal M (1991) The origin of chickpea (*Cicer arietinum* L.) karyotype and nuclear DNA amount. Heredity 66:367–372

Ozturk M, Duran A, Hakki EE (2013) Cladistic and phylogenetic analysis of the genus *Cicer* in Turkey. Plant Syst Evol 299:1955–1966

Palomino C, Fernandez-Romero MD, Rubio J, Torres A, Moreno MT, Millan T (2009) Integration of new CAPS and dCAPS-RGA markers into a composite chickpea genetic map and their association with disease resistance. Theor Appl Genet 118:671–682

Patil PB, Vrinten PL, Scoles GJ, Slinkard AE (1995) Variation in the ribosomal RNA units of the genera *Lens* and *Cicer*. Euphytica 83:33–42

Popov MG (1929) The genus *Cicer* and its species. Trudy Prikl Bot 21:1–254

Pundir RPS, Mangesha MH (1995) Cross compatibility between chickpea and its wild relative *Cicer echinospermum* Davis. Euphytica 83:241–245

Pundir RPS, Mangesha MH, Reddy GV (1993) Morphology and cytology of *Cicer canariense* a wild relative of chickpea. Euphytica 69:73–75

Radhika P, Gowda SJM, Kadoo NY, Mhase LB, Jamadagni BM, Sainani MN, Chandra S, Gupta VS (2007) Development of an integrated map of chickpea (*Cicer arietinum* L.) using two recombinant inbred line populations. Theor Appl Genet 115:209–216

Rajesh PN, Sant VJ, Gupta VS, Muehlbauer FJ, Ranjekar PK (2002) Genetic relationships among annual and perennial wild species of *Cicer* using inter simple sequence repeat (ISSR) polymorphism. Euphytica 129:15–23

Rakshit S, Winter P, Tekeoglu M, Juarez M, Pfaff J, Benko-Iseppon AM, Muehlbauer FJ, Kahl G (2003) DAF marker tightly linked to a major locus for ascochyta blight resistance in chickpea (*Cicer arietinum* L.). Euphytica 132:23–30

Redden RJ, Berger JD (2007) History and taxonomy of chickpea. In: Yadav SS, Redden RJ, Chen W, Sharma B (eds) Chickpea breeding and management. CABI, Wallingford, pp 1–13

Rizvi H, Babu BK, Agrawal PK (2014) Molecular analysis of kabuli and desi types of Indian chickpea (*Cicer arietinum* L.) cultivars using STMS markers. J Plant Biochem Biotechnol 23:52–60

Robertson LD, Ocampo B, Singh KB (1997) Morphological variation in wild annual *Cicer* species in comparison to the cultigens. Euphytica 95:309–319

Roorkiwal M, von Wettberg EJ, Upadhyaya HD, Warschefsky E, Rathore A, Varshney RK (2014) Exploring germplasm diversity to understand the domestication process in *Cicer* spp. using SNP and DArT markers. PLoS ONE 9:e102016

Ruperao P, Chan C-KK, Azam S, Karafiatova M, Hayashi S, Cizkova J, Saxena RK, Simkova H, Song C, Vrana J, Chitikineni A, Visendi P, Gaur PM, Millan T, Singh KB, Taran B, Wang J, Bately J, Dolezel J, Varshnet RK, Edwards D (2014) A chromosomal genomics approach to assess and validate the desi and kabuli draft chickpea genome assemblies. Plant Biotechnol J 2014:1–9

Salimath PM, Bahl PM, Mehra RB (1984) Genetic diversity in chickpea (*Cicer arietinum* L.). Z Pflanzenzucht 92:52–60

Santra DK, Tekeoglu M, Ratnaparkhe M, Kaiser WJ, Muehlbauer FJ (2000) Identification and mapping of QTLs conferring resistance to ascochyta blight in chickpea. Crop Sci 40:1606–1612

Sethy NK, Shokeen B, Edwards KJ, Bhatia S (2006) Development of microsatellite markers and analysis of intraspecific genetic variability in chickpea (*Cicer arietinum* L.). Theor Appl Genet 112:1416–1428

Shan F, Clarke HC, Plummer JA, Yan G, Siddique KHM (2005) Geographical patterns of genetic variation in the world collections of wild annual *Cicer* characterized by amplified fragment length polymorphisms. Theor Appl Genet 110:381–391

Sharma S, Upadhyaya HD, Roorkiwal M, Varshney RK, Gowda CLL (2013) Chickpea. In: Singh M, Upadhyaya HD, Bisht I (eds) Genetic and genomic resources of grain legume improvement. Elsevier Inc., London, pp 81–104

Singh A, Devarumath RM, RamaRao S, Singh VP, Raina SN (2008) Assessment of genetic diversity, and phylogenetic relationships based on ribosomal DNA repeat unit length variation and internal transcribed spacer (ITS) sequences in chickpea (*Cicer arietinum*) cultivars and its wild species. Genet Res Crop Evol 55:65–79

Singh KB, Ocampo B (1997) Exploitation of wild *Cicer* species for yield improvement in chickpea. Theor Appl Genet 95:418–423

Singh A, Singh NP (2012) Development of interspecific hybrid between *Cicer arietinum* and *C. judaicum* and characterization of interspecific derivatives for economic traits. Indian J Genet 72:174–182

Staginnus C, Desel C, Schmidt T, Kahl G (2010) Assembling a puzzle of dispersed retrotransposable sequences in the genome of chickpea (*Cicer arietinum* L.). Genome 53:1090–1102

Staginnus C, Winter P, Desel C, Schmidt T, Kahl G (1999) Molecular structure and chromosomal localization of major repetitive DNA families in the chickpea (*Cicer arietinum* L.) genome. Plant Mol Biol 39:1037–1050

Stamigna C, Crino P, Saccardo F (2000) Wild relatives of chickpea: multiple disease resistance and problems to introgression in the cultigens. J Genet Breed 54:213–219

Stebbins GL (1958) Longevity, habitat and release of genetic variability in hugher plants. Cold Spring Harb Symp Plant Biol 23:365–378

Sudupak MA (2004) Inter and intra-species inter simple sequence (ISSR) repeat variation in the genus *Cicer*. Euphytica 135:229–238

Sudupak MA, Kence A (2004) Genetic relationships among perennial and annual *Cicer* species growing in Turkey as revealed by allozymes. Genet Res Crop Evol 51:241–249

Sudupak MA, Akkaya MS, Kence A (2002) Analysis of genetic relationships among perennial and annual *Cicer* species growing in Turkey using RAPD markers. Theor Appl Genet 105:1220–1228

Sudupak MA, Akkaya MS, Kence A (2004) Genetic relationships among perennial and annual *Cicer* species growing in Turkey assessed by AFLP fingerprinting. Theor Appl Genet 108:937–944

Talebi R, Jelodar NB, Mardi M, Fayaz F, Furman BJ, Bagheri NA (2009) Phylogenetic diversity relationship among annual *Cicer* species using random amplified polymorphic DNA. Gener Appl Plant Physiol 35:3–12

Tanno K, Wilcox G (2006) The origins of cultivation of *Cicer arietinum* L. and *Vicia faba* L.: early finds from Tellel-Kerkh, north-west Syria, late 10th millennium B.P. Veget Hist Archeobot 15:197–204

Taran B, Warkentin TD, Tullu A, Vandenberg A (2007) Genetic mapping of ascochyta blight resistance in chickpea (*Cicer arietinum* L.) using a simple sequence repeat linkage map. Genome 50:26–34

Tayyar RI, Waines JG (1996) Genetic relationships among annual and perennial *Cicer* (Fabaceae) using isozyme variation. Theor Appl Genet 92:245–254

Tayyar RI, Lukaszewsky AJ, Waines JG (1994) Chromosome banding patterns in the annual species of *Cicer*. Genome 37:656–663

Tekeoglu M, Rajesh PN, Muehlbauer FJ (2002) Integration of sequence tagged microsatellite sites to the chickpea genetic map. Theor Appl Genet 105:847–854

Thudi M, Bohra A, Nayak SN, Varghese N, Shah TM, Penmesta RV, Thirunavukkarasu N, Gudipati S, Gaur PM, Kulwal PM, Upadhyaya HD, Kavikishor PB, Winter P, Kahl G, Town CD, Kilian A, Cook DR, Varshney RK (2011) Novel SSR markers from BAC-end sequences, DArT arrays and a comprehensive genetic map with 1291 marker loci for chickpea (*Cicer arietinum* L.). PLoS ONE 6:e27275

Toker C (2009) A note on the evolution of kabuli chickpea as shown by induced mutations in *Cicer reticulatum* Ladizinsky. Genet Res Crop Evol 56:7–12

Toker C, Uzun B, Ceylan FO, Ikten C (2014) Chickpea. In: Pratap A, Kumar J (eds) Alien gene transfer in crop plants. Achievements and impacts, vol 2. Springer, Heidelberg, pp 121–151

Udupa SM, Sharma A, Sharma RP, Pai RA (1993) Narrow genetic variability in *Cicer arietinum* L. as revealed by RFLP analysis. J Plant Biochem Biotechnol 2:83–86

Valarik M, Bartos J, Kovarova P, Kubalakova M, de Jong JH, Dolezel J (2004) High-resolution FISH on super-stretched flow-sorted plant chromosomes. Plant J 37:940–950

van der Maesen LJG (1972) *Cicer* L. a monograph of the genus, with special reference to the chickpea (*Cicer arietinum* L.). Veenman & Sons, Wageningen, pp 9–133

van der Maesen LJG (1987) Origin, history and taxonomy of chickpea. In: Saxena MC, Singh KB (eds) The chickpea. CABI, Wallingford, pp 139–156 van der Maesen LJG, Maxted N, Javadi F, Coles S, Davies AMR (2007) Taxonomy of the genus *Cicer* revisited. In: Yadav SS, Redden RJ, Chen W, Sharma B (eds) Chickpea breeding and management. CABI, Wallingford, pp 14–47

Varshney RK, Song C, Saxena RK, Azam S, Yu S, Sharpe AG, Cannon SB, Baek J, Taran B, Millan T, Zhang X, Rosen B, Ramsey LD, Iwata A, Wang Y, Nelson W, Farmer AD, Gaur PM, Soderlund C, Penmetsa RV, Xu C, Bharti AK, He W, Winter P, Zhao S, Hane JK, Carrasquilla-Garcia N, Condie JA, Upadhyaya HD, Luo M, Singh NP, Lichtenzveig J, Gali KK, Rubio J, Nadarajan N, Thudi M, Dolezel J, Bansal KC, Xu X, Edwards D, Zhang K, Kahl G, Gil J, Singh KB, Datta SK, Jackson SA, Wang J, Cook D (2013) Draft genome sequence of chickpea (*Cicer arietinum*) provides a source for trait improvement. Nat Biotechnol 31:240–246

Varshney RK, Thudi M, Nayak SN, Gaur PM, Kashwagi J, Krishnamurthy L, Jaganathan D, Koppolu J, Bohra A, Tripathi S, Rathore A, Jukanti AK, Jayalakshmi V, Vemula A, Singh SJ, Yasin M, Sheshshayee MS, Vishwanatha KP (2014) Genetic dissection of drought tolerance in chickpea (*Cicer arietinum* L.). Theor Appl Genet 127:445–462

Venora G, Ocampo B, Singh KB, Saccardo F (1995) Karyotype of the kabuli-type chickpea (*Cicer arietinum* L.) by image analysis system. Caryologia 48:147–155

Vlacilova K, Ohri D, Vrana J, Cihalikova J, Kubalakova M, Kahl G, Dolezel J (2002) Development of flow cytogenetics and physical genome mapping in chickpea (*Cicer arietinum* L.). Chromosome Res 10:695–706

Winter P, Pfaff T, Udupa SM, Huttel B, Sharma PC, Sahi S, Arreguin-Espinoza R, Wigand F, Meuhlbauer FJ, Kahl G (1999) Characterization and mapping of sequence tagged microsatellite sites in the chickpea genome. Mol Gen Genet 262:90–91

Winter P, Benko-Iseppon AM, Huttel B, Ratnaparkhe M, Tullu A, Sonnante G, Pfaff T, Tekeoglu M, Santra D, Sant VJ, Rajesh PN, Kahl G, Meuhlbauer FJ (2000) A linkage map of the chickpea (*Cicer arietinum* L.) genome based on recombinant inbred lines from a *C. arietinum* × *C. reticulatum* cross: localization of resistant genes for fusarium wilt races 4 and 5. Theor Appl Genet 101:1155–1163

Wood JA, Grusak MA (2007) Nutritional value of chickpea. In: Yadav SS, Redden RJ, Chen W, Sharma B (eds) Chickpea breeding and management. CABI, Wallingford, pp 101–142

Yadav SS, Kumar J, Turner NC, Berger J, Redden R, McNeil D, Materne M, Knights EJ, Bahl PN (2004) Breeding for improved productivity, multiple resistance and wide adaptation in chickpea (*Cicer arietinum* L.). Plant Genet Res 4:181–187

Zatloukalova P, Hribova E, Kubalakova M, Suchankova P, Simkova H, Adoracion C, Kahl G, Millan T, Dolezel J (2011) Integration of genetic and physical maps of the chickpea (*Cicer arietinum* L.) genome using flow-sorted chromosome. Chromosome Res 19:729–739

Zohary D, Hopf M (2000) Domestication of plants in the old world, 3rd edn. Oxford University Press, New York

Chapter 4
Genetic Improvement of Cotton

S. Manickam and A.H. Prakash

Abstract Cotton (*Gossypium* spp.) is the most important natural fiber crop cultivated in nearly 80 countries globally. India, China, and USA are the leading countries cultivating cotton on large scale. The cultivated and wild cotton species are differentiated mainly based on the type of seed hairs borne on the seeds. Of the 50 species of the genus *Gossypium*, four are cultivated commercially in different parts of the world. In this chapter, classification of cotton species, its morphology, floral biology etc. have been detailed. Origin of both Old World as well as New World cotton is discussed. A detailed account of gene pools, i.e., primary, secondary, and tertiary gene pools available in cotton is discussed and crossability among various species and their usefulness for commercial exploitation through introgression breeding is detailed. Genome grouping, genetic map for various useful traits have also been discussed. Cotton has been the most exploited crop through genetic engineering tools and a detailed discussion is presented on the genetic enhancement of cotton through biotechnological interventions for both biotic and abiotic stress management apart from enhancing the nutritional status of cotton seed. In most of the cotton growing countries, it is cultivated mainly under rain-fed situations, where water is the most limiting factor for productivity. Hence, detailed information on developing drought-resistant cultivars through both conventional and by using advanced QTL information has been furnished. Molecular markers serve as an important tool for the modern day breeders and information pertaining to marker technology published in cotton is also discussed. Information pertaining exploitation of male sterility system in cotton is detailed.

Keywords Cotton · Taxonomy · Evolution · Genetic modification · Marker-assisted selection · Male sterility

S. Manickam · A.H. Prakash (✉)
ICAR-Central Institute for Cotton Research, Regional Station, Coimbatore, India
e-mail: prakashcicr@gmail.com

S. Manickam
e-mail: manickam.cicr@gmail.com

V.R. Rajpal et al. (eds.), *Gene Pool Diversity and Crop Improvement*,
Sustainable Development and Biodiversity 10,
DOI 10.1007/978-3-319-27096-8_4

4.1 Introduction

Cotton (*Gossypium* spp.) is the world's leading natural textile fiber crop and a significant contributor of oilseed. Cotton is one of the best gifts that nature bestowed on mankind. Cotton fabric is the most skin friendly of all natural fibers available on earth. Cotton is cultivated in nearly 80 countries and contributes to 35.0 % of the global fabric.

Cotton (*Gossypium* spp.) belongs to the Malvaceae family and is one of the most important and earliest domesticated plants in the world. It was domesticated independently both in the Old World and in the New World. The word "cotton" originated from the Arabic term *al qutn*, which became *algodón* in Spanish and cotton in English (Brite and Marston 2013). Among the different domesticated species, the most widespread are *G. arboreum* L. and *G. herbaceum* L. domesticated in the Old World; and *G. hirsutum* and *G. barbadense* domesticated in the New World (Plate 4.1).

The global production of all fibers including cotton increased from 52.0 M tons in 2000 to 72.5 M tons in 2010. Though 60 % of the global fabrics are made from man-made synthetic fibers, and the proportion might further increase, cotton, with a share of 35.8 % in fabrics, continues to remain as the most skin friendly of all

Plate 4.1 Cultivated species of cotton. **a** *Gossypium arboreum.* **b** *Gossypium herbaceum.* **c** *Gossypium hirsutum.* **d** *Gossypium barbadense*

apparel available to mankind (Kranthi et al. 2011) and its demand for cloth will continue as long as civilization exists.

The global cotton production increased significantly over the past 5 years. During 2014–2015, a total of 25.90 million tons of cotton lint was produced. Among the six major cotton growing countries, Australia (2038 kg/ha) holds the highest productivity level followed by Brazil (1524 kg/ha), China (1484 kg/ha), USA (891 kg/ha), Pakistan (750 kg/ha), Uzbekistan (678 kg/ha), and India (523 kg/ha).

Cotton production: Global Scenario 2014–2015

	Acreage (Million hectares)	Production (Million tons)	Productivity -Lint Yield (kg/ha)	Area under rainfed (%)
Australia	0.24	0.49	2038	26
Mexico	0.18	0.26	1476	12
Brazil	1.00	1.50	1524	98
China	4.40	6.53	1484	6
USA	3.93	3.50	891	60
Uzbekistan	1.29	0.87	678	10
Pakistan	3.05	2.28	750	0
India	12.70	6.64	523	60
World	34.14	25.90	760	27

Source USDA—Cotton: World Markets and Trade. March 2015

India ranks first in terms of cultivated area occupying 32 % of the world cotton area followed by China, USA, and Pakistan. Before 5 years, global cotton production had reached a plateau of 19–20 million metric tons (111.7–117.6 M bales) during 1990–2002. Brazil, China, and India rank the best among countries which made significant progress during 1999–2009. China made spectacular progress with an impressive increase to 47.5 M bales in 2007 from 22.9 M bales in 1999. Brazil, which produced 3.0 M bales in 1998, increased its production to 9.4 M bales in 2007. Similarly, India doubled its production from a stagnating 15.8 M bales in 2001 to 31.5 M bales in 2007. The area under cotton in USA and Australia has been declining significantly over the past few years. The area in USA declined to 3.3 M hectares in 2008 from 5.7 M hectares in 2005. Similarly, the cotton area in Australia was 0.55 M hectares in 1999, but decreased to a mere 63,000 hectares in 2007.

4.2 Taxonomic Description of Cotton

Cotton belongs to the genus *Gossypium*, a member of the natural order *Malvales*, family *Malvaceae*, series *Hibisceae* of the dicotyledonous group of plants. The genera of *Hibisceae* are distinguished from each other on the basis of the following factors: (i) type of inflorescence; (ii) number, size and persistence of bracteoles;

(iii) form of the calyx; (iv) nature of the style: clavate or branched; (v) distribution of the oil glands; (vi) character of the fruit; (vii) nature of the seed; and (viii) chromosome number. It is widely distributed in the tropical and subtropical regions of America, Africa, Asia, and Australia. The cultivated species are the outcome of human selection made during various stages of their domestication under particular environment.

Several attempts have been made to classify the genus based on morphological characters, but were not successful. With the advance in genetic architecture, cytological differences, geographical distribution, and variations in morphological characters, a comprehensive classification was made possible.

The genus has been described by Hutchinson et al. (1947) as follows: "Haploid chromosome number of 13 or 26. Annual sub-shrubs, perennial shrubs or small trees. Branches terete or slightly angled, tomentose, hairy or glabrous, of two kinds, i.e., monopodial vegetative branches and sympodial fruiting branches, the latter sometimes reduced to jointed peduncles or flowering spures. The whole plant irregularly dotted with black oil glands. Bracteoles 3, usually foliar and persistent, sometimes small, or minute, rarely caducous. Calyx cup-shaped, truncate, undulate or 5-pointed. Stamens—numerous and lower part of the filaments united into a tube, the upper free, bearing unilocular anthers. Styles clavate, or furrowed, rarely divided at the tip. Ovary 3–5 locular, ripening to a dry, brittle, loculicidally dehiscent capsule. Loculi with seeds indefinite (rarely two only). Seeds covered with one or two coats of long unicellular hairs, or in some species almost naked."

The cultivated and wild cotton species are differentiated mainly based on the type of seed hairs borne on the seeds. Cotton seed hairs are of two types: unconvoluted and convoluted (Balasubramaniam 1963). In the former, the secondary thickening proceeds until the lumen is practically obliterated and when the capsule opens, the hairs dry without collapsing or forming convolution. In convoluted hairs, on the other hand, the secondary thickening leaves a considerable portion of lumen free, such that the hairs on drying, collapse on the central lumen, and form flat ribbons which twist spirally to give the characteristic convolutions. Unconvoluted hairs remain flat against the surface of the seed, whereas, the convoluted hairs expand, forming a fluffy mass. All species under cultivation bear true lint (partially thickened, convoluted hairs) on the seeds at maturity, while all other wild species except, *G. tomentosum*, from Hawaii have seeds bearing short brown fully thickened, unconvoluted hairs which are not spinnable.

Taxonomists from Linnaeus onward divided the genus into a number of species. The actual number proposed by each investigator naturally depended on the then existing concept of species, which in turn was based on the knowledge of the heritable variation and the plant material available at that time (Gadkari 1960).

4.2.1 Classification by Early Workers

Linnaeus distinguished five to six species in *Gossypium* on the basis of material from the cultivated types that he studied. De-Candole accepted 13 species with the genus, while Parlatore recognized only seven species. Todaro considered that there were 54 species within the genus. Watt (1907) divided the wild and cultivated cottons into five sections on the basis of presence or otherwise of fuzz and lint on seed, nature of bracteoles and other plant characteristics. Each section was further subdivided into a number of species, the total being 29 with 16 varieties.

With regard to the genus as cultivated in India, Gammie (1907) recognized six groups among the indigenous cultivated types besides New World imported a group of Dharwad American cotton. Among the Asiatic groups, 8 species comprising 12 varieties and subvarieties were recognized. On the other hand, only one species, *G. hirsutum* was mentioned in the New World group.

Leake and Ram Prasad (1940) attempted the classification of the cotton that they were handling for genetic studies and stressed the importance of monopodial and sympodial plant habits in the classification of cotton.

4.2.2 Classification by Recent Workers

The classification proposed by early workers was broadly based on purely morphological characters. Subsequently, with the advance of knowledge regarding the basis of variation and evolution, the concept of species was broadened and simplified, so that only those groups which differed in genetic architecture or between which cytological barriers existed, were regarded as separate species. As such, it became clear that the scientific classification must be based on morphological, physiological, genetical, ecological, cytological, and serological knowledge of the genus.

Efforts in this direction were made by Zaitzev (1928) and Harland (1932). All these workers divided the genus *Gossypium* primarily into two groups, one with 13 haploid chromosomes, and the other with 26 as the basic number. Zaitzev (1928) further subdivided these groups into two each on the basis of geographical distribution.

Zaitzev (1928) was of the opinion that the wild cottons should be removed from the genus *Gossypium* and he recognized 16 species within the genus came into the conclusion that Zaitzev's four subgroups were in essence four well-differentiated species. Harland (1932) disagreed with Zaitzev (1928) regarding the transfer of the wild cottons from the genus *Gossypium*. He proposed the division of the genus into 16 species, of which 5 were cultivated. Hutchinson and Ghose (1937) accepted the classification of the genus into 16 species suggested by Harland.

4.2.3 Modern System of Classification

On the basis of chromosome behavior, geographical description, morphological characters, and crossing behavior of different species, Hutchinson (1954) suggested that the genus *Gossypium* to be divided into 20 species, grouped under 8 sections. The related groups usually possess similar genetic potentialities and therefore, show a considerable parallelism in variation. The cultivated species of *Gossypium* belong to sections Herbacea and Hirsuta, and are *G. arboreum* L. and *G. herbaceum* L. belonging to the former and *G. hirsutum* L. and *G. barbadense* L., to the latter (Gadkari 1960). Based on chromosome, the cultivated linted species are classified into two groups as (i) Old World Asiatic diploid cottons with somatic chromosome number 26 ($2n = 2x = 26$) comprising *G. arboreum* L. and *G. herbaceum* L. and (ii) New World allotetraploid cottons with somatic chromosome number 52 ($2n = 4x = 52$) comprising *G. hirsutum* L. and *G. barbadense* L.

The wild species of *Gossypium*, i.e., those characterized by the lack of lint on seed and having capsules with hairs on their sutures, are very uniform and have not been subdivided into varieties except in case of the Californian wild species *G. klotzschianum.* A wild type called *davidsonii*, which was originally considered as a separate species, is now believed to be a variety of *G. klotzschianum.* The linted species have, however, undergone a greater degree of differentiation due to their spread by man in different habitats. These cultivated linted species of *Gossypium* have been subdivided into different taxonomic units such as varieties, forms and races by various taxonomists in the light of considerations that each one of them had adopted for his classification. As already mentioned, the cultivated cotton may be divided into four species: two Old World and two New World.

As per Silow (1944), the cultivated *G. arboreum* species include races viz., *Indicum, Burmanicum, Cernuum, Bengalense* and *Sinense*. Similarly, *G. herabceum* have races like *Persicum, Kuljianum, Acerifolium, Wightianum* and *Africanum* (Hutchinson 1950). The races of *G. hirsutum* include *Morrilli, Richmondii, Palmeri, Punctatum, Yucatenese, Marie-glante* and *Latifolium* (Hutchinson 1951).

The cotton genus (*Gossypium*) provides a model system for studying molecular evolution of genes duplicated by allopolyploidy. The five tetraploid *Gossypium* species ($n = 26$) are a monophyletic assemblage derived from a single allopolyploidization event that occurred approximately 1–2 MYA (Wendel 1989; Seelanan et al. 1997; Small et al. 1998).

Diploid *Gossypium* species (all $n = 13$) have been divided into genomic groups (A–K) based on differences in chromosome size and pairing behavior in interspecific hybrids (Endrizzi et al. 1985; Stewart 1995). The two diploid species that gave rise to the allotetraploids were from the A-genome and D-genome groups and are best represented by the extant species *G. herbaceum* L. and *G. raimondii* Ulbr., respectively (Endrizzi et al. 1985; Wendel et al. 1995; Small et al. 1998). Tetraploid species are therefore termed the AD-genome group, and their two constituent genomes are referred to as the A- and D-subgenomes (Fig. 4.1).

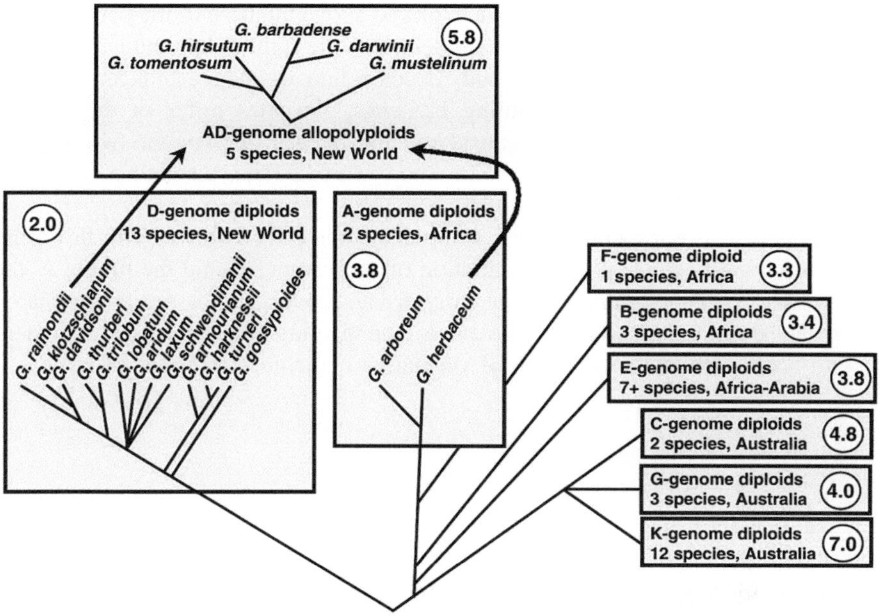

Fig. 4.1 Genome evolution in Cotton (Adopted from Wendel and Cronn 2003)

4.2.3.1 Plant Morphology

As the range of form exhibited by the plant in different species and the countries growing it is very great, details about its various parts is dealt in a broad way as given below (Balasubramanyan 1960).

Shoot

In the wild state, cotton is a perennial plant and may reach a height of 15–20 feet. Most of the cultivated types, however, attain heights ranging from 2–6 ft. The leaves are cordate, petiolate, three to nine lobed, and palmately veined. The size, texture, shape, and hairiness vary a great deal. In the New World cotton, the leaves above the cotyledons occur in a regular 3/8 spiral arrangement while in the cotton of the Old World series, the arrangement is 1/3 spiral. Glands occur on leaves, bracts, petioles, stems, and cotyledons. Nectaries are present on leaf calyx and bracts.

The primary axis bears two types of branches. The main stem arises from the terminal growing point, while the lateral branches arise from the axils of the leaves of main stem. The latter group consists of two types, viz., vegetative and fruiting. The former is a replica of the main stem but the latter is of a distinctive nature.

Each leaf has two buds at its axil. One is placed above the base of the petiole in the middle. The other bud is situated laterally either to the right or left and is, therefore, extra-axillary in position. Axillary buds develop into vegetative branches and the extra-axillary ones grow into fruiting branches. Varieties differ in the relative activity of the two buds. There is considerable difference between the two types of branches. The vegetative branches are more vertical and ascending while the fruiting branches are nearly horizontal.

The economic importance of the sympodial branching is great. The flowering and fruiting are dependent on the initiation of such branches and the timing of the crop for harvest is determined by the early or late production of such sympodia on the plant body. Very early varieties have their fruiting branches even at first or second node to the total exclusion of vegetative branching from leaf axils.

Root

The cotton plant has a slender tap root whose size and depth depend on the type of soil, the ranges of soil temperature and moisture, and the variety. The number and the extent of lateral roots arising from the tap root vary. In general, they originate at about half a foot below the soil surface and extend radially to a distance of 4 ft, branching and rebranching profusely so as to create a network of roots. In some cases, a second system of laterals develop at lower regions where the tap root touches the water logged saturated soil zone.

The lateral roots appear in four to five shallow longitudinal grooves which may develop a spiral appearance due to twisting of the tap root. Subterranean shoots are reported to develop freely in the Egyptian cottons, especially in the depressions adjacent to roots. These underground shoots look like galls in the beginning and grow to various sizes before coming out as a vegetative branch.

Stem

The primary axis is erect and branched. Its height varies according to varieties, season of planting and type of cultivation. The stem is irregularly ridged with three to five large bundles and many smaller ones located in the vascular cylinder. The vegetative branches and the main stem develop as a result of the terminal meristem of each axis. Such growth named monopodial is in contrast to the sympodial one found in fruiting branches. Early in the development of fruiting branch, when the first fruiting axis is being formed, it is pushed to one side so that the developing structures are projected to one side of the leaf on the main stem. Thus, a fruiting branch is more nearly horizontal with the primary axis and grows out either to the right or left of the subtending leaf on the main stem.

Leaf

The leaves are spirally arranged on main stem and the vegetative branches. The arrangement on the sympodia is in the form of alternate rows. The phyllotaxy of the American cottons is three-eighths while that of the Asiatic varieties is one-third. The leaves are petiolate and stipulate. The petioles are glabrous, pubescent, or fully hirsute depending on the variety. The shape of the leaf is generally cordate (entire in many wild species) and the lobing is palmate. The number and depth of sinus in lobes vary. In *barbadense* varieties, the leaves are large with lobes deeply incised (Plate 4.2a); in *hirsutum* group, they are big with lobes cut to half to one-third the length (Plate 4.2b); and in Asiatic types, the leaves are small and the lobes are rounded or very deeply cut. The deepest lobe is found in the Okra leaf (Plate 4.2c). The leaf surface may be hairy or glabrous. Nectaries are found on the undersides of the main veins. The common number is one to three though it may vary from one to five. They have the appearance of shallow pear-shaped naked pits, sometimes canopied by stellate hairs arising from the surrounding surface. Papillae made up of several glandular cells arise from the floor of the pit.

Multicellular hairs of epidermal origin have been recorded on the leaf surface. The color of the leaf may vary from light green to full green. Some varieties develop red pigment masking the full green as in red *arboreum* types. A pink spot of variable intensity at the junction of the petiole and the leaf is the feature of many *arboreum* and *hirsutum* varieties.

Stomata occur on both sides of the leaf, but they are more numerous on the underside. Two conspicuously thickened guard cells surround each stoma. Stomata are also found in cotyledons and hypocotyl.

Fruit

The development of the fruit (boll) begins with the fertilization, and shedding of withered floral organs enclosing it. An increase of 1 mm in diameter per day and the full growth of the boll are attained by the 25th day and the dehiscence by the 48th day. However, the number of days may vary according to the species and the varieties in addition to the environmental factors. Bolls developing under falling

Plate 4.2 Leaf shapes in cotton. **a** *G. barbadense* leaf. **b** *G. hirsutum* leaf. **c** Okra leaf

temperature will need more days to mature than those growing under rising temperature. These factors induce a difference of nearly 5 days in maturation. In general, the first half period of maturation of boll is spent in growth and the second half in internal development without any change in the boll size. The ripened boll contains seeds varying from one to nine in each loculus. A fair percentage of seeds remain undeveloped due to non-fertilization, heredity, and environment. These are called "motes."

Seed

The full grown seed is irregularly pear-shaped, varying in size depending on the variety and conditions of growing. It may be naked or bear short hairs called "fuzz." All cultivated cottons bear long fiberfibers called "lint" and a majority of them have also fuzz on the same seed. The lint can easily be separated from seed mechanically while, the fuzz remains attached firmly. The color of fibers may be white, brown, or green and that of the seed is usually gray, brownish, or black.

The mature seed has two cotyledons folded up and occupying the entire portion of its cavity. They are broad and kidney shaped. The seeds when planted in soil take about four to six days to germinate depending upon prevalent temperature and moisture. The primary root pushes its way through micropyle into the soil and arched hypocotyl, in the process of straightening up, lifts the convoluted cotyledons out of the ground. The cotyledons on being freed from the seed coat expand rapidly and growth of the epicotyl elongates the axis.

Seed Hairs

Lint and fuzz represents the outgrowth of epidermal cells on seeds. Some cells continue to lengthen while others stop growing after some time. The former are called lint while the latter the fuzz. Environment has been cited as the main factor of influence. The lint hair is unicellular and its development is phased in two stages: the first is a period of elongation and the second in thickness. A lint cell bulges first, the protoplasm inside turns granular, and the nucleus moves toward the bulge. The swelling enlarges until it is twice the diameter of the original cell and the nucleus moves to or near the tip. At this stage, the cells have large vacuoles and the walls are still thin. During a period of 24 days, the elongation continues and thereafter ceases. There is no change in thickness. The growth is irregular, slow at first but fast from about the 15th day. The rate slackens during days and quickens during nights.

The cell wall thickens during the second half of boll maturation. Deposits of cellulose are formed on the inside of primary wall. They are laid in layers as seen from some fibers showing as many as 25 concentric ones. During the first 10 days, the elongating lint is firmly attached to the seed and thereafter it becomes weak until the 30th day, regaining some but not all of the firmer attachment at its thickening phase. As soon as the boll dehisces, the hairs dry, collapse, and flatten the cylindrical form, assuming the ribbon-like shape go into spirals or convolutions. The twisting of fiber causes the lock of seed cotton to expand.

The mature hair is uniform in diameter up to three quarters in length and then gradually tapers to a point. It is slightly narrow at the end. A sample lint collected at maturity contains three types: ripe, half ripe, and unripe. The ripe fibers have thickened walls and good convolutions; the unripe fibers, known also as dead fibers, have thin walls, lack twist, and are weak, with a tendency to break up during manufacture; the half ripe fibers are intermediate between the two.

4.2.3.2 Floral Biology

Cotton flowers are extra-axillary, terminal, and solitary and are borne on the sympodial branches. The flower is subtended by an involucre of usually three unequal leaf-like bracts. Bracteoles, alternating with the bracts on the inside of the involucre or standing on either side of the small bract, may be present. The calyx, consisting of five undiverged sepals, is persistent and shaped as a shallow cup (Fig. 4.2). The calyx adheres tightly to the base of the boll as it develops.

The corolla is tubular, consisting of five obcordate petals alternating with calyx lobes and overlapping the next one in the series in a convolute manner. In some varieties, a spot of purple, sometimes called "eye," is found on the claw (base) of the petals. On the first day after anthesis, the corolla changes into pinkish hue and then into red during succeeding days. It withers and falls off on the third day, together with the staminal column and stigma leaving the ovary, calyx, and involucre intact.

The stamens are numerous and united to form a tubular sheath which surrounds the pistils except for the exposed portion of style and stigma at the tip. Under normal conditions, the pollen grains are viable up to 24 h and thereafter loses potency and fail to effect fertilization. The time taken by the pollen tube to traverse the style varies according to the variety and environment. In general, 10–13 h duration is required to traverse the entire length of style. Generally, the interval between pollination and fertilization varies from 36–40 h. The pollen grains of

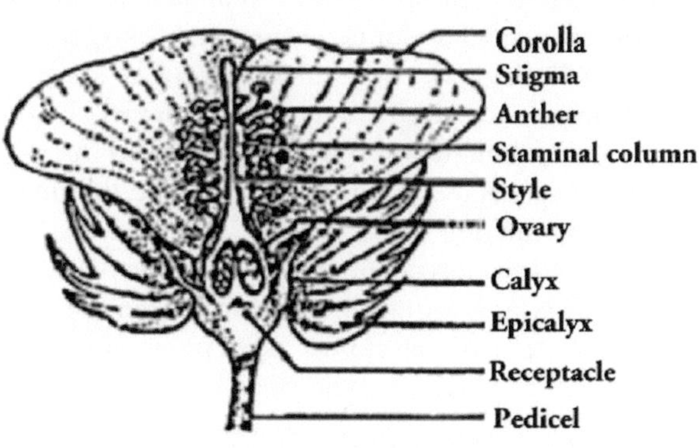

Fig. 4.2 Longitudinal section of a cotton flower

cotton are heavy, sticky, and warty leaving little chance for wind pollination. Insects are the natural agents for the pollen transfer.

The pistil consists of 3–5 undiverged carpels corresponding to the locular composition of a fully mature dehisced boll. The ovules are attached to parietal placenta of each locule. The style varies in length and splits near the apex into three, four, or five parts depending on the number of carpels. The dehiscence of the boll is along the dorsal sutures.

4.3 Origin of Cotton

Origin of *Gossypium* is estimated around 5–15 million years ago (mya) with a rapid early diversification of the major genome groups (Wendel and Cronn 2003). Cotton is unique among crop plants in that four separate species were independently domesticated for the specialized single-celled trichomes, or fibers, that occur on the epidermis of the seeds. This parallel domestication process involved four species, two from Americas, *Gossypium hirsutum and G. barbadense*, and two from Africa–Asia, namely *G. arboreum* and *G. herbaceum*. In each of these four cases, aboriginal people discovered several thousand years ago that the unique properties of cotton fibers made them useful for ropes, textiles, and other applications. As a consequence, cotton cultivation became increasingly widespread, such that over the millennia cotton became firmly established as the world's most important fiber crop and an important source of seed oil and protein meal (Wendel and Cronn 2003).

4.3.1 Old World Cotton

Cotton was first domesticated in the Old World about 7,000 years ago; the earliest archaeological evidence for cotton is from the Neolithic occupation of Mehrgarh, during the sixth millennium BC. The two main species, *G. arboreum* and *G. herbaceum* are genetically very different and probably diverged well before domestication (Pearsall 2008). Abundant archaeological evidence exists for the domestication and use of *G. arboreum*. Mehrgarh, the earliest agricultural village of the Indus Valley, presents evidence of cotton seeds and fibers dating to ca 6000 BC. At Mohenjo-daro, the famous archaeological site on the Indus river, fragments of cloth and cotton textiles have been dated to the fourth millennium BC, and archaeologists agree that most of the trade that made the city grow was based on cotton exportation.

G. herbaceum type of cotton is much less well known than *G. arboreum*. Traditionally, it is known to grow in African open forests and grasslands. However, the distribution of its closest wild progenitor suggests a northward distribution toward North Africa, and the Near East.

4.3.2 New World Cotton

Among the American species, *G. hirsutum* was apparently cultivated first in Mexico, and *G. barbadense* in Peru. The oldest evidence of *Gossypium hirsutum* in Mesoamerica comes from the Tehuacan valley and has been dated between 3400 and 2300 BC. In different caves of the area, archaeologists affiliated to the project of Richard MacNeish found remains of fully domesticated examples of this cotton. Recent studies have allowed the comparison of bolls and cotton seeds from excavation in Guila Naquitz Cave, Oaxaca, with living examples of wild and cultivated *G. hirsutum punctatum*, showing that they might come from the same species, originally domesticated in the Yucatan Peninsula.

The first clear evidence of domestication of *G. barbadense* type of cotton comes from Ancon, a site on the Peruvian coast where archaeologists found remains of cotton bolls dating to 4200 BC. By 1000 BC Peruvian cotton bolls were indistinguishable from modern cultivars of *G. barbadense*. Archaeological excavations showed these cotton has been found in different sites of Peru and Ecuador, especially Ancón, in the central coast of Peru.

Although all four cotton species spread far beyond their ancestral homes during the last several millennia, one species, *G. hirsutum*, recently has come to dominate world cotton commerce, having supplanted the vast majority of cultivation of the other three species, having spread from its original home in Mesoamerica to over 50 countries in both hemispheres (Wendel and Cronn 2003).

Two of the allotetraploid species, *G. hirsutum* L. and *G. barbadense* L., were independently domesticated within the last 5,000 years for their seed fiber (reviewed in Wendel 1995). The genetic consequences of domestication of *G. hirsutum*, the species that presently dominates world cotton commerce, have been explored in depth at the isozyme and restriction fragment length polymorphism (RFLP) levels (Wendel et al. 1992; Brubaker and Wendel 1994). The conclusions of these studies is that genetic diversity in *G. hirsutum* is very low and is especially restricted in the gene pool represented by modern annualized cultivars. *Gossypium hirsutum* was probably first domesticated in the Yucatan peninsula. The only extant form of *G. hirsutum* that arguably is wild, race "yucatanense," is found here as a common component of the indigenous beach strand vegetation (Stephens 1958; Sauer 1967), where it exists as a sprawling, perennial shrub. Evidence suggests that following initial domestication, the original perennial cultivated forms became widely dispersed throughout the Yucatan peninsula. Later, localized derivatives developed, the most important of which was the annualized race "latifolium," which is suggested to have spread to Guatemala and southern Mexico, where further agronomic development took place, leading to cultivated forms that spread via human-mediated diffusion throughout Mesoamerica. Molecular marker evidence (Wendel et al. 1992; Brubaker and Wendel 1994) shows that most of the gene pool of the modern annual forms of *G. hirsutum*, including the Upland cotton cultivars that are common in the cotton belt of the United States, traces to Mexican stocks that had been transported there from Guatemala and southern Mexico (Niles and Feaster 1984). This history of sequential genetic bottlenecks and rapid population expansion is thought to be responsible for the constrained levels of genetic diversity that are often observed in crop plant gene pools (Doebley 1989, 1992).

One generalization that has emerged from the recent massive effort in genome sequencing and mapping in a diversity of organisms is that genome doubling through polyploidy is a prominent process in plant evolution and has played a major role in the evolution of eukaryotic nuclear genomes. Polyploidization has been especially active and ongoing in higher plants, with up to 70 % of all angiosperms having experienced a relatively recent episode of genome doubling. Although there are various types of polyploidy (Grant 1981; Stebbins 1971), the most common is allopolyploidy, whereby two differentiated genomes, usually from different species, become reunited in a common nucleus as a consequence of a hybridization event. In the simplest case, allopolyploids have one complete diploid set of chromosomes derived from each parental species, and thus contain a doubled complement of genes (homoeologues). Examples of such polyploids abound and include many of the world's most important agricultural commodities (Hilu 1993) including cotton.

The establishment of a new allopolyploid species is not a trivial feat. First, all allopolyploids face several immediate genomic challenges, including the merger of divergent genomes, the resolution of potentially conflicting developmental signals and new or possibly accidental interactions with organellar genomes, in addition to overcoming the reproductive barriers associated with polyploidy (Wendel 2000; Comai 2005). Following this, and owing to their redundant genomic architecture, allopolyploid genomes then face several interesting and potentially dramatic evolutionary resolutions. These include the genomic decay of duplicate genes either in the form of genomic fragment loss (Shaked et al. 2001; Tate et al. 2009) or mutational obliteration (pseudogenization), genomic partitioning of ancestral functions (subfunctionalization; Force et al. 1999) or the possibility of a chance beneficial mutation conferring new functionality (neofunctionalization; Ohno 1970). These outcomes are not mutually exclusive (Conant and Wolfe 2008), and most probably require evolutionary time scales, and can be distorted by additional genomic disruptions, such as further hybridization and/or polyploidization leading to the accumulation of additional genomic content, yielding higher ploidies and additional genomic complexity.

A detailed restriction fragment length polymorphism map was used to determine the chromosomal locations and subgenomic distributions of quantitative trait loci (QTLs) segregating in a cross between cultivars of allotetraploid (AADD) *Gossypium hirsutum* ("Upland" cotton) and *Gossypium barbadense* ("Sea Island," "Pima," or "Egyptian" cotton) that differ markedly in the quality and quantity of seed epidermal fibers (Jiang et al. 1998). Most QTLs influencing fiber quality and yield are located on the "D" subgenome, derived from an ancestor that does not produce spinnable fibers. D subgenome QTLs may partly account for the fact that domestication and breeding of tetraploid cottons has resulted in fiber yield and quality levels superior to those achieved by parallel improvement of "A" genome diploid cottons. The merger of two genomes with different evolutionary histories in a common nucleus appears to offer unique avenues for phenotypic response to selection. This may partly compensate for reduction in quantitative variation associated with polyploid formation and be one basis for the prominence of polyploids among extant angiosperms.

Analysis of single-celled fiber transcriptomes from four wild and five domesticated accessions from two developmental time points revealed that at least one-third and likely one half of the genes in the genome are expressed at any one stage during cotton fiber development. Among these, ~5,000 genes are differentially expressed during primary and secondary cell wall synthesis between wild and domesticated cottons, with a biased distribution among chromosomes. Transcriptome data implicate a number of biological processes affected by human selection, and suggest that the domestication process has prolonged the duration of fiber elongation in modern cultivated forms. Functional analysis suggested that wild cottons allocate greater resources to stress response pathways, while domestication led to reprogrammed resource allocation toward increased fiber growth, possibly through modulating stress–response networks.

Tetraploid cottons are thought to have formed about 1–2 million years ago, in the New World, by hybridization between a maternal Old World "A" genome taxon resembling *Gossypium herbaceum* ($2n = 2x = 26$) and paternal New World "D" genome taxon resembling *Gossypium raimondii* or *Gossypium gossypioides* (Wendel et al. 1995) (both $2n = 2x = 26$). The antiquity of this New World event precludes human involvement in polyploid formation.

Wild A-genome diploid and AD tetraploid *Gossypium* taxa produce spinnable fibers that were a likely impetus for domestication (Fryxell 1979). Domesticated tetraploid cottons existed in the New World by 3500–2300 B.C. (Stephens and Moseley 1974) and have been widely distributed by humans throughout the world's warmer latitudes. Domesticated A-genome diploids existed in the Old World by 2700 B.C. (Chowdhury and Burth 1971), and one (of only two extant) species, *Gossypium arboreum*, remains intensively bred and cultivated in Asia. Its close relative and possible progenitor, the other extant A-genome diploid species *G. herbaceum* also produces spinnable fiber.

Although the seeds of D-genome diploids are pubescent, none produce spinnable fibers (Lee 1984). There is no evidence that domestication of D-genome *Gossypium* taxa has ever been attempted, although their geographic distribution overlaps that of several wild tetraploids. No taxa from the other recognized diploid *Gossypium* genomes (B, C, E, F, and G) have been domesticated.

Intense directional selection by humans has consistently produced AD tetraploid cottons that have superior yield and or quality characteristics compared to the A genome diploid cultivars. Selective breeding of *G. hirsutum* (AADD) has emphasized maximum yield, whereas *G. barbadense* (AADD) is prized for its fibers of superior length, strength, and fineness.

Nascent fiber evolution, before allopolyploidy, is elucidated by comparison of spinnable-fibered *Gossypium herbaceum* A and nonspinnable *Gossypium longicalyx* F genomes to one another and the outgroup D-genome of nonspinnable *Gossypium raimondii*. The sequence of a *G. hirsutum* A_tD_t (in which 't' indicates tetraploid) cultivar reveals many non-reciprocal DNA exchanges between subgenomes that

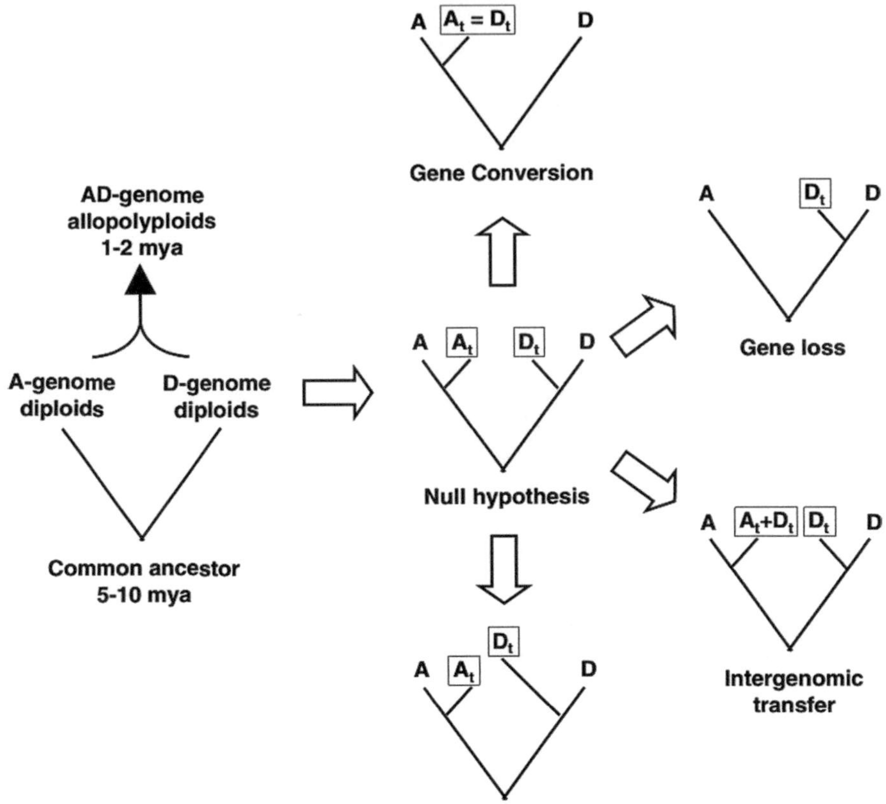

Fig. 4.3 Phylogenetic history of diploid and allopolyploid *Gossypium* species (*left*) and the various possibilities for the evolution of duplicated genes (Adopted from Wendel and Cronn 2003)

may have contributed to phenotypic innovation and/or other emergent properties such as ecological adaptation by polyploids (Fig. 4.3). Most DNA-level novelty in *G. hirsutum* recombines alleles from the D-genome progenitor native to its New World habitat and the Old World A-genome progenitor in which spinnable fiber evolved. Co-ordinated expression changes in proximal groups of functionally distinct genes, including a nuclear mitochondrial DNA block, may account for clusters of cotton fiber quantitative trait loci affecting diverse traits (Paterson et al. 2012).

4.4 Gene Pools in Cotton

The genus *Gossypium* consists of approximately 50 species, of which, 45 are diploid and five are allopolyploid. The species of *Gossypium* are classified into genome groups A to G in diploid and (AD) in allotetraploids group. The genus is distributed in Asia, Africa, America, Australia, Arabia and Pacific islands. The tetraploid cultivated cotton arose from a diploid 'A' genome species *G. herbaceum* var. *africanum* and a 'D' genome progenitor species *G. raimondii* by evolutionary process of migration, interspecific hybridization, polyploidy, some mutations for lintedness and selection over a period of time in nature as hypothesized from various studies. The list of currently recognized *Gossypium* species, organized by germplasm pools are furnished below.

Primary Germplasm pool

Species	Genome	Notes
hirsutum	AD_1	Current and obsolete cultivars, breeding stocks, primitive and wild accessions.
barbadense	AD_2	Current and obsolete cultivars, breeding stocks, primitive and wild accessions.
tomentosum	AD_3	Wild, Hawaiian Islands.
mustelinum	AD_4	Wild, NE Brazil
darwinii	AD_5	Wild, Galapagos Islands

Secondary Germplasm pool

Species	Genome	Notes
herbaceum	A_1	Cultivars and land races of Africa and Asia Minor, one wild from Southern African.
arboreum	A_2	Cultivars and land races of Asia Minor to SE Asia, and China, some African.
anomalum	B_1	Wild, two subspecies from Sahel and SW Africa
triphyllum	B_2	Wild, Cape Verde Islands
capitis-viridis	B_3	Wild, Cape Verde Islands
trifurcatum	(B)	Wild, Somalia
longicalyx	F_1	Wild, trailing shrub, East Central Africa
thurberi	D_1	Wild, Sonora Desert
armourianum	$D_{2\text{-}1}$	Wild Baja California
harknessii	$D_{2\text{-}2}$	Wild, Baja California
davidsonii	$D_{3\text{-}d}$	Wild, Baja California
klotschianum	$D_{3\text{–}k}$	Wild, Galapagos Islands
aridum	D_4	Wild, arborescent, Pacific slopes of Mexico
raimondii	D_5	Wild, Pacific slopes of Peru
gossypioides	D_6	Wild, South central Mexico
labatum	D_7	Wild, arborescent, SW Mexico
trilobum	D_8	Wild, West central Mexico

Tertiary Germplasm pool

Species	Genome	Notes
sturtianum	C_1	Wild, Ornamental, Central Australia
robinsonii	C_2	Wild, Western Australia
Bickii	G_1	Wild, Central Australia
australe	(G)	Wild, North Transaustralia
nelsonii	(G)	Wild, Central Australia
costulatum	(K)	Wild, decumbent, North Kimberleys of W. Australia
cunninghamii	(K)	Wild, ascending, Northern tip of NT, Australia
enthyle	(K)	Wild, erect, N Kimberleys, Australia
exgiuum	(K)	Wild, prostrate, N Kimberleys, Australia
londonerriense	(K)	Wild, ascending, N Kimberleys, Australia
marchantii	(K)	Wild, decumbent, Australia
nobile	(K)	Wild, erect, N Kimberleys, Australia
pilosum	(K)	Wild, ascending N Kimberleys, Australia
populifolium	(K)	Wild, ascending N Kimberleys, Australia
pulchellum	(K)	Wild erect, N Kimberleys, Australia
Rotundifolium	(K)	Wild, prostrate, N Kimberleys, Australia
Anapoides	(K)	Wild, erect, N Kimberleys, Australia
Stocksii	E1	Wild, Arabian Peninsula and Horn of Africa
Somalense	E2	Horn of Africa and Sudan
Areysianum	E_3	Arabian Peninsula
Incanum	E_4	Arabian Peninsula
Bricchettii	(E)	Somalia
Benadirense	(E)	Somalia, Ethiopia, Kenya
Vollensenii	(E)	Somalia

Diversity and Geographic Distribution of the Major Lineages of *Gossypium*

Genome group	Number of species	Geographic distribution
A	2	Africa, possibly Asia
B	3	Africa, Cape Verde Islands
C	2	Australia
D	13	Primarily Mexico; also Peru, Galapagos Islands, Arizona
E	7+	Arabian Peninsula, Northeast Africa, Southwest Asia
F	1	East Africa
G	3	Australia
K	12	NW Australia
AD	5	New World tropics and subtropics including Hawaii

Adopted from Wendel and Cronn (2003)

Present day cotton is the outcome of rigorous selection for production of desired type of fibers for the improved harvesting and processing. The side effect of this selection was the narrow genetic diversity for some specific traits including resistance against drought (Rosenow et al. 1983).

Presently, cotton yield improvement trends show the effect of a variety of biotic and abiotic threats on cotton due to meager genetic diversity. This is due to repeated utilization of a few genetic backgrounds for new cultivars development (May et al. 1995; Rahman et al. 2002, 2005, 2008). Moreover, farmers have planted the similar germplasm on large areas which lead to high level of genetic homogeneity in field (Rahman et al. 2012).

It is also noteworthy that exotic chromatin is lost from advanced-backcross progenies more quickly than would be expected (Jiang et al. 2000b; Liu et al. 2000). This suggests that the few classical studies of cotton introgression that have been done have under-estimated the potential benefits of introgression due to the rapid loss of introgressed chromatin. This potential artifact is added to the problem of confounding the benefits of introgression with the penalties, due to past inability to evaluate the effects of individual chromosome segments (lacking DNA markers). The use of a detailed molecular map will resolve these difficulties, identifying rare individuals retaining the exotic chromatin, thus permitting the independent evaluation of the positive and negative effects of each introgressed chromosome segment.

4.5 Interspecific Hybridization

Hybridization between species is resorted to for securing genes or gene combinations that are not normally available within the limits of a species. In addition, it may be possible to obtain increase or improvements in certain characters through what Stephens (1944) termed "transgressive breeding" since genes favorable for intensification of a particular character may occupy independent loci in the parental species and may also act independently of one another. In cotton, there are four cultivated species and 46 other wild species, thus a wide scope for experimental work on the possibilities of incorporating desired characteristics from cultivated or wild forms into the species of commerce.

4.5.1 Crosses Between Cultivated Species

Among the cultivars, *barbadense* possesses by far the best quality of lint. This species, on the other hand, has been used in improving fiber properties of *hirsutum* (Sikka and Joshi 1960). On the other hand, *barbadense* itself has been improved in the Sudan with regard to disease and pest resistance by the incorporation of the necessary genes from *arboreum* and *hirsutum* (Knight 1956). Several *hirsutum* cultivars have been developed in India by transferring general adaptability traits from *arboreum* and *herbaceum* (Pandya and Patel 1956). Further, they have reported transgressive value in *hirsutum* cultivars derived from *hirsutum-arboreum* hybridization. Some other *hirsutum* genotypes resistant to blackarm and jassids and possessing superior fiber quality have also been isolated from similar hybrids (Jagannatha Rao et al. 1953). Pandya and Patel (1958) reported improvement of ginning performance of *herbaceum* genotypes by hybridization with *arboreum*.

4.5.2 Wild Species as Source for Novel Genes and Traits

Wild species are a repository of a wide range of genes for economically important traits. The genes can be utilized to impart desired traits into cultivated elite genotypes for genetic enhancement of quality traits or resistance to biotic or abiotic stresses. The genes from wild species can be utilized to create a range of genotypes in the four cultivated species to create new variability and to enhance the available genetic diversity (Mehetre et al. 2004b). In cotton, 50 species belonging to the genus *Gossypium* distributed over three gene pools (Primary, Secondary and Tertiary) are classified into 12 genome groups viz., A to K and AD. The wild diploid species/races of cotton is found to have several useful characters which can be exploited to improve the cultivated cotton. A brief note of useful characters available in wild diploid species is furnished in Table 4.1.

Impressive work has been carried in India by cotton breeders by introgressing useful traits from wild species into cultivated species to enhance fiber traits and to combat biotic and abiotic stresses (Rajendran and Jain 2004; Gopalakrishnan et al. 2010). Some of the varieties developed in India by wide hybridization in cotton are presented below (Table 4.2).

Utilization of wild species for qualitative and quantitative crop improvement can be advanced by precision breeding approaches using the recently developed markers that are located on specific chromosome bins. It is important to overcome incompatibility barrier to obtain more number of useful derivatives of wide hybrids, through conventional as well as novel biotechnological approaches. With the assistance of new molecular tools, useful genes can now be introgressed into elite genotypes with adaptable agronomic background and also to reduce photosensitivity in majority of the genetic stocks.

4.6 Cross Compatibility Relationship Between Species

Any comprehensive program of interspecific hybridization will have to take into consideration the nature of obstacles met with and the ways and means of overcoming them. As a result of investigations carried out by several workers, valuable information has been collected. The known relationship between the various *Gossypium* species have been presented by Stephens (1945), Brown and Menzel

Table 4.1 Wild species as source of economic traits

Wild gene source	Special/contributing character	Remark
harknessi and *aridum*	Male sterile cytoplasm and Restorer genes	Cytoplasmic Male Sterile and Restorer lines, drought resistance
anomalum	Raised gossypol glands, hairiness, narrow bract	Resistance to blackarm, bollworms, leaf roller, semi- loopers, jassids, mites, *Earias*
	High fiber strength, Extreme fineness, high maturity percentage	Improved fiber quality
sturtii	–	Healthy foliage and luxuriant vegetative growth
stocksii	–	Resistance to drought
somalense		Resistance to bollworms
armourianum	Healthy leaf growth, fiber strength (latent feature), caducous bracteoles	Resistance to blackarm, spotted bollworm, jassids, aid in clean picking
raimondii	Vigorous root system, luxuriant vegetative growth,	Resistance to drought, black arm, spotted bollworm, leaf roller, jassids, improved fiber quality
tomentosum	Fine and strong fiber	Resistance to drought and jassids, improved fiber quality
australe and *bickii*	Pubescent bolls, high gossypol	Resistance to Bollworms
cernuum	High gossypol, long bolls, hard rind	Resistance to Pink bollworm
bengalense	Red plant, high gossypol	Resistance to Pink bollworm and *Helicoverpa*
palmeri	Palmate leaf	Resistance to Sucking pests, bollworms
thurberi	Deeply dissected leaves, more glands, prolific boll bearing, fiber strength (latent feature)	Resistance to Bollworms, frost, wilt, gummosis
mariegalantee	–	Resistance to Drought
africanum	–	Resistance to Drought
kondapathi	–	Fiber strength

(1952) and Menzel and Brown (1955). Based on several cross combinations attempted in *Gossypium* species, Sikka and Joshi (1960) concluded that some crosses did not give viable seeds of fertile offspring. Broad conclusions drawn by them are: (i) crosses between the Old World species, excluding some of those involving *sturtii*, produce viable seeds; (ii) with the exception of some of the crosses involving *davidsonii* and *gossypoides*, the crosses among the New World diploid species set viable seeds if fertilization is affected; (iii) in crosses between the New World and the Old World diploid species, excluding a few, fertilization is readily effected but many or all the seed, excluding a few, fertilization is readily affected but many or all the seeds obtained prove inviable. Incompatibility is then

Table 4.2 Utilization of traits from wild species for varietal improvement in India

Species involved in crossing	Variety developed (*G. hirsutum*)	Useful character transferred
G. tomentosum	B.1007, Khandwa-2	Jassid resistance
G. arboreum	Deviraj	Drought resistance
G. herbaceum	Devitej	Drought resistance
G. mexicanum	VRS 7	Verticillium wilt resistance
G. anomalum	Arogya	Bacterial blight resistance
G. anomalum	AKA 0841 (*G. arboreum*)	Fiber strength
(*G. anomalum x G. thurberi*)	PKV-Rajat	High yield
G. harknessii	PKV- Hy.3	CMS and Restorer genes
G. barbadense	MCU.5	Fiber quality.

apparently due to some postfertilization disharmony; (iv) tetraploid species hybridise successfully with a majority of the diploid species excluding *klotzchianum* and its variety *davidsonii*; and (v) interspecific crosses within the tetraploid group are also readily made.

There are several levels at which the hybridization between two *Gossypium* species may fail (Brown and Menzel 1952).

1. Flowers may drop soon after pollination without perceptible enlargement of the ovary, indicating that probably fertilization did not occur. In *hirsutum*, however, ovaries of emasculated and unpollinated flowers also enlarge for several days before the formation of abscission layers.
2. The ovary may enlarge, and the young boll remain on the plant several days to several weeks; in these cases fertilization has probably occurred.
3. The bolls may mature and contain either abortive seeds or large empty seeds. In rare instances, a hybrid may mature, which may be smaller than normal one.
4. Hybridization may fail after the development of an embryo; the seed germination, and the seedling dies in the cotyledon stage.
5. Death of seedlings may take place after several leaves are formed; in some cases plants may even reach the flowering stage.

4.7 Cytogenetic Stocks

Aneuploid substitution stocks have been developed from tetraploid TM-1 (GH) × 3–79 (GB) (Endrizzi and Ramsay 1979) and TM-1 × GT (Saha et al. 2006). Three of the chromosomes remained unidentified while 23 of the 26 chromosomes were identified by monosomics and telosomics together. Numerous SSRs and restriction fragment length polymorphism (RFLPs) have been assigned to respective chromosomes utilizing these resources.

4.8 Genome Size Variation and the Concept of Genome Group

As the genus diversified and spread, it underwent extensive chromosomal evolution, which has been studied by many researchers (reviewed by Endrizzi et al. 1985). Chromosome morphology is similar among closely related species, and this is reflected in the ability of related species to form hybrids that display normal meiotic pairing and high F1 fertility. In contrast, crosses among more distant relatives are often difficult or impossible to effect, and those that are successful are characterized by meiotic abnormalities. The collective observations of pairing behavior, chromosome sizes, and relative fertility in interspecific hybrids led to the designation of single-letter genome symbols (Beasley 1941) for related clusters of species. At present, eight diploid genome groups (A–G and K) are recognized (Endrizzi et al. 1985; Stewart 1995; Wendel et al. 1999).

Although all diploid *Gossypium* species share the same chromosome number (n = 13), there is more than a threefold variation in DNA content per genome (Bennett et al. 1997; Bennett et al. 1982; Edwards et al. 1974; Edwards and Mirza 1979; Kadir 1976; Michaelson et al. 1991), with 2C contents ranging from approximately 2 pg per 2C nucleus in the New World, D-genome diploids to approximately 7 pg per cell in Australian K-genome species. A-genome species have intermediate values of approximately 3.8 pg per 2C nucleus. The range in genome sizes is even greater when other diploid members of the tribe are considered, nearly seven-fold variation in DNA content is reported (Wendel et al. 2002) between the largest (*T. populnea*; 2C = 8.2 pg) and smallest (*Gossypioides kirkii* and *K. drynarioides*, each with 2C = 1.2 pg) genomes measured till date in the *Gossypieae*. This extraordinary variation in DNA content is widely believed to be caused by modification of the repetitive DNA fraction, with relatively little change in the absolute amounts of single-copy DNA (Geever et al. 1989). Increasing evidence indicates that both genome size expansion and contraction are common in evolution, not only in *Gossypium* and *Gossypieae* but also in angiosperms as a whole (Wendel et al. 2002). Polyploidy often confers emergent properties, such as the higher fiber productivity and quality of tetraploid cottons than diploid cottons bred for the same environments (Jiang et al. 1998).

4.9 Genetic Maps

There exists at many published genetic maps to map useful traits and markers which includes ~5,000 markers in public database including 3,300 RFLP, 700 amplified fragment length polymorphism (AFLP) 1,000 SSR, and 100 single nucleotide polymorphism (SNP). The STS-based genetic maps of diploid (D) and tetraploid (AtDt[1)] can serve as a source of probes for many of the other maps. Here "At" stands for A subgenome while "Dt" stands for D subgenome in tetraploid cotton

Gossypium genomes. The reference maps include 2,584 loci at 1.72 cM (~600 kb) intervals comprised on 2,007 probes (AtDt); and 1,014 loci at 1.42 cM (~600 kb) intervals identified by 809 probes (D) (Rong et al. 2004, 2005a). Respective genomes show great collinearity with each other (Rong et al. 2004) to get benefit of this, the gene order was concluded of a proposed mutual ancestor of the At, Dt, and D genomes. In total, 3,016 loci recognized by 2,337 probes were included in this map, and it spanned 2,324.7 cM. With the help of three different populations derived from interspecific crosses supplementary marker rich maps were developed which have been extensively utilized as reference maps for studying quantitative trait loci (QTLs) (Lacape et al. 2003; Guo et al. 2007; Lacape et al. 2007; Yu et al. 2007).

Diagnostic DNA markers linked with fiber traits have also been identified with the help of cotton molecular maps (Rahman et al. 2009, 2011; Cai et al. 2010). Study of genomics of drought tolerance has become vital due to heavy water use of cotton (Saranga et al. 2001, 2004; Zhao et al. 2008; Ullah 2009). During the last few years, major focus has also been the disease resistance with published information for *Xanthomonas* (Wright et al. 1998; Rungis et al. 2002), *Verticillium* (Bolek et al. 2005; Wang et al. 2008; Yang et al. 2008), root knot nematode (Shen et al. 2006; Wang et al. 2006a; Wang and Roberts 2006; Ynturi et al. 2006; Niu et al. 2007), *Thielaviopsis* (Niu et al. 2008), and cotton leaf curl disease (CLCuD; Rahman et al. 2005, 2006).

Nuclear restorer of a cytoplasmic male sterility has been a focus of R&D efforts because of increasing interest in growing hybrid cotton (Guo et al. 1998; Lan et al. 1999; Zhang and Stewart 2004; Wang et al. 2007). Different traits of defense umbrella such as leaf morphology (Jiang et al. 2000a; Song et al. 2005; Waghmare et al. 2005; Mei et al. 2004; Hao et al. 2008) and leaf color (Ali et al. 2009b) along with the pubescence, a specific characteristic of *Gossypium hirsutum* (Wright et al. 1999; Lacape and Nguyen 2005; Desai et al. 2008; Ali et al. 2009a), have also been the focus of study. Physical characteristics of the seed and its nutritional value are also being mapped due to the worth of cotton seed (Song and Zhang 2007).

A more profound method for the study of genetic control of a trait is the alignment and evaluation of several QTL mapping experiments using a mutual reference map rather than single study; it elucidates the genomic arrangement of trait distinction. A dense genetic map containing 432 QTLs (for fiber quality and its yield, leaf and flower morphology, trichome density, and their distribution etc.) and 3,475 loci were identified in a total of 11 populations (Jiang et al. 1998, 2000a; Wright et al. 1999; Saranga et al. 20012004; Paterson et al. 2003; Chee et al. 2005a, b; Draye et al. 2005; Rong et al. 2005b; Waghmare et al. 2005). Later, these QTLs were mapped on a consensus map which was presumed to be comparable to the DNA marker organization of the hypothetical ancestors of the two "subgenomes" of the tetraploids (Rong et al. 2005b). In another meta-analysis study of over 1,000 QTLs obtained from BC and recombinant inbred line (RIL) populations derived from the same parents, most consistent meta-clusters were reported for fiber color, fineness, and length (Lacape et al. 2010).

Assumption of cotton–Arabidopsis synteny relationships has become convenient due to consensus maps and it assists the study of correspondence between fiber or trichome related Arabidopsis genes and the cotton QTLs. This CMap resource was generated and can be accessed at http://chibba.agtec.uga.edu/cgi-bin/cmap/viewer for using in various genomic studies

4.10 Genetic Enhancement Through Biotechnological Interventions

The recent advances made in transgenic research over the past decade and the advent of insect-resistant transgenic (genetically modified) crops, have opened up exiting possibilities, new areas of research and new avenues in eco-sustainable pest management. Genetically modified crops have emerged as important components of modern eco-friendly high yielding agriculture. Thus far a total number of 20 transgenic crop plants incorporating 42 genes with 97 transgenic events were developed by 28 commercial companies including public funded institutions and have been released for commercial cultivation in more than 25 countries. Since, the first introduction in 1996, the area under transgenic crops increased to 148 m hectares by 2010 (Kranthi et al. 2011).

4.10.1 Genetically Modified Cotton for Insect and Disease Resistance

Genetically Modified (GM) cotton, popularly known in India as "Bt-cotton" was first developed and commercialized by the US multinational company, Monsanto, and later, by several other multinational companies such as Syngenta, Bayer, Dow, also by CAAS (Chinese Academy of Agricultural Sciences), and Indian companies such as JK seeds and Metahelix. A bacterial gene *cry1Ac* was isolated from a soil bacterium *Bacillus thuringiensis*, and introduced into the cotton genome through genetic engineering using a bacterium, *Agrobacterium tumefaciens* to develop commercial GM cotton called "Bollgard." The Bollgard-Bt-cotton with *cry1Ac* was first cultivated in the US during 1996 and was released in China and Australia in 1997. Later it was released in Mexico, Colombia, Indonesia, Argentina, South Africa, and India. Currently, an estimated 14.5 M hectares are under Bt cotton in 13 countries. This accounts for 42 % of the total global cotton area. Recently, several other genes such as protease inhibitors, *vip3A, cry1C, cry2Ab*, and *cry1F* have been used for the development of GM cotton and are being used for the control of cotton bollworms and other leaf-feeding caterpillars. Extensive studies in accredited laboratories in India and across the globe showed that Bt genes are specifically toxic to insect pests and have high level of safety to nontarget organisms such as beneficial insects, birds, fish, animals, and human beings.

After intensive biosafety studies and extensive field trials under the regulatory system of RCGM (Review Committee on Genetic Manipulation) and GEAC (Genetic Engineering Approval Committee), Bt cotton technology was first approved in 2002 by the GEAC for commercial cultivation in central and south Indian cotton–growing zones in India, and later in 2005 for cultivation in north India (Karihaloo and Kumar 2009).

An exploratory search for insecticidal plant and microbial species should be carried out to diversify the insect resistance in crop plants. A few recent examples deal with the use of allatotropins, allatostatins, proctolin etc. that have a significant effect on several lepidopteran species when consumed. Peptide phage display technology may be used to identify inhibitors for key target sites in insects. The potential of such peptides and neuropeptides for bollworm control has not yet been explored anywhere. The search for the insecticidal proteins includes plant sources (leaves, seeds, roots etc.), microbial organisms (*Bacillus*, *Xenorhabdus*, *Photorhabditis* etc.) and neurohormones from insect species (*Helicoverpa*, *Pectinophora,* and *Earias*).

The possibilities of discovering new genes for pest management have expanded into infinity, with introduction of new concepts such as gene silencing through RNA interference (RNAi). RNAi deploys double stranded RNA (dsRNA) to silence specific endogenous genes in the target organism, which can be specific to the class, genus, or even the specific target species. Thus, crucial species specific genes of a species such as the pink bollworm, *Pectinophora gossypiella* can be identified and the dsRNA expressed in plants to control the pink bollworm by silencing the specific target gene. Gene silencing has been used recently to develop a new biotech cotton variety that specifically controls bollworms by silencing a gossypol degrading enzyme called CYP6AE14 which otherwise enables bollworms survive on cotton (Mao et al. 2007). When bollworm eats the double stranded RNA (dsRNA) of the CYP6AE14gene, the enzyme is silenced and undigested gossypol remains in the stomach and kills larvae. The technology has immense potential in pest management that can be sophisticated to the extent of being extremely specific for the control of target pests alone. The RNAi technology is in the forefront of all the "state of art" technologies for pest management. Ever since the publication in Nature 1998 and the Nobel Prize awarded to Drs. Andrew Fire and Craig Mellow in 2006, for their discovery of dsRNA-based silencing of specific genes through RNAi (RNA interference), the technology has fired the imagination of researchers all over the world.

Globally, attempts are being made to introduce alternative genes (new Cry genes, lectins, protease inhibitors, genes from nematodes etc.) and RNAi (RNA interference-based gene silencing)-based crop protection through GM cotton for more effective pest management. However, insect-resistant GM crops that serve as alternate host plants of cotton bollworms (example, pigeon pea, chick pea, tomato, and other vegetables, which are hosts of the cotton bollworm, *Helicoverpa armigera*) should be developed with genes that are not used in GM cotton. Use of the same Cry genes in all crops will enhance the chances of resistance development in insects to the genes used. There are several sources in nature that have been used to isolate insecticidal genes. Genes from endosymbiotic bacteria of nematodes, *Xenorhabdus* and *Photorhabdus* are being actively considered for the development of transgenic crops (Kranthi et al. 2011).

Among animal sources, antichymotrypsin, antielastase, chitinase, cholesterol oxidase, and antitrypsin were isolated from the tobacco hornworm, *Manduca sexta* and used to develop biotech cotton resistant to sucking pests and lepidopteran insects. Trypsin inhibitors and spleen inhibitors isolated from cattle, protease inhibitors from plants (Soybean, barley, cowpea, squash, mustard, rice, potato, tomato), amylase inhibitor genes from beans and cereals, and lectins from plant sources have been used to develop biotech crops resistant to insect pests. Other genes include chitinases, glucanases, peroxidase, and tryptophan decarboxylase from various plant sources to develop insect- and disease-resistant cotton. Replicase genes and coat protein genes have been used to develop leaf curl virus resistant varieties through overexpression of the proteins or silencing of the genes through RNAi, especially for countries in Africa, India, and Pakistan, where the cotton leaf curl virus (CLCuV) problem can cause severe economic losses.

RNAi is an useful tool to develop insect- and disease-resistant varieties. Important diseases such as the cotton leaf curl virus and bacterial blight can be effectively managed by GM technologies including RNAi by pyramiding native resistance available in the country. Efforts should be made to identify "pathogen species specific" genes present in the pathogen species and "insect-species-specific" genes present in the insect gut which are functionally important for feeding, digestion, and other biological activities. There is a need to identify effective siRNAs and/or miRNAs and their targets. Gene sequences and the novel structures must be explored for their utility for crop protection through conventional or transgenic approaches for the management of cotton insect pests such as the bollworms, jassids, whiteflies, and new pests.

It is now been proven that new biotech crops that scare insects can be developed. Insects release chemicals called alarm pheromones when they are scared by their enemies. This warns their colonies to escape. New biotech crops express alarm pheromones that scare the specific insect pests. The alarm pheromone for many species of aphids, which causes dispersion in response to attack by predators or parasitoids, consists of the sesquiterpene (E)-farnesene (Ef). High levels of expression in *Arabidopsis thaliana* plants of an Efsynthase gene cloned from *Mentha piperita* were used to cause emission of pure Ef (Beale et al. 2006). These plants elicited potent effects on behavior of the aphid *Myzus persicae* (alarm and repellent responses) and its parasitoid *Diaeretiella rapae* (an arrestant response). Insect injury causes signal transduction. The signal transduction pathways leading to the release of plant volatiles have been found to alert other plants in the neighborhood. Jasmine scent reduced populations of jassids, aphids, *H. armigera,* and enhances populations of predators and parasitoids in cotton fields. Some plants have been found to help cotton crop to fight pests.

4.10.2 Genes to Enhance Nutritional Status of Cotton Seed

Low gossypol seed can be possible through biotech cotton expressing cytochrome P450 CYP6AE14 genes from pink bollworm and *Helicoverpa* to be expressed specifically in cotton seeds. The gene sequences are known and seed-specific promoters are available. These can be used to develop low gossypol seed varieties. Sunil Kumar et al. (2006) utilized RNA interference to inhibit the expression of the δ-cadinene synthase gene in a seed-specific manner, thereby disrupting a key step in the biosynthesis of gossypol in cotton. Compared to an average gossypol value of 10 μg/mg in wild type seeds, seeds from RNAi lines showed values as low as 0.2 μg/mg. Importantly, the levels of gossypol and related terpenoids that are derived from the same pathway were not diminished in the foliage and floral parts of mature plants and thus remain available for plant defense against insects and diseases. Further, it has been reported that the germinating RNAi seedlings are capable of launching terpenoid-based defense pathway when challenged with a pathogen. Thus, the silenced state of the δ-cadinene synthase gene that existed in the seed does not leave a residual effect that can interfere with the normal functioning of the cotton seedling during germination. Apart from silencing of δ-cadinene synthase, there are several innovative strategies that can be used to reduce gossypol specifically in seeds and also to increase monounsaturated fatty acids to make cotton seed oil more acceptable nutritionally.

4.10.3 Resistance to Drought

Drought stress or water deficit is a complex phenomenon affecting the physiology (Grimes and El-Zik 1990), growth, and productivity of cotton plant (Chu et al. 1995). Earlier (Guinn and Mauney 1984) reported that water stress decreased seed cotton yield, in part because of decreased flowering and in part because of decreased boll retention. The detrimental effects of drought can be minimized by the development of limited reports on this aspect due to complex nature of drought-tolerant mechanisms explaining slow progress in drought-prone areas (Cattivelli et al. 2008). Although conventional breeding has been successful in the past in developing drought-tolerant cotton cultivars by modifying the morphological and physiological traits, yet the approach is time consuming and labour intensive.

4.10.4 Mechanism of Drought Tolerance in Cotton

Drought tolerance is a multigenic trait (Mussell and Staples 1979; Cushman and Bohnert 2000; Ahmad et al. 2009) associated with morphophysiological characters

(Singh 2004) moderating the genetic improvement on morphophysiological-based selection of crop plants. These are mainly distance of 1st main lateral root from transition zone, seedling vigor, rate of increase in root system, root/shoot ratio (Cook 1985), increased size of tap roots (Pace et al. 1999), lower excised leaf water loss, lower transpiration rate, lower stomatal size and frequency and high-relative leaf water contents (Malik and Wright 1997; Parida et al. 2008), stomatal conductance, rate of photosynthesis (Nepomuceno et al. 1998), and canopy temperature (Lacape et al. 1998).

In genetic sense, drought tolerance mechanisms can be grouped into three categories viz., drought escape, drought avoidance/postponement, and dehydration tolerance (Mitra 2001). Drought escape is the capability of the plant to complete its lifecycle, i.e., vegetative and reproductive growth before the onset of drought season. Plants that are better prepared to escape drought tend to close their stomata, thus reducing CO diffusion but this ultimately leads to decreased rate of photosynthesis resulting in reduced growth and economic yield. Only those plants adopt this strategy that are surviving under extremely stressful condition that allow them to redirect absorption and energy which are usually used in metabolic processes of growth and production for the synthesis of protective molecules (Zhu 2002; Chaves et al. 2003).

In drought avoidance/postponement, the plant becomes capable of maintaining tissue water potential; this is done by increasing the water uptake from soil through their deeper and vigorous root system and reducing transpiration by stomatal closure (Izanloo et al. 2008; Agbicodo et al. 2009).

Dehydration tolerance is the ability of crop plants to maintain a favorable water balance through the expression of morphological traits, i.e., earliness, reduced leaf area, leaf rolling, wax content, efficient rooting system etc., physiological traits, i.e., reduced transpiration, high water-use efficiency, stomatal closure, and osmotic adjustment and biochemical traits, i.e., accumulation of proline, polyamine, trehalose, etc. traits (Brito et al. 2011).

4.11 Conventional Breeding for Drought Tolerance in Cotton

Drought stress is a complex phenomenon that affects the physiology of cotton plant. Cotton is classified as a drought-sensitive crop as it is not an efficient water consumer. Significant efforts have been made in the past to improve drought tolerance in cotton cultivars through conventional breeding. Most of the drought-related breeding programs concentrate on selection of those cotton cultivars that yield well under drought stress. This selection is generally based on identifying those traits that can be utilized for screening to drought tolerance. Many of such attributes including anatomical traits (root characters), physiological traits (gaseous exchange, osmotic adjustment), and plant water status measurement (leaf water potential, relative water contents, cell membrane stability) are recognized as important components of drought tolerance in cotton (Steele et al. 2006; Basu et al. 2007). But for a plant breeder to evaluate a large population by these parameters is too tedious and time consuming.

Exotic tetraploid cottons are relatively more tolerant to drought and heat. Progenies generated after making an interspecific cross between *Gossypium tomentosum* and *Gossypium hirsutum* showed relatively high tolerance under water stress (Gotmare and Singh 2004). Due to better adaptation to drought stress, an exotic strain of *Gossypium hirsutum*, var. *marie-galante* remained under cultivation in the beginning of twentieth century in Brazil (Boulanger and Pinheiro 1971).

4.11.1 QTLs Associated with Drought Tolerance in Cotton

There are many minor genes (poly genes) with additive effect in their expression which control drought stress (Zhao 2002; Mohammadi et al. 2005; Thi and Chi 2008). Natural variability that exist in the crop can be utilized for selection under stress environment, it might be natural or stimulated using mapping of QTLs (polygenes), and following the marker-assisted selection (Ashraf et al. 2008).

Quantitative trait loci (QTLs) which are responsible for improved productivity of crops under water stress condition have been identified via genetic mapping techniques (Agrama and Moussa 1996; Tuinstra et al. 1996; Ribaut et al. 1997).

It is also reported that physiological variation due to QTLs is considered to be coupled with tolerance against stress such as osmotic adjustment (Roberto and Silvio 2006).

The so-called QTLs providing the path toward marker-assisted selection (MAS) (Morgante and Salamini 2003) and following their cloning are further manipulated through genetic engineering (Salvi and Tuberosa 2005). QTL mapping is very much important as it paves the way for the mechanism of gene action, identification of location and number, and total phenotypic effect.

QTLs analysis play a key role in controlling drought stress implementing their wide assessment via traditional ways in past and through advanced mapping approaches and DNA marker based examination in detail (Humphreys and Humphreys 2005).

However, due to very low genetic variability in cotton, there are few reports about genes conferring drought tolerance (Quisenberry et al. 1981; Basal et al. 2003).

Under high water-deficit conditions, HSPCB gene responsible for peptide synthesis become activated in leaves of cotton drought-tolerant genotypes (Voloudakis et al. 2002). An alpha crystalline small heat shock protein gene (GHSP26) was identified from *Gossypium arboreum* L. GHSP26 gene activated in stress condition and improves drought tolerance by regulating cellular metabolism as it assists in protein folding and prevent protein denaturation (Maqbool et al. 2007). Selvam et al. (2009) identified a novel drought-resistant gene KC3. Physiological and biochemical studies proved that KC3 gene improves drought tolerance in cotton.

Under drought stress condition in cotton, dehydration-responsive element binding gene (DREB) I and II were activated and their function is to assists DNA binding. It also act as a transcription activator which controls the expression of a number of stress-related genes; thus it improves the stress tolerance of crop plants by interacting with the specific cis-acting element named DRE/CRT (Liu et al. 1998). By using a chromosome-walking technique, a TPS (trehalose-6- phosphate synthatase) gene is identified during stress which participates in the biosynthesis of trehalose. Trehalose is an osmolyte that help in protection of proteins under water stress (Kosmas et al. 2006).

Using differential display of PCR, A4B1, a novel *Gossypium* transcript of 855 base pairs was reported that was induced by drought stress which had a very similar function as small heat shock proteins (SHSPs) of plants (Maqbool et al. 2007).

From multiple tissues and organs under different conditions 185,000 *Gossypium* ESTs were organized from 30 cDNA libraries which contain more than 94,800,000 nucleotides including drought stress tolerance (Udall et al. 2006). It has been reported that in drought stress condition, photo system IPsaH protein, and H (+)-ATPase-related gene expressed reducing evaporation adjusting photosynthesis, hence help to utilize water efficiency. Increase expression level of glyceraldehyde-3-phosphate dehydrogenase, alcohol dehydrogenase, and drought-induced cysteine protease (Koizumi et al. 1993; Harrak et al. 2001) was analyzed under drought conditions (These enzymes assists in cellular functions during stress and maintain chloroplast membrane and chlorophyll contents by catabolic actions).

Microarrays have been developed from cotton ESTs to facilitate cotton genomics research. Udall et al. (2006) obtained 51,107 unigenes by analyzing approximately 185,000 ESTs from both fiber/ovule (124,299 ESTs) and non-fiber source (608,99 ESTs) of *Gossypium hirsutum*, *Gossypium arboreum,* and *Gossypium raimondii*. In cotton, several additional batches of ESTs or cDNA-based microarray have reported (Udall et al. 2006; Shi et al. 2006; Wu et al. 2006) that there is no question that these microarrays of cotton will act as a key tool for genomic studies in cotton.

A drought-tolerant cotton having "stay green" like phenotype was obtained by introducing *Arabidopsis* gene GF14 lambda encoding a 14-3-3 protein which have good stomatal conductance, response well under drought condition by higher rate of photosynthesis (67). Lv et al. (2009) reported that drought stress was obtained by transferring H (+)-PPase gene TSVP from *Thellugiella halophilla* in cotton. Transgenic cotton have better shoot and root growth, higher photosynthesis, greater stability of cell membrane and higher relative water content (RWC) under water-deficit conditions. This research was quite similar with the work of (Pasapula et al. 2010) who reported that increase efficiency of cotton under drought stress condition, when Arabidopsis vacular H (+)-pyrophosphaatase gene (AVP1) was introduced in cotton. That facilitated ions and sugar movement into roots resulting greater absorption of water and reducing water potential hence greater fiber yield than wild type cotton.

Now, several drought-related genes have been cloned and characterized in recent times. Zhang et al. (2009) reported on the nine ESTs including photosystem I psaH protein, and H + -ATPase-related genes which were upregulated at different levels in drought stress cotton seedlings. These genes are responsible for the absorption and utilization of water through adjusting the photosynthesis process. Under drought stress the two genes were found to be highly induced. cDNAs differentially expressed in response to drought stress also revealed the role of CaLEALI gene in response to various abiotic stresses. Transgenic approach provided proof of concept of the relevance of many genes such as P5CS, Glyoxalase AHK1/ATHK1, DREBs, PDH45Helicase, NPK1, DREB2 like small protein such as CAP2, GmDREB2, AtHARDY, ARAG etc., which mainly addressed cellular level tolerance in model plant species, need to be utilized in cotton. Maintenance of positive carbon balance during stress is of significant importance to avoid yield penalty.

4.11.2 Resistance to Other Abiotic Stress and Climate Change

Climate change can adversely affect adaptability levels of varieties that were carefully selected and developed by farmers and plant breeders over the past several years to suit specific agro-ecological conditions of specific cropping zones. Salinity problems in irrigated regions are increasing. Erratic distribution of rainfall that has become a more common phenomenon recently has also become more detrimental to the crop that is cultivated in more than 60 % area in India (Kranthi et al. 2011). It is necessary to identify, classify, and categorize germplasm lines that have unique traits to withstand cold, heat, high CO_2, and other stresses that are envisaged to occur with climate change.

Cotton is sensitive to photoperiod and thermal conditions and does not adjust easily to new environments. Cotton varieties from particular latitudes are known to take inordinately long time to adapt to unfamiliar latitudes across the globe. Genetic engineering can help to develop cotton varieties that can grow anywhere in the world. Genetic manipulation of Rubisco activase can alter photoperiod and thermal sensitivity to enhance the adaptability of cotton to a wide range of environments. Cotton susceptibility to abiotic stress especially drought, water logging, and salinity adds to the adjustment complications.

Drought-responsive element binding proteins (DREB) rd29A genes for drought, high-salt, and cold stress have been identified and used in several crops including cotton. Superoxide dismutase (SOD) confers chilling stress and is being explored for its utility in cotton. A few years ago attempts were made to develop biotech cotton for abiotic stress-tolerance, through the deployment of genes that are responsible for modification of a single metabolite that would confer increased tolerance to salt or drought stress. Stress-induced proteins with known functions such as water channel proteins, key enzymes for osmolyte biosynthesis of betaine, proline, trehalose, and polyamines were the initial targets of plant transformation.

Cotton being C3 crop, genetic manipulation to maintain positive carbon balance either by increasing carboxylation reaction or by decreasing photo respiration will enhance the water use efficiency (WUE) and nutrient use efficiency (NUE), thereby enhancing yields. Single-cell C4 mechanism suggests possibilities to express C4 genes in C3, recent progress in cloning and expression provides leads for the co-ordinated expression of relevant genes in target organelles. Recent successful demonstration of increasing CO_2 (CCM) and biomass in *Arabidopsis* by utilizing decarboxylation of glycolate (a pathway that exists in bacteria) acts as an option for improvement of C3 crops such as cotton. Along with the above improvement of primary constitutive traits such as root growth (Alfin, AUX1, PIN-1, NAC-1), Wax (SHINE/WIN1, WXP1/WXP2)-associated traits using transgenic approach. Finally, combination of the above traits in coordinated manner will help us to obtain cotton genotypes with better adaption to climate change particularly abiotic stresses.

Substantial differences in heat tolerance, root growth, WUE, and dry matter accumulation among exotic *Gossypium hirsutum* lines were shown (Quisenberry et al. 1982). However, different degrees of sterility and distinct segregation distortions, and propensity to preserve the parental haplotypes of interspecific hybrids are unusual features which hinder the progress of enriching the genomes of cultivated cotton species (Jiang et al. 2000b; Waghmare et al. 2005).

GM cotton varieties for traits such as drought and disease (leaf curl virus) management have not yet been released commercially and have immense potential in many countries. Herbicide-resistant GM cotton in small-scale production systems should find a useful place with careful planning and the design of alternative placement of intercrops to avoid the direct effect of herbicide on them and also to ensure that cotton does not become the sole crop in the production systems because of the new weed management GM technology.

4.12 Molecular Markers

Molecular markers are the firm landmarks in the genome of an organism rather than the normal genes because mostly they do not have the biological impacts and may or may not relate with phenotypic expression of a trait (Agarwal et al. 2008). The development of the DNA markers is simple due to the availability of large-scale genomic database (Andersen and Lubberstedt 2003). In plant breeding, these markers are very helpful in recognition, characterization, identification of genetic variations, marker-assisted selection (MAS), linkage mapping, and genomic fingerprinting (Kalia et al. 2011), to remove linkage drag in backcrossing, and to identify the traits which are not easy to measure by visual observation (Appleby et al. 2009). The comparison of different aspects of generally used molecular markers is given in Table 4.3. A narrow genetic base is reported in cotton by several workers using different molecular markers (Tatineni et al. 1996; Brubaker and Wendel 1994; Multani and Lyon 1995; Pillay and Myers 1999; Ullah et al. 2012).

Table 4.3 Comparison of marker systems in cotton

Marker	Template DNA quantity	Template DNA quality	Genetics	Cost	Reliability
RFLPs	High	High	Codominant	High	High
RAPDs	Low	High	Dominant	Low	Low
ISSRs	Low	Medium	Dominant	Low	Medium
SSRs	Low	Moderate	Codominant	Low	High
AFLPs	Medium	Moderate	Dominant	Moderate	High
SNPs	Low	High	Codominant	Low	High
GBS	Low	High	–	Low to moderate	High

(Adopted from Malik et al. 2014)

Narrow genetic base and complex allotetraploid genome of cotton (*Gossypium hirsutum* L.) is stimulating efforts to avail required polymorphism for marker based breeding. The availability of draft genome sequence of *G. raimondii* and *G. arboreum* and next generation sequencing (NGS) technologies facilitated the development of high-throughput marker technologies in cotton (Malik et al. 2014). Tetraploid genome of cotton is relatively large and contains about 2200–3000 Mb of DNA (Arumuganathan and Earle 1991; Paterson and Smith 1999). The intraspecific DNA polymorphism is low in this species (Tatineni et al. 1996; Brubaker and Wendel 2001), which makes it a challenging crop for development of molecular markers. There is an undeniable need for highly polymorphic molecular markers if progress in plant breeding is to be made using marker-assisted breeding programs. A comparison of molecular marker systems in cotton is made in Table 4.3.

4.12.1 Use of Molecular Markers for Studying Genetic Diversity in Cotton

The success of any breeding program mainly depends on the availability of the genetic diversity in the germplasm resources. Understanding of the genetic relationships among plant genotypes is significant to know the complexity of available germplasm, to discover the differences in available genotypes and to build up useful conservation plans (Dahab et al. 2013). Thus, evaluation based on the molecular markers can give valuable insight into the genetic structure of a plant population, which helps in the development of new varieties (Russell et al. 1997).

RAPD and ISSR techniques have been utilized to analyze genetic diversity and hybridization and for the incident of somaclonal variations in various crops involving cotton (Vafaie-Tabar et al. 2003; Mehetre et al. 2004a; Rana et al. 2007; Sheidai et al. 2008, 2010). Five prominent studies were conducted to evaluate genetic diversity using RAPD markers during 1990s. Genetic diversity of 16 elite homozygous genotypes obtained from the interspecific hybridization was studied using 80 RAPD markers (Tatineni et al. 1996). RAPD markers were used to differentiate the *G. hirsutum* lines from the *G. arboreum* (Iqbal et al. 1997). Similarly, 25 short duration genotypes of cotton were analyzed using arbitrary primers (Wang et al. 1997). Later, (Khan et al. 2000) studied genetic diversity of 31 *Gossypium* species, 3 subspecies, and 1 interspecific hybrid using 45 RAPD primers and the results showed that genetic relationship of many species is related to the center of origin. Recently, genetic diversity in 18 cotton genotypes of Pakistan studied by 5 RAPD primers showed that two diverse genotypes of cotton (CIM-240 and CIM-443) have resistance against cotton leaf curl virus (Mumtaz et al. 2010).

AFLP technique was also used to distinguish the differences among diploid and tetraploid species of cotton by utilizing the variations in ribosomal RNA genes (Pillay and Myers 1999). The genetic diversity between the upland cotton, wild species (*G. raimondii*, *G. thurberii*, and *G. sturtianum*), and their BC_3 progenies was evaluated using AFLP markers (Vroh et al. 1999). Intra- and interspecific relatedness of the *G. barbadense*, *G. arboreum*, *G. raimondii*, and *G. hirsutum* are determined by AFLPs which demonstrated its usefulness for genetic relatedness across wide range of species (Abdalla et al. 2001). The relationship between the parents and 4-day neutral backcross generations of cotton was determined using 43 AFLP markers (Zhong et al. 2002). Comparative study was conducted to evaluate AFLP and RAPD techniques using 16 diploid cotton genotypes and it was concluded that AFLP markers are more efficient for polymorphism detection and for analyzing of genetic diversity as compared to RAPDs (Rana and Bhat 2004). Similarly, genetic diversity of 26 Tanzanian cotton genotypes (*Gossypium hirsutum* L.) was studied using the AFLP markers (Lukonge et al. 2007). The results of this study indicated the high values of genetic similarity which show the lower genetic diversity among Tanzanian cotton cultivars. Reference (Myers et al. 2009) mapped 98 AFLP markers and assigned 22 distinctive chromosomal positions using cytogenetic deletion stocks. Mapping information enhanced the utilization of AFLPs and can be used to saturate the existing marker frequency over different chromosomes.

In cotton, SSRs are considered as a new class of DNA markers which hastened cotton genetic diversity and mapping studies (Preetha and Raveendren 2008) and are important source to observe the transcribed genes (Saha et al. 2003). There are multiple reports about using the SSR markers for genetic diversity. Liu et al. (2000) identified 71 SSR loci with 65 primer pairs and placed them on distinctive chromosomes of cotton. Genetic diversity among U.S. and Australian cultivars, and day neutral lines of *G. hirsutum* was also analyzed by SSR markers (Gutierrez et al. 2002).

Further, saturation of SSR markers was extended by the addition of 204 markers which exhibited 261 segregating bands giving rise to 233 mapped loci in cotton (Nguyen et al. 2004). Interspecific polymorphism between *G. barbadense* and *G. hirsutum* was also studied using SSR markers and results showed that polymorphism between species was high but it was low within species (Rungis et al. 2005). Frelichowski Jr. et al. (2006) developed new SSR markers, analyzed the status of

23 chromosomes and found that the interloci distance was 4.9 cM. Diversity among 52 different *G. hirsutum* cultivars was studied by 31 SSR primer pairs and successfully discriminated the 52 cultivars through broader allelic coverage (deMagalhaes Bertini et al. 2006). Similarly, genetic diversity of 43 upland cotton varieties (Chen and Du 2006), 56 sea island cotton accessions, (Wang et al. 2011) 19 Bt cotton genotypes (Ullah et al. 2012), 50 representative Pakistani genotypes (Dahab et al. 2013), and 193 upland cotton cultivars (Fang et al. 2013) were evaluated using 36,237,104,70, and 448 SSR markers, respectively. SSRs have also been used to assess the genetic purity of the cotton hybrids (Selvakumar et al. 2010) and demonstrated as an effective tool for hybrid identification.

Recent developments in next generation sequencing (NGS) and RNA-seq technology have generated high-throughput sequence data which facilitated the identification of SNPs as effective and highly saturated markers for genetic studies in cotton. Genetic variations within and between the different species of cotton have been characterized by 1000 SNPs and 279 In-Dels from the 270 and 92 loci segregating in *G. barbadense* and *G. hirsutum* to provide mapped molecular markers for crosses within species and introgression of foreign germplasm in cotton (van Deynze et al. 2009). A genome reduction experiment based on the restriction site conservation (GR-RSC) and previously generated assembly of express sequence tags (ESTs) were used to discover the SNPs in four accessions of *G. hirsutum* and *G. barbadense*. A total of 11,834 and 1,679 nongenic and 4,327 genic SNPs were identified in the GR-RSC and EST assemblies using high conservation parameters. The KASPar assays were used to target the 1,052 (704 nongenic and 348 genic) genome-specific SNPs between the *G. hirsutum* accessions (Byers et al. 2012). The assay then tested for the Mendelian segregation ratio in the F_2 population derived from a cross of upland cotton (*G. hirsutum*) cultivars.

4.13 QTLs Mapping for Important Economic Traits in Cotton

The regions in genomes to have genes linked with a quantitative trait are known as quantitative trait loci, QTLs (Collard et al. 2005), and the process of developing linkage maps and performing QTL analysis is referred to as QTL mapping (Paterson 1996). QTL analysis stands on the principal of identifying a connection among phenotype and genotype of markers (Collard et al. 2005).

RFLPs have been widely used to map genes of economic interest in cotton. Previously, the RFLP map of *G. hirsutum* and *G. barbadense* was used to map 14 QTLs for fiber-related traits (Jiang et al. 1998). Similarly, genes influencing density of stem and leaf trichomes (Wright et al. 1999), high gossypol plant, and low seed gossypol contents (Vroh et al. 1999) were confined by RFLP markers. Jiang et al. (1998) developed an RFLP map of 261 markers distributed among 26 linkage groups using F2 plants from an interspecific cross. Another genetic linkage map was developed using RFLP markers, and 26 QTLs were recognized for agronomic and fiber quality traits (Ulloa and Meredith Jr. 2000). Later on, RFLP-based QTL mapping was extended to leaf chlorophyll contents (Saranga et al. 2001). Backcross population of *G. hirsutum* and *G. barbadense* was used to map 28, 9, and 8 QTLs for fiber length, length uniformity, and short fiber contents, respectively, using the 262 RFLP markers (Chee et al. 2005).

RAPDs have also been widely used for QTL mapping in cotton; however, lack of reproducibility and unknown chromosomal positions remained main disadvantages which restricted the use of RAPDs in advanced studies. Kohel et al. (2001) used 85 RAPD markers and identified 13 QTLs associated to the fiber quality in the F_2 population derived from the *G. hirsutum* and *G. barbadense* cross. There are numerous studies on using the RAPDs for QTL mapping along with other molecular markers. An extensive SSR genotyping was conducted over F_2 populations from three diverse upland cotton genotypes using 1378 markers and 39 fiber-related QTLs were identified (Shen et al. 2005). Recombinant inbred lines (RILs) are also important mapping populations and several QTLs related to plant architecture (Wang et al. 2006b), yield (Zhang et al. 2013), and fiber quality (Zhang et al. 2005) have been identified in upland cotton using RILs. About 31 QTLs linked to the yield and fiber quality traits are detected by wide array of SSR and EST-SSR markers (6123) in four-way cross populations developed from the four inbred lines of *G. hirsutum* (Qin et al. 2008). A genetic linkage map of the tetraploid cotton was developed using 1601 pairs of SSR and 247 SNP markers (Yu et al. 2012). The genetic map consisted of the 2072 loci covering 3380 cm of the cotton genome. Two F_2 populations were generated by the crosses of upland cotton cultivars and 4083 SSR markers were used for QTL analysis, which detect 54 QTLs linked to early maturity (Li et al. 2013).

A total of 144 primer combinations of AFLPs and 150 of SSRs were used to detect 28 QTLs related to the fiber traits (Mei et al. 2004). To know the significant threshold for the LR statistics, permutation tests were carried out after which 7 QTLs remain significant. RIL lines developed from the intraspecific cross of upland cotton are used to detect the 12 epistatic and 4 main QTLs related to the plant architectural traits by 2130 SSR, 2 RAPD and 1 SRAP markers (Wang et al. 2006b).

Studies in *Gossypium* have focused on QTL which are involved in fiber strength, length, uniformity, micronaire, color, disease resistance, fruiting nodes, boll weight and number, yield, seed oil and protein content, leaf morphology, and various seed related traits (Lacape et al. 2010; Rong et al. 2007). Numerous independent studies have reported QTL pertaining to all of these traits using independent or updated linkage maps from different or the same segregating populations (Said et al. 2013).

Said et al. (2013) utilized the published QTL information for various traits and used a mixture of more advanced mapping techniques including composite interval mapping (CIM) and multiple QTL mapping (MQM) for comprehensive meta QTL analysis. In their study, a total of 1,223 QTL from 42 different QTL studies in *Gossypium* were surveyed and mapped using Biomercator V3 based on the *Gossypium* consensus map from the Cotton Marker Database. They have identified putative QTL clusters via meta-analysis which may be useful for breeding programs

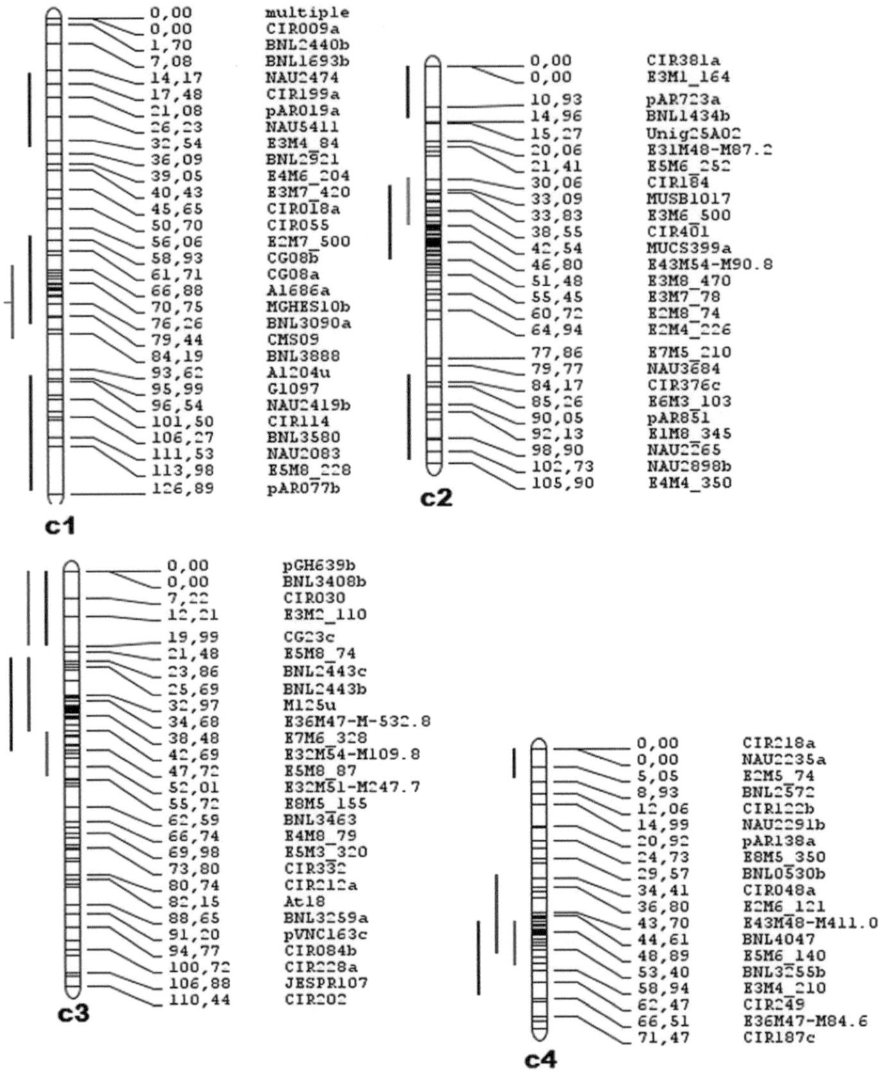

Fig. 4.4 QTL Clusters and Hotspots for various characters identified in Cotton by Said et al. (2013)

and future studies involving *Gossypium* QTL. The presence of QTL clusters and hotspots indicated consensus regions across cultivated tetraploid *Gossypium* species, environments, and populations which contain large numbers of QTL, and in some cases multiple QTL associated with the same trait termed a hotspot. Their study combined two previous meta-analysis studies and added all other currently available QTL studies, making it the most comprehensive meta-analysis study in cotton to date (Fig. 4.4).

Fang et al. (2014) identified 131 fiber QTLs and 37 QTL clusters in a random-mated recombinant inbred (RI) population. These QTLs were responsible for the combined phenotypic variance ranging from 62.3 % for short fiber content to 82.8 % for elongation. Two major QTL clusters were observed on chromosomes 7 and 16. Comparison of these 131 QTLs with the previously published QTLs indicated that 77 were identified before, and 54 appeared novel.

Conclusively, huge arrays of QTLs have been identified using multiple molecular marker technologies. Description of stable QTL from diverse generations, common QTL from various populations and homologous QTLs raises the information on the genetic base. Information about distribution of important QTLs in the genome of cotton is very important and promises the future strategy for marker-assisted breeding. Cotton Gen serves as an important database for such information and currently this database has 988 QTLs for 25 different traits (http://www.cottongen.org/data/qtl) which can be surveyed according to objectivity.

4.14 Genome Wide Association Studies (GWAS) in Cotton

Association mapping, also known as linkage disequilibrium (LD) mapping, has appeared as a tool to determine the variation in complex traits using historical and evolutionary recombination actions at the population level (Nordborg and Tavare 2002). In association mapping, nonstructured populations are phenotyped and genotyped to identify the trait associated with marker (Myles et al. 2009). This results into capture of wider recombination and higher resolution mapping as compared to linkage mapping (Zhu et al. 2008). The applications of association mapping for cotton assist extensive employment of natural genetic diversity conserved within the worldwide collections of cotton germplasm (Abdurakhmonov 2007), as in other plant germplasm resources. Turning the efforts of gene-tagging from biparental QTL mapping to LD-based association study promise the productive employment of ex situ conserved genetic diversity of global germplasm resources of cotton (Abdurakhmonov et al. 2008). The cotton genome may need few numbers of markers for productive associating mapping of complex traits, which is also reported for other crops (Barnaud et al. 2006). Regarding the tetraploid genome of cotton with a total recombination length of about 5,200 cm and an average 400 kb per cm (Paterson and Smith 1999), the LD block sizes of ~5–6 cm distance is sufficient to conduct an association mapping of different traits that would require a maximum of ~1,000 polymorphic markers for successful and reliable association mapping (Abdurakhmonov et al. 2009). Extent of genome-wide LD and association mapping of fiber quality traits were reported using 95 SSR markers in 285 exotic accessions of *G. hirsutum* comprised of 208 landraces and 77 varieties (Abdurakhmonov et al. 2008). Similarly, LD-based association mapping was conducted for fiber quality traits in 335 *G. hirsutum* germplasm using 202 SSR markers (Abdurakhmonov et al. 2009). Progress in genome sequencing technology provides an opportunity to produce large size genotypic data, which supports association mapping over QTL mapping and because of this association mapping is becoming more common (Edwards and Batley 2010).

4.15 Marker-Assisted Selection (MAS) in Cotton

Marker-assisted selection (MAS) is a procedure by which a phenotype is selected on the basis of genotype of a marker (Collard et al. 2005). Selecting the plants in the segregating population that have the suitable genes combinations is the important component of plant breeding (Weeden et al. 1993). Once the markers tightly linked to the genes have been detected, breeders may use particular DNA marker to identify the plants carry the genes (Young 1996). The effectiveness and cost of MAS are influenced by the marker technique; therefore, it must be selected carefully (Coryell et al. 1999). During the past two decades, RAPD techniques have been used for MAS for getting the glanded plants and glandless seeds in the interspecific population of *G. sturtianum* and other species (Mergeai et al. 1998). It was exposed that the DNA markers connected to the major QTL (QTLFS1) for fiber strength could be utilized in MAS to increase fiber strength of commercial varieties in segregating populations (Zhang et al. 2003). Some RAPD markers were developed into locus specific sequence characterized amplified region (SCAR) markers to screen the BC1F4 upland cotton. For example, SCAR 1920 marker for the major fiber strength QTL was developed and has been used for selecting desirable genotypes (Guo et al. 2003). Screening of the SNPs which are mapped on chromosome 10 recognized extra 3 SNP markers that were associated with blue disease resistance gene (Cbd) which were employed to efficiently characterize a trait allowing MAS for strong levels of blue-disease-resistance in cotton breeding programs (Fang et al. 2010).

Numerous QTLs and markers linked with traits have been identified in cotton for utilizing in MAS. For example, QTLs for dense leaf pubescence in cotton (Wright et al. 1999) and QTLs related with drought (Saranga et al. 2001; Paterson et al. 2003; Ullah 2009) have been identified. Similarly, DNA markers associated with nectariless, hairiness, and red leaf color traits were reported (Rahman et al. 2002; Ali et al. 2009a, b). Markers linked with a gene which restores male-fertility in upland cotton have further been mapped (Lan et al. 1999). Useful markers linked with two restorer genes (Rf1 and Rf2) for developing restorer parental lines were detected in two cotton lines of D2 genome (Zhang and Stewart 2004; Feng et al. 2005). Markers associated with genes conferring resistance to CLCuD (Rahman et al. 2002, 2006) were used in probing cotton plants containing the genes in succeeding generations, resulted into the development of two resistant cotton lines NIBGE-2 (Rahman et al. 2008) and NIBGE-115 (Rahman et al. 2008). QTLs for Fusarium wilt are a handy tool for conducting MAS (Wang et al. 2009).

DNA markers associated with fiber quality traits are useful for MAS in cotton (Zhang et al. 2003; Asif 2010). A total of nine DNA markers (three SSRs and six random amplified polymorphic DNA (RAPDs)) mapped on one linkage group which were found to be linked with two QTLs for fiber strength. In cotton, molecular breeding programs screening for QTLs linked with fiber strength has been successfully conducted with markers (Guo et al. 2003). Chen et al. (2009) further mapped this major fiber strength QTL on Chr-24 (D8) recently. In MAS for high-quality lint production, markers linked with the QTLs identified in *G. barbadense* are more helpful (Chee et al. 2005b; Mumtaz 2007). Introgression of loci and or genomic regions derived from *Gossypium barbadense* into *G. hirsutum* was tracked by SSRs, resulted in a 2–3-mm increment in fiber length (Mumtaz 2007). Similarly, AFLPs linked with fiber and agronomic traits have also utility in cotton breeding (Jixiang et al. 2007). In another study, a total of 56 QTLs (LOD > 3.0) linked with 14 different agronomic and fiber traits and one QTL linked with fiber elongation were found which can be helpful in MAS (Wu et al. 2009).

4.15.1 Fiber Genes—Exploring Unique Genetic Pathways

The cotton fiber is an outstanding single-celled model system helpful in studying processes such as cellulose biosynthesis and cell expansion. They differentiate from the protoderm of developing seeds, and are unicellular and unbranched. It has been reported that greater than one-half million quasi-synchronously elongating fibers usually develop in each boll or ovary (Kim and Triplett 2001). Genes involved in fiber development and secondary wall depositions are helpful in understanding many genetic pathways (Haigler et al. 2005).

4.16 Male Sterility Systems in Cotton

The conventional method of hybrid seed production through hand emasculation and pollination is labour intensive and expensive. Hence, utilization of male sterile lines (Plate 4.3) will help to reduce cost of seed production. In cotton, two types of male sterility viz., genetic male sterility (GMS) and cytoplasmic male sterility (CMS) have been explored thus far (Manickam et al. 2004). Both the systems have been utilized extensively for the commercial hybrid seed production in India and other countries.

4.16.1 Genic Male Sterile System (GMS)

There are various male sterility genes; both dominant as well as recessive, identified in *G. hirsutum*, *G. barbadense*, and *G. arboreum* cotton (Table 4.4). There are more number of recessive genes controlling male sterility in cotton has been identified

Plate 4.3 Flower part showing male sterile (*left*) and fertile (*right*) anthers

Table 4.4 Genic male sterile lines identified in cotton

Gene Symbol	Species	Fertility status
Ms1	*G. hirsutum*	Partially sterile
Ms2	*G. hirsutum*	Completely sterile
Ms3	*G. hirsutum*	Partially sterile
Ms4	*G. hirsutum*	Completely sterile
Ms5 Ms6	*G. hirsutum*	Completely sterile
Ms7	*G. hirsutum*	Completely sterile
Ms8 Ms9	*G. hirsutum*	Completely sterile
Ms10	*G. hirsutum*	Completely sterile
Ms11	*G. barbadense*	Completely sterile
Ms12	*G. barbadense*	Completely sterile
Ms13	*G. barbadense*	Completely sterile
Ms14	*G. hirsutum*	Completely sterile
Ms15	*G. hirsutum*	Completely sterile
Ms16	*G. hirsutum*	Partially sterile
Ms17	*G. hirsutum*	Completely sterile
Ms18	*G. barbadense*	Completely sterile
Ms19	*G. barbadense*	Completely sterile
ams 1	*G. arboreum*	Completely sterile
ar. Ms	*G. arboreum*	Completely sterile

rather than dominant genes. Recessive sterility is controlled by both single as well as duplicate genes, whereas, the dominant sterility is controlled by single gene only. Gregg MS 399 conditioned by duplicate recessive ms5ms5ms6ms6 is the most stable and promising one (Weaver 1968). This is comparatively least affected by environment. The male sterility due to ms3 can be manipulated environmentally; it is completely sterile in hot and dry environment and partially fertile in cool and moist environment. GMS lines developed in tetraploid cotton are listed in Table 4.4.

In India and Pakistan, many *G. hirsutum* genotypes have been converted into GMS background using Gregg MS 399 line. The GMS lines are maintained by sib mating with fertile plants (heterozygous for single loci). In the following generation, the plants segregate into 1:1 ratio for sterile and fertile plants. In the hybrid seed production plot, the fertile plants are rouged out at the time of flowering. The procedure of identifying male fertile plants at flowering is tedious and time consuming. Therefore, it should be helpful to develop a GMS line with indicative character appearing in seed or seedling stage. Even though several marker characters linked with male sterility have been reported by several workers (Quisenberry and Kohel 1968; Leport 1970; Feng 1988; Zhang et al. 1992), none of them have been adopted on a commercial scale due to some drawbacks or the other in its practical utilization.

4.16.2 Cytoplasmic-Genetic Male Sterility Systems (CGMS)

In cotton, CGMS system is developed by introgressing cytoplasm of one species into the nuclear background of another species. From 1970 onward, Meyer in USA introgressed several cytoplasms viz., *G. barbadense, G. tomentosum, G. arboreum, G. herbaceum, G. anomalum, G. longicalyx,* and *G. harknessii* into *G. hirsutum* and developed a CMS system with cytoplasm of $D_{2\text{-}2}$ *G. harknessii* and its fertility restorer dominant 'F' (Meyer 1973). This system was developed by repeated backcrossing, testing, and selection in *G. hirsutum* genome. Genetic analysis proved that its fertility can be restored by one or two dominant genes. One RAPD marker linked with this fertility-restoring gene has been identified (Guo et al. 1998). The stable male sterile *G. hirsutum* cotton developed by Meyer (1975) through introgression of *Gossypium harknessii,* Brandg., has been most commonly used for the development of CMS lines. A single dominant gene (Rf) from *G. harnessii* has been transferred to good combining male parents. However, the system suffers due to yield suppression, low ginning and fiber fineness, induction of female sterility in certain interspecific crosses due to deleterious effects of *harknessii* cytoplasm. Hence, search is on to use more wild species cytoplasm to develop new CMS sources with better restorers (Table 4.5). Considerable success has been achieved using *G. aridum* cytoplasm with better restorers, which are under evaluation. Comparison of GMS and CGMS system in cotton is presented in Table 4.6.

Table 4.5 List of cytoplasmic male sterile lines developed in cotton

Name	Cytoplasm	Fertility expression
C9	*G. anomalum*	Partially sterile
–	*G. arboreum*	Partially sterile
P24-6A	*G. arboreum*	Partially sterile
HAMS 16 277	*G. harknessii*	Sterility, restorer developed
–	*G. hirsutum*	Sterility, restorer developed
104-7A	*G. hirsutum*	Sterility, restorer developed
–	*G. trilobum*	Gametophyte sterile
Jin-A	Unknown	Completely sterile
–	G. *aridum*	Completely sterile

4.16.3 Identification of New Male Sterility Source

In India, two new male sterility sources each in upland and arboreum cottons have been identified since 1993 (Manickam et al. 2004). In upland cotton, wild species *G. aridum* has been identified as the new source of cytoplasmic genetic male sterility at Akola Centre through interspecific hybridization. In upland cotton, a new GMS has been identified through induced mutation from the cultivar Abadhita (10 kR gamma rays + 0.2 % EMS combination). In *G. arboreum*, two male sterility loci (ams1—as spontaneous mutant of cultivar DS 5 and ar.ms from *G. anomalum*) have been identified (Table 4.7).

4.16.4 Chemosterilants

To identify superior hybrid combinations, we have to mobilize a broad range of elite genetic diversity. Flexibility is not possible with fewer GMS, CGMS, and 'R' lines. Overdependence on too limited genetic base will lead to genetic vulnerability. Hence, other methods of inducing male sterility should be explored. Male gametocides lead to artificial, nongenetic male sterility in female parent and provides an answer to this to produce hybrid seed with less labour requirement and cost of seed production.

Few gametocides were tried in India in 1960s like FW-450, Maleic Hydrazide (MH) etc., but were not successful (Bhale 1999). USA and China have made considerable efforts to provide low-cost practical pollination control. Use of 2, 3-dichloroisobutanoate (commercial name FW-450) as a selective gametocides to produce hybrids for the first time in cotton was reported by Eaton (1957). Spraying FW-450 once @ 1.2 % solution in green house on Empire cotton resulted in male sterility. However, extensive testing under field condition revealed lack of selectivity causing both male and female sterility.

Table 4.6 Comparison of two male sterile system in cotton

	CGMS	GMS
Source of male sterility	Only one source of stable cytoplasm, i.e., of *G. harknessii*	Several male sterile genes are available
Stability of male sterility	Highly stable because it is not affected by environmental factors such as temperature and day length	Less stable, and highly influenced by environment factors. Sterile plants become fertile especially under low temperature, i.e., below 16 °C
Fertility of hybrids	Restoration capacity of R lines is poor and hence some of the F_1 plants are found to be sterile	All the F_1 plants are fertile.
Availability of diverse male parents	Restricted because of lack of diversity in source of restorer genes	Unlimited, any genotypes of *G. hirsutum* with dominant Rf genes can be used
Conversion of parents	Both female and male parents are to be converted. Female parents are to be converted into CMS 'A' lines, whereas, male parents are to be converted into 'R' lines	Only female parent of good hybrid combination needs to be converted into GMS lines
Conversion of female parents into MS lines	Relatively easy, simple, and quick.	Difficult, cumbersome, and time consuming. Requires back crossing and selfing in alternate generations and hence more number of generations required for conversion.
Maintenance of male sterile lines	By crossing with isogenic 'B' lines	By sib mating with fertile sibs
Rouging of fertile plants in hybrid seed production plots	Not necessary, all plants are sterile	Plants segregate in 1:1 ratio for sterile and fertile plants, needs to be rouged at the time of flowering
Cost of hybrid seed production	Cheaper than GMS system	Costlier than CGMS system
Amount of hybrid seed produced	More per unit area	Less per unit area than CGMS because of reduction in sterile plant population due to rouging of fertile plants
Adverse effect of sterility genes	Cytoplasmic genes have adverse effect. For example, *harknessii* cytoplasm is susceptible to sucking pests	No adverse effect on yield, quality, or other parameters reported so far

In China, use of dichloropropianic acid, Dalapan, and FW-450 were reported but due to the instability, it was difficult to control the spray dose and phytotoxic effect also noticed to large extent. Since 1981, Potassium 3, 4-dichloro-5-isothiazolecarboxylate (TD-1123) has been tested (Olvey et al. 1981; Loper et al. 1987) in cotton. This chemical is systematically accumulated in bracts and developing bolls and has male

Table 4.7 New male sterility sources of cotton identified in India

Species	Type of male sterility	Source	Method used
G. hirsutum	CMS	*G. aridum*	Interspecific hybridization
	GMS	Abadhita	Induced mutation
G. arboreum	GMS	ams1	Spontaneous mutant (DS5)
		G. anomalum	Interspecific hybridization

gametocidal properties. It also causes some amount of phytotoxicity. The frequency and rate of application of TD-1123 to maintain male sterility depends on the cultivar and temperature (Olvey et al. 1981). L-O-methylthreonine (OMT) also had selective gametocidal activity at 0.45 kg/ha but had phytotoxic effect (Ladyman et al. 1990).

4.16.5 Thermosensitive Genetic Male Sterility (TGMS)

Regarding thermosensitive genetic male sterility system in diploids (*G. arboreum*), five lines have been stabilized in India after raising boll to row progenies and observing for flower fertility/sterility at Nagpur as well as Dharwad (Unpublished). The critical fertility temperatures have been found to be around 18 °C based on correlation of minimum temperature with percentage boll set. TGMS in tetraploids was studied at Mudhol and Nagpur which revealed 40 °C as critical sterility temperature. The inheritance study done in these showed the trait to be monogenic recessive and segregated in 3:1 ratio.

References

Abdalla AM, Reddy OUK, El-Zik KM, Pepper AE (2001) Genetic diversity and relationships of diploid and tetraploid cottons revealed using AFLP. Theor Appl Genet 102(2–3):222–229

Abdurakhmonov IY (2007) Exploiting genetic diversity. Proceedings of the World cotton research conference, vol 1. Lubbock, Tex, USA, pp 10–14

Abdurakhmonov IY, Kohel RJ, Yu JZ et al (2008) Molecular diversity and association mapping of fibre quality traits in exotic *G. hirsutum* L. germplasm. Genomics 92(6):478–487

Abdurakhmonov IY, Saha S, Jenkins JN et al (2009) Linkage disequilibrium based association mapping of fibre quality traits in *G. hirsutum* L. variety germplasm. Genetica 136(3):401–417

Agarwal M, Shrivastava N, Padh H (2008) Advances in molecular marker techniques and their applications in plant sciences. Plant Cell Rep 27(4):617–631

Agbicodo EM, Fatokun CA, Muranaka S et al (2009) Breeding drought tolerant cowpea: constraints, accomplishments and future prospects. Euphytica 167:353–370

Agrama HAS, Moussa ME (1996) Mapping QTLs in breeding for drought tolerance in maize (*Zea mays* L.). Euphytica 91:89–97

Ahmad RT, Malik TA, Khan IA, Jaskani MJ (2009) Genetic analysis of some morpho-physiological traits related to drought stress. Int J Agric Biol 11:235–240

Ali I, Kausar A, Rahman M et al (2009a) Development of genetic linkage map of leaf hairiness in *Gossypium hirsutum* (cotton) using molecular markers. Pak J Bot 41(4):1627–1635
Ali I, Ashraf M, Rahman M et al (2009b) Development of genetic linkage map of leaf red colour in cotton (*Gossypium hirsutum*) using DNA markers. Pak J Bot 41(3):1127–1136
Andersen JR, Lubberstedt T (2003) Functional markers in plants. Trends Plant Sci 8(11):554–560
Appleby N, Edwards D, Batley J (2009) New technologies for ultra-high throughput genotyping in plants. Methods Mol Biol 513:19–39
Arumuganathan K, Earle ED (1991) Nuclear DNA content of some important plant species. Plant Mol Biol Rep 9(3):208–218
Ashraf M, Athtar HR, Harris PJC, Kwon TR (2008) Some prospective strategies for improving crop salt tolerance. Adv Agron 97:45–110
Asif M (2010) Genomic analysis for quality traits in cotton (*G. hirsutum* L) by DNA fingerprinting technology. B Z Univ Multan Pakistan
Balasubramanyan R (1960) Morphology. In: Sethi BL, Sikka SM, Dastur RH (eds) Cotton in India – A Monograph, vol I. Indian Central Cotton Committee, Bombay, India
Balasubramaniam R (1963) *Gossypium*—Taxonomy—Diagnostic description of cultivated species. In: Monograph on Cotton in Madras State. Government of Madras, India
Barnaud A, Lacombe T, Doligez A (2006) Linkage disequilibrium in cultivated grape vine, *Vitis vinifera* L. Theor Appl Genet 112(4):708–716
Basal H, Bebeli P, Smith CW, Thaxton PS (2003) Root growth parameters of converted race stocks of upland cotton and two BC_2 F_2 populations. Crop Sci 43:1983–1988
Basu PS, Ali M, Chaturvedi SK (2007) Osmotic adjustment increases water uptake, remobilization of assimilates and maintains photosynthesis in chickpea under drought. Ind J Exp Biol 45:261–267
Beasley JO (1941) Hybridization, cytology, and polyploidy of Gossypium. Chronica Botanica 6:394–395
Beale HM, Birkett MA, Bruce TJA et al (2006) Aphid alarm pheromone produced by transgenic plants affects aphid and parasitoid behavior. PNAS 103(27):10509–10513
Bennett, MD, Cox AV, Leitch IJ (1997) Angiosperm DNA c-values Database, http://www.rbgkew.org.uk/cval/database1.html
Bennett MD, Smith JB, Heslop-Harrison JS (1982) Nuclear DNA amounts in angiosperms. Proc R Soc Lond B216:179–199
Bhale NL (1999) Heterosis Breeding. In: Sundaram V, Basu AK, Narayanan SS et al (eds) Handbook of cotton in India. Indian Society for Cotton Improvement, Mumbai, pp 57–78
Bolek Y, El-Zik KM, Pepper AE et al (2005) Mapping of *Verticillium* wilt resistance genes in cotton. Plant Sci 168:1581–1590
Boulanger J, Pinheiro D (1971) Evolution de la production au nord-est Brasil. Coton et Fibres Tropicales 26:319–353
Brite EB, Marston JM (2013) Environmental change, agricultural innovation, and the spread of cotton agriculture in the Old World. J Anthropol Archaeol 32(1):39–53
Brito GGD, Sofiatti V, Lima MMDA et al (2011) Physiological traits for drought phenotyping in cotton. Acta Sci Agron 33(1):117–125
Brown MS, Menzel MY (1952) The cytology and crossing behaviour of *Gossypium gossypoides*. Bull Torrey Bot Cl 79:110–125
Brubaker CL, Wendel JF (1994) Reevaluating the origin of domesticated cotton (*Gossypium hirsutum*; Malvaceae) using nuclear restriction fragment length polymorphisms (RFLPs). Am J Bot 81:1309–1326
Brubaker CL, Wendel JF (2001) RFLP diversity in cotton. Genetic improvement of cotton, emerging technologies. Science Publishers, Enfield, pp 81–101
Byers RL, Harker DB, Yourstone SM et al (2012) Development and mapping of SNP assays in allotetraploid cotton. Theor Appl Genet 124(7):1201–1214
Cai Y, Xie Y, Liu J (2010) Glandless seed and glanded plant research in cotton. Agron Sustain Dev 30:181–190

Cattivelli L, Rizza F, Badeck FW et al (2008) Drought tolerance improvement in crop plants: An integrative view from breeding to genomics. Field Crop Res 105:1–14
Chaves MM, Maroco JP, Pereira JS (2003) Understanding plant response to drought from genes to the whole plant. Funct Plant Biol 30:239–264
Chee P, Draye X, Jiang CX et al (2005a) Molecular dissection of phenotypic variation between *Gossypium hirsutum* and *Gossypium barbadense* (cotton) by a backcross-self approach: III. Fibre length. Theor Appl Genet 111:772–781
Chee P, Draye X, Jiang CX et al (2005b) Molecular dissection of interspecific variation between *Gossypium hirsutum* and *Gossypium barbadense* (cotton) by a backcross-self approach: I. Fiber elongation. Theor Appl Genet 111:757–763
Chen G, Du XM (2006) Genetic diversity of source germplasm of upland cotton in China as determined by SSR marker analysis. Acta Genetica Sinica 33(8):733–745
Chen H, Qian N, Guo WZ et al (2009) Using three overlapped RILs to dissect genetically clustered QTL for fibre strength on chro.24 in Upland cotton. Theor Appl Genet 119:605–612
Chowdhury KA, Burth GM (1971) Cotton seeds from the Neolithic in Egyptian Nubia and the origin of Old World cotton. Biol J Linn Soc Lond 3:303–312
Chu C, Henneberry TJ, Radin JW (1995) Effect of irrigation frequency on cotton yield in short season production systems. Crop Sci 35(4):1069–1073
Collard BCY, Jahufer MZZ, Brouwer JB, Pang ECK (2005) An introduction to markers, quantitative trait loci (QTL) mapping and marker-assisted selection for crop improvement: the basic concepts. Euphytica 142(1–2):169–196
Comai L (2005) The advantages and disadvantages of being polyploid. Nat Rev Genet 6:836–846
Conant GC, Wolfe KH (2008) Turning a hobby into a job: how duplicated genes find new functions. Nat Rev Genet 9:938–950
Cook CG (1985) Identifying root traits among MAR and non-MAR cotton, *Gossypium hirsutum* L. cultivars that relate to performance under limited moisture conditions. Master thesis. Texas A&M University, USA
Coryell VH, Jessen H, Schupp JM et al (1999) Allele-specific hybridization markers for soybean. Theor Appl Genet 98(5):690–696
Cushman JC, Bohnert HJ (2000) Genomic approaches to plant stress tolerance. Curr Opin Plant Biol 3:117–124
Dahab AA, Saeed M, Mohamed BB et al (2013) Genetic diversity assessment of cotton (*Gossypium hirsutum* L.) genotypes from Pakistan using simple sequence repeat marker. Austral. J Crop Sci 7(2):261–267
de Magalhaes Bertini CHC, Schuster I, Sediyama T et al (2006) Characterization and genetic diversity analysis of cotton cultivars using microsatellites. Genet Mol Biol 29(2):321–329
Desai A, Chee PW, May OL, Paterson AH (2008) Correspondence of trichome mutations in diploid and tetraploid cottons. J Hered 99:182–186
Doebley JF (1989) Isozymic evidence and the origin of crop plants. In: Soltis DE, Soltis PS (eds) Isozymes in plant biology. Dioscorides Press, Portland, Oreg, pp 165–186
Doebley JF (1992) Molecular systematics and crop evolution. In: Soltis PS, Soltis DE, Doyle JJ (eds) Molecular systematics of plants. Chapman and Hall, New York, pp 202–222
Draye X, Chee P, Jiang CX et al (2005) Molecular dissection of interspecific variation between *Gossypium hirsutum* and *G. barbadense* (cotton) by a backcross-self approach: II. Fibre fineness. Theor Appl Genet 111:764–771
Eaton FM (1957) Selective gametocide opens way to hybrid cotton. Science 126:1174–1175
Edwards D, Batley J (2010) Plant genome sequencing: applications for crop improvement. Plant Biotechnol J 8(1):2–9
Edwards GA, Mirza MA (1979) Genomes of the Australian wild species of cotton. II. The designation of a new G genome for *Gossypium bickii*. Can J Genet Cytol 21:367–372
Edwards GA, Endrizzi JE, Stein R (1974) Genomic DNA content and chromosome organization in *Gossypium*. Chromosoma 47:309–326
Endrizzi J, Ramsay G (1979) Monosomes and telosomes for 18 of the 26 chromosomes of *Gossypium hirsutum*. Can J Genet 21:531–536

Endrizzi JD, Turcotte EL, Kohel RJ (1985) Genetics, cytology, and evolution of *Gossypium*. Adv Genet 23:271–375
Fang DD, Xiao J, Canci PC, Cantrell RG (2010) A new SNP haplotype associated with blue disease resistance gene in cotton (*Gossypium hirsutum* L.). Theor Appl Genet 120(5):943–953
Fang DD, Hinze LL, Percy RG et al (2013) A microsatellite-based genome-wide analysis of genetic diversity and linkage disequilibrium in Upland cotton (*Gossypium hirsutum* L.) cultivars from major cotton-growing countries. Euphytica 191(3):391–401
Fang DD, Jenkins JN, Deng DD et al (2014) Quantitative trait loci analysis of fiber quality traits using a random-mated recombinant inbred population in Upland cotton (*Gossypium hirsutum* L.). BMC Genom 15:397
Feng FS (1988) An introduction to a male sterile germplasm in cotton. China Cottons 15:15–16
Feng CD, Stewart JMD, Zhang JF (2005) STS markers linked to the Rf(1) fertility restorer gene of cotton. Theor Appl Genet 110:237–243
Force A, Lynch M, Pickett FB et al (1999) Preservation of duplicate genes by complementary, degenerative mutations. Genetics 151:1531–1545
Frelichowski JE Jr, Palmer MB, Main D et al (2006) Cotton genome mapping with new microsatellites from Acala "Maxxa" BAC-ends. Mol Genet Genomics 275(5):479–491
Fryxell PA (1979) The natural history of the cotton tribe. Texas A&M University Press, College Station
Gadkari PD (1960) Taxonomy. In: Sethi BL, Sikka SM, Dastur RH (eds) Cotton in India—A monograph, vol I. Indian Central Cotton Committee, Bombay
Gammie GA (1907) The Indian cottons. Mem Dept Agric India Bot Ser 11:2
Geever RF, Katterman FRH, Endrizzi JE (1989) DNA hybridization analyses of a *Gossypium* allotetraploid and two closely related diploid species. Theor Appl Genet 77:553–559
Gopalakrishnan N, Manickam S, Sabesh M et al (2010) All India coordinated cotton improvement project—In the service of Indian cotton farmers. Central Institute for Cotton Research, Regional Station, Coimbatore, India
Gotmare V, Singh P (2004) Use of wild species for cotton improvement in India. ICAC Rec XXII:12–14
Grant V (1981) Plant speciation. Columbia University Press, New York
Grimes DW, El-Zik KM (1990) Cotton. In: Stewart BA, Nielsen DR (eds) Irrigation of agricultural crops. Agronomy series no 30. ASA- CSSASSSA, Madison, WI, USA, pp 741–773
Guinn G, Mauney JR (1984) Fruiting of cotton. I. Effects of moisture status of flowering. Agron J 76:90–94
Guo W, Zhang T, Pan J, Kohel RJ (1998) Identification of RAPD marker linked with fertility-restoring gene of cytoplasmic male sterile lines in upland cotton. Chin Sci Bull 43:52–54
Guo W, Zhang T, Shen X et al (2003) Development of SCAR marker linked to a major QTL for high fibre strength and its usage in molecular-marker assisted selection in upland cotton. Crop Sci 43(6):2252–2256
Guo W, Cai CP, Wang CB et al (2007) A microsatellite-based, gene-rich linkage map reveals genome structure, function and evolution in *Gossypium*. Genetics 176:527–541
Gutierrez OA, Basu S, Saha S et al (2002) Genetic distance among selected cotton genotypes and its relationship with F_2 performance. Crop Sci 42(6):1841–1847
Haigler CH, Zhang DH, Wilkerson CG (2005) Biotechnological improvement of cotton fibre maturity. Physiol Plant 124:285–294
Hao JJ, Yu SX, Dong ZD et al (2008) Quantitative inheritance of leaf morphological traits in upland cotton. J Agric Sci 146:561–569
Harland SC (1932) The genetics of *Gossypium*. Bib Genet 9:107
Harrak H, Azelmat S, Tabaeizadeh Z (2001) Isolation and characterization of a gene encoding a drought induced cysteine protease in tomato (*Lycopersicon esculentum*). Genome 44:368–374
Hilu KW (1993) Polyploidy and the evolution of domesticated plants. Am J Bot 80:1494–1499
Humphreys MO, Humphreys MW (2005) Breeding for stress resistance: general principles. In: Harris PJC (ed) Ashraf M. Abiotic stresses, Plant resistance through breeding and molecular approaches, pp 19–46

Hutchinson JB (1950) Some geographic races of Asiatic cottons. Emp Cott Gr Rev 27:123–127

Hutchinson JB (1951) Intra-specific differentiation in *Gossypium hirsutum*. I. Heredity 5:161–163

Hutchinson JB (1954) New evidence on the origin of the old world cottons. Heredity 8:225–241

Hutchinson JB, Ghose RLM (1937) Studies on crops ecology. I. The composition of cotton crops of Central India and Rajputana. Indian J Agric Sci 7:1

Hutchinson JB, Silow RA, Stephens SG (1947) The evolution of *Gossypium* and the differentiation of the cultivated cottons. Oxford Univ Press, London

Iqbal MJ, Aziz N, Saeed NA et al (1997) Genetic diversity evaluation of some elite cotton varieties by RAPD analysis. Theor Appl Genet 94(1):139–144

Izanloo A, Condon AG, Langridge P et al (2008) Different mechanisms of adaptation to cyclic water stress in two South Australian bread wheat cultivars. J Exp Bot 59(12):3327–3346

Jagannatha Rao C, Marar KS, Santhanam V (1953) A brief review of cotton breeding problems in Madras State with special reference to improvement in yield and quality. Indian Cott Gr Rev 7:48–50

Jiang C, Wright RJ, El-Zik KM, Paterson AH (1998) Polyploid formation created unique avenues for response to selection in *Gossypium* (cotton). Proc Natl Acad Sci USA 95(8):4419–4424

Jiang C, Wright RJ, Woo SS, DelMonte TA, Paterson AH (2000a) QTL analysis of leaf morphology in tetraploid Gossypium (cotton). Theor Appl Genet 100:409–418

Jiang C, Chee P, Draye X, Morrell P, Smith C, Paterson A (2000b) Multi-locus interactions restrict gene flow in advanced generation interspecific populations of polyploid *Gossypium* (cotton). Evolution 54:798–814

Jixiang W, Johnie NJ, McCarty JC et al (2007) AFLP marker associations with agronomic and fibre traits in cotton. Euphytica 153:153–163

Kadir ZBA (1976) DNA evolution in the genus *Gossypium*. Chromosoma 56:85–94

Kalia RK, Rai MK, Kalia S et al (2011) Microsatellite markers: an overview of the recent progress in plants. Euphytica 177(3):309–334

Karihaloo JL, Kumar A (2009) Bt cotton in India—Status Report, 2 e. Asia-Pacific Consortium on Agricultural Biotechnology (APCoAB) and Asia-Pacific Association of Agricultural Research Institutions (APAARI), New Delhi, India

Khan SA, Hussain D, Askari E et al (2000) Molecular phylogeny of *Gossypium* species by DNA finger printing. Theor Appl Genet 101(5–6):931–938

Kim HJ, Triplett BA (2001) Cotton fibre growth in plant and in vitro: models for plant cell elongation and cell wall biogenesis. Plant Physiol 127:1361–1366

Knight RL (1956) The genetical approach to disease resistance in plants. Emp Cott Gr Rev 33:191–196

Kohel RJ, Yu J, Park YH, Lazo GR (2001) Molecular mapping and characterization of traits controlling fibre quality in cotton. Euphytica 121(2):163–172

Koizumi M, Yamaguchi-Shinozaki K, Shinzaki K (1993) Structure and expression of two genes that encode distinct drought induce able cysteine proteinases in *Arbidobsis thaliana* L. Gene 129:175–182

Kosmas SA, Argyrokastritis A, Laikas MG et al (2006) Isolation and characterization of drought related trehalose-6-phosphate-synthase gene from cultivated cotton (*Gossypium hirsutum* L). Planta 223(2):326–339

Kranthi KR, Venugopalan MV, Sabesh M, Yadav MS (2011) Vision 2030. Central Institute for Cotton Research, Nagpur, India

Lacape JM, Nguyen TB (2005) Mapping quantitative trait loci associated with leaf and stem pubescence in cotton. J Hered 96:441–444

Lacape JM, Wery J, Annerosa DJM (1998) Relationship between plant and soil water status in five field-growing cotton (*Gossypium hirsutum* L.) cultivars. Field Crops Res 57:29–48

Lacape JM, Nguyen TB, Thibivilliers S et al (2003) A combined RFLP-SSR-AFLP map of tetraploid cotton based on a *Gossypium hirsutum* x *Gossypium barbadense* backcross population. Genome 46:612–626

Lacape JM, Dessauw D, Rajab M, Noyer JL, Hau B (2007) Microsatellite diversity in tetraploid Gossypium germplasm: assembling a highly informative genotyping set of cotton SSRs. Mol Breed 19:45–58

Lacape JM, Llewellyn D, Jacobs J et al (2010) Meta-analysis of cotton fibre quality QTLs across diverse environments in a *Gossypium hirsutum* × *G. barbadense* RIL population. BMC Plant Biol 10:132

Ladyman JAR, Goss JR, Walhood VT, Chai EY (1990) Activity of a novel chemical hybridizing agent in field-grown cotton. Field Crops Res 23:177–186

Lan T-H, Cook C, Paterson A (1999) Identification of a RAPD marker linked to a male-fertility restoration gene in cotton (*Gossypium hirsutum* L.). J Agric genomics 4:1–5

Leake HM, Prasad Ram (1940) Studies in Indian Cotton. Mem Dept Agric India Bot Ser. 6:115–150

Lee J (1984) Cotton. In: Kohel RJ, Lewis CF (eds) ASA/CSSA/SSSA Publishers. Madison, WI, pp 6–24

Leport PL (1970) Complementary studies on the localization of the ms_3 gene of parental pollen sterility, its linkage with yg. Cotton Fibres Tropicales 25:311–314

Li C, Wang X, Dong N et al (2013) QTL analysis for early-maturing traits in cotton using two upland cotton (*Gossypium hirsutum* L.) crosses. Breeding Sci 63(2):154–163

Liu Q, Kasuga M, Sakuma Y et al (1998) Two transcription factors DREB1 and DREB2 with EREBP/AP2 DNA binding domain separate two cellular signal transduction pathways in drought and low temperature-responsive gene expression respectively in *Arabidopsis*. Plant Cell 10:1391–1406

Liu S, Cantrell RG, McCarty JC, Stewart JM (2000) Simple sequence repeat-based assessment of genetic diversity in cotton race stock accessions. Crop Sci 40:1459–1469

Loper GM, Olvey J, Berdel RL (1987) Concentration of the systemic gametocide, TD-1123, in cotton nectar, and honeybee response. Crop Sci 27:558–561

Lukonge K, Hersalman L, Labuschagne MT (2007) Genetic diversity of Tanzanian cotton revealed by AFLP analysis. In: Proceedings of the African crop science conference, vol 8, pp 773–776

Lv SL, Lian LJ, Tao PL et al (2009) Over expression of *Thellungiella halophilla* H(+)- PPase (TSVP) in cotton enhance drought stress resistance of plants. Planta 229(4):899–910

Malik TA, Wright D (1997) Use of net photosynthesis and water use-efficiency in breeding wheat for drought resistance. Pak J Bot 29(2):337–346

Malik W, Ashraf J, Iqbal MZ et al (2014) Molecular markers and cotton genetic improvement: current status and future prospects. Sci World J. doi:10.1155/2014/607091

Manickam S, Gururajan KN, Rajendran TP (2004) Status of hybrid cotton in India. In: Rai M, Singh M, Kumar S (eds) Souvenir on National symposium on harnessing heterosis in crop plants. Indian Institute of Vegetables Research, Varanasi, India 13–15th March 2004, pp 78–86

Mao YB, Cai WJ, Wang JW et al (2007) Silencing a cotton bollworm P450 monoxygenase gene by plant-mediated RNAi impairs larval tolerance of gossypol. Nat Biotechnol 25:1307–1313

Maqbool A, Zahur M, Irfan M et al (2007) Identification, characterization and expression of drought related alpha-crystalline heat shock protein gene (GHSP26) from *Desi* cotton. Crop Sci 47:2437–2444

May OL, Bowman DT, Calhoun DS (1995) Genetic diversity of U.S. upland cotton cultivars released between 1980 and 1990. Crop Sci 35:1570–1574

Mehetre SS, Gomes M, Susan E et al (2004a) RAPD and cytomorphological analyses of F_1, F_2 and amphidiploid (A1) generations of *Gossypium arboreum* × *Gossypium capitis-viridis*. Cytologia 69(4):367–379

Mehetre SS, Patil SD, Aher AR (2004b) Improvement of fibre properties through interspecific hybridization—A review. J Indian Soc Cotton Impr 29:55–64

Mei M, Syed NH, Gao W et al (2004) Genetic mapping and QTL analysis of fibre-related traits in cotton (*Gossypium*). Theor Appl Genet 108:280–291

Menzel MY, Brown MS (1955) Isolation mechanisms in hybrids of *Gossypium gossypoides*. Am J Bot 42:49–57

Mergeai G, Baudoin JP, Vroh BI (1998) Production of high gossypol cotton plants with low gossypol seed from trispecific hybrids including *Gossypium sturtianum*. In: Proceedings of the World cotton research conference, vol 2, pp 206–210, Athens, Greece, September 1998

Meyer VG (1973) A Study of reciprocal hybrids between upland cotton (*Gossypium hirsutum* L.) and experimental lines with cytoplasms from seven other species. Crop Sci 13:439–444

Meyer VG (1975) Male sterility from *G. harknessii*. J Hered 66:23–27

Michaelson MJ, Price HJ, Ellison JR, Johnston JS (1991) Comparison of plant DNA contents determined by Feulgen microspectrophotometry and laser flow cytometry. Am J Bot 78:183–188

Mitra J (2001) Genetics and genetic improvement of drought resistance of crop plants. Curr Sci 80:758–763

Mohammadi M, Taleei A, Zeinali H et al (2005) QTL analysis for phonotypic traits in double haploid population of barley. Indian J Agric Biol 7(5):820–823

Morgante M, Salamini F (2003) From plant genomics to breeding practice. Curr Opin Biotechnol 14:214–219

Multani DS, Lyon BR (1995) Genetic fingerprinting of Australian cotton cultivars with RAPD markers. Genome 38(5):1005–1008

Mumtaz H (2007) Identification of structural and functional genomic markers for fibre quality traits in cotton using interspecific population (*G. hirsutum* x *G. barbadense*). M Phil thesis, QA Univ Islamabad Pakistan

Mumtaz AS, Naveed M, Shinwari ZK (2010) Assessment of genetic diversity and germination pattern in selected cotton genotypes of Pakistan. Pak J Bot 42(6):3949–3956

Mussell H, Staples RC (1979) Stress physiology in crop plants. John and Sons, New York

Myers GO, Jiang B, Akash MW et al (2009) Chromosomal assignment of AFLP markers in upland cotton (*Gossypium hirsutum* L.). Euphytica 165(2):391–399

Myles S, Peiffer J, Brown PJ et al (2009) Association mapping: critical considerations shift from genotyping to experimental design. Plant Cell 21(8):2194–2202

Nepomuceno AL, Oosterhuis DM, Stewart JM (1998) Physiological responses of cotton leaves and roots to water deficit induced by polyethylene glycol. Environ Exp Bot 40(1):29–41

Nguyen TB, Giband M, Brottier P et al (2004) Wide coverage of the tetraploid cotton genome using newly developed microsatellite markers. Theor Appl Genet 109(1):167–175

Niles GA, Feaster CV (1984) Breeding. In: Kohel RJ, Lewis CF (eds) Cotton. American Society of Agronomy, Madison, pp 201–231

Niu C, Hinchliffe DJ, Cantrell RG et al (2007) Identification of molecular markers associated with root knot nematode resistance in upland cotton. Crop Sci 47:951–960

Niu C, Lister HE, Nguyen B, Wheeler TA, Wright RJ (2008) Resistance to *Thielaviopsis basicola* in the cultivated A genome cotton. Theor Appl Genet 117:1313–1323

Nordborg M, Tavare S (2002) Linkage disequilibrium: what history has to tell us. Trends Genet 18(2):83–90

Ohno S (1970) Evolution by gene duplication. Springer, New York

Olvey JM, Fisher WD, Patterson LL (1981) TD-1123: a selective male gametocide. Proc Beltwide Cotton Prod Res, Conf

Pace PF, Cralle HT, El-Halawany SHM et al (1999) Drought induced changes in shoot and root growth of young cotton plants. J Cotton Sci 3:183–187

Pandya PS, Patel CT (1956) Review of further progress in interspecific hybridization work at Surat. In: 7th Conference on cotton growing problems India, ICCC, Bombay, pp 9–14

Pandya PS, Patel CT (1958) Cotton improvement through interspecific hybridization and the role of wild germplasm. In: 8th Conference on cotton growing problems India, ICCC, Bombay

Parida AK, Dagaonkar VS, Phalak MS, Aurangabadkar LP (2008) Differential responses the enzymes involved in proline biosynthesis and of degradation in drought tolerant and sensitive cotton genotypes during drought stress and recovery. Acta Physiol Plant 30:619–627

Pasapula V, Shen G, Kuppu S et al (2010) Expression of an Arabidopsis vascular H(+)-Pyrophosphatase gene AVP1 in cotton improves drought and salt tolerance and increase fibre yield in field conditions. Plant Biotechnol J 9(1):1–12

Paterson AH (1996) Making genetic maps. In: Genome Mapping in Plants. RG Landes Company, Austin, Tex, USA; Academic Press, San Diego, Calif, USA, pp 23–39
Paterson AH, Smith RH (1999) Future horizons: Biotechnology of cotton improvement. Cotton: origin, history. technology and production. Wiley, New York, pp 415–432
Paterson AH, Saranga Y, Menz M et al (2003) QTL analysis of genotype x environment interactions affecting cotton fibre quality. Theor Appl Genet 106:384–396
Paterson AH, Wendel JF, Gundlach H et al (2012) Repeated polyploidization of *Gossypium* genomes and the evolution of spinnable cotton fibres. Nature 492:423–427
Pearsall DM (2008) Plant domestication and the shift to agriculture in the Andes. In: Silverman H and Isbell W(eds) The Handbook of South America Archaeology, Springer
Pillay M, Myers GO (1999) Genetic diversity in cotton assessed by variation in ribosomal RNA genes and AFLP markers. Crop Sci 39(6):1881–1886
Preetha S, Raveendren TS (2008) Molecular marker technology in cotton. Biotechnol Mol Biol Rev 3(2):32–45
Qin H, Guo W, Zhang YM, Zhang T (2008) QTL mapping of yield and fibre traits based on a four-way cross population in *Gossypium hirsutum* L. Theor Appl Genet 117(6):883–894
Quisenberry JE, Kohel RJ (1968) A leaf abnormality for the identification of a genetic male sterile in *Gossypium hirsutum* L. Crop Sci 8:369–370
Quisenberry JE, Jorden WR, Roark BA, Fryrear DW (1981) Exotic cotton as genetic sources for drought resistance. Crop Sci 21:889–895
Quisenberry JE, Roark B, McMichael BL (1982) Use of transpiration decline curves to identify drought-tolerant cotton germplasm. Crop Sci 22:918–922
Rahman M, Shaheen T et al (2012) Cotton genetic resources. A review. Agron Sustain Dev Springer Verlag (Germany) 322):419–432. doi:10.1007/s13593-011-0051-z.<hal-00930526>
Rahman M, Hussain D, Zafar Y (2002) Estimation of genetic divergence among elite cotton (*Gossypium hirsutum* L.) cultivars/genotypes by DNA fingerprinting technology. Crop Sci 42:2137–2144
Rahman M, Hussain D, Malik TA, Zafar Y (2005) Genetics of resistance to cotton leaf curl disease in *Gossypium hirsutum*. Plant Pathol 54:764–772
Rahman M, Ahmed N, Asif M, Zafar Y (2006) Identification of DNA markers linked with cotton leaf curl disease (CLCD). In: International cotton genome initiative (ICGI) workshop, Brasilia Brazil, pp 77–78
Rahman M, Yasmin T, Tabassum N et al (2008) Studying the extent of genetic diversity among *Gossypium arboreum* L. genotypes/cultivars using DNA fingerprinting. Genet Resour Crop Evol 55:331–339
Rahman M, Zafar Y, Paterson AH (2009) *Gossypium* DNA markers types, number and uses. In: Paterson AH (ed) Genomics of cotton. Springer, Dordrecht
Rahman M, Asif M, Shaheen T et al (2011) Marker-assisted breeding in Higher Plants. In: Lichtfouse Eric (ed) Sustainable agriculture reviews, vol 6; Alternative farming systems, biotechnology, drought stress and ecological fertilisation. Springer, Dordrecht, pp 39–76
Rajendran TP, Jain KC (2004) Achievements in cotton research in all india coordinated cotton improvement project. Central Institute for Cotton Research, Regional Station, Coimbatore, India
Rana MK, Bhat KV (2004) A comparison of AFLP and RAPD markers for genetic diversity and cultivar identification in cotton. J Plant Biochem Biotechnol 13(1):19–24
Rana MK, Singh S, Bhat KV (2007) RAPD, STMS and ISSR markers for genetic diversity and hybrid seed purity testing in cotton. Seed Sci Technol 35(3):709–721
Ribaut JM, Jiang C, Gonzalez-DE-Leon D et al (1997) Identification of quantitative trait loci under drought condition in tropical maize 2: yield components and marker assisted selection strategies. Theor Appl Genet 94:887–896
Roberto T, Silvio S (2006) Genomics-based approaches to improve drought tolerance of crops. Trends Plant Sci 11(8):405–412

Rong J, Abbey C, Bowers JE (2004) A 3347-locus genetic recombination map of sequence-tagged sites reveals features of genome organization, transmission and evolution of cotton (*Gossypium*). Genetics 166:389–417

Rong J, Bowers JE, Schulze SR et al (2005a) Comparative genomics of *Gossypium* and *Arabidopsis*: Unravelling the consequences of both ancient and recent polyploidy. Genome Res 15:1198–1210

Rong J, Pierce G, Waghmare V et al (2005b) Genetic mapping and comparative analysis of seven mutants related to seed fibre development in cotton. Theor Appl Genet 111:1137–1146

Rong J, Feltus EA, Waghmare VN et al (2007) Meta-analysis of polyploid cotton QTL shows unequal contributions of subgenomes to a complex network of genes and gene clusters implicated in lint fiber development. Genetics 176(4):2577–2588

Rosenow D, Quisenberry J, Wendt C, Clark L (1983) Drought tolerant sorghum and cotton germplasm. Agric Water Manag 7:207–222

Rungis D, Llewellyn D, Dennis ES, Lyon BR (2002) Investigation of the chromosomal location of the bacterial blight resistance gene present in an Australian cotton (*Gossypium hirsutum* L.) cultivar. Aust J Agric Res 53:551–560

Rungis D, Llewellyn D, Dennis ES, Lyon BR (2005) Simple sequence repeat (SSR) markers reveal low levels of polymorphism between cotton (*Gossypium hirsutum* L.) cultivars. Aust. J Agric Res 56(3):301–307

Russell JR, Fuller JD, Macaulay M et al (1997) Direct comparison of levels of genetic variation among barley accessions detected by RFLPs, AFLPs, SSRs and RAPDs. Theor. Appl Genet 95 (4):714–722

Saha S, Karaca M, Jenkins JN et al (2003) Simple sequence repeats as useful resources to study transcribed genes of cotton. Euphytica 130(3):355–364

Saha S, Raska D, Stelly DM (2006) Upland (*Gossypium hirsutum* L.) x Hawaiian cotton (*G. tomentosum* Nutt. ex Seem.) F_1 hybrid hypoaneuploid chromosome substitution series. J Cotton Sci 10:263–272

Said JI, Lin Z, Zhang X et al (2013) A comprehensive meta QTL analysis for fiber quality, yield, yield related and morphological traits, drought tolerance, and disease resistance in tetraploid cotton. BMC Genom 14:776

Salvi S, Tuberosa R (2005) To clone or not to clone plant QTLs: Present and future challenges. Trends Plant Sci 10:297–304

Saranga Y, Menz M, Jiang CX (2001) Genomic dissection of genotype x environment interactions conferring adaptation of cotton to arid conditions. Genome Res 11:1988–1995

Saranga Y, Jiang CX, Wright RJ et al (2004) Genetic dissection of cotton physiological responses to arid conditions and their inter-relationships with productivity. Plant Cell Environ 27:263–277

Sauer J (1967) Geographic reconnaissance of seashore vegetation along the Mexican Gulf Coast. Coastal Studies Series No. 21. Louisiana State University Press, Baton Rouge, La

Seelanan T, Schnabel A, Wendel JF (1997) Congruence and consensus in the cotton tribe. Syst Bot 22:259–290

Selvakumar P, Ravikesavan R, Gopikrishnan A et al (2010) Genetic purity analysis of cotton (*Gossypium* spp.) hybrids using SSR markers. Seed. Sci Technol 38(2):358–366

Selvam, JN, Kumaravadivel N, Gopikrishnan A, *et al* (2009) Identification of a novel drought tolerance gene in (*Gossypium hirsutum* L.cv) KC3. Commun Biometry Crop Sci 4(1):9–13

Shaked H, Kashkush K, Ozkan H et al (2001) Sequence elimination and cytosine methylation are rapid and reproducible responses of the genome to wide hybridization and allopolyploidy in wheat. Plant Cell 13:1749–1759

Sheidai M, Yahyazadeh F, Farahanei F, Noormohammad Z (2008) Genetic and morphological variations induced by tissue culture in tetraploid cotton (*Gossypium hirsutum* L.). Acta Biologica Szegediensis 52(1):33–38

Sheidai M, Bahrami L, Majd A et al (2010) Genetic diversity in F_2 back-cross progenies of cotton. Gene Conserve 9:167–187

Shen X, Guo W, Zhu X et al (2005) Molecular mapping of QTLs for fibre qualities in three diverse lines in Upland cotton using SSR markers. Mol Breed 15(2):169–181

Shen XL, Van Becelaere G, Kumar P et al (2006) QTL mapping for resistance to root-knot nematodes in the M-120 RNR upland cotton line (*Gossypium hirsutum* L.) of the Auburn 623 RNR source. Theor Appl Genet 113:1539–1549

Shi YH, Zhu SW, Mao XZ et al (2006) Transcriptome profiling, molecular, biological, physiological studies reveals a major role ethylene in cotton fibre cell elongation. Plant Cell 18 (3):651–664

Sikka SM, Joshi AB (1960) Breeding. In: Sethi BL, Sikka SM, Dastur RH (eds) Cotton in India—A monograph, vol I. Indian Central Cotton Committee, Bombay

Silow RA (1944) Genetics and species development in old world cottons. J Genet 46:62

Singh P (2004) Cotton breeding 2 e. Kalyani Publishers, New Delhi

Small RL, Ryburn JA, Cronn RC et al (1998) The tortoise and the hare: choosing between noncoding plastome and nuclear Adh sequences for phylogeny reconstruction in a recently diverged plant group. Am J Bot 85:1301–1315

Song XL, Zhang TZ (2007) Identification of quantitative trait loci controlling seed physical and nutrient traits in cotton. Seed Sci Res 17:243–251

Song XL, Guo WZ, Han ZG, Zhang TZ (2005) Quantitative trait loci mapping of leaf morphological traits and chlorophyll content in cultivated tetraploid cotton. J Integr Plant Biol 47:1382–1390

Stebbins GL (1971) Chromosomal evolution in higher plants. Edward Arnold, London

Steele KA, Price AH, Shashidhar HE, Witcombe JR (2006) Marker-assisted selection to introgress rice QTLs controlling root traits into an Indian Upland rice variety. Theor Appl Genet 112:208–221

Stephens SG (1944) The application of genetics to plant breeding. Trop Agric (Trin) 21:126–129

Stephens SG (1945) Colchicine produced polyploids in *Gossypium*. II. Old World triploid hybrids. J Genet 46:303–312

Stephens SG (1958) Salt water tolerance of seeds of *Gossypium* species as a possible factor in seed dispersal. Am Nat 92:83–92

Stephens SG, Moseley ME (1974) Early domesticated cottons from archaeological sites in central coastal Peru. Am Antiquity 39:109–122

Stewart JM (1995) Potential for crop improvement with exotic germplasm and genetic engineering. In: Constable GA, Forrester NW (eds) Challenging the future, proceedings of the world cotton research conference-1. CSIRO, Melbourne, pp 313–327

Sunil Kumar G, Campbell LM, Puckhaber L et al (2006) Engineering cottonseed for use in human nutrition by tissue-specific reduction of toxic gossypol. PNAS 103(48):18054–18059

Tate JA, Joshi P, Soltis KA et al (2009) On the road to diploidization? Homoeolog loss in independently formed populations of the allopolyploid *Tragopogon miscellus* (Asteraceae). BMC Plant Biol 9:80

Tatineni V, Cantrell RG, Davis DD (1996) Genetic diversity in elite cotton germplasm determined by morphological characteristics and RAPDs. Crop Sci 36(1):186–192

Thi LN, Chi BB (2008) Fine mapping for drought tolerance in rice (*Oryza sativa* L.) Omonrice 16:9–15

Tuinstra MR, Grote EM, Goldsbrouh PB, Ejeta G (1996) Identification of quantitative loci associated with pre-flowering drought tolerance in sorghum. Crop Sci 36:1337–1344

Udall JA, Swanson JM, Haller K et al (2006) Global assembly of cotton ESTs. Genome Res 16(3):441–450

Ullah I (2009) Molecular genetic studies for drought tolerance in cotton. PhD thesis, Quaid-i-Azam University, Islamabad

Ullah I, Iram A, Iqbal MZ et al (2012) Genetic diversity analysis of Bt cotton genotypes in Pakistan using simple sequence repeat markers. Genet Mol Res 11(1):597–605

Ulloa M, Meredith WR Jr (2000) Genetic linkage map and QTL analysis of agronomic and fibre traits in an intraspecific population. J Cotton Sci 4(3):161–170

Vafaie-Tabar M, Chandrashekaran S, Singh RP, Rana MK (2003) Evaluation of genetic diversity in Indian tetraploid and diploid cotton (*Gossypium* spp.) by morphological characteristics and RAPDs. Indian J Genet Plant Breed 63(3):230–234

van Deynze A, Stoffel K, Lee M et al (2009) Sampling nucleotide diversity in cotton. BMC Plant Biol 9 Article 125

Voloudakis AE, Kosmas SA, Tsakas S et al (2002) Expression of selected drought-related genes and physiological responses of Greek cotton varieties. Funct Plant Biol 29:1237–1245

Vroh Bi I, Maquet A, Baudoin JP, (1999) Breeding for low-gossypol seed and high-gossypol plants in upland cotton. Analysis of tri species hybrids and backcross progenies using AFLPs and mapped RFLPs. Theor Appl Genet 99(7–8):1233–1244

Waghmare VN, Rong JK, Rogers CJ et al (2005) Genetic mapping of a cross between *Gossypium hirsutum* (cotton) and the Hawaiian endemic, *Gossypium tomentosum*. Theor Appl Genet 111:665–676

Wang C, Roberts PA (2006) Development of AFLP and derived CAPS markers for root-knot nematode resistance in cotton. Euphytica 152:185–196

Wang XY, Guo WZ, Zhang TZ, Pan JJ (1997) Analysis of RAPD fingerprinting on short-seasonal cotton cultivars in China. Acta Agronomica Sinica 23:669–676

Wang C, Ulloa M, Roberts PA (2006a) Identification and mapping of microsatellite markers linked to a root-knot nematode resistance gene (rkn1) in Acala NemX cotton (*Gossypium hirsutum* L.). Theor Appl Genet 112:770–777

Wang BH, Wu YT, Huang NT et al (2006b) QTL mapping for plant architecture traits in upland cotton using RILs and SSR markers. Acta Genetica Sinica 33(2):161–170

Wang F, Stewart JM, Zhang J (2007) Molecular markers linked to the Rf(2) fertility restorer gene in cotton. Genome 50:818–824

Wang HM, Lin ZX, Zhang XL et al (2008) Mapping and quantitative trait loci analysis of *Verticillium* wilt resistance genes in cotton. J Integr Plant Biol 50:174–182

Wang P, Su L, Qin L et al (2009) Identification and molecular mapping of a *Fusarium* wilt resistant gene in upland cotton. Theor Appl Genet 119(4):733–739

Wang XQ, Feng CH, Lin ZX, Zhang XL (2011) Genetic diversity of sea-island cotton (*Gossypium barbadense*) revealed by mapped SSRs. Genet Mol Res: GMR 10(4):3620–3631

Watt G (1907) The wild and cultivated plants of the world. Longmans, London

Weaver JB Jr (1968) Analysis of a double recessive completely male sterile cotton. Crop Sci 8:597–600

Weeden NF, Timmerman GM, Lu J (1993) Identifying and mapping genes of economic significance. Euphytica 73(1–2):191–198

Wendel JF (1989) New World tetraploid cottons contain old world cytoplasm. Proc Natl Acad Sci USA 86:4132–4136

Wendel JF (1995) Cotton. In: Simmonds N, Smartt J (eds) Evolution of crop plants. Longman, London, pp 358–366

Wendel JF (2000) Genome evolution in polyploids. Plant Mol Biol 42:225–249

Wendel JF, Cronn RC (2003) Polyploidy and the evolutionary history of cotton. Adv Agron 78:139–186

Wendel JF, Brubaker CL, Percival AE (1992) Genetic diversity in *Gossypium hirsutum* and the origin of upland cotton. Am J Bot 79:1291–1310

Wendel JF, Schnabel A, Seelanan T (1995) An unusual ribosomal DNA sequence from *Gossypium gossypioides* reveals ancient, cryptic, inter genomic introgression. Mol Phylogenet Evol 4: 298–313

Wendel JF, Small RL, Cronn RC, Brubaker CL (1999) Genes, jeans, and genomes: Reconstructing the history of cotton. In: van Raamsdonk LWD, den Nijs JCM (eds) Proceedings of the VII International symposium of the international organization of plant biosystematists. Plant Evolution in Man-made Habitats Hugo de Vries Laboratory, Amsterdam, pp 133–161

Wendel JF, Cronn RC, Johnston JS, Price HJ (2002) Feast and famine in plant genomes. Genetica 115:37–47

Wright R, Thaxton P, El-Zik K, Paterson AH (1998) D-subgenome bias of Xcm resistance genes in tetraploid *Gossypium* (Cotton) suggests that polyploid formation has created novel avenues for evolution. Genetics 149:1987–1996

Wright R, Thaxton P, Paterson AH, El-Zik K (1999) Molecular mapping of genes affecting pubescence of cotton. J Hered 90:215–219

Wu Y, Machado AC, White RG et al (2006) Expression profiling identifies gene expressed early during lint fibre initiation in cotton. Plant Cell Physiol 47:107–127

Wu J, Gutierrez OA, Jenkins JN, McCarty JC, Zhu J (2009) Quantitative analysis and QTL mapping for agronomic and fibre traits in an RI population of upland cotton. Euphytica 165:231–245

Yang C, Guo WZ, Li GY et al (2008) QTLs mapping for Verticillium wilt resistance at seedling and maturity stages in *Gossypium barbadense* L. Plant Sci 174:290–298

Ynturi P, Jenkins JN, McCarty JC et al (2006) Association of root-knot nematode resistance genes with simple sequence repeat markers on two chromosomes in cotton. Crop Sci 46:2670–2674

Young ND (1996) QTL mapping and quantitative disease resistance in plants. Ann Rev Phytopathol 34:479–501

Yu JW, Yu SX, Lu CR et al (2007) High-density linkage map of cultivated allotetraploid cotton based on SSR, TRAP, SRAP and AFLP markers. J Integr Plant Biol 49:716–724

Yu JZ, Kohel RJ, Fang DD et al (2012) A high-density simple sequence repeat and single nucleotide polymorphism genetic map of the tetraploid cotton genome. Genes Genomes Genetics 2(1):43–58

Zaitzev GS (1928) A contribution to the classification of *Gossypium*. Trans Turkestan Plant Br Stat No. 12

Zhang JF, Stewart JM (2004) Identification of molecular markers linked to the fertility restorer genes for CMS-D8 in cotton. Crop Sci 44:1209–1217

Zhang TZ, Feng YZ, Pan JJ (1992) Genetic analysis of four genic male sterile line in upland cotton. Acta Gossypii Sinica 4:1–8

Zhang T, Yuan Y, Yu J et al (2003) Molecular tagging of a major QTL for fibre strength in upland cotton and its marker-assisted selection. Theor Appl Genet 106(2):262–268

Zhang ZS, Xiao YH, Luo M et al (2005) Construction of a genetic linkage map and QTL analysis of fibre-related traits in upland cotton (*Gossypium hirsutum* L.). Euphytica 144(1–2):91–99

Zhang L, Li F, Liu C et al (2009) Construction and analysis of cotton (*Gossypium arboreum* L.) drought-related cDNA library. BMC Res Notes 2:120–128

Zhang T, Qian N, Zhu X, *et al* (2013) Variations and transmission of QTL alleles for yield and fibre qualities in upland cotton cultivars developed in China. PLoS ONE 8(2) Article ID: e57220

Zhao J (2002) QTLs for oil content and their relationship to other agronomic traits in a Eurpeon × Chinese oilseed rape population. Diss Grorg-August. University of Goettingen, Germany

Zhao XQ, Xu JL, Zhao M et al (2008) QTLs affecting morph-physiological traits related to drought tolerance detected in overlapping introgressionlines of rice (Oryza sativa L.). Plant Sci 174:618–625

Zhong M, McCarty JC, Jenkins JN, Saha S (2002) Assessment of day-neutral backcross populations of cotton using AFLP markers. J Cotton Sci 6(2):97–103

Zhu JK (2002) Salt and drought stress signal transduction in plants. Ann Rev Plant Biol 53:247–273

Zhu C, Gore M, Buckler ES, Yu J (2008) Status and prospects of association mapping in plants. Plant Genome 1(1):5–20

Chapter 5
Genetic Diversity and Germplasm Patterns in *Brassica juncea*

S.S. Banga and Shashi Banga

Abstract Crop Brassicas, also known by their trade name of rapeseed-mustard, belong to the family Brassicaceae. This family includes about 380 genera and over 3,000 species (Mabberley 1993). Crop Brassicas are the most important *Brassiceae* species that were domesticated as sources of vegetables, condiments, and edible oils. These crops include oilseed forms of *Brassica napus*, *B. juncea*, *B. carinata*, and three ecotypes of *B. rapa*. As per United States Department of Agriculture (USDA) estimates, global rapeseed production during 2014/2015 was estimated to be 71.24 million metric tons (www.worldrapeseedproduction.com). European Union (24.07 MT), Canada (15.55 MT), China (14.70 MT), and India (6.85 MT) were major contributors. Different forms of oilseed brassicas are cultivated throughout the world. Winter type of *B. napus* predominates in Europe, China, and eastern United States. Spring forms of *B. napus* are grown in Canada and Australia. Winter *B. juncea* once occupied large tracks in China, until higher yielding forms of *B. napus* replaced it. Spring types of *B. juncea* are cultivated in the Indian subcontinent. It is also grown as a condiment crop in Europe (Vaughan and Hemingway 1959) and is at present an option for the drier ecologies of Canada, Australia, and even in northern United States. *B. rapa*, once an oilseed crop of global distribution, is now restricted to limited geographies with winter type of *B. napus* replacing it in Europe and higher yielding and disease resistant *B. juncea* replacing it in the Indian subcontinent. *B. juncea* is the principal winter oilseed crop in India. It covers over 90 % of the area under rapeseed-mustard crops. Despite its huge economic significance for Indian subcontinent, yield levels in *B. juncea* are stagnating. This is in spite of large-scale hybridizations and selection programs. Selection efficiencies of the breeding programs are also declining and there is growing concern over the absence of resistance sources for insect and diseases. Superior alleles from primary gene pools and the related species are being explored

S.S. Banga (✉) · S. Banga
Department of Plant Breeding and Genetics, Punjab Agricultural University, Ludhiana, India
e-mail: nppbg@pau.edu

S. Banga
e-mail: shashibanga@yahoo.com

V.R. Rajpal et al. (eds.), *Gene Pool Diversity and Crop Improvement*,
Sustainable Development and Biodiversity 10,
DOI 10.1007/978-3-319-27096-8_5

for transfer into the superior agronomic base in cultivated germplasm. These are needed to impart stability, wider adaptability, stress tolerance, oil quality, and increase productivity to meet the existing and emerging nutritional and industrial uses. This chapter describes the current status of genetic diversity in *B. juncea*.

Keywords Crop Brassicas · Genetic diversity · Germplasm · Genomes · Breeding

5.1 Introduction

5.1.1 Brassica Genomes

Crop Brassicas comprise three diploid species namely, *B. rapa* (AA, $n = 10$), *B. nigra* (BB, $n = 8$) and *B. oleracea* (C genome, $n = 9$). These diploid species diverged from a common ancestor after a triplication event that took place 7.9–14.6 million years ago (Lysak et al. 2005). Of the three primary diploids, *B. rapa,* followed by *B. nigra*, is the most ancient (Gomej-Campo and Prakash 1999). A and C genome diverged less than four million years ago (Inaba and Nishio 2002). Despite existing genome homologies between diploid progenitors (Attia and Robbelen 1986; Parkin et al. 2003), more recent studies converge on the concept of two lineages "oleracea"/"rapa" evolving from one ancestor and "nigra" originating from another (Warwick and Black 1991; Pradhan et al. 1992). The "nigra-oleracea" clades separated 20 Ma. The core "oleracea" lineage that led to evolution of *B. oleracea* and *B. rapa*arose ≈3 Ma in the northeastern Mediterranean, from where ancient forms of *B. oleracea* spread across Europe and *B. rapa* to Asia (Arias et al. 2014). Three allopolyploids: *B. juncea* (AABB), *B. napus* (AACC), and *B. carinata* (BBCC) arose from multiple natural hybridizations between genome donor diploid species (Olsson 1960; Prakash 1973a, b, 1974; Song and Osborn 1992). The evolution of participating genomes in the natural Brassica allo-tetraploids had been influenced strongly by the cytoplasmic background (Prakash et al. 2009). The role of the cytoplasm in mediating transmission frequency of meiotic-driven genetic alterations in the neo-polyploids stands demonstrated in *B. napus* (Szadkowski et al. 2010) as well as *Brassica* hybrids or allopolyploids resulting from crosses between any two of the three diploids crop brassica species (Cui et al. 2012). In *B. juncea,* there is contradiction regarding A-genome having been modified during evolution (Prakash et al. 2009) or B genome (Liu and Wang 2006). In *B. carinata*, B genome is relatively unchanged, but the C genome is considerably altered. In *B. napus,* both A and C genomes seemed to have co-evolved. Synthetic allotetraploids involving any two of the three Brassica genomes show aberrant meiotic behavior and chromosome aberrations during initial generations after polyploidy (Prakash et al. 2009). Natural hybridization events have been the basis of genome evolution of *Brassica* and interspecific crosses. This enabled gene exchanges to contribute significantly to the differentiation within the genus by generating new type or species and also allowing gene exchange across species boundaries. During the domestication and breeding process of each species, divergent selection may have also supplemented the diversity of the corresponding cultivars and crops.

5.2 Origin

Brassica juncea is a crop of antiquity. Its seeds were recovered during archaeological excavations dated to 2300 BC (Parkash 1980). Taxonomic investigations based on morphological diversity and molecular markers, have shown that *B. juncea* arose from multiple independent hybridization events between wild forms of *B. rapa* and *B. nigra* in the Middle East or adjoining areas with sympatric distribution of diploid donors (Burkill 1930; Olsson 1960; Vaughan et al. 1963; Axelsson et al. 2000; Prakash et al. 2009). There is also a view that polyploidization events led to the current agronomic types of *B. juncea* (Prakash and Hinata 1980; Kaur et al. 2014). Vavilov (1949) considered Afghanistan and bordering areas as the primary center of origin of *B. juncea*, with Asian Minor, central or western China and eastern India as secondary centers of diversity. Spect and Diederichsen (2001) considered the area between eastern Europe and China, having parental sympatry as the place of origin. A separate origin of Chinese foliage types and Indian oleifera types were also proposed (Vaughan et al. 1963; Vaughan 1977). Some Indian researchers opined that Indian *B. juncea* types have originated within India (Duhoon and Koppar 1998). However, existence of high allelic diversity in land races from Afghanistan bordering areas of Pakistan (Banga unpub.) brings out the need to re-investigate Afghanistan and the adjoining regions were the primary center of origin as was originally proposed by Vavilov (1949).

5.3 Botanical Classification

B. juncea possess enormous morphotype diversity and end product usage. Five subspecies are recognized. These are: *juncea, crispifolia, foliosa, integrifolia, and napiformis*. In contrast, Spect and Diederichsen (2001) recognized four subspecies, based on varied morphology, quality characteristics and uses. These are: (i) ssp. *integrifolia*, used for leaf vegetable in Asia; (ii) ssp. *juncea*, cultivated primarily for its seed or sometimes as fodder; (iii) ssp. *napiformis*, used as a root-tuber vegetable; and (iv) ssp. *taisai*, stalks and leafs are consumed as vegetables in China. Two races of B. *juncea*, namely Indian and the Oriental races have been described based on the geographic distribution. These were found closer to the *B. rapa* and B. *nigra* progenitors, respectively (Vaughan et al. 1963; Vaughan 1977; Prakash et al. 2009). Sequence variation of nuclear internal transcribed spacer regions of ribosomal DNA (ITS1, 5.8SrRNA and ITS2) from Chinese vegetable types (AB-genome) and its genome donor parents (Xiao-Hua et al. 2007) helped to decipher the molecular evolution patterns of Chinese vegetable forms. Studies revealed two major clades, one carrying four accessions of vegetable mustard, showing close affinity with "*B. nigra*" lineage and the other containing 12 accessions of *B. juncea.* This classification contradicts the suggestions of *B. juncea* as being closer to only the A-genome species. It is likely that *B. juncea* forms derived from China have evolved through different polyploidy events and unidirectional concerted evolution. Existence of two SSR defined groups of *B. juncea,* occurring in overlapping regions of India and China has been reported (Chen et al. 2013; Kaur et al. 2014).

5.4 Cytoplasmic Variation

Despite appreciating the importance of cytoplasm in molding responses of nuclear genomes (Szadkowski et al. 2010; Cui et al. 2012), very few attempts are available that document cytoplasmic variation in a broader assemblage of *B. juncea* landraces and historical cultivars. Until recently, the natural hybridizations between the parental genome donor species were considered unidirectional, with *B. rapa* as the cytoplasmic donor to *B. juncea* (Uchimiya and Wildman 1978; Banga et al. 1983; Erickson et al. 1983; Palmer et al. 1983; Warwick and Black 1991; Pradhan et al. 1992). Most of conclusions were, however, drawn from genotyping small germplasm collections by using neutral markers. The availability of microsatellite markers specific to cytoplasmatic organelles, now permit characterization of larger sets of germplasm to understand phylogenetic relationships. Occurrence of two major plasmotypes in *B. juncea* was recently reported (Kaur et al. 2014). One group primarily included genotypes/species from the "nigra" lineage, and the second group possibly belonged to "rapa/oleracea" lineage. Unexpectedly, the East European *B. juncea* grouped with *B. nigra,* implying *B. nigra* to be the cytoplasmic donor species for test East European collections. It was also suggested that bulk of current *B. juncea* genotypes belonging to the *B. rapa* lineage carried cytoplasm that was not present in any of the *B. rapa* germplasm included in the investigations. It is possible that ancestral *B. rapa* forms that participated in the ancient hybridization events have become extinct. These results also supported the proposition of two distinct but geographically co-existing SSR-defined groups in *B. juncea,* (Chen et al. 2013). Group I comprised East European and some Indian *B. juncea* germplasm, and group II carried most of the germplasm from India, China, and Australia. Two geographic races of oilseed *B. juncea* (Indian and the Oriental races), closer to the *B. rapa* and *B. nigra,* respectively, was proposed earlier (Vaughan et al. 1963; Vaughan 1977). Kaur et al. (2014), have proposed that groups carrying *B. rapa* or *B. nigra* cytoplasms be named as the group-A and group-B, respectively, to recognize their cytoplasmic backgrounds. The coexistence of two plasmon lineages in India and China may explain differential interpretations about A-genome having remained mostly intact, as compared to considerably altered B genome (Prakash et al. 2009) or the A-nuclear genome considerably altered, and. relatively a conserved B genome (Liu and Wang 2006). Such contradictions may result from the type of germplasm resources used.

5.5 Possible Migration Route to India

Prain (1898) had originally proposed that Rai (oilseed mustard) migrated into India from China through a northeastern frontier. This proposition, though supported by Sinskaja (1928), was contradicted by Burkill (1930) and Sun (1970). Due to the absence of any historical data, a very large study was recently conducted (Banga, unpublished). This study was based on the proposition that both plasmotypes of *B. juncea* originated in Middle East or Afghanistan, and migrated to India through ancient trade routes. Over 125 *B. juncea* land races were collected from the point of entry of ancient trade routes to India (Fig. 5.1). Also included were the land races (inc. historical varieties) of *B. juncea*, *B. rapa,* and *B. nigra* from Pakistan, Kashmir and Punjab in NW India and also from NE India. These were genotyped using chloroplast specific SSR primers and nuclear SSR primers, developed using sequence information from adaptation-related candidate genes within or from genomic regions closely linked to flowering and dispersal-related genes. The diversity tree generated from DNA polymorphism obtained from chloroplast specific SSR's allowed clear discrimination between land races collected from NW India and NE India as both formed distinct groups. NW group aligned with *B. nigra* and a B lineage species *B. tournefortii*. NE group comprising *B. juncea* and *B. rapa* was distinct with almost no overlap with NW group.

Coupled with the information from past studies (Chen et al. 2013; Kaur et al. 2014), it is proposed that one group of *B. juncea* land races from north and eastern India; carried *B. nigra* cytoplasm. It has cytoplasmic and genetic affinities with the *B. juncea* genotypes from central-western China. Another groups comprising genotypes from central and western plains of India carried *B. rapa* cytoplasm. It may correspond to the germplasm from eastern China (Fig. 5.2).

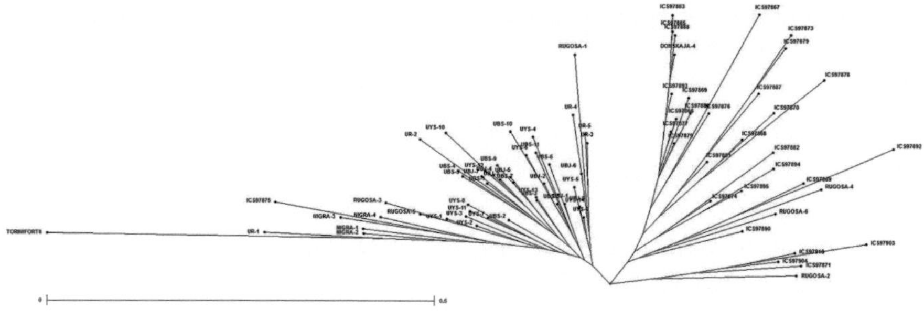

Fig. 5.1 Diversity tree for *B. juncea* landraces collected from NW and NE parts of India

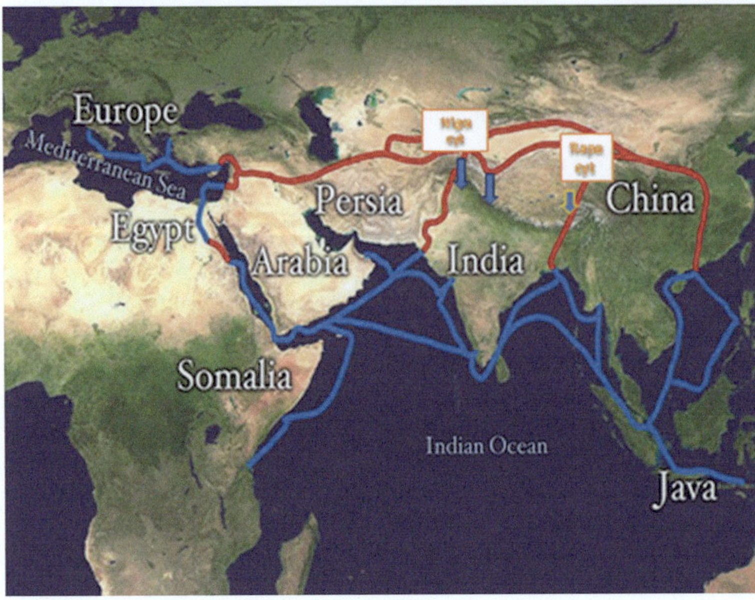

Fig. 5.2 Possible routes of migration of *B. juncea* to India

We propose that NW group of *B. juncea* carrying *B. nigra* cytoplasm, migrated to Indian mainland along the ancient silk route from Afghanistan or adjoining Tibet and Xinjiang area of China, through NWFP (Pakistan), Kashmir and Uttrakhand. The NE group of land races of leafy *B. juncea* that carried *B. rapa* cytoplasm may have arrived in NE from adjacent areas of Tibet, Sichuan or Yunnan provinces of China. We also believe that both forms of *B. juncea,* carrying *nigra* or *rapa* cytoplasm originated in Afghanistan or Middle East, before their migration to China, India and Europe, where these continue to coexist. *B. juncea* carrying "nigra cytoplasm" seems to have greater adaptation to cooler climatic regions. Xian-Hui et al. (1999) have shown that the geographical distribution and ecological environment were important factors affecting genetic diversity. Various landraces tended to cluster together, in line with their eco-geographic distribution. Genetic differences existed between winter and spring land races in *B. juncea*. Greater genetic diversity was found within winter *B. juncea* than within spring types.

5.6 Genetic Diversity and Germplasm Patterns

Genetic diversity can be broadly defined in terms of both the genetic resources and in the context of evolution and domestication. It is a critical resource for improving selection efficiency and hybrid performance for crop improvement. Genetic diversity may include quantitative genetic or allelic variation at gene level. Aside morphological assessments, pedigree information used to compute pairwise coefficients of parentage (COP) as a reliable indicator of genetic diversity in self-pollinating crops (Souza and Sorrells 1989). However, COP estimates are limited by pedigree errors, besides the assumptions of unrelated ancestry and absence of any selection biases (Cox et al. 1986; Souza and Sorrells 1989). Establishing genetic diversity and inter germplasm relatedness allow transforming germplasm constellations into potential gene pools and unravel underlying evolutionary forces. It may also promote conservation and optimal deployment of variation to maximize breeding gains. Till early 1990s; ecogeographic diversities, quantitative genetic assessments or isozyme variations were considered as primary descriptors of genetic diversity. However, with the development of new bio-techniques, molecular markers have come to occupy the central stage. At molecular level, genetic diversity is measured in terms of genotypic or allele frequencies, a proportion of polymorphic loci, besides observed and expected heterozygosities (Petit et al. 1998).

In the context of structured populations, molecular measures of differentiation are based on genetic distances among populations (Nei 1987). Isozymes (Sekhon and Gupta 1995), random amplified polymorphic DNA (RAPD; Jain et al. 1994), inter-simple sequence repeat (ISSR; Quiros et al. 1995; Bornet and Branchard 2004), amplified fragment length polymorphisms (AFLPs; Vos et al. 1995; Negi et al. 2000), restriction fragment length polymorphisms (RFLPs; Song 1995; Cavell et al. 1998) and microsatellites, or simple sequence repeats (SSRs; Tautz 1989) have been routinely used (Zhou et al. 2006; Chen et al. 2008) in crop Brassica species. SSR markers are especially useful due to their multiallelic nature, reproducibility and automation. Knowledge of their chromosome location may be particularly useful in genetic diversity studies. AFLPs offer distinctive advantages due to their high multiplex ratio, especially when complete genome coverage is required (Pejic et al. 1998). Molecular markers reflect variation at DNA level and cover coding and none-coding regions of genome. They are environment, development stage, tissue and organ neutral. The relationship between genetic relatedness estimated based on the neutral markers and that on the basis of the quantitative traits is generally low.

5.7 Morphological Characterization

Measuring morphological traits provide a simple way to quantify genetic variation and at the same time assessing genotype performance. Such estimates of quantitative genetic variation are generally based on characterization of phenological and productive traits. Analysis of data from breeding families is also used to estimate additive genetic variance or heritability for the polygenic traits (Falconer and Mackay 1996). However, such assessments are limited by poor polymorphism, low heritability, and genotype x environmental interactions. D^2 multivariate analysis is the method of choice, especially if, the numbers of genotypes for estimating genetic divergence are large. Association of genetic diversity with heterosis is routinely attempted, although past studies are not unambiguous regarding the correlation between genetic and geographic diversities (Gupta et al. 1991) or between genetic diversity and heterosis (Gupta et al. 1991; Krishnapal and Ghose 1992). Thakur and Zarger (1989) showed that high magnitude of desirable heterosis was not directly related to extreme parental divergence. This was also suggested by Ali et al. (1995). One of the most significant findings from assessment of morphological diversity in *B. juncea* was the discovery of two major gene pools, namely the Indian and the East European gene pools. Pradhan et al. (1993) assessed genetic divergence in *B. juncea* at the morphological level using 25 accessions of Indian and East European origins and some resynthesized *B. juncea* lines produced earlier (Prakash 1973a, b). The accessions were grouped into different clusters based on nine agronomic and yield traits. Maximum divergence was observed between Indian and the East European types. East European lines were found to be the sources of several useful traits like branching, pod number, resistance to white rust, seed quality, oil content, and yellow seed coat (Pradhan et al. 2003). Indian germplasm pool is characterized by early flowering, shorter stature, longer pods, bolder seeds, and superior resistance to pod shattering (Ramchiary et al. 2007). Pradhan et al. (1993) also studied heterosis in 10 × 10 diallel crosses where the parents were selected on the basis of genetic divergence. Maximum heterosis was recorded between the genetically diverse Indian × East European crosses. These findings led to the development of the first commercial hybrid, DMH 1 in *B. juncea*. It was based on a cross between an Indian gene pool line, Pusabold, and an East European genotype, EH-2 (Sodhi et al. 2006). In another study (Shalini et al. 2000), morphological assessment of 81 genotypes helped to establish genetic diversity using D^2 statistic. Days to 50 % flowering, plant height, and the number of siliques per plant were found to be the best contributors to genetic divergence and these were proposed as phenotypic indicators of divergence. In another study involving D^2 statistics (Goswami and Behl 2006); days to 50 % flowering, days to maturity, plant height, main shoot length, siliques on main shoot, seed yield/plant, and oil content were considered primary contributors to the genetic diversity. Morphological characterization of 90 genotypes of Indian mustard for 15 different traits enabled their grouping into nine clusters based on statistics (Lodhi et al. 2013). Number of primary branches/plant, number of secondary branches/plant, silique angle, siliques on main shoot, primary branch angle, 1000-seed weight, and main shoot length contributed the maximum to genetic diversity. Almost similar results were obtained by Shekhawati et al. (2014), who grouped 60 genotypes into 13 clusters based on D^2 analysis. The traits like seed yield per plant (27.33 %), number of silique per plant (24.13 %), and length of main branch (21.17 %) were major determinatorsof genetic divergence.

In a very interesting study, Singh et al. (2013) grouped 62 commercial mustard cultivars, bred in India, into eight clusters on the basis of 18 morphological and six computed variables. Cluster 1 included 50 genotypes. The narrow genetic base of mustard varieties in India was attributed to their close ancestry (Chauhan et al. 2011; Singh and Chauhan 2010). Acharya and Swain (2003) did not consider heterosis as a function of genetic diversity on an absolute scale. However, parents from clusters separated by a medium range of intercluster D^2 values favored higher heterotic for seed yield, pods per plant and seeds per pod. Apparently, intermediate diversity was more important for heterotic combination (Ghosh and Gulati 2002). There are also reports those widely diverse genotypes which may also reside in the same geographical region (Verma et al. 2000). Kumar et al. (2007) grouped 25 genotypes into seven clusters. Varietal hybridization between genotypes belonging to different clusters was advocated to increase the probability of getting promising recombinants.

5.8 Molecular Characterization of Genetic Diversity

Although most of the loci affecting the yield and its components are still to be fully mapped, the average allele differences between target germplasm can be estimated by molecular markers. Characterization of the genetic variations permits breeders to narrow the search for new alleles within a germplasm of interest. Genotypic information from germplasm collections also helped to assign lines and populations to heterotic groups, monitor changes in allele frequencies during varietal development, understand parallel variation in germplasm collections from different eco-geographies, study the evolutionary history of wild relatives and recreating selection history (Warburton and Hoisington 2001).

Sekhon and Gupta (1995) used isozymes to document genetic diversity and investigated the use of isozyme derived genetic distance along with pedigree-based diversity between parents for predicting F_1 heterosis in Indian mustard. The results indicated that isozyme patterns were better predictors of heterosis in crosses based on morphological diversity and pedigree information. DNA polymorphism generated by RAPD markers was used to document genetic variation among 12 Indian and 11 East European *B. juncea* accessions Jain et al. (1994). Cluster analysis identified two groups, group-A comprised only the East European germplasm and group-B included all the Indian and four introduced genotypes. The correlation between genetic distance and heterosis was not consistent. These results clearly suggested that genetic diversity is useful for understanding relationships among *Brassica* genotypes despite the absence of any direct correlation between the genetic distance and heterosis. Burton et al. (2004) used AFLP markers to assess the genetic diversity of 77 canola-quality *B. juncea* breeding lines derived from the Canadian and the Australian breeding programs and 15 non-canola-quality genotypes from India, Russia, Canada, and China. Polymorphism generated by AFLP markers led to the formation of two groups, with Indian genotypes being distinct, from rest of the genotypes. Hybrids between the two pools showed yield heterosis. Crosses made between genetically divergent Indian × East European lines also showed significant level of heterosis; on the other hand, crosses made within the Indian or the East European genotypes recorded either negative or insignificant yield heterosis. Maintaining the overall divergence of the two pools is necessary for heterosis breeding in *B. juncea*. Therefore, a systematic genetic analysis using molecular markers and subsequent marker-assisted precision introgressions needs to be undertaken for the manipulating complex traits. Genotyping with AFLP markers allowed grouping of 21 natural *B. juncea* accessions originating from India, Russia, Australia, and Canada and nine resynthesized lines (Srivastava et al. 2001) into three distinct diversity groups. All the Indian types formed one group; the resynthesized lines formed a separate cluster and the lines from Australia, Canada, and Russia (all East European types) formed the third cluster. Genetic diversity was also established between 50 genotypes of Indian genotypes of mustard using agronomic data and RAPD genotyping (Singh et al. 2013). Maximum variability was observed for biological yield followed by seed yield. Agronomic traits and molecular markers allowed discrimination of genotypes into different clusters, however, clustering pattern varied. There was no relationship between geographic origin and clustering of genotypes. (Vinu et al. 2013) studied the genetic diversity in a set of 44 mustard genotypes, sourced from India and abroad. They used genome specific SSR markers and quantitative data for yield and component traits. Estimates of genetic distances allowed grouping of test genotypes into four clusters. The clustering patterns allowed discrimination of genotypes as per their origin and pedigree records. Grouping of genotypes on the basis of SSR marker data appeared more convincing than the phenotypic data. Although, there was no correlation between the distance matrices developed based on the genotypic and phenotypic data, SSR markers appeared better than quantitative data for discriminating *B. juncea* genotypes.

RAPD markers helped to characterize a collection of *B. juncea* germplasm, comprising 41 accessions from Pakistan, six oilseed cultivars/lines and five Japanese vegetable cultivars. Clustering placed most of the collected germplasm and oilseed cultivars/lines close to each other, suggesting a narrow genetic base of the germplasm in Pakistan (Rabbani et al. 1998). This was attributed to the selection for similar agronomic traits and horticultural uses. However, the oilseed collections and cultivars clustered separately from the vegetable types. Khan et al. (2008) fingerprinted 37 germplasm collections, five advance breeding lines and three crop cultivars using RAPD markers. Advanced breeding lines were relatively divergent from other test genotypes and these formed an independent cluster. Clustering based on molecular data did not mirror geographic diversities. In another study involving *B. juncea* germplasm from Pakistan, Tahira et al. (2013) explored 30 lines and varieties using RAPD markers. On an average 84 % similarity, matrix (SM) was observed between all thirty genotypes. RBJ-97001 was the most diverged genotype with average genetic diversity of 29 % while RBJ-02017 showed the highest average genetic SM (87 %). The genotypes RBJ-07017 and RL-18 were closest to each other with genetic SM of 95 %. In contrast, Raya 49/2 and RBJ-97001 were most distinct genotypes with genetic SM of 71 %. Cluster analysis categorized 30 genotypes into two major groups (A and B).

High diversity of *B. juncea* was also identified in southwest and western China through RAPD analysis (Li et al. 1997; An et al. 1999). Grouping of *B. juncea* landraces (73) from southwest China was largely related to agro ecological adaptations (Pu et al. 2007). The association between molecular genetic diversity seemed to be associated with geological and biological conditions prevailing at collection sites of 101 accessions from western China. The genetic diversity was based on AFLP, sequence-related amplified polymorphism (SRAP), and simple sequence repeat (SSR) markers, (Xu et al. 2008). Genetic diversity in winter, type of *B. juncea* was higher than that in spring types, and genetic diversity of *B. juncea* in Shaanxi and Xinjiang provinces was higher than that observed among genetically distinct genotypes in Tibet (Xu et al. 2008). Using SRAP markers, Wu et al. (2009) assessed 67 oilseed forms of *B. juncea* from China and 10 accessions from other countries, along with 18 non-oilseed types. Genetic diversity of the test genotypes largely reflected their growth habit (spring or winter).

Recent studies on molecular characterization of very large global collections of *B. juncea* have helped to establish Germplasm patterns of this crop. In a very elaborate study, involving 119 oilseed forms of *B. juncea,* Chen et al. (2013) reported two distinct groups as identified by SSR markers for both the A and B genome. Group I was reported to include germplasm from central and western India, besides that from eastern China. Group II included accessions from northern and eastern India, along with those from central and western China. European and Australian accessions also clustered in Group II. Chinese germplasm lines appeared very diverse (Group, I) and carried higher frequency alleles as compared to those from India. Population structure analysis based on the polymorphism generated by nuclear markers from 122 germplasm lines (Kaur et al. 2014), also confirmed the occurrence of two groups. The group one in this case included *B. juncea* accessions from East Europe and India. Second group carried a bulk of genotypes from India, China, and Australia.

Generally speaking, the genetic diversity evaluated by various molecular markers was different. Yu et al. (2005) used morphological characters, isozymes, proteins, and RAPD markers for estimating genetic diversity (GD) and associating it with hybrid performance. Association between different estimate of GD and F_1 performance were significant for many agronomic traits but not the seed yield. Further, the genetic diversity was not found to be significantly associated with geographic diversity. Contrasting results, using RAPD markers were, however, reported by Hu et al. (2003). They could establish parallelism between geographic and genetic divergence.

5.9 Expanded *Brassica juncea* Gene Pool

There is now a broad agreement that *B. juncea* possess a narrow genetic base due to dual bottlenecks of polyploidy and domestication. Intensive plant breeding activities have further eroded its genetic base (Chauhan et al. 2011; Singh et al. 2013). The genetic base can be expanded by resynthesis of *B. juncea* from diploid progenitors, which may help to mobilize genetic variation available in diploid genome donor species. Another option is to introgress variation available in wild and weedy crucifers.

5.10 Resynthesized *B. juncea*

Interspecific hybridization, followed by chromosome doubling is arguably the most potent evolutionary process that precipitates gross restructuring of participating genomes. Deletions (Ma and Gustafson 2006) gene conversions (Kovarik et al. 2005), transposon activation (Madlung et al. 2005), chromosomal rearrangements (Udall et al. 2005) and DNA methylation (Madlung et al. 2005; Lukens et al. 2006). Inter genomic exchange (Prakash 1973b) or increased meiotic recombination (Pecinka et al. 2011) can precipitate changes in neo-polyploids. Resynthesis of *B. juncea* is now being viewed as a major option to augment germplasm by mobilizing novel alleles from extant progenitors and also to benefit from de novo variation accruing from polyploidization.

Nagaharu (1935) is credited with the first experimental resynthesis of *B. juncea.* It was also resynthesized by several other workers (Ramanujam and Srinivasachar 1943; Frandsen 1943, 1947). Objectives of the resynthesis research varied from basic information (Srivastava et al. 2001, 2004), to applied aspects of developing novel phenotypes (Olsson 1960; Prakash 1973a). It was also possible to resynthesize *B. juncea* with both *B. rapa* (Olsson 1960; Prakash 1973a, b) and *B. nigra* (Campbell 1991; Song et al. 1993) as cytoplasm donor parents. Sexually obtained synthetics show reduced pollen and seed fertility in early generations, sometimes as low as 6 % in *B. juncea* (Olsson 1960). In the advanced generations, however, meiotic stability and seed fertility tended to improve significantly. In general, resynthesized genotypes are lower yielding than natural forms. Nevertheless, these constitute rich reservoir of diversity, as was revealed by allozyme and molecular marker studies (Srivastava et al. 2004). Standout outcome was the demonstration of heterosis for seed yield expressed by including resynthesized lines into experimental hybrids in *B. juncea* (Bansal et al. 2009, 2012).

Despite enormous efforts, breeding value of newly created *B. juncea* has remained poor due to genetic instabilities and associated linkage drag that *B. nigra* used for resynthesis did not experience any selection as an oilseed crop during domestication, or subsequent breeding may be a significant reason for its poor breeding value. Attempts have been made to enhance the genetic base of allotetraploids through hybridization between allotetraploid and diploid species or among allotetraploid *Brassica* species. In all these cases, recipient allotetraploid species were the primary targets for improvement by homeologous introgressions from a donor species. Such a mobilization of useful variation from progenitor species has not been very successful in *B. juncea*. We recently developed a new concept of derived amphiploidy to overcome these limitations. It is based on the hypothesis that the resynthesis can be pre-tuned for improved agronomic performance by bringing together diploid genomes from different but related high-value allotetraploids. Implicit is the proposition that the repeat polyploidization of already diplodized genomes with a new nuclear genome partner will set in motion fast-paced and far-reaching genetic consequences. We created *B. juncea* (A^nA^n B^cB^c), by hybridizing *B. napus* ($A^nA^nC^nC^n$) cv. Surpass with *B. carinata*

($B^cB^cC^cC^c$) cv. PC5. Synthesis of octoploid ($A^nA^nC^nC^nB^cB^cC^cC^c$) was the critical step in this method. This ensured preferred elimination of the chromosomes in the tetrasomic dose ($C^cC^cC^nC^n$) as these largely formed multivalents. There was irregular chromosome segregation, leading to an unequal chromosome migration to the poles and micro-nuclei during both meiotic divisions. It is well known that if homologous partners are present, chromosomes would preferentially pair within the same genome (e.g., A^nA^n/B^cB^c) to form bivalents as in the allotetraploids and to form multivalents (e.g., $C^cC^cC^nC^n$) as in autopolyploid. Multivalents invariably result in aberrant meiosis. Genomic instabilities, meiotic irregularities and reduced fertility are evolutionary disadvantageous due to decreased fitness. Therefore, the gametic sieve, especially through pollen, favors a shift back to diploid-like meiosis to stabilize polyploid lineages.

Development of *B. juncea* by re-associating A and B genomes from *B. napus* and *B. carinata,* respectively, precipitated far-reaching genetic and phenotypic changes (Gupta et al. 2015). Stable meiosis with euploid chromosome number and high pollen fertility confirmed the genetic viability of the concept of derived amphiploidy. Chromosome substitutions, A/C substitutions, were recognized in eleven progenies (Gupta et al. 2014). Population structure analysis allowed identification of three diversity pools. Progenies with determinate inflorescence were identified, and these were genetically closer to B genome carrying species like *B. nigra* and *B. carinata.* The indeterminate group was genetically closer to the extant *B. juncea.* Derived genotypes were agronomically superior, and these showed high heterosis in crosses with natural *B. juncea.*

5.11 Alien Genetic Introgressions

Extensive genetic diversity occurs in the wild and weedy germplasm of *Brassica* allies for nuclear and cytoplasmic genes. Wild Brassicaceous species occupy very diverse ecological landscapes (Tsunoda 1980) in the areas stretching from western Mediterranean to the eastern end of the Sahara desert and further to the northwest of India. Due to evolution under diverse eco-geographical habitats, wild Brassiceae constitutes a rich repository of genetic resources primarily for the defensive and adaptive traits (Warwick 2011). The desirable traits include the C3–C4 photosynthetic system (*Moricandia arvensis*), high nervonic acid content (*Thlaspi perfoliatum*), sclerotinia resistance (*Erucastrum cardaminoides*), alternaria leaf spot resistance (*Sinapisalba, Camelina sativa),* aphid resistance (*B. fruticulosa*) etc.

Harnessing alien genetic diversity is an important step to widen the germplasm base of *B. juncea.* To facilitate this, hybridization between wild and crop brassicas is the first step. As the wild germplasm belong to the secondary and tertiary gene pool, several kinds of hybridization barriers exist. These operate either at the time of fertilization or after fertilization, depending on the extent of reproductive isolation. Pre-fertilization barriers include pollen germination, pollen tube entry in the stigma and growth of the pollen tube through the style. Several crosses show unilateral incompatibility, i.e., pollination is effective only in one direction while the reciprocal crosses show strong pre-fertilization barriers. In general, success is greater when wild species are used as female parents (Shivanna 1996). Post-fertilization barriers cause embryo abortion leading to formation of shriveled or rudimentary seeds, without embryo. In recent years in vitro, fertilization has been effectively used to raise several intergeneric hybrids (Zenkteller 1990). Embryo rescue has been the most effective technique, pioneered by Japanese scientists to overcome post-fertilization barriers and is very widely used to raise hybrids. This procedure covers all the techniques, which are used to promote the growth of hybrid embryo and includes ovary culture, ovule culture and sequential culture. Understanding the phylogenetic relationships between cultivated Brassicas and wild crucifers (Prakash et al. 2009), through the interpretation of the chromosome pairing patterns or through genetic linkage maps anchored to the Arabidopsis genome, have opened up enormous possibilities of directed gene exchange across taxonomic domains. Though somatic hybridization is the technique of choice in overcoming fertilization barriers, modifications of the embryo rescue techniques continue to be employed due to their simplicity and operational ease. Though, a large number of interspecific and intergeneric hybrids involving *B. juncea* and wild crucifers have been developed (Prakash et al. 2009), with major emphasis on establishing genomic relatedness. However, only limited attempts have been made toward introgression of desirable nuclear encoded genes. Recently, resistance to 2, 4-D was transferred from *B. kaber* to *B. juncea*, by traditional backcross method Mithila and Christopher (2013). Plant screening of hybrids was carried out with dicamba.

5.12 Cytoplasmic Male Sterility-Fertility Restoration Systems

Major emphasis of alien introgressions in *B. juncea* has been towards development of cytoplasmic male sterility-fertility restoration (CMS-FR) systems using alloplasmic variation. Cytoplasmic male sterile lines could be developed by backcross substitution of *B. juncea* genomes in the cytoplasmic background of wild crucifers. To facilitate that sexually synthesized allopolyploids or somatic hybrids between wild and crop species were used as the bridging species. Wild species were invariably used as the female parent of such crosses. As expected, CMS originating from sexual hybridizations possess unaltered organelle genomes because of exclusive maternal inheritance. Since organelle assortment and intergenomic mitochondrial recombinant is of frequent occurrence in Brassicaceae, the cytoplasmic constitution is entirely different in those originating from somatic hybrids, and different combinations of mitochondrial and chloroplast genomes have been reported in different CMS lines (Prakash et al. 2009). A number of such CMS systems are now available (Table 5.1).

Table 5.1 Alloplasmic male sterile systems developed for *B. juncea*

CMS system	Cytoplasm donor	Reference
Ogura	*Raphanus sativus*	Kirti et al. (1995a)
Oxyrrhina	*B. oxyrrhina*	Prakash and Chopra (1988, 1990), Kirti et al. (1993)
Siifolia	*Diplotaxis siifolia*	Rao et al. (1994), Rao and Shivanna (1996)
Trachystoma	*Trachystoma ballii*	Kirti et al. (1995b)
Moricandia	*Moricandia arvensis*	Prakash et al. (1998), Kirti et al. (1998), Kaur et al. (2014)
Erucoides	*D. erucoides*	Malik et al. (1999), Prakash (2001), Bhat et al. (2006)
Berthauti	*D. berthauti*	Malik et al. (1999), Bhat et al. (2008)
Canariense	*Erucastrum canariense*	Prakash et al. (2001)
Catholica	*D. catholica*	Pathania et al. (2007)
Lyratus	*Enarthrocarpus lyratus*	Deol et al. (2003), Janeja et al. (2003)
Fruticulosa	*Brassica fruticulosa*	Banga (unpub)

Alloplasmic CMS plants, in general, general, were similar to euplasmic plants in development and morphology. However, many of them exhibited developmental and floral abnormalities as a consequence of altered nucleo-cytoplasmic interactions. Varying degrees of leaf chlorosis were associated with *Raphanus*/Ogu, *Oxyrrhina*, *Tournefort*ii, *Moricandia*, and *Enarthrocarpus* systems. Floral abnormalities in male sterile plants included: petaloid anthers (*nigra, muralis, trachystoma, raphanus, tournefortii, canariense*); poor or absent nectarines (*tournefortii and raphanus*); crooked style (*tournefortii, raphanus*); thick pistil (*raphanus*); and low seed fertility (*raphanus, tournefortii, enarthrocarpus* and *trachystoma*). Fertility restorers for *moricandia, ogura, catholica, and erucoides* and *lyratus* CMS systems could be developed by introgressing gene(s) for fertility restoration from cytoplasm donor wild species (Banga and Banga 2009; Prakash et al. 2009). Fertility restorer for the *mori* CMS could also restore fertility of *eru* CMS system (Bhat et al. 2005). At present, *ogura* and *mori* cytoplasmic male sterility systems are being used to develop hybrids.

5.13 Disease Resistance

Several fungal diseases cause heavy losses to seed yield in *Brassica* crops. The major ones are white rust caused by *Albugo candida*, alternaria leaf spot (*Alternaria* spp.) blackleg or stem canker (*Leptosphaeria maculans*) and *Sclerotinia* stem rot (*Sclerotinia sclerotiorum).* Nuclear genes conferring resistance to these diseases and for other desirable agronomic traits have been incorporated from related sources exploiting nonhomologous recombination following sexual/somatic hybridization and also by generating alien chromosome addition lines. Examples for successful gene introgressions in crop Brassicas include: black leg resistance from *B. juncea to B. napus* (Roy 1984) and from *B. nigra* to *B. napus* (Chevre et al.

1996, 1997); clubroot resistance from *B. napus* to cabbage (Chiang et al. 1977); self-incompatibility alleles from *B. rapa* to forage rape, earliness to oil rape leading to release of cultivars like Norin 16 and Asahi-Natane in Japan (Shiga 1970; Namai et al. 1980); resistance to pod shattering from *B. juncea to B. napus* (Parkash and Chopra 1990); resistance against *Phoma lingam* from *B. juncea* to *B. napus* (Sacristan and Gerdemann 1986) and from *B. nigra* to *B. napus* (Struss et al. 1996). Examples of introgression through somatic hybridization include beet cyst nematode resistance (*Heterodera schachtii)* from *Sinapis alba* and *Raphanus sativus* to *B. napus* (Lelivelt et al. 1993; Lelivelt and Krens 1992); alternaria leaf spot resistance from *S. alba* to *B. napus* (Primard et al. 1988); clubroot resistance (*Plasmodiophora brassicae*) from *Raphanus sativus* to *B. oleracea* var. botrytis. Eight species (*Brassica desnottesii, Diplotaxis berthautii, D. catholica, D. cretacea, D. erucoides, Erucastrum gallicum* and *Camelina sativa, Coincya pseuderucastrum*) were found completely resistant to alternaria pathogen (Sharma et al. 2002).

There are only a few reports regarding successful introgression of genes for disease resistance in *B. juncea*. Four complete sets of introgression lines were developed following hybridization of four wild crucifers (viz. *Erucastrum cardaminoides, Diplotaxis tenuisiliqua, E. abyssinicum and Brassica fruticulosa*) with *B. juncea* (Banga unpub). These four sets are named as Intro-cardaJ, Intro-tenuiJ, Intro-abbyJ, and Intro-frutiJ, respectively. Their resistance responses to sclerotinia stem rot were characterized finally in BC_1S_5 and BC_1S_7 generation by using standard stem inoculation test, using the *Sclerotinia* isolates (Garg et al. 2010). The introgession lines could be classified as hypersensitive, resistant (lesion size < 1 cm) and susceptible (lesion size > 1 cm), based on their resistance responses. Majority of the lines fell into the susceptible categories. This is expected as introgression of a desired trait is a chance event in wide crosses due to rarity of homoeologous pairing. The frequency of lines with the hypersensitive response in the present context was 0.08 (Intro-frutiJ), 0.10 (Intro-cardaJ), 0.10 (Intro-tenuiJ) and 0.02 (Intro-abbyJ). The resistant lines of each introgression set had much higher level of resistance than that found in a representative set of *Brassica juncea* germplasm.

5.14 Insect Resistance

Aphid (*Lipaphiserysimi*) is a major pest of oilseed *Brassicas*. Under epiphytotic conditions, this pest can cause up to 70 % yield losses. A large number of wild and weedy crucifers were assessed for resistance responses to mustard aphid {*Lipaphiserysimi* (Kaltenbach) Homoptera} infestation (Kumar et al. 2011). A wild crucifer; Brassica *fruticulosa* was found resistant to mustard aphid. An allopolyploid, (*B. fruticulosa* × *B. rapa* var. brown *sarson*) was first synthesized for use as a bridging species to transfer *fruticulosa* resistance to *B. juncea*. Backcrossing and selfing allowed development of a large number of introgression lines, carrying resistance to mustard aphid. Resistance levels of the introgression lines were established through a series of choice and no-choice experiments as well as field evaluations. *B. rapa* cv. BSH-1 was the most preferred host; B. *fruticulosa* was the least preferable one along with select introgression lines. This was evident from the number of aphids that settled on circular leaf bits of these genotypes, 24 and 48 h after aphid release in a choice experiment. *Brassica fruticulosa* and select introgression lines exhibited antibiosis against the mustard aphids in no-choice experiment, and the aphids died within 5–8 days after their release. Maximum survival (76.7 %) was recorded on BSH-1. Other demographic parameters like development time, fecundity and longevity also reflected a similar trend. In the screen house studies, *B. fruticulosa* and identified introgression lines suffered no seedling mortality after 30 days of aphid release. However, 80 % mortality was observed on BSH-1. Field-level studies also confirmed the introgressed resistance, emphasizing heritable nature of *fruticulosa* resistance. Most of the introgression lines had normal chromosome number ($2n = 36$) and were fully fertile. These introgression lines were also characterized using well distributed and transferable markers. Average proportions of recipient and donor genome in the substitution lines were 49.72 and 35.06 %, respectively (Atri et al. 2012). Minimum alien parent genome presence (27.29 %) was observed in the introgression line, Ad3 K-280.

References

Acharya NN, Swain D (2003) Hybrid performance in relation to genetic divergence in Indian mustard [*Brassica juncea* (L.) Czern&Coss]. Crop Res 25:312–315

Ali M, Copeland LO, Elias SG, Kelly JD (1995) Relationship between genetic distance and heterosis for yield and morphological traits in winter canola (*Brassica napus* L.). Theor Appl Genet 91:118–121

An XH, Chen BY, Fu TD, Liu HL (1999) Genetic diversity of Chinese landraces in *Brassica juncea* was analysed by RAPD markers. J Huazhong Agric U 18:524–527

Arias T, Beilstein MA, Tang M, McKain MR, Pires JC (2014) Diversification times among *Brassica* (Brassicaceae) crops suggest hybrid formation after 20 million years of divergence. Am J Bot 101:86–91

Atri C, Kumar B Kumar H, Kumar S,Sharma S, and Banga SS (2012) Development and characterization of Brassica juncea—fruticulosa introgression lines exhibiting resistance to mustard aphid (Lipaphis erysimi Kalt). BMC Genetics 13:104. doi:10.1186/1471-2156-13-104

Attia T, Robbelen G (1986) Meiotic pairing in haploids and amphidiploids of spontaneous versus synthetic origin in rape, *Brassica napus* L. Can J Genet Cytol 28:330–334

Axelsson T, Bowman CM, Sharpe AG, Lydiate DJ, Lagercrantz U (2000) Amphidiploid *Brassica juncea* contains conserved progenitor genomes. Genome 43:679–688

Banga SS, Banga SK (2009) Crop improvement strategies in rapeseed-mustard. In: Hegde DM (ed) Vegetable oils scenario: approaches to meet the growing demands. ISOR, Hyderabad, pp 13–35

Banga SS, BangaSK, LabanaKS (1983) Nucleo-cytoplasmic interaction in Brassica. In: Proceedings of the 6th International Rapeseed Congress, Paris, 17–19 May 1983, 602–606 pp

Bansal P, Kaur P, Banga SK, Banga SS (2009) Augmenting genetic diversity in *Brassica juncea* through its resynthesis using purposely selected diploid progenitors. Int J Plant Breed 3:41–45

Bansal P, Banga S, Banga SS (2012) Heterosis as investigated in terms of polyploidy and genetic diversity using designed *Brassica juncea* amphiploid and its progenitor diploid species. PLoS ONE 7:e29607

Bhat SR, Prakash S, Kirti PB, Kumar DV, Chopra VL (2005) A unique introgression from *Moricandiaarvensi* confers male fertility upon two different cytoplasmic male-sterile lines of *Brassica juncea*. Plant Breed 124:117–120

Bhat SR, Priya V, Ashutosh Dwivedi KK, Prakash S (2006) *Diplotaxiserucoides* induced cytoplasmic male sterility in *Brassica juncea* is rescued by the *Moricandiaarvensis* restorer: genetic and molecular analyses. Plant Breed 125:150–155

Bhat SR, Kumar P, Prakash S (2008) An improved cytoplasmic male sterile (*Diplotaxisberthautii*) *Brassica juncea*: identification of restorer and molecular characterization. Euphytica 159:145–152

Bornet B, Branchard M (2004) Use of ISSR fingerprints to detect microsatellites and genetic diversity in several related *Brassica* taxa and *Arabidopsis thaliana*. Hereditas 140:245–248

Burkill IH (1930) The Chinese mustards in the Malaya Peninsula. Gardners Bull 5:99–117

Burton WA, Ripley VL, Potts DA, Salisbury PA (2004) Assessment of genetic diversity in selected breeding lines and cultivars of canola quality *Brassica junceaand* their implications for canola breeding. Euphytica 136:181–192

Campbell DR (1991) Effects of floral traits on sequential components of fitness in *Ipomopsisaggregata*. Am Nat 137:713–737

Cavell AC, Lydiate DJ, Parkin IAP, Dean C, Trick M (1998) Collinearity between a 30-centimorgan segment of *Arabidopsis thaliana* chromosome 4 and duplicated regions within the *Brassica napus* genome. Genome 41:62–69

Chauhan JS, Singh KH, Singh VV, Satyanshu K (2011) Hundred years of rapeseed-mustard breeding in India: accomplishments and future strategies. Indian J Agric Sci 81:1093–1109

Chen S, Nelson MN, Ghamkhar K, Fu T, Cowling WA (2008) Divergent patterns of allelic diversity from similar origins: the case of oilseed rape (*Brassica napus* L.) in China and Australia. Genome 51:1–10

Chen S, Wan Z, Nelson MN, Chauhan J, Redden R, Burton WA, Lin P, Salisbury PA, Fu T, Cowling WA (2013) Evidence from genome-wide simple sequence repeat markers for a polyphyletic origin and secondary centers of genetic diversity of *Brassica juncea* in China and India. J Hered 104:416–427

Chevre AM, Eber F, This P, Barret P, Tanguy X, Brun H, Delseny M, Renard M (1996) Characterization of *Brssicanigra* chromosomes and of backleg resistance in *B. napus-B. nigra* addition lines. Plant Breed 115:113–118

Chevre AM, Barret P, Eber F, Dupuy P, Brun H, Tanguy X, Renard M (1997) Selection of stable *Brassica napus- B. juncea* recombinant lines resistant to backleg (*Leptosphaeriamaculans*).I. Identification of molecular markers, chromosomal and genomic origin of the introgression. Theor Appl Genet 95:1104–1111

Chiang MS, Chiang BY, Grant WF (1977) Transfer of resistance to race 2 of *Plasmodiophora brassicae* from *Brassica napus* to cabbage (*B. oleracea* var. *capitata*). I. Interspecific hybridization between *B. napus* and *B. oleracea* var. *capitata*. Euphytica 26:319–336

Cox TS, Murphy JP, Rodgers DM (1986) Changes in genetic diversity in the red winter wheat regions of the United States. Proc Natl Acad Sci 83:5583–5586

Cui C, Ge X, Gautam M, Kang L, Li Z (2012) Cytoplasmic and genomic effects on meiotic pairing in Brassica hybrids and allotetraploids from pair crosses of three cultivated diploids. Genetics 191:725–738

Deol JS, Shivanna KR, Prakash S, Banga SS (2003) Enarthrocarpuslyratus based cytoplasmic male sterility and fertility restorer system in *Brassica rapa*. Plant Breed 122:438–440

Duhoon SS, Koppar MN (1998) Distribution, collection and conservation of biodiversity in cruciferous oilseeds in India. Genet Resour Crop Evol 45:317–323

Erickson LR, Straus NA, Beversdorf WD (1983) Restriction patterns reveal origins of chloroplast genomes in *Brassica* amphiploids. Theor Appl Genet 65:201–206

Falconer DS, MacKay TFC (1996) Introduction to quantitative genetics. Longman, London

Frandsen KJ (1943) The experimental formation of *Brassica juncea* Czern&Coss. Dansk Bot Ark 11:1–17

Frandsen KJ (1947) The experimental formation of *Brassica napus* L. var. *oleifera* DC. and *Brassica carinata* Braun. Dansk Bot Ark 12:1–16

Garg H, Atri C, Sandhu PS, Kaur B, Renton M, Banga SK, Singh H, Singh C, Barbetti MJ, Banga SS (2010) High level of resistance to *Sclerotiniasclerotiorum* in introgression lines derived from hybridization between wild crucifers and the crop Brassica species *B. napus* and *B. juncea*. Field Crops Res 117:51–58

Ghosh SK, Gulati SC (2002) Parental diversity to realize maximum hetrosis in Indian mustard (*Brassica juncea* (L.) Czern&Coss). Indian J Genet 62:25–28

Gómez-Campo C, Prakash S (1999) Origin and domestication. In: Gómez-Campo C (ed) Biology of Brassica coenospecies. Elsevier Science, Amsterdam, pp 59–106

Goswami PK, Behl RK (2006) Genetic divergence in Indian mustard. Ann Agric Res 27:187–190

Gupta VP, Sekhon MS, Satija DR (1991) Studies on genetic diversity, heterosis and combining ability in Indian mustard. Indian J Genet 51:448–453

Gupta M, Atri C, Banga SS (2014) Cytogenetic stability and genome size variations in newly developed derived *Brassica juncea* allopolyploid lines. J Oilseed Brassica 5:118–127

Gupta M, Gupta S, Kumar H, Kumar N, Banga SS (2015) Population structure and breeding value of a new type of *Brassica juncea* created by combining A and B genomes from related allotetraploids. Theor Appl Genet 128:221–234

Hu SW, Ovesná J, Kucera L et al (2003) Evaluation of genetic diversity of *Brassica napus* germplasm from China and Europe assessed by RAPD markers. Plant Soil Environ 49:106–113

Inaba R, Nishio T (2002) Phylogenetic analysis of the *Brassiceae* based on the nucleotide sequences of the S-locus related gene SLR1. Theor Appl Genet 105:1159–1165

Jain A, Bhatia S, Banga SS, Prakash S, Laxmikumaran M (1994) Potential use of random amplified polymorphic DNA (RAPD) technique to study genetic diversity in Indian mustard (*B. juncea*) and its relatedness to heterosis. Theor Appl Genet 88:116–122

Janeja HS, Banga SK, Bhasker PB, Banga SS (2003) Alloplasmic male sterile *Brassica napus* with *Enarthrocarpuslyratus* cytoplasm: introgression and molecular mapping of an *E. lyratus* chromosome segment carrying a fertility restoring gene. Genome 46:792–797

Kaur P, Banga S, Kumar N, Gupta S, Akhtar J, Banga S (2014) Polyphyletic origin of *Brassica juncea* with *B. rapa* and *B. nigra* (Brassicaceae) participating as cytoplasm donor parents in independent hybridization events. Am J Bot 101:1–10

Khan MA, Rabbani MA, Munir M, Ajmal SK, Malik MA (2008) Assessment of genetic variation within Indian mustard (*Brassica juncea*) germplasm using random amplified polymorphic DNA markers. J Integr Plant Biol 50:385–392

Kirti PB, Narasimhulu SB, Mohapatra T, Prakash S, Chopra VL (1993) Correction of chlorosis in alloplasmic male sterile *Brassica juncea* through recombination between chloroplast genomes. Genet Res 62:11–14

Kirti PB, Banga SS, Prakash S, Chopra VL (1995a) Transfer of Ogura cytoplasmic male sterility to *Brassica juncea* and improvement of the male sterile line through somatic cell fusion. Theor Appl Genet 91:517–521

Kirti PB, Mohapatra T, Baldev A, Prakash S, Chopra VL (1995b) A stable cytoplasmic malesterile line of *Brassica juncea* carrying restructured organelle genomes from the somatic hybrid *Trachystomaballii*+ *B. juncea*. Plant Breed 114:434–438

Kirti PB, Prakash S, Gaikwad K, Kumar VD, Bhat SR, Chopra VL (1998) Chloroplast substitution overcomes leaf chlorosis in a *Moricandiaarvensis*-based cytoplasmic male sterile *Brassica juncea*. Theor Appl Genet 97:1179–1182

Kovarik A, Pires JC, Leitch AR, Lim KY, Sherwood AM, Matyasek R, Rocca J, Soltis DE, Soltis PS (2005) Rapid concerted evolution of nuclear ribosomal DNA in two *Tragopogon* allopolyploids of recent and recurrent origin. Genetics 169:931–944

Krishna P, Ghose SK (1992) Heterosis in relation to genetic divergence in rapessed and mustard. J Oilseeds Res 9:169–174

Kumar M, Singh KP, Singh D (2007) Genetic divergence analysis for seed yield, its parameters and oil content in Indian mustard [*Brassica juncea*(L.) Czern&Coss]. Res Crops 8:183–185
Kumar S, Atri C, Sangha MK, BangaSS (2011) Screening of wild crucifers for resistance to mustard aphid, *Lipaphiserysimi* (Kaltenbach) and attempt at introgression of resistance gene(s) from *Brassica fruticulosa* to *Brassica juncea*. Euphytica 179:461–470
Lelivelt CLC, Krens FA (1992) Transfer of resistance to the beet cyst nematode (*HeteroderaschachtiiSchm.)* into the *Brassica napus* L. gene pool through introgeneric somatic hybridization with *Raphanussativus* L. Theor Appl Genet 83:887–894
Lelivelt CLC, Leunissen EHM, Frederiks HJ, Helsper JPFG, Krens FA (1993) Transfer of resistance to the beet cyst nematode (*Heteroderaschachtii* Scgn.) from *Sinaps alba* L. (white mustard) to the *Brassica* napus gene pool by means of sexual and somatic hybridization. Theor Appl Genet 85:688–696
Li RG, Zhu L, Wu NF, Fan YL, Wu XM, Qian XZ (1997) Genetic diversity among oilseed cultivars of *Brassica juncea* Czern. Coss in China. J Biotechnol 5:26–31
Liu A, Wang J (2006) Genomic evolution of *Brassica* allopolyploids revealed by ISSR marker. Genet Resour Crop Evol 53:603–611
Lodhi B, Thakral NK, Singh D, Avtar R, Bahadur R (2013) Genetic diversity analysis in Indian mustard (*Brassica juncea*). J Oilseed Brassica 4:57–60
Lukens LN, Pires JC, Leon E, Vogelzang R, Oslach L, Osborn T (2006) Patterns of sequence loss and cytosine methylation within a population of newly resynthesized *Brassica napus* allopolyploids. Plant Physiol 140:336–348
Lysak MA, Koch MA, Pecinka A, Schubert I (2005) Chromosome triplication found across the tribe *Brassicaceae*. Gen Res 15:516–525
Ma XF, Gustafson J (2006) Timing and rate of genome variation in triticale following allopolyploidization. Genome 49:950–958
Mabberley DJ (1993) The plant-book: a portable dictionary of the vascular plants. Cambridge University Press, London
Madlung A, Tyagi AP, Watson B, Jiang HM, Kagochi T, Doerge RW, Martienssen R, Comai L (2005) Genomic changes in synthetic *Arabidopsi* polyploids. Plant J. 41:221–230
Malik M, Vyas P, Rangaswamy NS, Shivanna KR (1999) Development of two new cytoplasmic male-sterile lines in *Brassica juncea* through wide hybridization. Plant Breed 118:75–78
Mithila J, Christopher JH (2013) Transfer of auxinic herbicide resistance from *Brassica kaber* to *Brassica juncea* and *Brassica rapa* through embryo rescue. In Vitro Cell Dev Biol-Plant 49:461–467
Nagaharu U (1935) Genomic analysis in *Brassica* with special reference to the experimental formation of *B. napus* and peculiar mode of fertilization. Jpn J Bot 7:389–452
Namai H, Sarashima M, Hosoda T (1980) Interspecific and intergeneric hybridization breeding in Japan. In: Tsunoda S, Hinata K, Gomez-Campo C (eds) *Brassica* crops and wild allies: biology and breeding. Japan Scientific Societies Press, Tokyo, pp 191–204
Negi MS, Devic M, Delseny M, Lakshmikumaran M (2000) Identification of AFLP fragments linked to seed coat colour in *Brassica juncea* and conversion to a SCAR marker for rapid selection. Theor Appl Genet 101:146–152
Nei M (1987) Molecular evolutionary genetics. Columbia University Press, New York
Olsson G (1960) Species crosses within the genus *Brassica* II. Artificial *Brassica napus* L. Hereditas 46:171–222, 351–386
Palmer JD, Shields CR, Cohen DB, Orton TJ (1983) Chloroplast DNA evolution and the origin of amphiploid *Brassica* species. Theor Appl Genet 65:181–189
Parkin IAP, Sharpe AG, Lydiate DJ (2003) Patterns of genome duplication within the *Brassica napus* genome. Genome 46:291–303
Pathania A, Kumar R, Kumar DV, Asutosh Dwivedi KK, Kirti PB, Prakash S, Chopra VL, Bhat SR (2007) A duplication of *coxI* gene is associated with CMS (*Diplotaxiscatholica*) *Brassica juncea* derived from somatic hybridization with *Diplotaxiscatholica*. J Genet 86:93–101

Pecinka A, Fang W, Rehmsmeier M, Levy AA, MittelstenScheid O (2011) Polyploidization increases meiotic recombination frequency in Arabidopsis. BMC Biol 9:24–30

Pejic I, Marsan AP, Morgante M, Kozumplick V, Castiglioni P, Taramino G, Motto M (1998) Comparative analysis of genetic similarity among maize inbred lines detected by RFLPs, RAPDs, SSRs, and AFLPs. Theor Appl Genet 97:1248–1255

Petit RJ, Demesure B, Dumolin S (1998) cpDNA and mtDNA primers in plants. In: Karp A, Isaac PG, Ingram DS (eds) Molecular tools for screening biodiversity. Chapman and Hall, London, pp 256–261

Pradhan AK, Prakash S, Mukhopadhyay A, Pental D (1992) Phylogeny of *Brassica* and allied genera based on variation in chloroplast and mitochondrial DNA patterns: molecular and taxonomic classifications are incongruous. Theor Appl Genet 85:331–340

Pradhan AK, Sodhi YS, Mukhopadhyay A, Pental D (1993) Heterosis breeding in Indian mustard (*Brassica juncea* L. Czern&Coss): analysis of component characters contributing to heterosis for yield. Euphytica 69:219–229

Pradhan AK, Gupta V, Mukhopadhyay A, Arumugam N, Sodhi YS, Pental D (2003) A high density linkage map in *Brassica juncea* (Indian mustard) using AFLP and RFLP markers. Theor Appl Genet 106:607–614

Prain D (1898) The mustards cultivated in Bengal. Agric Ledger 5:1–180

Prakash S (1973a) Haploidy in *Brassica nigra* Koch. Euphytica 22:613–614

Prakash S (1973b) Artificial synthesis of *Brassica juncea* Coss. Genetica 44:249–263

Prakash S (1974) Probable basis of diploidisation of *B. juncea* Coss. Can J Genet 16:232–234

Prakash S (1980) Cruciferous oilseeds in India. In: Tsunoda, Hinata SK,Gomez-Campo C (eds) *Brassica* crops and wild allies. Jpn Sci Soc, Tokyo, Japan, p 151–163

Prakash S (2001) Utilization of wild germplasm of *Brassica* allies in developing cytoplasmic male sterility-fertility restoration systems in Indian mustard *Brassica juncea*. In: Liu H, Fu T (eds) Proc Int Symp Rapeseed Sci. Science Press, New York, pp 63–67

Prakash S, Chopra VL (1988) Introgression of resistance to shattering in *Brassica napus* from *Brassica juncea* through nonhomologous recombination. Plant Breed 101:167–168

Prakash S, Chopra VL (1990) Reconstruction of alloploid Brassicas through non-homologus recombination: introgression of resistance to pod shatter in *Brassica napus*. Genet Res Camb 56:1–2

Prakash S, Hinata K (1980) Taxonomy, cytogenetics and origin of crop *Brassicas*—a review. Oper Bot 55:1–57

Prakash S, Kirti PB, Bhat SR, Gaikwad K, Kumar VD, Chopra VL (1998) A *Moricandiaarvensis*-based cytoplasmic male sterility and fertility restoration system in *Brassica juncea*. Theor Appl Genet 97:488–492

Prakash S, Bhat SR, Quiros CF, Kirti PB, Chopra VL (2009) *Brassica* and its close allies: cytogenetics and evolution. Plant Breed Rev 31:21–187

Primard C, Vedel F, Mathieu C, Pelletier GD, Chevre AM (1988) Interspecific somatic hybridization between *Brassica napus* and *Brassica hirta* (*Sinapis alba* L). Theor Appl Genet 75:546–552

Pu X, Wang M, Luan L, Wang X, Zhang J, Li H, Zhang Q, Li Y, Cao Y, Jiang L (2007) Genetic diversity analysis of *Brassica juncea* landraces in southwest China. Sci Agric Sin 40:1610–1621

Quiros CF, Truco MJ, Hu J (1995) Sequence comparison of two codominant RAPD markers in *Brassica nigra* deletions, substitutions and microsatellites. Plant Cell Rep 15:268–270

Rabbani MA, IwabuchiA Murakami Y, Suzuki T, Kenji Takayanagi K (1998) Genetic diversity in mustard (*Brassica juncea* L.) germplasm from Pakistan as determined by RAPDs. Euphytica 103:235–242

Ramanujam S, Srinivasachar D (1943) Cytogenetical investigations in *Brassica* and the artificial synthesis of *Brassica juncea*. Indian J Genet 3:73–88

Ramchiary N, Padmaja KL, Sharma S, Gupta V, Sodhi YS, Mukhopadhyay A, Arumugam N, Pental D, Pradhan AK (2007) Mapping of yield influencing QTL in *Brassica juncea*:

implications for breeding of a major oilseed crop of dryland areas. Theor Appl Genet 115:807–817

Rao GU, Shivanna KR (1996) Development of a new alloplasmic CMS *Brassica napus* in the cytoplasmic background of *Diplotaxissiifolia*. Eucarpia Cruciferae Newsl 18:68–69

Rao GU, Sarup BV, Prakash S, Shivanna KR (1994) Development of new cytoplasmic male-sterility system in *Brassica juncea* through wide hybridization. Plant Breed 112:171–174

Roy NN (1984) Interspecific transfer of *Brassica juncea*-type high blackleg resistance to *Brassica napus*. Euphytica 33:295–303

Sacristan MD, Gerdemann M (1986) Different behaviour of *Brassica juncea* and *B. carinata* as sources of *Phoma lingam* resistance in experiments of interspecific transfer to *B. napus*. Plant Breed 97:304–314

Sekhon MS, Gupta VP (1995) Genetic distance and heterosis in Indian mustard: developmental isozymes as indicators of genetic relationships. Theor Appl Genet 91:1148–1152

Shalini S, Sheriff RA, Kulkarni RS, Venkantarmana P (2000) Correlation and path analysis of Indian mustard germplasm. Res Crops India 1:226–229

Sharma G, Kumar DV, Haque A, Bhat SR, Prakash S, Chopra VL (2002) *Brassica coeno* species: a rich reservoir for genetic resistance to leaf spot caused by *Alternariabrassicae*. Euphytica 125:411–417

Shekhawat N, Jadeja GC, Singh J, Ramesh (2014) Genetic diversity analysis in relation to seed yield and its component traits in Indian mustard (*Brassica juncea* Czern&Coss). Bioscan 9:713–717

Shiga T (1970) Rape breeding by interspecific crossing between *Brassica napus* and *Brassica campestris* in Japan. Jpn Arg Res Q 5:5–10

Shivanna KR (1996) Incompatibility and wide hybridization. In: Chopra VL, Prakash S (eds) Oilseed and vegetable Brassicas: indian perspective. Oxford and IBH Publishing Co, New Delhi, pp 77–102

Singh KH, Chauhan JS (2010) Morphological descriptor of rapeseed mustard varieties. Directorate of Rapeseed Mustard Research, Sewar, Bharatpur, Rajasthan, India, p 61

Singh KH, Shakya R, Thakur AK, Chauhan DK, Chauhan JS (2013) Genetic Diversity in Indian Mustard (*Brassica juncea* (L.) Czernj&Cosson) as Revealed by Agronomic Traits and RAPD Markers. Nat Acad Sci Lett 36:419–427

Sinskaja EN (1928) Theoleiferous plants and root crops of the family *Cruciferae*. Bull Appl Bot Genet Plant Breed 19:3–648

Sodhi YS, Chandra A, Verma JK, Arumugam N, Mukhopadhyay A, Gupta V, Pental D, Pradhan AK (2006) A new cytoplasmic male sterility system for hybrid seed production in Indian oilseed mustard *Brassica juncea*. Theor Appl Genet 114:93–99

Song KM (1995) Rapid genome change in synthetic polyploids of *Brassica* and its implications for polyploid evolution. Proc Natl Acad Sci 92:7719–7723

Song KM, Osborn TC (1992) Polyphyletic origins of *Brassica napus*: new evidence based on nuclear and RFLP organelle analysis. Genome 35:992–1001

Song KM, Tang K, Osborn TC (1993) Development of synthetic *Brassica* amphidiploids by reciprocal hybridization and comparison to natural amphidiploids. Theor Appl Genet 86:811–821

Souza E, Sorrells ME (1989) Pedigree analysis of North American oat cultivars released from 1941 to 1985. Crop Sci 29:595–601

Spect CE, Diederichsen A (2001) Brassica. In: Hanelt. P. (ed) Mansfeld's encyclopedia of agricultural and horticultural crops. Springer, Berlin, pp 1453–1456

Srivastava A, Gupta V, Pental D, Pradhan AK (2001) AFLP-based genetic diversity assessment amongst agronomically important natural and some newly synthesized lines of *Brassica juncea*. Theor Appl Genet 102:193–199

Srivastava A, Mukhopadhyay A, Arumugam M, Gupta V, Verma JK, Pental D, Pradhan AK (2004) Resynthesis of *Brassica juncea* through interspecific crosses between *B. rapa* and *B. nigra*. Plant Breed 123:204–206

Struss D, Quiros CF, Plieske J, Robbelen G (1996) Construction of *Brassica* B-genome synteny groups based on chromosomes extracted from three different sources by phenotypic isozymes and molecular markers. Theor Appl Genet 93:1026–1032

Sun VG (1970) Breeding plants of *Brassica*. J Agric Assoc China 71:41–52

Szadkowski E, Eber F, Huteau V et al (2010) The first meiosis of resynthesized *Brassica napus*, a genome blender. New Phytol 186:102–112

Tahira R, Ullah I, Saleem M (2013) Evaluation of genetic diversity of raya (*Brassica juncea*) through RAPD markers. Int J AgricBiol 15:1163–1168

Tautz D (1989) Hypervariability of simple sequences as a general source for polymorphic DNA markers. Nuc Acids Res 17:6463–6471

Thakur HL, Zarger MA (1989) Heterosis in relation to genetic diversity and specific combining ability in Indian mustard. Indian J Genet 51:448–453

Tsunoda S (1980) Ecophysiology of wild and cultivated forms in *Brassica* and allied genera. In: Tsunoda, S, Hinata K, Gomez-Campo C (eds) *Brassica* crops and wild allies, biology and breeding. Jpn Sci Soc, Tokyo, pp 109–120

Uchimiya H, Wildman SG (1978) Evolution of fraction 1 protein in relation to origin of amphidiploid *Brassica* species and other members of the *Cruciferae*. J Hered 69:299–303

Udall JA, Quijada PA, Osborn TC (2005) Detection of chromosomal rearrangements derived from homeologous recombination in four mapping populations of *Brassica napus* L. Genetics 169:967–979

Vaughan G (1977) A multidisciplinary study of the taxonomy and origin of *Brassica* crops. Bioscience 27:35–40

Vaughan JG, Hemingway JS (1959) The utilization of mustards. Econ Bot 13:196–204

Vaughan JG, Hemingway JS, Schofield HJ (1963) Contributions to a study of variation in *Brassica juncea* Coss.&Czern. J Linn Soc Bot 58:435–447

Vavilov NI (1949) The origin, variation, immunity and breeding of cultivated plants. Chron Bot 13:1–364

Verma OP, Kushwaha GD, Singh HP (2000) Heterosis in relation to genetic diversity in Indian mustard. Cruciferae Newsl 22:93–94

Vinu V, Singh N, Vasudev S, Yadava DK, Kumar S, Naresh S, Bhat SR, Prabhu KV (2013) Assessment of genetic diversity in *Brassica juncea* (Brassicaceae) genotypes using phenotypic differences and SSR markers. Rev Biol Trop 61:1919–1934

Vos P, Hogers R, Bleeker M, Reijans M, Lee TVD, Hornes M, Friters A, Pot J, Paleman J, Kuiper M, Zabeau M (1995) AFLP: a new technique for DNA fingerprinting. Nucl Acids Res 23:4407–4414

Warburton M, Hoisington D (2001) Applications of molecular marker techniques to the use of international germplasm collections. In: Henry RJ (ed) Plant Genotyping: DNA fingerprinting of plants. CABI Publishing, Wallingford, pp 83–93

Warwick SI (2011) Brassicaceae in agriculture. In: Schmidt R, Bancroft I (eds) Genetics and genomics of the Brassicaceae. Plant genetics and genomics: crops and models. Springer, New York, pp 33–65

Warwick SI, Black LD (1991) Molecular systematics of *Brassica* and allied genera (Subtribe *BrassicinaeBrassicae*) chloroplast genome and cytodeme congruence. Theor Appl Genet 82:81–92

Wu X, Chen B, Lu G, Wang H, Xu K, Guizhan G, Song Y (2009) Genetic diversity in oil and vegetable mustard (*Brassica juncea*) landraces revealed by SRAP markers. Genet Resour Crop Evol 56:1011–1022

Xian-Hui A, Bao-yuan C, Ting-dong F, Hou-li L (1999) Genetic diversity analysis among chinese landraces in *Brassica juncea* using RAPD markers. In: Proceedings of the 10th international rapeseed congress, Canberra, Australia

Xiao S, Xu J, Li Y, Zhang L, Shi S, Wu J, Liu K (2007) Generation and mapping of SCAR and CAPS markers linked to the seed coat color gene in *Brassica napus* using a genome walking technique. Genome 50:611–618

Xu A, Ma C, Xiao E, Quan J, Ma C, Tian G, Tu J, Fu T, Zhang G (2008) Genetic diversity of *Brassica juncea* from western China. Acta Agron Sin 34:754–763

Yu CY, Hu SW, Zhao HX, Guo AG, Sun GL (2005) Genetic distances revealed by morphological characters, isozymes, proteins and RAPD markers and their relationships with hybrid performance in oilseed rape (*Brassica napus* L.). Theor Appl Genet 110:511–551

Zenkteller M (1990) *In vitro* fertilization of ovules of some species of Brassicaceae. Plant Breed 105:221–228

Zhou WJ, Zhang GQ, Tuvesson S, Dayteg C, Gertsson B (2006) Genetic survey of Chinese and Swedish oilseed rape (*Brassica napus*L.) by simple sequence repeats (SSRs). Genet Resour Crop Evol 53:443–447

Chapter 6
Potato Diversity and Its Genetic Enhancement

Avinash Srivastava, Vinay Bhardwaj, BP Singh and SM Paul Khurana

Abstract Potato is the world's third largest food crop after rice and wheat widely grown across all continents. It belongs to the genus Solanum and section Petota that contain approximately 2000 species that are distributed from the South-western United States (38°N) to Chile (41°S) between 2000 to 4000 m altitudes. Potato has 6 cultivated species, 225 wild relatives and 110 wild tuber-bearing species. The main cultivated potato species *Solanum tuberosum* L., a tetraploid ($2n = 4x = 48$) originated from Andes of Peru and Bolivia in South America over 10,000 years ago. The ploidy of potatoes varies from diploid ($2n = 24$) to hexaploid ($2n = 72$) with majority being diploids. Potatoes were introduced to Europe in 1570s and by beginning of seventeenth century they spread to the other parts of the world. Systematic potato breeding started in 1807 in England followed by other parts of Europe, North America, India, International Potato Centre, Peru and China. There are two basic approaches to conserve potato genetic resources, viz. in situ and ex situ. Currently, cryo-conservation is being tapped for long-term conservation. Seven major potato gene banks are present worldwide to conserve existing diversity. Although more germplasm are being evaluated, the use of genetic resources has been much poorer to their evaluations mainly due to undesirable tuber traits of the wild species and crossability barriers. This has led to narrow genetic base of the cultivated potatoes. The 'Irish famine' of 1840s depicts the devastating effect of

A. Srivastava
Central Potato Research Station, 5th Mile Upper Shillong, Shillong 793 009, Meghalaya, India
e-mail: bhu.avinash@gmail.com

V. Bhardwaj · B. Singh
Central Potato Institute Shimla, Shimla 171001, HP, India
e-mail: vinaycpri@gmail.com

B. Singh
e-mail: directorcpri@gmail.com

S.P. Khurana (✉)
Faculty of Science, Engineering & Technology, University Science Instrumentation Centre, Amity University, Gurgaon 122413, Haryana, India
e-mail: smpaulkhurana@gmail.com

V.R. Rajpal et al. (eds.), *Gene Pool Diversity and Crop Improvement*, Sustainable Development and Biodiversity 10,
DOI 10.1007/978-3-319-27096-8_6

growing large areas under a single variety. Cultivated potato exhibits complex tetrasomic inheritance and high heterozygosity. Dihaploids of tuberosum cross readily with many diploid species thus providing opportunity for introgression of useful traits from alien sources to cultivated background. The other well-exploited techniques in potato breeding, viz. somaclonal variations, somatic hybridization, molecular markers, genetic transformation and RNAi approaches. Potato is one of the rare crops where maximum tissue culture and genetic engineering interventions have been connoted. Today, potato genome is sequenced and it opens up new vistas for developing tailor-made varieties in future.

Keywords Potato diversity · Germplasm conservation · Genetic stability · Abiotic stresses · Quality traits

6.1 Introduction

Potato rightly called, 'the vegetable that changed history' provided both the spark and the fuel for centuries to the social change. While conquering the world, it was banned and lauded, cursed and praised, feared and loved until humanity welcomed it into its home and hearth. Today, as one of the world's major non-cereal food crop, potato is grown in more than 148 countries in a wide variety of soils and climates surpassed only by wheat, rice and maize in total production. Yet till sixteenth century it was unknown to the people of Europe, Asia, Africa and North America. The crop has a very fascinating history of its origin, evolution and spread in the world, stretching to nearly 7000–9000 years back. Some of it is well documented while other has been chronicled from the archaeological remains and historical evidence (Hawkes 1990).

6.2 Origin

The potato is believed to be originated in the South American continent where it grows as wild in nature and represents the widest diversity of forms in tuber shape, size, colour, taste, etc. The mainly cultivated potato species *S. tuberosum* L., a tetraploid ($2n = 4x = 48$) is believed to have originated from the basin of lake Titicaca on Peru–Bolivian borders from its wild diploid ancestors many of which may be extinct now. Two main centres of diversity of tuber-bearing *Solanum* species are Central America and Andean region of North-western Argentina, Peru and southern Bolivia. The species grow in a wide variety of habitats from semi-desert conditions of northern Argentina, southern Bolivia and Mexico to the high rainfall receiving subtropical forests of Central and South America exhibiting a wide adaptation to altitude right from the sea level to nearly 5,000 m.

Archaeological evidence suggests that the potters from the Moche cultures in northern Peru (c. AD 1–600) and the Chimu people (c. AD 900–1450), as well as Huari or the Nazca valley in southern Peru (c. AD 650–700) obtained potatoes by barter or other means from farmers in the highlands where potatoes were actually 'cultivated'. Actual remains of the potatoes were also recovered infrequently from tombs, dwellings and rubbish heaps including chuño or tunta, from some archaeological sites. Archaeological remains of potatoes from the Chilca valley near Lima have been radiocarbon-dated to 7000 years before present (Hawkes 1990). There is much later evidence for potatoes from rubbish heaps, graves and food stores in 4500–3500 BC (Ugent et al. 1982). The first historical record of potato can be traced back to 1537, when a band of Spaniards led by Jiménez de Quesada penetrated into the highlands of Colombia. Later potatoes were accounted by López de Gomara (1552) in southern Peru and by Pedro Cieza de León (1553) in the southern Colombia and northern Ecuador. Potatoes in Chile received the first mention by Sir Francis Drake in 1578 (Drake 1628).

6.3 Early History

In South America, where it originated, potato was the most productive source of main food for centuries for the people in the high Andes and southern Chile. Potatoes were dried by Andean Indians to make chuño, a freeze-dried potato powder of the bitter, frost-resistant potatoes grown at 3,600–4,400 masl during food shortage and between successive crops during periods of scarcity caused by frost or other unfavourable growing conditions. Following the conquest of Peru, the Spaniards introduced potatoes in Spain and further spread it to many European countries including Italy, Belgium, Germany, France, Switzerland and Holland by the end of the sixteenth century. Initially, potato was grown only as a curiosity in the Europe's botanical gardens and remained a shunned plant—at best a food for swine and country bumpkins for next two centuries. The crop remained a botanical curiosity till about the mid-eighteenth century, and was not grown in any Western European country barring Ireland, where potatoes became the most profitable new crop, mainly for human consumption, and for pigs thriving well on potatoes. Throughout the eighteenth century, none seems to have been aware of the danger to the economy of a nation dependent on a single crop. The warnings of Curwen (1818) went unheeded till August 1845, when suddenly one warm, rainy day in August, an unknown malady (late blight) struck the Irish potato fields. Potatoes quickly rotted in the fields, sending an unbearable stench across the countryside and repeating the same scene across whole of Europe. This was also true in 1846, 1847 and 1848 resulting in famous famine and death of nearly 2.5 million and migration of one million Irish including the famous Kennedys and Reagans to North America.

6.3.1 Spread in Europe

Potato was introduced into Europe, first in Spain in *c.* 1570 and second in England in *c.* 1590 (Hawkes 1990). The Spanish introduction rests on market records from the Hospital de la Sangre in Seville, whilst the English records are extremely complex; nevertheless, we know from Gerard that he grew potato in his garden and described and figured it in *Herbal* of 1597.

Hawkes (1994) reports that European herbalists obtained their potatoes from the Flemish herbalist, Carol Clusius (1601), his specimens being derived from Spain via Italy. The early European potato came from the Andes, perhaps from the northern Colombian part. It is interesting to note that these first European potatoes were adapted to short (12 h) day of the Andes and not to long (16–18 h) day as present day potatoes of Europe, a fact which is also confirmed from the evidence from the Spanish archives. These potatoes were probably the Andean form of the tetraploid potato (*S. tuberosum* subsp. *andigena*), which further evolved through several centuries of 'unconscious' selection in Europe to adapt it to the long summer days of northern Europe showing morphological correlates to this evolutionary change, including reduced top growth, shorter internodes, larger leaves and reduced flowering and fruiting. Hence it was not until the late eighteenth and early nineteenth centuries that this new-day-length adaptation was complete, allowing potato cultivation on a large scale to spread into Central and Eastern Europe.

From Spain the potato was taken to Italy by the Carmelite Friars, as Clusius (1601) mentions that it was grown in Italy before 1587. Clusius received potato from Italy and sent to botanists in many parts of Germany and Austria. The Swiss herbalists C. Bauhin and J. Bauhin obtained tubers from Clusius in the late sixteenth century and sent them to France by about 1600. The Slavic nations seem to have obtained their potatoes from Germany since the names of potato are derived from German ones, e.g. *Kartoffel, Grundbirne,* etc. The potato was brought to Russia by Peter the Great from Holland at the end of seventeenth century.

After the introduction into England in 1590, it was not until the mid-eighteenth century that it was grown on a large scale. The same time scale is recorded for Scotland and Wales. In Ireland, it was grown on field scale by the early seventeenth century. From Scotland the potato was taken to Norway in the mid-eighteenth century and then to Sweden and Denmark.

6.3.2 Spread in Asia, Africa

The potato's global voyage began in the seventeenth century. While stay-at-home Europeans may have had misgivings about the new crop, the sailors, soldiers, missionaries, colonial officials and explorers quickly carried it to their foreign outposts. Thus, Belgian, British, Dutch, French, Portuguese and Spanish sailors carried the potato first to ports in Asia and the South Pacific while trading, whaling and fishing and later inland to their homes.

Dutch settlers believed to have taken potatoes to the Penghu Islands in the Taiwan Strait as early as 1603 (Anonymous 2007). Belgian and French missionaries introduced them into Taiwan. Soon the crop had spread throughout China passing across Eastern Europe, over the Urals and into the steppes of Asia.

The potato arrived in Africa relatively late. A few grew in South Africa as early as 1830, but British and German colonists and missionaries did not introduce potatoes into East Africa until about 1880. In North and West Africa, the two world wars were the main stimulus for the crop's introduction. With supply lines from Europe cut, armies and colonial personnel were forced to grow their own *bombiderres*. While Africa is not a major producer in terms of volume, more African countries grow potatoes today than any other continent.

In the latter half of the twentieth century, the crop found a home in the arid Middle East, where it established itself as an important commodity in Jordan, Israel and other countries. It is even grown in climate-controlled facilities in the Gulf States.

In North America, potato was completely unknown until the early seventeenth century. It first received potatoes from England via Bermuda in 1621 where it was introduced in 1613. The first potatoes were grown in Virginia. Later in the century, there were more introductions from England and Ireland, but no records of an introduction were made from South America before Goodrich, in 1863, (Hawkes 1992) obtained some varieties in a Panama market.

6.3.3 Spread in India

In India, potato was introduced in the early years of the seventeenth century, most probably by either Portuguese sailors or by the Britishers to the hills of the north India and to Sri Lanka where it flourished in the colonial home gardens. The earliest reference of the potato occurs in account of the voyage of Edward Terry in 1655 (Upadhya 1974) who was chaplain to Sir Thomas Roe, British Ambassador to the court of the Mughal Emperor Jahangir from 1615 to 1619. Similarly, Fryer's travel records (1672–1681) mention the potato as a well-established garden crop in Surat and Karnataka in 1675.

By late eighteenth or early nineteenth century the potato was an important, established, vegetable crop in the hills and plains of India. Early introductions resembled the andigena potatoes and were adapted to short winter days having long dormancy and capable of withstanding higher temperatures under country stores. They were grown by various names in local dialects depicting some character, viz. *Phulwa*—flowering in the plains; *Gola*—round potatoes, *Satha*—maturing in 60 days, etc. and came to be known as *desi* varieties (Pushkarnath 1969).

Between 1924 and end of World War II, the state agricultural departments and other agencies introduced a large number of European potato varieties with a view to selecting those suitable for local conditions. These efforts, however, proved of little value mainly due to poor adaptation of temperate long day adapted European varieties in the subtropical short days available in plains of India. Only few foreign varieties approached the yield levels of local varieties under commercial culture, however, none proved to be a very good yielder. Fast degeneration of seed stocks was another important factor in non-establishment of European varieties in the subtropics. Lack of adequate cold storages for storing seed potatoes over long periods of hot summer also posed problems. The local old varieties could be kept well in country stores, whereas the imported varieties could not survive these conditions. The introductions from Europe thus made no impact on potato culture in subtropical plains of India. However, a few introductions such as Magnum Bonum, Up-to-Date, Royal Kidney, Great Scot, Craig's Defiance, etc. survived in the hills. In India till 1950, 16 each of desi and European varieties were identified to be mostly under cultivation.

6.4 Genetic Diversity Among Cultivated and Wild Potatoes

There are seven cultivated tuber-bearing *Solanum* species, viz. *S. stenotomum*, *S. ajanhuiri*, *S. phureja*, *S. chaucha*, *S. juzepczukii*, *S. tuberosum* ssp. *andigena*, *S. tuberosum* ssp. *tuberosum* and *S. curtilobum*, occurring in a polyploid series ranging from diploid to pentaploid. Several of them are fairly similar to each other probably because they were initially confined to cool temperate climatic region of the Andes of South America and the lowlands of southern Chile and for that reason

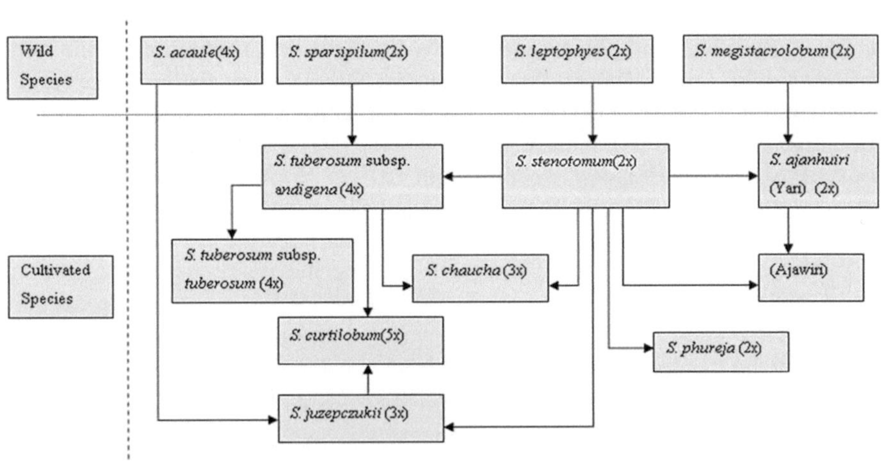

Fig. 6.1 Evolutionary relationships of cultivated potatoes and their ploidy levels (after Hawkes 1990)

were classified by Dodds (1962) as 'groups' of *S. tuberosum* rather than distinct species. Their probable evolutionary relationships are shown in Fig. 6.1.

The wild potatoes seem to have evolved by means of geographical and ecological isolation rather than by genetic incompatibility. The related wild species are much more widespread. The diploid species, *S. stenotomum* is grown from central Peru to central Bolivia and is believed to be the most primitive, probably having been derived from the diploid wild species, *S. leptophyes*, or possibly *S. canasense*, both of which still occur in the central part of its distribution area.

At least four wild potato species are widely believed to be involved in the process of evolution of the cultivated species of potato. Evidence indicates that hybridization of *S. stenotomum* with the weedy species *S. sparsipilum* and subsequent chromosome doubling produced the tetraploid *S. tuberosum* subsp. *andigena* in the central Andes (Cribb and Hawkes 1986). Some workers, however, consider that the tetraploid Andean potatoes are derived from *S. stenotomum* by simple chromosome doubling. This tetraploid subspecies was carried by ancient people into southern Chile, where it became adapted to the long-day length to evolve into subsp. *tuberosum*. A similar process in Europe caused the same development to take place under the long-day conditions. However, certain authors (Grun 1979) believe that subspecies of *tuberosum* from Chile and Europe differ from subspecies of *andigena* by certain cytoplasmic factors acquired from some wild diploid species, such as *S. chacoense*.

In pre-conquest days, the cultivated diploid species *S. phureja* evolved from *S. stenotomum* through a process of artificial selection by Andean farmers in lower, warmer eastern valleys and acquired shorter dormancy so that three crops could be grown in a year. In contrast, natural hybridization of *S. stenotomum* with the wild frost-resistant species *S. megistacrolobum* gave rise to the diploid *S. ajanhuiri*. The F_1 hybrid produced the 'Yari' group of varieties and a probable backcross to the cultivated parent gave rise to the 'Ajawiri' group of varieties. Similarly, the F_1 cross from a series of hybridizations between *S. stenotomum* and the wild tetraploid species *S. acaule* gave rise to a highly sterile triploid *S. juzepczuki*, which incorporated the strong frost resistance of *S. acaule*. A further natural cross between *S. juzepczukii* and *S. tuberosum* subsp. *andigena* produced the only slightly less frost-resistant pentaploid species *S. curtilobum*. This evidently involved a $2n$ gamete from *S. juzepczukii* and a normal gamete from *S. tuberosum* subsp. *andigena* (Hawkes 1990). A series of crosses between *S. stenotomum* and subsp. *andigena* have given rise to the triploid hybrids named *S. chaucha*.

We thus have a network of cultivated species or species groups, which evolved chiefly in the central Andes of Peru and Bolivia, involving four original wild species, viz. *S. acaule*, *S. sparsipilum*, *S. leptophyes* and *S. megistacrolobum*. All but two of these cultivated potatoes have always been confined to that central area. However, the diploid *S. phureja* has extended northwards into Ecuador, Colombia and Venezuela, whilst the tetraploid *S. tuberosum* spread into southern Chile.

A lot of studies have been done to assess genetic diversity among different cultivars grown across the world utilizing both conventional and molecular tools. Molecular markers have become important tools in studies of genetic diversity

(Bered et al. 2005), due to the high resolution and reliability in the identification of cultivars. They are also applied in the genetic characterization of potato (Ford and Taylor 1997, Schneider and Douches 1997). Random amplified polymorphic DNA (RAPD) markers have the advantage of detecting polymorphism simply and quickly (Demeke et al. 1996; Kujal et al. 2005), while simple sequence repeat (SSR) markers or microsatellites provide high reproducibility and genetic informativeness. Both the markers have been used in the molecular characterization of potato cultivars (Coombs et al. 2004) as well as of other species, e.g. soybean (Garcia et al. 2007). These studies have unanimously concluded that the cultivable varieties possess narrow genetic base. The European cultivated potato has been known to have arisen from a limited number of introductions (Glendinning 1983), resulting in a low level of genetic diversity, compared to the potato gene pool of the American countries. Moreover, the selection of genotypes which produced tubers under long-day conditions, combined with selection for superior agronomic traits, further narrowed the European gene pool (Provan et al. 1999; Spooner and Alberto 2006).

6.5 Maintenance of Diversity

6.5.1 Gene Banks

Up to early twentieth century, *S. tuberosum* L. was the only potato species known outside South America. It was only after the first expedition carried out in centre of origin by Prof. S.M. Bukasov and his co-workers in 1925–1926 that knowledge about the wealth of genetic diversity among the tuber-bearing *Solanum* started accumulating. Thereafter a number of N.I. Vavilov expeditions largely led by S.M. Bukasov and S.W. Juzepczuk were held in the Americas. This laid the foundation of the first gene bank in Leningrad starting in 1927. This gene bank still survives as N.I. Vavilov Institute of Plant Industry, Russia. Taking advantage of the pioneering Russian work, British Empire expedition held in 1939, collected materials in Mexico and the Andean countries of South America. As a result, Commonwealth Potato Collection was established in Cambridge and is now situated at Scottish Crop Research Institute, Pentlandfield, Scotland. Lately, the major bulk of collection has been mediated by the International Potato Centre (CIP), Lima, Peru, a CGIAR organization with the primary mandate of collection and conservation of vast potato genetic resources. Presently, CIP is the holder of the largest collection of the potato germplasm. In addition, large potato collections also exist at Inter-regional Potato Introduction Station (IR-1), Sturgeon-Bay, Wisconsin, USA; Instituto Nacional de Tecnologia Agropecuaria (INTA), Balcarce, Argentina; Chilean Potato Gene bank in Valdivia, Chile; Dutch-German Potato Collection, Braunschweig, Germany, and Institut fur Kartoffelforschung, Gross-Lusewitz, Germany. Some national potato programmes also maintain potato genetic resources, e.g. in India, Central Potato Research Institute, Shimla holds a modest collection of more than 4,100 accessions of elite potato varieties, parental lines as well as wild species imported from

Fig. 6.2 Variability in potato germplasm for tuber & flesh colour

40 countries with good variability in tuber skin and flesh colour (Fig. 6.2). This is the largest potato collection in Asia (Gopal and Gaur 1997). In spite of concerted efforts to collect the genetic variability, only about 130 out of the total known number of 235 potato species exist in gene banks. More efforts are needed to collect the others, which are known only as descriptors or dried specimens.

6.5.2 Conservation

There are two basic approaches for conservation of genetic resources: in situ conservation and ex situ conservation. In situ conservation refers to situations where the material is maintained in the natural habitat, within the community of which it forms a part. In situ conservation is based on the concept that it allows natural evolution to continue. In situ conservation is generally suggested only for wild relatives because they alone live in natural communities. It is carried out under the auspices of national governments through biospheres reserves, national parks, world heritage sites and other protected areas. It can also be carried out through farmers who manage land races and wild relatives in eco-geographic pockets of genetic diversity. In potato, there is no specific programme to follow this method of conservation because of the practical problems of maintenance, economic viability, etc. Further, in this population size should be large enough to avoid risk of inbreeding and of genetic drift as these lead to the decay of genetic diversity. It is not realistic to expect that all species and their component populations can be covered by natural reserves, national parks and other protected areas. Further this method cannot safeguard the species in face of unforeseen natural calamities.

Due to the practical limitations of in situ conservation, it is preferred to maintain variability under managed conditions (ex situ) in gene banks. This method, however, freezes the evolutionary process of species though it ensures conservation of its existing genetic variability. Potato genetic resources can be conserved as vegetative propagules or as true (botanical seeds). Potato is a highly heterozygous crop due to which sexual reproduction results in segregating populations. So when the objective is to maintain the exact genotype of an accession, vegetative propagation (in vitro or in vivo) is the only option. However, if the objective is not to maintain the exact genotype, but total gene pool, germplasm can be conserved as true seeds.

In vivo propagation is done through tubers in glass house, as well as in fields. Genetic identity is maintained by roguing mixtures and clones are protected from diseases and pests by using various protective measures. Still, there is risk of exposure of germplasm to viral and mycoplasmal diseases from year to year, resulting in degeneration of the germplasm stocks. Loss of material due to natural calamities and loosing identity due to mechanical mixtures or wrong labelling are the other risks associated with this method. Labour and maintenance costs are also high. This is the traditional method of conservation. This method, however, provides a continuous opportunity to evaluate and compare the characteristics of different genotypes. Further, seed/propagules can be readily made available to the users.

The ability to grow plants under aseptic conditions has allowed the development of in vitro preservation techniques for germplasm conservation and exchange. In vitro maintenance of germplasm has several advantages. A large number of accessions can be conserved in a small space under disease free conditions irrespective of the crop season. The risk of loss due to biotic and abiotic factors is minimized and the possibility of cross infection between accessions is eliminated. In vitro materials can be made free of systemic bacteria, fungi, viruses and mycoplasmas and maintained as pathogen-free stocks (Khurana and Garg 1998; Jeffries et al. 2006). This eases the quarantine regulations for international distribution of genetic materials. A major limitation of in vitro germplasm conservation is the likelihood of genetic instability in the process of culture. In vitro potato germplasm can be conserved through slow-growth conservation for medium-term storage or cryo-preservation for long-term storage.

6.5.2.1 Slow-Growth Conservation

This is based on the micropropagation of apical or axillary buds (nodal cuttings) on modified Murashige and Skoog medium. In order to avoid frequent subculturing of micropropagated plants, subculture period is enhanced by following the slow-growth conservation strategy. For this, a number of approaches like low temperature storage, reduced light intensity, high sucrose concentration, use of osmoticums or growth retardants in the medium, increase in volume of medium, sealing of the culture vessels, mineral oil layer on the medium, etc. are used. The most commonly used protocol for in vitro conservation of potato is based on combination of low temperature, reduced light intensity and use of osmoticums. By this method, in vitro plantlets can be conserved for 2–3 years depending on the genotype (Fig. 6.3). In subtropics, where maintenance of low temperature is problematic and expensive due to high demand on energy, plantlets could be conserved at normal propagation temperatures using MS medium with osmoticum (4 % sorbitol, 2 % sucrose) for 12 months without subculture (Khurana et al. 1998).

Fig. 6.3 In vitro conservation of potato germplasm

6.5.2.2 Cryo-Preservation

By this method plant material is frozen at ultra low temperature around −196 °C of liquid nitrogen. At ultra low temperatures, the cells are in state of metabolic inactivity. Due to inhibition of cell division this method allows storage of material with minimal risk of genetic instability. In this method, meristematic tissue of in vitro material is usually used. The technique can also be used for preserving embryos, callus, pollen, cell suspension, etc. These are, however, rarely used as potato germplasm due to their intrinsic genetic variability. This is followed by pretreatment with cryo-protectants with low molecular weight such as glycerol and dimethyl sulphoxide, which penetrates the cell with ease and high molecular weight compounds such as polyvinyl pyrrolidone and dextran, which penetrates slowly and can reduce cryo-damage significantly. They protect surface membranes by reducing growing rate and size of ice crystals, and by lowering the effective concentration of solutes in equilibrium with ice inside and outside the cell. They also help to increase membrane permeability which aids removal of water from the cell and facilitates protective dehydration in the early stages of freezing. By pretreatment with cryo-protectants some degree of dehydration is induced in cells and tissues thereby avoiding the damage caused by the formation of ice crystals during the freezing and defrostation process. Freezing is the most crucial step in the whole process of cryo-preservation. This can be achieved through slow freezing where cultures are frozen by slow cooling at freezing rate between 0.5 and 4 °C per min, starting from 0 °C until the temperature reaches −100 °C, and finally transferred to liquid nitrogen; by rapid freezing wherein the materials contained in vials are lowered directly into a tank filled with liquid nitrogen. The temperature decreases rapidly at the rate of 300–1,000 °C per min. The stepwise freezing method combines both the procedures of slow and rapid freezing. Initially plant material is cooled slowly and stepwise (ca 1–5 °C per min) to an intermediate temperature, maintained at that temperature for 30 min, and then rapidly cooled by plunging it into liquid nitrogen. In the initial slow freezing, ice is formed outside the cells and the unfrozen protoplasm losses water due to the vapour pressure deficit between the super-cooled protoplasm and the external ice. Vitrification is a process by which water undergoes a phase transition from a liquid to an amorphous 'glassy state'; in this form, water does not possess a crystalline structure. Vitrification occurs when the solute concentration becomes so high that ice formation is prevented and the water molecules form glass. For this, tissue is sufficiently dehydrated with a highly concentrated vitrification solution at 25 or 0 °C without causing injury prior to immersion in liquid nitrogen. Plant vitrification solutions like PVS2 and PVS3 have been developed which are glycerol based and less toxic. During encapsulation/dehydration, the cells/tissues are trapped into calcium alginate beads followed by incubation in 0.85 M sucrose (as sole source of cryo-protection) for 14–16 h, air-drying for 3–4 h in a laminar flow chamber and rapid freezing in liquid nitrogen. The above alternatives are also combined, e.g. encapsulation–vitrification, pregrowth–desiccation, droplet freezing, etc. to achieve the desired results. Thawing of the frozen material is achieved by transferring it to warm water at 37–40 °C. The optimal thawing rate is one that prevents ice formation by recrystallization in the process of warming. The recovery of thawed cultures can be improved by nursing them through an initial recovery period involving gradual dilution of cryo-protectants through several steps of washing and by keeping osmotic disruption to the minimum. Plants are regenerated from the recovered tissue by culturing on suitable tissue culture media (Khurana et al. 1998). To avoid somaclonal variations, shoot tips are directly regenerated into plantlets, without any adventitious growth or callus formation. This is ensured by developing suitable protocols of cryo-conservation and regrowth media for different cultivars/accessions.

6.5.2.3 Sexual Propagation

Potato true seed (botanical seed or TPS) is orthodox type, meaning it is able to tolerate a high degree of desiccation and storage at low temperature, i.e. between 5 and –20 °C without loss of viability. Due to this, preservation of potato germplasm through sexual propagation as true seed is less laborious and much cheaper than the preservation of vegetatively propagated material. In addition, it is easy to maintain the material free of pathogens this way, as only a few virus diseases are known to be seed transmitted. Seeds occupy relatively small space and their transport is also economical. However, majority of the wild species of potatoes are diploid and self-incompatible. In such species true seeds are required to be produced by sib-mating. The sample size should be large enough to maintain the heterozygosity and to preserve all alleles at the loci of interest. Sample size depends on the nature of genetic control of the character(s) concerned.

True seeds with low moisture content can be kept at low temperatures for many years. Successful seed storage depends on effective control of several factors including physiological maturity of the seed, temperature, seed moisture content, storage atmosphere, etc. It is desirable to harvest seeds at physiological maturity as germination, and vigour are maximum in fully mature seeds and longevity of mature seeds is more than in those harvested at other stages. The seeds extracted from the mature berries should be dried in shade or in seed dryers at temperature <30 °C. Drying in direct sunlight or at temperatures above 30 °C can adversely affect the viability of the seed. The moisture content of the seeds can be further reduced by keeping them in silica gel for a week or so. Within certain limits, viability of the seeds increases by drying and storing them at low temperature. For long-term storage, potato seeds should have moisture content 5 ± 1 % and stored at −10 to −20 °C. For short to medium-term storage samples should be stored at 0–10 °C after drying up to 8 ± 1 % moisture content. Before storage, the samples should be hermetically sealed in airtight containers.

6.5.3 Genetic Stability of Conserved Germplasm

Preservation of the genetic integrity of plants generally multiplied by vegetative means is fundamental to long-term conservation. Field gene banks where potato accessions are maintained by propagation through tubers are generally considered safe from this angle as natural mutations are rare, and if there is any off-types can be rogued out. However, apprehensions are expressed about the genetic integrity of the material conserved in vitro. In slow-growth in vitro conservation, plantlets are grown under suboptimal conditions. Due to this they show signs of stress and have bush like appearance with thin stems and reduced or nil leaves. Continuous maintenance under these conditions may lead to physiological as well as genetic changes in the material. In the cryo-preservation protocols, genetic variations could occur during tissue culture before or after cryo-preservation or during eventual plant regeneration. During the storage phase, some genetic variations may be induced by the accumulation of mutations caused by background ionizing radiations, even in explants which were previously genetically stable. Bi-nucleation and abnormal chromosome distribution in DMSO (cryo-protectant) treated cells has also been reported.

Accurate characterization of the clone is required to answer questions related to genetic stability. Morphological analysis provides one approach for the description of the genetic stability when phenotypic variation is demonstrably linked to genetic variation. The expression of morphological characters, however, can be subject to environmental or physical effects. This and other limitations have led to research on methods of analysis based on more stable features of the genome. Non-specific proteins and isozyme analyses are also used, but these too suffer from environmental and developmental effects. Recently, molecular markers such as RFLP, RAPD, AFLP and SSR have shown their usefulness in the genetic characterization of plant material after long-term storage. RAPD and AFLP are dominant markers that help to determine genetic distance in terms of providing estimates of genetic differences rather than of genetic similarities. The advantage of co-dominant markers, such as microsatellites, is the possibility to display all four alleles at a single locus. Hence by their nature, microsatellites are more informative in detecting genetic changes, if any.

6.5.3.1 Evaluation of Germplasm

Evaluation for different traits

The species evaluated for various characters of resistance or adaptation are even fewer than those collected. Important sources of resistance to the major potato diseases and pests, and adaptation to environmental extremes are listed in Table 6.1.

Table 6.1 Sources of resistance to the major biotic and abiotic stresses

Biotic/Abiotic stress	Causal organism	Source of resistance
Fungus resistance	*Phytophthora infestans* (late blight)	
	i. Race specific	*S. cardiophyllum, S.demissum, S. ediense, S. stoloniferum* and *S. verrucosum*
	ii. Non-race specific	*S. berthaultii, S. bulbocastanum, S. chacoense, S. circaeifolium, S. demissum, S. microdontum, S. phureja, S. pinnatisectum, S. polyadenium, S. stoloniferum, S. tarijense, S. tuberosum* ssp *andigena, S. vernei* and *S. verrucosum*
	Alternaria solani (early blight)	*S. bulbocastanum, S. chacoense* and *S. tarijense*
	Synchytrium endobioticum (wart)	*S. acaule, S. berthaultii, S. boliviense, S. demissum, S. gourlayi, S. sparsipilum, S. spegazzinii, S. sucrense, S. tuberosum* (both subspecies) and *S. vernei*
	Fusarium spp. (Fusarium wilt)	*S. acaule, S. kurtzianum* and *S. spegazzinii*
Bacterial resistance	*Pseudomonas (Ralstonia) solanacearum* (bacterial wilt)	*S. chacoense, S. microdontum, S. phureja, S. sparsipilum* and *S. stenotomum*
	Erwinia carotovora (soft rot; blackleg)	*S. acaule, S. brevidens, S. bulbocastanum, S. chacoense, S. demissum, S. hjertingii, S. leptophyes, S. megistacrolobum, S. microdontum, S. phureja, S. pinnatisectum, S. tuberosum* ssp. *andigena* and *S. vernei.*
	Streptomyces scabies (common scab)	*S. chacoense, S. commersonii, S. jamessi, S. tuberosum* ssp. *andigena, S. yungasense* and various cultivated varieties
Virus resistance	Potato virus X	*S. acaule, S. berthaultii, S. brevicaule, S. chacoense, S. commersonii, S. curtilobum, S. phureja, S. sparsipilum, S. sucrense, S. tarijense* and *S. tuberosum ssp. andigena*
	Potato virus Y	*S. acaule, S. chacoense, S. demissum, S. gourlayi, S. phureja, S. rybinii, S. stoloniferum, S. tuberosum* ssp. *andigena*
	Potato virus S	*S. tuberosum* ssp. *tuberosum* varieties like Great Scot, Kerries Pink, Red Skin, Eclipse, British Queen, Up-to-date, Duke of York and Foxlyna etc.
	Potato leaf roll virus	*S. acaule, S. brevidens, S. chacoense, S. demissum, S. etuberosum, S. raphanifolium, S. stolonifrum* and *S. tuberosum* ssp. *andigena*
	Spindle tuber viroid	*S. acaule, S. berthaultii, S. gurreroense, S. hjertingii* and *S. multidissectum*

(continued)

Table 6.1 (continued)

Biotic/Abiotic stress	Causal organism	Source of resistance
Insect resistance	*Leptinotarsa decemlineata* (Colorado beetle)	*S. berthaultii, S. chacoense, S. commersonii, S. demissum, S. jamesii, S. pinnatisectum, S. polyadenium* and *S. tarijense*
	Myzus persicae, Macrosiphum euphorbiae (aphids)	*S. berthaultii, S. bukasovii. S. bulbocastanum, S. chomatophilum, S. infundibuliforme, S. lignicaule, S. marinasense, S. medians, S. multidissectum, S. neocardenasii, S. stoloniferum*
	Phthorimaea operculella (Tuber moth)	*S. chacoense, S. stenotomum* and *S. tuberosum* ssp. *andigena*
Nematode resistance	*Globodera rostochiensis, G. pallida* (potato cyst nematode)	*S. acaule, S. berthaulti, S. boliviense, S. bulbocastanum, S. capsicibaccatum, S. cardiophyllum, S. demissum, S. gourlayi, S. kurtzianum, S. leptophyes, S. multidissectum, S. oplocense, S. sparsipilum, S. spegazzinii, S. sucrense, S. tuberosum* ssp. *andigena* and *S. vernei*
	Meloidogyne incognita (root-knot nematode)	*S. bulbocastanum, S. cardiophyllum, S. chacoense, S. curtilobum, S. hjertingii, S. kurtzianum, S. microdontum, S. phureja, S. sparsipilum and S. tuberosum* ssp. *andigena*
Physiological characters	Frost	*S. acaule, S. ajanhuiri, S. boliviense, S. brachistotrichum, S. brevicaule S. brevidens, S. canasense, S. chomatophilum, S. commersonii, S. curtilobum, S. demissum, S. etuberosum, S. juzepczukii, S. megistacrolobum, S. multidissectum, S. pumilum, S. raphanifolium, S. sanctae-rosae, S. toralapanum and S. vernei.*
	Heat and drought	*S. acaule, S. bulbocastanum, S. chacoense, S. commersonii, S. gourlayi, S. megistacrolobum, S. microdontum, S. ochoae, S. papita, S. pinnatisectum, S. spegazzinii* and *S. tarijense*
	Lack of tuber blackening	*S. hjertingii*
	High protein content	*S. phureja* and *S. vernei*
	High starch content	*S. vernei*
	Low reducing sugars at low temperature storage	*S. phureja, S spegazzinni* and *S. vernei*
Quality traits	Chip making directly from cold storage	*S. medians, S. okadae, S. pinnatisectum, S. raphanifolium, S. sogarandinum*
	Use in medicine	*S. siparunoides, S. sisymbrifolium, S. stramonifolium, S. tuberosum*
	High carotenoid content	*S. phureja, S. stenotomum*
	High starch content	*S. phureja, S. vernei*
	High ascorbic acid content	*S. phureja, S. estoloniferum*

Information on usefulness of various accessions and species is getting enriched and published as more and more germplasm is evaluated. Where such information is not available, passport data on characteristics of the natural habitats of species are of great importance, for example late blight resistant species are found in Mexican gene pool where late blight fungus has been found to reproduce sexually. Similarly frost resistance is found in species capable of growing at altitudes above 3,500 m. Species from dry and warm climate areas are tolerant to these stresses.

6.6 Utilization of Genetic Resources

Use of genetic resources has been even much poorer to their evaluation. Only 10 primitive cultivars or species had been used in the pedigrees of 62 % of the 627 cultivars listed in the Index of European Potato Varieties. Remaining 38 % cultivars involved only ssp. *tuberosum.* Only 13 species have been used so far in the variety improvement programmes of the world. Parentage of as many as 171 American varieties could be traced to a single variety Rough Purple Chilli introduced from Chiloe region of southern Chile. The wild species which occur in the pedigree of several cultivars are briefly described below.

S. demissum: A hexaploid, self-fertile spp. from Mexico. It is resistant to PVY and PLRV, hypersensitive and resistant to late blight, resistant to wart and both *Globodera* species; alkaloids in this confer resistance on Colorado beetle and other insects. It is moderately resistant to frost also. Genes of *S. demissum* have been incorporated into more than 50 % of the world's cultivars mainly for resistance to late blight and PLRV.

S. acaule: A tetraploid and also hexaploid, self-fertile spp. from Andes of Argentina to Peru. It is extremely resistant to PVX and resistant to PLRV, PSTVd, wart, both spp. of *Globodera* and frost.

S. chacoense: A diploid spp. from Argentina, Southern Brazil, Bolivia, Paraguay and Uruguay. It is extremely resistant to PVA and PVY, resistant to late blight, Colorado beetle, tuber moth and other insects. *S. chacoense* has been used mainly with *S. phureja* to produce a hybrid which served as a bridge between *S. demissum* and *S. tuberosum* to create tetraploid hybrids by the smooth euploid transfer of gene into ssp. *tuberosum.*

S. spegazzinii Bitt: A diploid spp, from Northwest Argentina. It is resistant to *Fusarium,* wart and *Globodera* spp and has low content of reducing sugars. This has been mainly used as a source of resistance to nematodes.

S. stoloniferum Schlechtd. et Beche: A tetraploid, self-fertile spp. from Mexico. It is extremely resistant to PVA and PVY, hypersensitive and field resistant to late blight. About 20 cultivars carry the gene *Ry* for extreme resistance to PVA and PVY derived from *S. stoloniferum.*

S. vernei Bitt. et Wittm.: A diploid spp. from Northwest of Argentina. It is outstanding for protein and starch content and resistant to late blight and both spp. of *Globodera.* It has low content of reducing sugars. It has been used in breeding mainly for resistance to nematodes.

In the utilization of gene pool, it is just as important to be aware of possible undesirable characters as it is to know of desirable characters from a specific source. Undesirable tuber traits of the wild species and crossability problem of certain species have acted as deterrents for their use by the breeders. Pre-breeding of the wild species for combining resistance to various diseases and insect pests with agronomic characters requires special attention. This will undoubtedly lead to a closer association between gene banks and gene bank users and help in turning 'gene bank collections' into 'working collections'.

6.7 Potato Genetic Enhancement

Even though wild species of potato have been known to harbour myriad of genes for various biotic and abiotic resistance traits, the associated genetic load of poor yield and wildness traits renders them unfit for any breeding programme. Limited effort has been done to improve the tuber traits of these wild species through pre-breeding. These traits can be those missing from the advanced populations with resistance to pests, viruses or bacterial wilt, complementary genes for more durable resistance against late blight, resistance genes for emerging man-made constraints like heat, salinity, etc. or genes for improved nutritional value such as zinc and iron. Despite lack of pre-breeding in most of the cases, wild species have been used enough for transfer of the resistance genes to cultivated species.

6.7.1 Utilization of Wild Species for Improvement of Cultivated Potato

Conventionally, potato breeding has been restricted to the members of the cultivated tuberosum accessions. Choice of a parent is most often based on its de facto phenotypic expression of the characters of interest. Normally breeders try to pick pairs of clones/varieties for inter-mating based on complementary sets of characters with desirable level of expression. The ideal strategy would be to identify the superior parents based on their general combining ability and inter-mating them in all feasible combinations to select a few top combinations based on progeny test. This is followed by selection over successive generation for yield and other desirable traits (Bradshaw and Mackay 1994).

S. tuberosum is known to have a narrow genetic base and little improvement is observed in tuber yield among the recombinants produced by crossing among *tuberosum* parental clones. To overcome these problems attempts were made in Europe and North America (where the crop is grown under temperate long days) to use *andigena* in potato breeding programmes. But the short-day requirement for tuberization of *andigena* resulted in poor performance of *tuberosum* × *andigena* progenies. However, the use of *andigena* pre-selected for long days, in crosses with *tuberosum* led to progenies yielding higher than *tuberosum* x *tuberosum* progenies. In 1970s, the name *neo-tuberosum* was proposed for advanced *andigena* material adapted to long days. However, in the plains of subtropics, potato is grown during winter under short days. Under these conditions, *andigena* adapted to short days can be used advantageously for exploiting heterosis. The progenies from such crosses, however, are generally late maturing. To overcome this drawback and avoid the undesirable tuber characters in t*uberosum x andigena* populations, there is a need to select early bulking/maturing *andigena* clones with desired tuber characters, like uniformly white or red skin and shallow eyes.

Utilization of wild species in genetic improvement of potato suffers from many impediments owing to its clonal propagation, reproductive barriers, EBN imbalances and complex inheritance pattern. To be able to make the desired cross for any genetic and breeding work, it is essential that the parental genotypes flower over a sufficient length of time and that the flowers do not drop, but develop into fruits. Genotype, day length and temperature are the major factors determining behaviour of flowering and fruiting in potato, though a number of other factors like inflorescence position, plant/stem density, competition between flower and tuber, precipitation, nutrients and date of planting are also known to influence production of flowers and fruits in potato.

In the tropics and subtropics, conditions conducive to flowering and fruiting are available only at high altitudes (>1500 m above sea level) where potato crop is grown during summer season. Even under such conditions, not all genotypes produce flowers and fruits. During the short days of autumn season when the potato crop is grown in the plains in subtropics, flowering genotypes are few. Attempts to induce flowering in non-flowering genotypes by planting on bricks, removal of tubers, grafting shoots on tomato stock, extending of photoperiod by artificial illumination and spray of hormones like gibberellic acid and 2,4-dichlorophenoxy acetic acid, etc. have been successful to certain extent.

Male sterility is a very serious constraint in potato breeding. Although both pollen and ovule sterility can occur, pollen sterility ranging from partial to complete absence of pollen grains is very common in potato. Almost one-third of the potato cultivars derived from *S. tuberosum* ssp. *tuberosum* do not form berries. Further potato suffers from crossability barriers. Both self- and cross-incompatibility conditioned by a series of allelomorphs are widely prevalent in potatoes. Most diploid species are obligate out-breeders due to self-incompatibility (Ross 1986).

Incompatibility barriers can be pre-zygotic or post-zygotic. Pre-zygotic barriers are expressed between pollination and fertilization as non-germination of pollen on stigma; germination, but no penetration into the stigma; penetration, but inhibition of pollen tube growth to different degrees and at different sites in the style, ovary or ovulum. Post-zygotic barriers are expressed during and after fertilization in ovules, or more specifically in embryo sacs and also during growth and flowering of plants (F1) produced upon hybridization. These barriers, like the pre-zygotic ones, may cause partial or complete failure of hybridization and introgression (Caligari 1992).

Diploid potato species are a rich source of many desirable traits. The dihaploid of the tetraploid cultivated potato can be utilized for mating with the diploid species. Dihaploids of potato are generally produced by their parthenogenetic development in $4x \times 2x$ crosses between S. *tuberosum* × *S. phureja* (pollinator). Anther culture has also been used to produce dihaploids in potatoes, but the application of this approach is limited due to the lack of response of many genotypes to another culture. Dihaploids of *tuberosum* cross readily with many diploid wild spp. and thus provide the opportunity to introgression of useful traits from alien sources to cultivated background. Dihaploids with disomic rather than tetrasomic inheritance, facilitate the genetics and breeding of potatoes. Smaller populations are required to detect recessive genes in $2x$ crosses than in $4x$ intercrosses.

The final goal of breeding at the diploid level is to develop 2x clones with specific attributes, good breeding value for agronomic traits and $2n$ gametes. Production of $2n$ (unreduced) gametes by diploid–dihaploid hybrids is important to revert to tetraploid level through $4x \times 2x$ or $2x \times 4x$ (unilateral) or $2x \times 2x$ (bilateral) polyploidization (Tai 1994). The natural production of $2n$ gametes in many tuber-bearing *Solanum* species is well established, although their frequency may vary greatly within and between species owing to genetic, environmental and physiological factors. Such $2n$ gametes may originate in different ways, but genetically the basic distinction is between $2n$ FDR (First Division Restitution) gametes (reduction of $2n$ to n chromosomes followed by restitution to $2n$ through incomplete first division) and $2n$ SDR (Second Division Restitution) gametes (reduction of $2n$ to n chromosomes followed by restitution to $2n$ through incomplete second division). It can be calculated that FDR gametes brought 80 % introgression of the intact parental genotype to the hybrid progeny and SDR gametes 40 %. This high percentage of intact gene transfer through FDR gametes is especially important when the parent plant carries many desirable genes for qualitative as well as quantitative characters. The analytical breeding scheme based on dihaploids recommends the use of $2n$ gametes for production of tetraploids with maximal heterozygosity.

Due to the plasticity of various causal organisms responsible for a number of maladies in potato, new pathotypes mainly through mutations keep on emerging in nature. The situation is further aggravated by the cultivation of resistant monocultures on a large scale, rendering the available resistant varieties infructuous. This phenomenon makes the breeding of resistant varieties against the biotic stresses a continuous process (Shekhawat et al. 2000). The breeding of new varieties, therefore, always requires search of new/appropriate genes obtaining resistance from wild/semi-cultivated species and their introgression in suitable agronomically superior background. In a number of instances, however, resistant genes are not available and in this context, biotechnological interventions in conjunction with conventional breeding techniques can be useful in transferring resistance across the species, genera, families, etc. Quantitative trait loci (QTLs) (Schafer- Pregl et al. 1998), a biotechnological tool, in potato, holds promise as most of the biotic/abiotic stresses are polygenic and inherited in quantitative manner. The QTL markers, if could be exploited in a manner as in case of Mendelian factors, can substantially reduce the time lag in breeding.

Despite the problems, conventional breeding has been successful to various extent in improving the yield and introgression of various desirable traits from diverse sources. A brief account of it is as follows.

6.7.1.1 Late Blight

Late blight is caused by *Phytophthora infestans* (Fig. 6.4) which is highly variable and therefore, breeding varieties against it has been a see-saw story. Pathological variants (races) could be detected universally after the resistance genes from *S. demissum* were transferred to the commercial cultivars. Towards the end of twentieth century, the racial complexity had reached its zenith in most of the countries. The fungus had developed virulences against almost all the known R-genes in potato. This may be due to the fact that A_2 mating type now occurs almost throughout Europe and even in some Asian countries and therefore, racial variability through sexual reproduction in these regions cannot be ruled out (Singh 2000).

In 1909, R.N. Salaman demonstrated the heritable nature of resistance to late blight in wild species *S. edinense* (Salaman 1929). Resistance to blight can occur both in foliage and tubers. However, breeders have largely neglected the latter. Broadly resistance can be grouped into two types: (i) race-specific resistance (also called vertical resistance or major gene resistance, qualitative or discontinuous resistance) and (ii) race non-specific (also called horizontal resistance or minor gene resistance, field resistance, polygenic resistance, quantitative resistance or partial resistance).

The race-specific resistance based on gene-for-gene relationship was initially identified in hexaploid ($2n = 6x = 72$), wild species *S. demissum*. It is expressed in the form of hypersensitive response of the tissue to all races of *P. infestans* that did not possess the corresponding virulence to the resistance genes (R-genes). Specific resistance is conditioned by a series of major dominant genes each of which is brought into action by distinct pathotypes; currently 11 such genes (ex-demissum) are recognized. *S. stoloniferum* has also been found to possess similar, if not identical, resistance genes.

However, the R-genes wherever deployed were defeated in due course of time. In India, the process of transfer of R-genes from *S. demissum* background started in mid-1950s and the first set of late blight resistant varieties was released for commercial cultivation in 1968. Of these, cv. Kufri Jyoti (possessing R-genes 3.4.7) became the most popular which is still grown in several parts of the country. Since

Fig. 6.4 Late blight infested potato leaf

then a lot of varieties carrying R-genes have been bred and deployed across the country. In cv. Kufri Jyoti, matching virulences (3.4.7) were detected immediately after 5–6 years of its cultivation both in North-eastern and North-western hills. Both frequency of matching virulences, their combinations and disease increased rapidly making it completely susceptible by 1988. In India, this problem has been avoided by making adjustments in screening methodology (Singh and Shekhawat 1999). The seedlings in F_1C_1 are challenge-inoculated with the most complex race (8–9 gene complexes) (Fig. 6.5). The seedlings showing either complete susceptibility or immunity are discarded. The selected seedlings possess both R-genes coupled with a high degree of field resistance.

Race non-specific resistance is a quantitative and multifaceted trait, probably governed by many genes, it is, therefore, difficult to analyse in Mendelian ratios. Field resistance to late blight operates mainly through four factors, viz. infection efficiency, incubation period, colonization rate and sporulation efficiency. Many host factors, environmental aspects, edaphic, nutritional and climatic have an effect on these four components of resistance. Besides, components of field resistance to tuber blight include the depth in the soil at which the tubers are produced, the ease with which the spores are washed down from the canopy into the soil, the rapidity of periderm formation and the resistance to wounding. Although, at the phenotypic level, both types of resistances can be easily identified, at the genotypic level these are almost similar. Genetic analysis of resistance to late blight using DNA markers showed that major genes for resistance (R-gene) are closely linked to the factors controlling quantitative resistance suggesting that there is no real difference between qualitative and quantitative resistance to late blight as far as the nature of the genes involved. The differences observed at the phenotypic level may be the result of various allelic and non-allelic interactions. Thus a complex picture emerges, which renders both selection for and evaluation of blight resistance a slow process.

Several wild *Solanum* species possess high degree of resistance to late blight. Species like *S. bulbocastanum*, *S. demissum*, and *S. stoloniferum* had clones, which

Fig. 6.5 Late blight screening chamber showing F1C1 seedlings inoculated with *P. infestans*

possessed low infection frequency. Besides clones of *S. bulbocastanum* and *S. demissum* developed only small lesions, whereas clones of *S. stoloniferum* possessed high degree of resistance to tissue colonization. Umaerus demonstrated that *S. demissum* is a treasure of resistance. Besides R-genes, it also possesses field resistance, which operates primarily through low infection frequency at seedling stage. In German-Dutch potato collection nothing encouraging was found in *S. tuberosum* ssp. *andigena* and the primitive cultivars except an accession of *S. phureja*. However, resistance was detected in wild diploid Mexican species like *S. pinnatisectum*, *S. bulbocastanum*, *S. polyadenium* and *S. verrucosum*. It was also detected in Bolivian and Argentinean species, including *S. chacoense*, *S. berthaulti*, *S. microdontum* and *S. vernei*. Although initially *S. tuberosum* ssp. *andigena* was thought to be not having any resistance to late blight, there has been renewed interest in it as andigena potatoes respond well to selection for the resistance and produced a number of highly resistant clones. Resistance has also been reported in the Russian diploid sources, including *S. polytrichon*, *S. simplicifolium* and *S. microdontum*, which showed resistance to tuber blight as well.

6.7.1.2 Viruses

Viral diseases are an important constraint for potato crop because of their systemic distribution in the host and are mainly responsible for the degeneration of seed stocks (Khurana 1999, 2008; Jeffries et al. 2006). At least 12 viral diseases are known to infect potato crop in India and elsewhere. Among them, PVX, PVY, PVS, PVA, PVM, leaf roll (PLRV) and apical leaf curl viruses (PALCV) are important (Figs. 6.6, 6.7 and 6.8). Introducing resistant cultivars is one of the most efficient ways of reducing the losses caused by viruses. Resistance genes to different potato viruses have been identified in many wild potato species. Some of these genes have been incorporated in many of the recently released potato cultivars.

Fig. 6.6 PVY infected potato plant

Fig. 6.7 PALCV infected potato plant

The nature of resistance against viruses is of several types: (i) tolerance, (ii) resistance to infection, (iii) hypersensitivity usually giving field immunity and (iv) extreme resistance or immunity. Tolerance to viruses in potatoes is usually considered a dangerous type of resistance. Resistance to infection can be defined as the type in which only a small percentage of infection appears in the field and is governed by polygenes. Hypersensitivity and immunity are on the other hand due to mostly single dominant genes. Out of the four types of resistance, immunity gives almost complete elimination of virus and is preferable over other types. However, in recent years, more emphasis is being given to vector resistance where the resistance sought is against the vectors (aphids and other vectors), the carrier of viruses and not against viruses themselves.

There are three main groups of strains of PVY, viz. PVY^O (common strains), PVY^N (tobacco veinal necrosis strains) and PVY^C (stipple streak strains). The strain-specific resistance is controlled by the resistance gene N_Y while the extreme resistance is controlled by the gene R_Y. *Ny* genes are found in large number of cultivars including Pentland Crown, Pentland Ivory, King Edward and Cana and in hybrids derived from wild species like *S. chacoense*, *S. demissum* and *S. microdontum*. The sources of *Ry* gene are *S. stoloniferum*, *S. hougasii* and

S. tuberosum ssp. *andigena*. Accordingly the genes are named as Ry_{sto}, Ry_{hou} and Ry_{adg}. *Ry* is inherited as a single dominant gene and hence easy to breed.

There are mild, moderate and severe strains of PVA. They differ in severity of symptoms produced in potato cultivars. The gene *Na*, present in many cultivars, protects the plant from infection under natural pressure from PVA by means of a hypersensitive response. The gene *Na* is linked to gene Nx_{tbr}, which controls the resistance to PVX.

PVX strains can be separated according to their serological reaction into two main pathotypes: 1 and HB. Cockerham (1970) identified several genes conferring hypersensitivity, viz. $Nx_{tbr}, Nb_{tbr}, Nx_{chc}$ *and* Rx^{n}_{acl}. Similarly, genes conferring extreme resistance are known to occur in several species (Rx_{adg}, Rx_{acl} and Rx_{scr}).

Strains of PLRV differ in the severity of the symptoms they produce on potato and on test plants. Resistance to PLRV has been found to be oligogenic and has been detected in *S. brevidens* and *S. tuberosum*. Transfer of resistance genes from *S. brevidens* has been achieved through protoplast fusion but from *S. tuberosum* it has been possible only through bridge crosses. Among the hexaploid somatic hybrids derived from *S. tuberosum* and *S. brevidens* by protoplast fusion, some hybrids with high PLRV resistance were obtained.

Potato apical leaf curl disease, first reported in northern India by Garg et al. (2001) has been associated with a geminivirus, which was confirmed to be a strain of tomato leaf curl New Delhi virus (ToLCNDV) belonging to the genus Begomovirus (Usharani et al. 2003). The virus is transmitted by whiteflies and the affected plants show curling/crinkling mosaic of apical leaves.

Solomon-Blackburn (2001) and Tiwari et al. (2012) have excellently reviewed on prevalence of resistance against different virus among potato species and the nature of genes controlling them. Both major genes and quantitative trait loci (QTLs) have been found to govern virus resistance. The two major QTLs, *Plrv.1* and *Plrv.4* confer resistance to PLRV. Major genes Rl_{adg} and Rlr_{etb} confer high resistance to PLRV infection. At present there are four different known *R* genes: Ry_{adg}, Ry_{sto}, Ry_{hou} and Ry_{chc}, which confer extreme resistance to PVY in potato. In addition, Ny_{chc}, Ny_{dms}, Nc_{tbr}, Ny_{adg} Ny_{tbr} and *Ny-1* confer hypersensitive resistance on PVY. The gene Ny_{adg} controlling hypersensitive resistance to PVY^{O} is epistatic to Ry_{adg}. As a result, the genotypes carrying both Ry_{adg} and Ny_{adg} exhibited extreme resistance to PVY. The *R* genes Rx_{adg} (*Rx1*), Rx_{tbr}, Rx_{acl} (*Rx2*) and $Rx^{scr}_{HB}/Rx^{scr}_{CP}$ confer extreme resistance to PVX. The *N* genes Nx_{acl}, Nx_{chc}, Nb_{tbr}/Nx_{tbr}, Nx^{spl}_{tbr}, Nx_{phu} confer hypersensitive resistance to PVX. A single dominant gene *Ns* conferring hypersensitive resistance to PVS. The gene *Rm* confers hypersensitive resistance while gene *Gm* confers resistance to PVM infection. The *R* genes Ry_{sto}, Ra_{sto}, Ra_{adg} and Ry_{hou} confer extreme resistance while *N* genes Na_{adg}, Na_{sto}, Ny_{chc}, Na_{dms}, Ny_{dms}, Na_{tbr} and Na^{tbr}_{KE} confer hypersensitive resistance to PVA.

The development of cultivars with multiple virus resistance, however, remains a challenge for the breeders. This may be because the breeder select for many

Fig. 6.8 PLRV infected potato plant

important characters, therefore, introducing even a few genes for resistance to viruses becomes a difficult task. At CPRI, parental lines having virus resistance in duplex/triplex/tetraplex form have been developed. The progeny of triplex/tetraplex parents are being crossed with nulliplex parents to produce almost immune clones. After the evaluation for viral resistance in early generations, evaluation for horticultural traits is being done in later clonal generations.

6.7.1.3 Bacterial Wilt

Bacterial wilt or brown rot caused by *Ralstonia solanacearum*, first reported from India in 1892, is the most destructive of all bacterial diseases. Incidence of bacterial wilt is wide spread in all mid hill regions of the country and pockets of Assam, Meghalaya and Maharashtra. The disease damages the crop in two different ways—premature wilting of standing crop and rotting of tubers in fields, transit and stores (Fig. 6.9). It is primarily tuber borne, but survives equally well in soil. Host resistance is hard to find because of lack of co-evolution of the host and the bacterium, high variability in the bacterium and instability of the host resistance.

Resistance to bacterial wilt is a partially dominant character and is more of a polygenic type. Inheritance of resistance and its expression is complex and both additive and non-additive gene actions are involved, but the latter component is more important. A gene-for-gene relationship is not applicable to bacterial wilt. Certain genes other than those 'for resistance alone' have turned out to have the novel (pleiotropic) effects in conferring the resistance once the potato plant has come into contact with pathogen under a certain set of environmental conditions.

Fig. 6.9 Bacterial wilt infected field and tubers

These genes were eventually called 'genes for resistance' once a certain level of resistance was detected. The major or minor status of these genes depends on the particular genotype of the pathogen, and the particular environmental conditions that influence their expression. Attempts to transfer resistance from wild *Solanum* spp. into common potato resulted in excessive recombination and in breakdown upon intercrossing. Non-strain specificity and race cultivar specificity are the common features required for resistance to bacterial wilt. Thus, the host genotype x pathogen genotype interaction in potato, *R. solanacearum* system, seems to be artifactual. Both the host and pathogen are sensitive to environmental changes. Therefore, host genotype *x* pathogen genotype interaction may also be a result of host genotype x environment and/or pathogen *x* environment interaction.

The resistance in the clones of species such as *S. phureja* mainly and a few other *Solanum* species have been exploited extensively in the South American countries. But the Indian isolates of the bacterium have proved to be highly virulent making these sources ineffective. Resistance in *S. phureja*, the only species where it has been studied in detail, is strain- and temperature-specific, and it breaks down under the warm climates. Nematode injury also leads to its break down. A collection of nearly 500 clones of *Solanum* species which carry low to moderate degree of resistance, i.e. *S. phureja*, *S. microdontum*, *S. canasense*, *S. stenotomum*, *S. pinnetisectum*, *S. sparsipilum*, *S. kurtzianum*, *S. jamesii*, *S. polytrichon*, *S. vernei*, *S. acaule* and *S. stoloniferum* and interspecific hybrids between a number of above species and also resistant varieties developed so far in other parts of the world, viz. Prisca, Cruza, Caxamarca, Molenera and Ampola were screened against different isolates of the pathogen. All these cultivars/cultures proved susceptible to Indian isolates except for the one clone of diploid *S. microdontum* showing a moderate level of resistance. The efforts made to transfer the useful resistance from this source into the *tuberosum* background via dihaploids resulted in the development of two promising meiotic tetraploids. However, in field tests these were also proved to be susceptible.

6.7.1.4 Wart

Wart disease, of potato caused by a phycomycetous fungus, *Synchytrium endobioticum* (Schilb), was first reported from India in 1953 in North Bengal Hills. To avoid its further spread to other parts of the country, this area was brought under domestic quarantine in 1959. This has helped in containing this dreaded disease in Darjeeling district only. The disease causes cauliflower like growths on tubers, stolons and stem bases (Fig. 6.10). The heavy infection of disease causes rotting of entire products and results in total loss of crop. Cultivation of wart-immune varieties on a long-term basis is the only viable alternative. The resistance genes are available in a number of varieties of *S. tuberosum*. Besides, a number of wild species such as *S. boliviense*, *S acaule*, *S microdontum, S. demissum, S. sparsipilum*, *S. polytrichon*, *S. simplicifolium, S. chacoence* f.sp. *boergerii, S. vernei* and *S. spegazzinii* are known to have resistance to the disease. Monogenic dominant mode of inheritance for at least the control of necrotic response has been proved. However, modifying genes are also present which condition the nature and extent of response. Systematic breeding programme for wart immunity started in 1964. Since the pathogen survives in soil over long period, the only plausible defence mechanism is to develop varieties immune to it and saturate the area with the same. The crosses between wart-immune *tuberosum* parents result in recovery of quite high percentage of resistant clones than between resistant x susceptible parents. Using Adina × Ultimus, several late blight resistant and wart-immune hybrids were developed. One of them was released for commercial cultivation under the name Kufri Sherpa in 1983, which did not become popular because of its poor keeping quality, unattractive dull white skin, round tubers with medium deep eyes. Indian cultivars, viz. Kufri Jyoti, Kufri Chamatkar, Kufri Muthu, Kufri Sheetman, Kufri Bahar, Kufri Khasigaro and Kufri Kumar are immune to the race of *S. endobioticum* prevalent in the Darjeeling hills. These cultivars except Kufri Jyoti did not establish in the area because of local preference for varieties with red skin tubers. To develop a red tuber variety, pimpernel was used as one of the parents. One hybrid from cross

Fig. 6.10 Wart infected potato plant

SLB/Z 405a x Pimpernel was selected and has since been released as cultivar Kufri Kanchan. This variety is immune to wart and possesses high degree of field resistance to late blight.

6.7.1.5 Nematodes

About 90 species of nematodes belonging to 38 genera have been reported to be associated with potatoes. Among these, the root-knot nematodes and potato cyst nematodes have been recognized as the major pests.

At least nine species of root-knot nematode (*Meloidogyne spp.)* are known to infect potatoes. Among these, *M. incognita* is the most important throughout the world followed by *M. javanica*. The dominant root-knot nematode species affecting potato both in hills and plains is *Meloidogyne incognita* while *M. javanica* infestation is restricted to mid hills and plains.

Heavily infested plants are stunted with yellowish leaves while no visible tuber symptoms of nematode injury are seen under low infestation levels. The galls on potato roots are small and often go unnoticed. Wart-like structures are formed due to tuber infestation, which reduces the commercial value and keeping quality of tubers. The second-stage juveniles (hatched out from the egg masses laid by females) infest the young roots resulting in the formation of giant cells and the formation of syncytium and development of the nematode in the roots causes formation of galls or root knots. The nematode infection on tubers is characterized by the formation of typical wart-like pimples on the outer skin.

Work on breeding potato varieties resistant to *M. incognita* was initiated at the CPRI way back in 1961. Out of 101 hybrid cultures tested in a nematode infested field, only one line HC-294, a cross between Kufri Red × (Gladstone × Taborky) proved resistant to root-knot nematode. Further tests confirmed that only a few larvae invaded the roots of this selection and giant cells were not well developed. The resistance to root-knot nematode in this variety was inferred to be polygenic in nature. Later, another selection HC 115 from a cross of HB-289 × Kufri Red was also reported to be resistant to *M. incognita.* Both these lines were tolerant to heat, drought and frost. However, crosses involving HC-294 did not possess any resistance to root-knot nematodes indicating that the resistance was controlled by recessive genes. Further screening of accessions of tuber-bearing wild *Solanum* species indicated that a high degree of resistance was available in *S. spegazinii* and *S. vernei*, which were used as parental material for producing several crosses.

Potato cyst nematodes, *Globodera* spp. also known as golden nematode or potato root eelworms, is considered as one of the major pests throughout the world. Quarantine or regulatory actions are imposed against them in most countries. In India, the potato cyst nematode was first detected in 1961 by Jones at Ootacamund in Tamil Nadu. The nematode is reported to be present in about 3,050 hectare areas of Nilgiri and about 200 hectares of Kodaikanal hills of Tamil Nadu. Both these species *G. rostochiensis* and *G. pallida* are prevalent in these hills singly and also as mixed populations. The hatching of larvae from cysts (Fig. 6.11) is initiated by the

Fig. 6.11 PCN cysts in potato roots

root diffusates of potato or the members of the family Solanaceae. The nematode takes about 35 and 40 days for completion of life cycle during the crop season (Krishna Prasad 1993).

Heavy infestations in the absence of control measures often result in total crop loss. When the population in soil is sufficiently high, small patches of poorly growing plants may appear in the field. Temporary wilting of plants occurs during hotter parts of the day. Typical symptoms of heavy infestations are stunted plants with unhealthy foliage, premature yellowing, poor development of root symptom, reduction in size and number of tubers and poor yields. Since control of potato cyst nematode through chemicals is inadequate, expensive and environmentally hazardous, breeding cultivars resistant to the pest is the most effective way to control it.

The efforts to locate the source of resistance to cyst nematodes began in 1968, but a systematic breeding programme at CPRI was started in 1971. Over 2,000 genotypes comprising group tuberosum, group andigena and tuber-bearing *Solanum* species were screened and resistance was located in 20 accessions of 14 wild tuber-bearing species, viz. *S. ehrenberjii*, *S. vernei*, *S. chacoense*, *S. phureja*, *S. demissum*, *S. gourlayi*, *S. microdontum*, *S. sucrense*, *S. tarijense*, *S. acaule*, *S. fendleri*, *S. multidissectum*, *S. oplocense*, *S. sparsipilum* and some accessions of *S. tuberosum* ssp. *adigena*. Simultaneously, efforts were also made to procure resistant breeding lines to both the species from the Netherlands and USDA. A parental line VTn^2 62.33.3 (*S. tuberosum* × *S. vernei* hybrid) received from the Netherlands, having resistance to both the species, was extensively used in crosses with late blight resistant cultivar Kufri Jyoti resulting in release of hybrid Kufri Swarna for cultivation in cyst nematode infested areas of Nilgiri hills. Later one more variety, Kufri Neelima was released in 2010 for commercial cultivation in these hills using the same parental line.

Besides the important resistant lines from the Netherlands, nine resistant cultures with ssp. andigena in their pedigree were also received from Dr. Howard of Plant breeding Institute, Cambridge in 1976. Of these, two cultures, viz. D 40/8 and D 42/9, proved resistant to both the species of cyst nematodes. These are still used in the breeding programme.

6.7.1.6 Heat Stress

The potato has long been considered a crop for cool and temperate climates. Higher temperatures inhibit yield by overall reduction of plant development due to heat stress or by reduced partitioning of assimilates to tubers. Tuberization is reduced at night temperatures above 20 °C with complete inhibition of tuberization above 25 °C. Exposure of potato plants to heat stress alters the hormonal balance in the plants. As a result most of assimilated carbon is partitioned to above ground vegetative parts at the cost of the tubers. Aspects of heat tolerance that are considered important and should be taken into account in breeding programme includes ability of the plants to tuberize at night temperature of 22 °C and above (Fig. 6.12), low shoot/root ratio at high temperature, and early maturity of the crop.To breed heat-tolerant genotypes for Indian conditions, crosses were made amongst known heat-tolerant and local high-yielding genotypes. The known heat-tolerant genotypes used in the breeding programme were LT-1, LT-2, LT-5, LT-7, LT-8, LT-9, DTO-28, DTO-33 (received from CIP) Katahdin, Desiree and Kufri Lauvkar. The progenies were screened for heat tolerance by their ability to form tubers within 1 month after shifting to high temperatures. The selected genotypes were multiplied and further selected at early planting in Indo-Gangetic plains at Modipuram and Jalandhar under heat stress. Kufri Surya, a heat-tolerant variety (Minhas et al. 2006) has been released for cultivation as early planting in north-western plains as well as in rabi and kharif crops in peninsular India. This variety yields excellent defect-free

Fig. 6.12 Impact of heat stress (24 °C) on leaf bud tuberization in potato cultivar, *K. Surya* (heat tolerant) and *K. Chandramukhi* (heat sensitive)

tubers with high proportion of large (>85 mm) tubers suitable for processing into high quality French fries and chips. The reducing sugar content of tubers of this variety is less than 100 mg/100 g fresh weight and the tuber dry matter content is 20–21 % at harvest.

6.7.1.7 Drought Stress

Drought is a major limiting factor for potato production in the world influencing yield as well as tuber quality. Drought may occur due to erratic rainfall, inadequate irrigation techniques and lack of water supply. Even with good irrigation practices, water stress may occur because of high transpiration rates especially during mid-day, when root system cannot completely meet the transpiration requirements of the plant. Drought may affect potato growth and production by reducing the amount of productive foliage, decreasing the rate of photosynthesis per unit of leaf area and shortening the vegetative period.

Drought resistance can be the result of drought avoidance (e.g. closure of stomata, large root system) or drought tolerance (capacity for osmotic adjustment, rapid resumption of photosynthesis activity, etc.). Aspects of drought resistance that are considered important and should be taken into account in breeding programme includes the effect of short periods of stress on productivity and tuber quality, survival and recovery of the plants after water stress and water use efficiency. Experiments on reduction of leaf extension rate upon exposure to water stress and its recovery showed that potato cultivars could be placed in three groups. Group 'A' was characterized by minimum growth reduction under stress and rapid recovery on re-watering with final increase in the leaf length exceeding that of the unstressed controls. In group 'B' plants stress created moderate reduction in growth and on recovery the increase in leaf length became comparable to that of controls. Group 'C' was characterized by large reduction in growth and re-watering did not result in final leaf length increases comparable to that of controls. Both of these characters can be used as selection criteria in the breeding programme for drought tolerance.

6.7.1.8 Frost Tolerance

The problem of frost is known in almost every country where potatoes are grown. Temperatures below –2 °C in the field can produce partial or complete loss of the crop. In temperate zones, frosts can occur during spring when the crop is establishing itself, or during autumn when the crop is maturing. Higher crop losses occur in tropical highlands and subtropical plains where frosts can occur any time during the crop growth period. In India, more than 80 % of the potatoes are grown during winter in plains and the crop is prone to frosts during the months of December and January. Based on the field observations, two types of frosts are often distinguished. 'White frost' occurs when there is a decrease in temperature and relative humidity is high. 'Black frost' occurs under low temperatures and much drier conditions, hence more damaging and severe, because plant tissue is darkened immediately. Acclimation or hardening may increase the resistance to frosts in many plants. Exposure of the plants to prolonged low temperature is effective in increasing resistance to frost injury in *S. tuberosum, S. multidisectum, S. chomatophilum, S. acaule and S. commersonii*. Genetic variability exists in the genus *Solanum* with respect to frost injury. *S. acaule* has the ability to withstand extracellular ice formation up to –5 °C, which gives this species frost tolerance. This species can be used in the breeding programmes to transfer frost tolerance to *S. tuberosum*. In the north of India, frost occurs during December and January in the plains of Punjab and Eastern UP. Cultivars Kufri Sheetman and Kufri Dewa released by the Central Potato Research Institute possess resistance to frost. High degree of frost resistance was observed in other 28 hybrids from crosses involving *S. acaule*.

6.7.1.9 Fertilizer Use Efficiency

The potato is considered to be heavy feeder on nutrients and requires high inputs of NPK and water for optimum production. This not only increases the cost of production but also causes environmental pollution. The application of high rates of N and K fertilizers and irrigation water on coarse-textured soils on which the shallow-rooted crop is often grown can result in loss of N and K, which represents an economic loss to the grower and may cause environmental degradation of groundwater. While numerous studies have explored N and K management practices as a strategy for minimizing N and K loss, there is potential for exploiting the genetic variability among cultivars of asexually propagated crop species for improved N and K uptake (Trehan 2009). A nutrient-efficient potato can produce higher yields per unit of nutrient, applied or absorbed even at a limited nutrient supply (Graham 1984). Such genotypes could reduce N fertilization and nitrate leaching (Duynisveld et al. 1988; Sharifi et al. 2007). The recovery of applied phosphorus by potato crop is not more than 15–20 % (Trehan et al. 2008). Numerous studies have demonstrated the existence of considerable variation for nutrient efficiency among crop species and cultivars within species, which suggests genetic control of inorganic plant nutrition (Errebhi et al. 1999; Trehan et al. 2005). Although plant breeders seldom select for nutrient use efficiency (NUE), breeding programmes that develop lines that produce high yields may result in unconscious selection of genotypes that use nutrients more efficiently (Batten 1993).

At CPRI, studies on nutrient use efficiency have shown that potato cultivars showed wide variation in agronomic use efficiency (AUE), nutrient uptake efficiency (NUE) and physiological use efficiency (PUE) with respect to nitrogen, phosphors and potassium. Kufri Pukhraj was the most N, P and K efficient cultivar among ten cultivars tested in the absence as well as the presence of green manure. The efficient cultivars gave higher tuber yield under nutrient stress (i.e. with less dose of N, P and K fertilizer) than less efficient cultivars. The main cause of higher nitrogen efficiency in the presence of green manure was the capacity of a genotype to use/absorb more N per unit green manured soil, i.e. the ability of the root system of a genotype to acquire more N from green manured soil (NUE). The variation in potassium and phosphorus efficiency of different potato cultivars was due to both their capability to use absorbed K and P to produce potato tubers (PUE) and to their capacity to take up more K and P per unit soil (NUE). Breeders should combine parameters/characters (NUE and PUE) responsible for high nitrogen, phosphorus and potassium efficiency to breed multi-nutrient efficient potato cultivar. Kufri Sindhuri, a red-skinned potato variety released in 1967, is termed poor man's potato owing to its capacity to produce higher yield even under poorly managed growing conditions (Pushkarnath 1976). Recently CPRI has released potato variety Kufri Gaurav (Fig. 6.13) having higher nutrient use efficiency. The variety requires lower

Fig. 6.13 Leaf, flower, sprout and tubers of nutrient-efficient potato cultivar, *Kufri Gaurav*

doses of N, P and K than other cultivars to produce a particular fixed yield in the same field.

6.7.1.10 Quality Traits

Potatoes represent a non-fattening, nutritious and wholesome food, which supply important nutrients to the human diet. Tubers contain significant concentrations of vitamin C and essential amino acids. They are also a valuable source of at least 12 essential vitamins and minerals. Besides being important in human diet, potatoes are also used as animal feed and as raw material for starch and alcohol production. Potato quality parameters change according to the specific market utilization types, and are often referred by two major categories. The first category groups 'external quality', aspects comprising skin colour, tuber size and shape, eye depth. These traits are deemed very important for fresh consumption where external traits are most likely to influence consumer's choice. The second category comprises 'internal quality' aspects including nutritional properties, culinary value, after-cooking properties or processing quality. Internal quality is given by traits such as dry matter content, flavour, sugar and protein content, starch quality, type and amount of glycoalkaloids. Although quality is one of the most important characteristics of potato, it is probably the most poorly defined and least researched at the genetic level (Dale and Mackay 1994). There are several factors affecting tuber quality. They include the genetic makeup of the cultivar, crop maturity, agronomic practices, environmental conditions, storage temperatures, the presence of pests and diseases Traits that are genetically controlled can be grouped as biological traits (proteins, carbohydrates, vitamins, minerals, reduced amounts of toxic glycoalkaloids), sensorial traits (flavour, texture, colour) and industrial traits (tuber shape and size, dry matter content, cold sweetening, oil absorption, starch quality).

Breeding potato for quality traits requires a continuous flow of new genes and allelic diversity into the *S. tuberosum* gene pool. The genetic improvement of this crop is hampered by its tetrasomic inheritance, high level of heterozygosity, and incompatibility barriers. However, recent advances in plant biotechnology have significantly improved the possibilities of producing novel genetic variability and efficiently perform selection, especially when biotechnologists pool resources with breeders. Equally important is the fact that basic studies have contributed to elucidate our knowledge on the genetics, biochemistry and physiology of several quality traits, making breeding efforts less empirical and more predictable.

Use of *Solanum* species is particularly important in potato in that this crop has more related and cultivated and wild relatives than any other crop plant (Pavek and Corsini 2001), so that almost any trait that is important for breeding can be found in this germplasm. Since most *Solanum* species are diploid ($2n = 2x = 24$), a simple and efficient approach for their use is based on analytic breeding scheme involving first the production of *S. tuberosum* haploids ($2n = 2x = 24$) prior to crossing with compatible $2x$ species. Once the resulting diploid hybrids that produce $2n$ gametes at acceptable frequencies are selected for traits of interest (e.g. good chipping ability, high dry matter content and resistance to diseases), the return to the tetraploid level of the cultivated potato may be achieved through sexual polyploidization schemes.

Lu et al. (2001) represents an example on the use of this approach. The authors produced *S. phureja–S. stenotomum* diploid hybrids and screened them for individual and total carotenoid content of tubers. They found a linear correlation between carotenoids and yellow-flesh intensity and selected hybrids containing 13 times more carotenoids than control cultivar 'Yukon Gold' (yellow flesh), and 22 times more carotenoids than 'Superior' (white flesh). The best hybrids also produced 2*n* pollen, and thus represent unique material for unilateral sexual polyploidization schemes aimed at producing 4*x* offspring for further selection.

Varieties obtained through sexual polyploidization approach are available worldwide. 'Yukon Gold', for example, is a yellow-flesh Canadian variety obtained through 4*x* × 2*x* crosses between 4*x* cultivar 'Norgleam' and a 2*x S. tuberosum–S. phureja* hybrid (Johnston and Rowberry 1981). Potato cultivars obtained through unilateral sexual polyploidization are also being released in China, now ranking first in the world potato production, (Jin et al. 2004). Up to now the 4*x* × 2*x* approach has been mainly used to transfer resistance traits. It also has a great potential for improving quality traits due to the high number of *Solanum* species with noteworthy quality characteristics (Table 6.1).

In potato genetic engineering techniques have been applied to produce routinely and several transformation protocols are currently available. Data published by Dunwell (2000) indicated that the potato ranks second, after corn, in the list of plant species for which field trials were carried out in the United States. In the past, most potato genotypes were transformed with genes for herbicide or insect resistance. However, much emphasis is now on quality traits. The production of starches with modified amylose-to-amylopectin ratio represents a good example of the possibilities offered by genetic engineering in improving potato quality traits. Lloyd et al. (1999) provided evidence that in transgenic potato lines where the activity of ADP-glucose pyrophosphorylase (AGPase) was reduced through antisense technology had a significant reduction of amylose. They also observed that in AGPase antisense plants, amylopectin accumulated shorter chains and that the size of starch granules was reduced. On contrast, the simultaneous antisense inhibition of two isoforms of starch-branching enzymes (SBE A and B) to below 1 % of the wild type activity increased amount of amylose in transgenic lines (Schwall et al. 2000). The amylose content of their transgenic lines was comparable to that of the high-amylose corn starch reported by Shi et al. (1998).

Great attention has been given to improve the essential amino acid composition of tubers and especially their lysine, tyrosine, methionine and cysteine content. Chakroborty et al. (2000) transformed a potato genotype with the gene *AmA1* from *Amaranthus hypocondriacus,* encoding a protein with a nutritionally balanced amino acid composition. The amino acid profile in tubers of both types of transgenic plants showed a 1.5- to 8-fold increase for all essential amino acids in the wild type.

Genetic engineering has been recently used to improve carotenoid content of tubers. In particular, to overcome the zeaxanthin deficiency of human diet, Römer et al. (2002) downregulated the synthesis of zeaxanthin epoxidase specifically in tubers through antisense technology and co-suppression approaches. Both strategies achieved a decreased conversion of zeaxanthin to violaxanthin in transgenic tubers with a corresponding increase of zeaxanthin content of 4- to 130-fold. Due to the use of a tuber-specific promoter, leaf carotenoid content of all transformants was very similar to the control plants, and thus photosynthesis was not negatively affected by lack of violaxanthin.

One main constraint in the use of wild species is that, together with useful traits, they can transfer characteristics that are undesired from the commercial standpoint. In the case of *Solanum* species, traits such as long stolons, deep eyes, which are negative quality traits, can be transmitted. As reported recently by Pavek and Corsini (2001), transmission of undesired traits has very much limited the use of potato genetic resources. Therefore, after interspecific crosses, time-consuming evaluation and selection are necessary to eliminate unwanted wild type genes and restore the cultivated improved phenotypes.

Marker-assisted selection is perhaps the most powerful approach that uses DNA markers efficiently for selection of interspecific hybridization by reducing the linkage drug in terms of time and space. The use of DNA markers can be ascribed not only to the use of markers tightly linked to target genes (positive-assisted selection), but also to the use of markers specific for the wild donor parent to perform selection against the wild genome (negative-assisted selection; Barone 2004). For example, *S. commersonii*, a diploid wild species possesses several useful traits like frost resistance, acclimation capacity, high dry matter content of tubers, resistance to *Ralstonia solanacearum* along with high content of demissine, tomatine and commersonine, glycoalkaloids which are extremely toxic to humans and animals. To efficiently identify desirable *S. tuberosum–S. commersonii* hybrids, a negative-assisted selection approach was followed to estimate estimating the wild genome content of each hybrid by using *S. commersonii* specific AFLP markers with assumption that hybrids with low wild genome content may show only in minimal part the negative traits associated to the genome of the wild parent. This approach helped to identify hybrids combining low wild genome content with resistance and quality traits from *S. commersonii* (Barone et al. 2001; Carputo et al. 2002).

A very exciting development in the context of efficient selection has been the generation of a molecular-linkage map based on functional gene markers involved in carbohydrate metabolism and transport (Chen et al. 2001). Using diploid mapping populations for which molecular maps were already available, the authors performed CAPS, SCAR and RFLP marker assays for 69 functional genes previously studied and identified (among the others *AGPase*, *SssI*, *GbssII*, *Dbe*, *UGPase*, *Ppc*, and *Cis*). This work allowed the identification of 85 genetic loci covering a considerable amount of the potato genome. The availability of this molecular function map allowed a candidate gene approach to be used for studying starch- and other sugar-related agronomic traits in potato. Chen et al. (2001) compared the QTL map for starch content previously published (Schäfer-Pregl et al. 1998) with the molecular function map, and various correlations between the map positions of 14 QTLs for tuber starch content and function-related loci were found.

A candidate gene approach has been also used by Menendez et al. (2002) to study cold sweetening in potato. Using RFLP and AFLP markers, they generated a QTL and linkage map of two segregating diploid populations previously evaluated for sugar content after cold storage. The authors mapped ten potato genes with known map position and unknown function in carbon metabolism or transport, and tested them for their effects on sugar content. Results displayed linkage between glucose, fructose and sucrose QTLs and all of eight candidate gene loci (*AGPaseS*, *AGPaseB*, *SbeI*, *GapC*, *Invap*, *Ppa1*, *Sut1*, *Sut2*). The authors pointed out that their results provide a basis for performing marker-assisted selection using allelic variants of candidate genes in the *Solanum* gene pool.

6.8 Biotechnology in Potato Genetic Enhancement

Potatoes are highly amenable to tissue culture and have been used widely for the development of biotechnology techniques. These techniques are now becoming handy in the potato variety improvement programmes. Various techniques have been used in potato breeding programmes includes somaclonal variations for creating useful variability, somatic hybridization for overcoming barriers to normal crossing, molecular markers for marker-assisted selection and genetic transformation for transfer of single, specific genes from any source to the existing varieties.

Use of biotechnology in potato improvement is already a reality, but these techniques are likely to remain as adjuncts to the more classical methods of breeding, which themselves will continue to evolve in efficiency as the underlying mechanisms and properties of genetic systems become more clearly known.

We are not dealing herewith about potato improvement through biotechnology for want of space, and also volume of data generated beyond the scope of this chapter and the readers are therefore suggested to refer to recent papers and more specific review chapters/books on the subject.

References

Anonymous (2007) Potato and potato products cultivation, seed production, manuring, harvesting, organic farming, storage and processing, NIIR Project Consultancy Services, pp 1–12

Barone A (2004) Molecular marker-assisted selection for potato breeding. Am J Potato Res 81:111–117

Barone A, Sebastiano A, Carputo D, Della Rocca F, Frusciante L (2001) Molecular marker-assisted introgression of the wild *Solanum commersonii* genome into the cultivated gene pool. Theor Appl Genet 102:900–907

Batten GD (1993) A review of phosphorus efficiency in wheat. In: Randall PJ et al (eds) Genetic aspects of plant mineral nutrition. Kluwer Academic Publishers, Dordrecht, pp 215–220

Bered F, Terra TF, Spellmeier M, Neto JFB (2005) Genetic variation among and within sweet corn populations detected by RAPD and SSR markers. Crop Breed Appl Biotechnol 5:418–425

Bradshaw JE, Mackay GR (1994) Breeding strategies for clonally propagated potatoes. In: Bradshaw JE, Mackay GR (eds) Potato genetics. CAB International, Wallingford, pp 467–498

Caligari PDS (1992) Breeding new varieties. In: Harris PM (ed) The potato crop, 2nd edn. Chapman & Hall, London, New York, pp 334–372

Carputo D, Frusciante L, Monti L, Parisi M, Barone A (2002) Tuber quality and soft rot resistance of hybrids between *Solanum tuberosum* and the incongruent wild relative *S. commersonii*. Am J Potato Res 79:345–352

Chakaborty S, Chakaborty N, Datta A (2000) Increased nutritive value of transgenic potato by expressing a nonallergenic seed albumin gene from *Amaranthus hypochondriacus*. Proc Natl Acad Sci USA 97:3724–3729

Chen X, Salamini F, Gebhardt C (2001) A potato molecular-function map for carbohydrate metabolism and transport. Theor Appl Genet 102:284–295

Clusius C (1601) Rariorum planatarum historia. Moretus, Plantin, Antwerp

Cockerham G (1970) Genetical studies on resistance to potato viruses X and Y. Heredity 25: 309–348

Coombs JJ, Frank LM, Souches DS (2004) An applied fingerprinting system for cultivated potato using simple sequence repeats. Am J Potato Res 81:243–250

Cribb PJ, Hawkes JG (1986) Experimental evidence for the origin of *S. tuberosum* subsp. *andigena*. In: D'Arcy WG (ed) Solanaceae: biology and systematics. Columbia University Press, New York, pp. 383–404

Curwen JC (1818) Observations on the state of Ireland, vol I. Baldwin, Craddock and Brown, London

Dale MFB, Mackay GR (1994) Inheritance of table and processing quality. In: Bradshaw JE, Mackay GR (eds) Potato genetics. CAB International, Wallingford, pp 285–315

de Cieza De León P (1553) *La Crónica Del Perú*. Seville. English edition: de Cieza De León P (1553) The Chronicles of Peru (trans: Sir Clements R). Markham in Hakluyt Society 33 and 68, 1864 and 1883, London

Demeke T, Lynch DR, Kawchuck LM, Kozub GC, Armstrong JD (1996) Genetic diversity of potato determined by random amplified polymorphic DNA analysis. Plant Cell Rep 15:662–667

Dodds KS (1962) Classification of cultivated potatoes. In: Correl D (ed) The potato and its wild relatives. Texas Research Foundation, Renner, Texas, pp. 517–539

Drake SF (1628) The world encompassed, vol 16. Hakluyt Society, London, p 1854

Dunwell JM (2000) Transgenic approaches to crop improvement. J Exp Bot 51:487–496

Duynisveld WHM, Strebel O, Bottcher J (1988) Are nitrate leaching from arable land and nitrate pollution of ground water avoidable. Ecol Bull. 39:116–125

Errebhi M, Rosen CJ, Lauer FI, Martin MW, Bamberg JB (1999) Evaluation of tuber-bearing Solanum species for nitrogen use efficiency and biomass partitioning. Am J Potato Res 76: 143–151

Ford R, Taylor PWJ (1997) The application of RAPD markers for potato cultivar identification. Aust J Agric Res 48:1213–1217

Garcia AF, Alberini JL, Zucchi MI, De Souza AP (2007) Microsatellite molecular markers in the cultivar identification of Brazilian soybean for human consumption. Crop Breed Appl Biotechnol 7:155–164

Garg ID, Khurana SMP, Kumar S, Lakra BS (2001) Association of a gemi-nivirus with potato apical leaf curl in India and its immuno-electron microscopic detection. J Indian Potato Assoc 28:227–232

Glendinning DR (1983) Potato introductions and breeding up to the early 20th century. New Phytol 94:479–505

Gomara LD (1552) Historia General de las Indias (General History of the Indies)

Gopal J, Gaur PC (1997) Potato genetic resources in India. Ind J Plant Genetic Res 10:205–215

Graham RD (1984) Breeding for nutritional characteristics in cereals. Adv Plant Nutr 1:57–102

Grun P (1979). Evolution of the cultivated potato: a cytoplasmic analysis. In: Hawkes JG, Lister, RN, Skelding AD (eds) The biology and taxonomy of *Solanaceae*. London, Academic Press, p 655–665

Hawkes JG (1990) The potato, evolution, biodiversity and genetic resources. Belhaven Press, London, 259 p

Hawkes JG (1992) History of the potato. In: Harris PM (ed) The potato crop. Chapman & Hall, London, pp 2–12

Hawkes JG (1994) Origins of cultivated potatoes and species relationships. In: Bradshaw JE, Mackay GR (eds) Potato genetics. CAB International, UK, pp 3–42

Jeffries C, Barker H, Khurana SMP (2006) Viruses & viroids. In: Gopal J, Paul Khurana SM (eds) Handbook of potato production, improvement and post harvest management. The Howarth Press Inc., USA, pp 387–448

Jin LP, Qu DY, Xie KY, Bian CS, Duan SG (2004) Potato germplasm, breeding studies in China. In: Proceedings of the Fifth world potato congress, Kunming, China, 26–31 March 2004, pp. 175–177

Johnston GR, Rowberry RG (1981) Yukon gold: a new yellow-fleshed, mediumearly, high quality table and French-fry cultivar. Am Potato J 58:241–244

Khurana SMP (1999) Potato viruses and viral diseases. Tech Bull No 35 (Revised), CPRI, Shimla, 94 p

Khurana SMP (2008) Potato viruses. In: Rao GP et al (ed) Characterization, diagnosis & management of plant viruses. Volume I: industrial crops. Studium Press LLC, Texas, pp 1–45

Khurana SMP, Garg ID (1998) Present status of controlling non-persistently aphid-transmitted potato viruses. In: Hadidi A et al (eds) Plant virus disease control. APS Press, St. Paul, pp 583–615

Khurana SMP, Chandra R, Upadhya MD (eds) (1998) Comprehensive potato biotechnology. MPH, New Delhi, 370 p

Krishna Prasad KS (1993) Nematodes-distribution, biology and management. In: Chadha KL, Grewal JS (eds) Advances in Horticulture volume 7-potato. Malhotra Publishing House, New Delhi, pp 635–647

Kujal S, Chakrabarti SK, Pandey SK, Khurana SMP (2005) Genetic divergence in tetraploid potato (*Solanum tuberosum* s spp *tuberosum*) as revealed by RAPD vis a vis morphological markers. Potato J 32(1–2):17–27

Lloyd JR, Springer F, Buleon A, Müller-Röber B, Willmitzer L, Kossmann J (1999) The influence of alterations in ADP-glucose pyrophosphorylase activities on starch structure and composition in potato tubers. Planta 209:230–238

Lu W, Haynes K, Wiley E, Clevidence B (2001) Carotenoid content and color in diploid potatoes. J Am Soc Hort Sci 126:722–726

Menéndez CM, Ritter E, Schäfer-Pregl R, Walkemeier B, Kalde A, Salamini F, Gebhardt C (2002) Cold sweetening in diploid potato: mapping quantitative trait loci and candidate genes. Genetics 162:1423–1434

Minhas JS, Devendra Kumar TA, Joseph BT, Raj SMP, Khurana SK, Pandey SV, Singh BPS, Naik PS (2006) Kufri Surya: a new heat tolerant potato variety suitable for early planting in north-western plains, peninsular India and processing into French fries and chips. Potato J 33:35–43

Pavek JJ, Corsini DL (2001) Utilization of potato genetic resources in variety development. Am J Potato Res 78:433–441

Provan J, Powell W, Dewar H, Bryan G, Machray GC, Waugh R (1999) An extreme cytoplasmic bottleneck in the modern European cultivated potato (*Solanum tuberosum*) is not reflected in decreased levels of nuclear diversity. Proc R Soc Lond B Biol Sci 266:633–639

Pushkarnath (1969) Potato in India-varieties. New Delhi, Indian Council of Agricultural Research, p 493

Pushkarnath (1976) Varieties and varietal problems. In: Pushkarnath (ed) Potato in sub tropics. Orient Longman Limited, New Delhi, pp 173–216

Römer S, Lübeck J, Kauder F, Steiger S, Adomat C, Sandman G (2002) Genetic engineering of a zeaxanthin-rich potato by antisense inactivation and cosuppression of carotenoid epoxidation. Metab Eng 4:263–272

Ross H (1986) Potato breeding-problems and perspectives. Springer, Berlin, p 132

Salaman RN (1929) Genetic studies in potatoes: abnormal segregation in families arising from the cross *S. utile* X *S. tuberosum*. J Genet 20:311–343

Schäfer-Pregl R, Ritter E, Concilio L, Hesselbach J, Lovatti L, Walkemeier B, Thelen H, Salamini F, Gebhardt C (1998) Analysis of quantitativetrait loci (QTLs) and quantitative trait alleles (QTAs) for potato tuber yield and starch content. Theor Appl Genet 97:834–846

Schneider K, Douches DS (1997) Assessment of PCR-based simple sequence repeats to fingerprint North American potato cultivars. Am Potato J 74:149–160

Schwall GP, Safford R, Westcott RJ, Jeffcoat R, Tayal A, Shi Y-C, Gidley MJ, Jobling SA (2000) Production of very-high-amylose potato starch by inhibition of SBE A and B. Nat Biotechnol 18:551–554

Sharifi M, Zebarth BJ, Coleman W (2007) Screening for nitrogen-use efficiency in potato with a recirculating hydroponic system. Commun Soil Sci Plant Anal 38:359–370

Shekhawat GS, Khurana SMP, Singh BP (2000) Important diseases of potato and their management. In: Khurana SMP, Shekhawat GS, Singh BP, Pandey SK (eds) Potato global research and development, vol 1. Indian Potato Association, Shimla, pp 281–303

Shi YC, Capitani T, Trzasko P, Jeffcoat R (1998) Molecular structure of a lowamylopectin starch and other high-amylose maize starches. J Cereal Sci 27:289–299

Singh BP (2000) Status of late blight in sub-tropics. In: Khurana SMP, Shekhawat GS, Singh BP, Pandey SK (eds) Potato global research and development, vol 1. Indian Potato Association, Shimla, pp 525–533

Singh BP, Shekhawat GS (eds) (1999) Potato late blight in India. Tech Bull No. 27 (Revised). CPRI, Shimla, 85 p

Solomon-blackburn RM, Barker H (2001) Breeding virus resistant potatoes (*Solanum tuberosum*): a review of traditional and molecular approaches. Heredity 86:17–35

Spooner D, Alberto S (2006) Structure biosystematics and genetic resources. In: Gopal J, Khurana SMP (eds) Handbook of potato-production improvement, and postharvest management. Haworth Press, Inc. New York, pp 1–40

Tai GCC (1994) Use of 2n gametes. In: Bradshaw JE, Mackay GR (eds) Potato genetics. CAB International, Wallingford, pp 109–132

Tiwari JK, Gopal J, Singh BP (2012) Marker-assisted selection for virus resistance in potato: options and challenges. Potato J 39(2):101–117

Trehan SP (2009) Improving nutrient use efficiency by exploiting genetic diversity of potato. Potato J 36(3–4):121–135

Trehan SP, Dessougi HEL, Claassen N (2005) Potassium efficiency of ten potato cultivars as related to their capability to use non-exchangeable soil potassium by chemical mobilization. Commun Soil Sci Plant Anal 36:1809–1822

Trehan SP, Upadhayay NC, Sud KC, Kumar M, Jatav MJ, Lal SS (2008) Nutrient management in potato. Technical Bulletin No. 90. Central Potato Research Institute, Shimla, India, p 64

Ugent D, Pozorski S, Pozorski T (1982) Archaeological potato remains from the Casma valley of Peru. Econ Botony 36:181–192

Upadhya MD (1974) Potato. In: Hutchinson J (ed) Evolutionary studies in world crops. Cambridge University Press, Cambridge, pp 139–149

Usharani KS, Surendranath B, Khurana SMP, Garg ID, Malathi VG (2003) Potato leaf curl—a new disease of potato in northern India caused by a strain of Tomato leaf curl New Delhi virus. New Dis Rep 8:2

Chapter 7
Genome Plasticity in Buckwheat

Nikhil K. Chrungoo, Lashaihun Dohtdong and Upasna Chettry

Abstract Plasticity is the ability of a plant genotype to respond to different environmental conditions by producing different phenotypes. Classic examples of phenotypic plasticity in plants include response of leaves to sun, heterophylly, environmental control of cleistogamy, responses to herbivory, inter- and intra-specific competition, allelopathy. True plastic responses to variations in environment have just as firm a genetic basis as other plant characters. As a parameter which is determined by those genetic systems that control development, plasticity can be considered as an epigenetic phenomenon. Thus, plastic responses represent changes in 'typical' developmental sequences due to the interaction of the organism's genotype with the environment. Even though the diversity of genetic resources is fundamental for ecosystem functioning, sustainable agricultural production and attainment of food and nutritional security, yet only a few crop species are utilized for food production throughout the world. Further, erosion of genetic resources is having serious consequences, both on the genetic vulnerability of crops to changes in environmental factors as well as in their plasticity to respond to changes in climate or agricultural practices. Since a crop's ability to tolerate the vagaries of environment is dependent on a complex combination of responses and mechanisms, an understanding of morphological, physiological, and genetic mechanisms involved in the responses of these crops assumes significance. As a source of agronomic traits for breeding and adaptability to changing environments genetic diversity in agricultural crops have tangible values. However, the shrinkage of agricultural basket due to "agricultural simplification," is having a significant impact on sustainability of farm agroecosystems. Of particular concern, the cultivation of traditional crops has declined and continues to decline globally, yet such crops offer greater genetic diversity, and have the potential to improve food and nutritional security. Among these, the International Plant Genetic Resources Institute (IPGRI) and Consultative Group on International Agriculture (CGIAR) have identified buckwheat (*Fagopyrum* spp.), grain amaranth (*Amaranthus* spp.),

N.K. Chrungoo (✉) · L. Dohtdong · U. Chettry
Plant Molecular Biology Laboratory, UGC-Centre for Advanced Studies in Botany, North-Eastern Hill University, Shillong 793 022, India
e-mail: nchrungoo@gmail.com

V.R. Rajpal et al. (eds.), *Gene Pool Diversity and Crop Improvement*,
Sustainable Development and Biodiversity 10,
DOI 10.1007/978-3-319-27096-8_7

and (*Chenopodium* spp.) as crops of potential for future. Common buckwheat (*Fagopyrum esculentum* Moench), a diploid (2*n* = 16) annual crop plant, is widely cultivated in Asia, Europe and America. Due to short growth span, capability to grow at high altitudes and the high-quality protein content of its grains, it is an important crop in mountainous regions of India, China, Russia, Ukraine, Kazakhstan, parts of Eastern Europe, Canada, Japan, Korea, and Nepal. The plant is known to have three viz. summer, intermediate, and late summer ecotypes. While the late-summer ecotypes are low altitude cultivars, the summer ecotypes are cultivated at high altitudes. The summer ecotypes have been suggested to have been evolved from late summer ecotypes through selection of early flowering plants under long-day conditions; the selection being a part of the domestication process in buckwheat for climatic adaptation.

Keywords Genome plasticity · Buckwheat · Protein content · Ecotypes · Agro-ecosystems · Fagopyrum spp. · Plasticity · Genome size · Genetic diversity

The genus *Fagopyrum* (Polygonaceae) consists of 19 species that are divided into two monophyletic groups viz. the *cymosum* group which includes *F. esculentum*, *F. tataricum*, and *F. homotropicum* and the *urophyllum* group which includes wild species that have small lustrous achenes completely covered with a persistent perianth. More than 7500 accessions, belonging to different species of buckwheat have been recorded from different regions of the world. On the basis of its achene morphology, the genus *Fagopyrum* has been divided into (i) the *cymosum* group which comprises of *F. esculentum, F. tartaricum, F. cymosum, F. homotropicum, F. lineare, and F. pilus,* all of which have large lusterless achenes that are incompletely covered with a persistent perianth, and the (ii) *urophyllum* group which comprises of *F. urophyllum* and other wild species which have small lustrous achenes which are completely covered with a persistent perianth.

Even though buckwheat cultivation has been going on for a long time, the crop still has a long way to go in the direction of greater popularity and importance. Some of the inherent limitations in buckwheat like low plasticity and unstable yields, indeterminate growth habit, seed shattering, lodging, presence of allergenic compounds, bitter taste, and low shelf life of its flour restrict its popularity. Advancing the utilization of buckwheat would require an integrated approach involving marker assisted selection of genotypes for higher plastic responses and quality traits.

7.1 Introduction

Plant genetic resources, representing the sum total of diversity accumulated through years of cultivation under domestication and natural selection, are considered as one of the most important gifts of nature to mankind. These resources constitute important sources of food, feed, and shelter for mankind. Given, the constraints of growth because of their structural complexity and autotrophic nature which necessitates the plants drawing water essentially from the rhizosphere, plants have developed a certain extent of plasticity so that a single plant genotype is able to give rise to a wide range of phenotypes depending on the prevailing environmental conditions. The plastic responses of plants have have received attention in two interlinked contexts. First, developmental plasticity has profound implications for plant evolution. If a single genotype can result in multiple phenotypes, this weakens the link between selection at the level of the phenotype and changes in allele frequency at the level of the genotype. However, the plastic responses of plants also make it possible to investigate the genetic control mechanisms underlying the phenomenon. Although adaptive plasticity is generally considered to be of most evolutionary and ecological interest, the plasticity in plant responses to varying conditions of growth has a tremendous bearing on the survival and colonization potential of a plant. Nonadaptive plasticity reduces the likelihood of persistence in a new environment that is stressful and thus increases the strength of selection. Grether (2005) has suggested that whenever a species encounters stiff environmental conditions that trigger phenotypic changes with reduced fitness, selection will favor appropriate genetic changes that counteract the phenotypic changes so as to restore the phenotype to its ancestral state. Such an adaptive evolution, referred to as "genetic compensation," leads to genetic differences among populations so that the mean trait values of populations in different environments may appear to be more similar when measured in their native habitat than when grown under common environmental conditions. Thus, while genetic compensation would tend to reduce phenotypic variability it would promote genetic divergence in a species. However, stressful environments could also increase phenotypic variation via the expression of cryptic genetic variations. Thus, while the phenotypic diversity under a set of environmental conditions may represent only a fraction of the diversity that is actually available, cryptic genetic variation could be revealed but under stressful environments.

Even though the diversity of genetic resources is fundamental for ecosystem functioning, sustainable agricultural production, and attainment of food and nutritional security, yet only a few crop species are utilized for food production throughout the world. Currently, only 30 plant species are known to provide 95 % of the world's food energy needs. These plants have been selected from a large basket of agrobiodiversity which comprises of more than 7,000 species which is approximately 1/10 of the estimated number of edible species present in nature. A consequence of increased reliance on major food crops is the shrinking of food basket which humankind has been relying upon for generations (Prescott-Allen and Prescott-Allen 1990). This "nutritional paradox" as it is called (Ogle and Grivetti 1995) has its roots in "agricultural simplification", a process that favored some crops instead of others on the basis of their comparative advantages for growing in a wider range of habitats, simple cultivation requirements, easier processing and storability, nutritional properties, taste, etc. Of particular concern, the cultivation of traditional crops has declined and continues to decline globally, yet such crops offer greater genetic diversity, and have the potential to improve food and nutritional security. While the twentieth century witnessed a systematic approach toward rescuing the genetic resources of staple crops (Pistorius 1997), the twenty first century has started with the awareness on the necessity of rescuing and improvement of crops left aside by research, technology, marketing systems as well as conservation efforts. Among these, the International Plant Genetic Resources Institute (IPGRI) and Consultative Group on International Agriculture (CGIAR) have identified buckwheat (*Fagopyrum* spp.), grain amaranth (*Amaranthus* spp.) and (*Chenopodium* spp.) as crops of potential for use in crop improvement programs.

7.2 Adaptive Plasticity: A Selective Advantage

Despite the long history of both theoretical and empirical research on plasticity, the mechanisms that regulate plasticity of plant responses are still poorly understood. While natural selection is expected to favor locally adapted plants when the environment is constant, phenotypic plasticity would be more beneficial in spatially or temporally heterogeneous environments. Since climate variability has always been a part of risk associated with agriculture, farmers have developed many ways of managing that risk. While farm communities have adopted maintenance of genetic diversity as one of the livelihood strategies, vagaries and the rates of climate change have been observed to be much higher than that required for landraces to evolve and adapt for changing environments. Hence, the plasticity of genetic resources will remain important for such situations. Pigliucci et al. (2006) have defined phenotypic plasticity as a "property of a genotype to produce different phenotypes" in response to changing environment. While the notion of "phenotypic plasticity as an evolving trait" has received much focus (Scheiner 1993, 2013; Alpert and Simms 2002; DeWitt and Scheiner 2004) the fact that phenotypic plasticity can be a useful paradigm to understand the interactions of genetics, development, ecology, and evolution can not be underestimated.

Although plastic responses are quite often considered to be adaptive, the response can be considered as just one of several factors which may cause variation in amounts and patterns of plastic response between taxa. In addition, the amount and pattern of plastic response can evolve independently of the character mean (Schlichting 1984). For example, in the genus *Portulaca*, the character shoot/root ratio has a very similar mean value for *P. grandiflora* (20.4) and *P. oleracea* (18.5), but their amounts (CV = 30 and 12, respectively) and directions (Spearman rho = 0.00) of plastic response differ markedly (Zimmerman 1976). Forces responsible for evolutionary changes in phenotypic plasticity can be assigned to three general categories *viz.* selection, drift, and disruption of the genetic system. While the phenotypic plasticity due to selection would include situations where advantageous plastic responses are related to specific environmental regimes, plasticity may also evolve through random changes not driven by selection including factors such as inbreeding, hybridization, and polyploidy (Gama and Hallauer 1980; Pooni and Jinks 1980; Garbutt and Bazzaz 1983; Levin 1983; Schlichting and Levin 1986). While the most common misconception regarding phenotypic plasticity is that it represents a nongenetic means of dealing with a change in the environment, well-defined plastic responses to environmental variation have just as firm a genetic basis as other plant characters (Khan et al. 1976; Jain 1978). Thus plastic responses can be conceived as 'typical' developmental sequences due to the interaction of the organism's genotype with the environment. While seed selection by farmers over seasons exerts selection pressure on populations of genotypes thereby reducing the plasticity at population level, introduction of new varieties or new selection and introgression of genes from hybridization with wild species or varieties would lead to enhanced level of plasticity and thereby a higher capability to withstand environmental changes. In this context, the neglected and underutilized crop genetic resources assume significance for sustainable agriculture (Eyzaguirre et al. 1999; Mal 2007). Many underutilized species occupy important niches, adapted to risky and fragile conditions of rural communities, and have a comparative advantage in marginal lands as they can withstand stress. They also contribute to the diversity and stability of agro-ecosystems and are potential crops for the diversification of agriculture. Most of the these crops do not require high inputs and can be successfully grown in marginal, degraded, and wastelands with minimal inputs and at the same time can contribute to increased agricultural production, enhanced crop diversification, and improved environment and have the potential to contribute useful genes to breed better varieties capable of withstanding and sustain the climate change scenario (Mal and Joshi 1991; Mal 2007; Padulosi et al. 2009).

7.3 Cultivation, Utilization and Diversity of Buckwheat

Genetic diversity which is currently underutilized may become more attractive to farmers as a result of climate change. Many neglected and underutilized species which are currently maintained through in situ conservation on farm could be the crops of future. Their adaptability, plasticity and resilience to stresses provide farmers with much needed coping strategies to confront with climate changes. Because of changes in shift in rainfall pattern and temperature deviations from normal, community-based management of a wide portfolio of plant and animal genetic diversity is required to allow adaptive capacity. The suitability of current crop genotypes to local conditions will change in both positive and negative ways, depending upon the crop and region, but will affect many production systems. The processes of in situ/on-farm management of agricultural biodiversity carried out by millions of farmers in the world have developed a range of genetic diversity that helps to diversifying incomes and livelihoods of people in such changing situations. On-farm management of genetic diversity has traditionally allowed farmers to cope with adversity and this process will continue to serve that function in future too.

Even though buckwheat is considered to be a minor crop, it is an indispensable food in the temperate and hill regions of East Asia and Europe. Buckwheat is a multipurpose crop used for food, feed, medicine, and manure (Li and Zhang 2001; Zeller 2001). Common buckwheat (*Fagopyrum esculentum* Moench), a diploid ($2n = 16$) annual, is widely cultivated in Asia, Europe, and America. Due to short growth span, capability to grow at high altitudes, and the high quality protein of its grains it is an important crop in mountainous regions of India, China, Russia, Ukraine, Kazakhstan, parts of Eastern Europe, Canada, Japan, Korea, and Nepal. The plant is a rich source of Zn, Cu, Mn, Se, vitamin B_1, B_2, E, and dietary proteins for gluten sensitive individuals (Wei et al. 2003; Stibilj et al. 2004). Buckwheat leaves and flowers are a rich in source of rutin, catechins, and other polyphenols that are potential antioxidants (Luthar 1992; Oomah and Mazza 1996; Watanabe 1998). Buckwheat proteins have also been reported to have anticancer, hypoglycaemic, and antihypertension properties (Kayashita et al. 1999; Park and Obha 2004; Kayashita et al. 1995a, b, 1996, 1997; Tomotake et al. 2000, 2006). As a green manure crop, buckwheat produces only modest biomass but offers rapid growth, improves soil and makes phosphorous more available. Quick, aggressive growth accounts for its success as a smother crop for suppressing weeds, particularly in late summer.

Buckwheat is an erect annual profusely branched herb which grows in height from 0.5 to 1.5 m (Fig. 7.1). The plant has hollow stems which vary in color from green to red and brown at maturity. Buckwheat has a shallow tap root system, with numerous laterals extending from 0.9 to 1.2 m in depth. Having an indeterminate growth habit, the plants begin to flower 5–6 weeks after sowing and mature in 80–110 days. The genus *Fagopyrum* consists of about 19 species of which only two species, namely, *F. esculentum* and *F. tartaricum* are cultivated. On the basis of its achene morphology, the genus *Fagopyrum* has been divided into two phylogenetic groups viz. the *cymosum* group and the *urophyllum* group. The *cymosum* group comprises of *F. esculentum* and *F. tartaricum* which are cultivates and *F. cymosum, F. homotropicum, F. lineare, F. pilus* which grow wild. Plants belonging to this group have large lusterless achenes that are incompletely covered with a persistent perianth. The *urophyllum* group, on the other hand, comprises of *F. urophyllum* and other wild species which have small lustrous achenes which are completely covered with a persistent perianth (Ohnishi and Matsuoka 1996; Yasui and Ohnishi 1998; Ohsako and Ohnishi 2000). The genus *Fagopyrum* is comprised of both perennial as well as annual species with diploid ($2n = 2x = 16$) as well as tetraploid ($2n = 4x = 32$) cytotypes.

While the genus *Fagopyrum* is comprised of both self- as well as cross-pollinated species, the occurrence of dimorphic heterostyly renders some of the species self incompatible. Joshi (1999) have reported a wide range of variation in 11 morphological characters including plant height, number of branches, number of internodes, number of leaves, leaf length, leaf width, days to flower, no. of seeds per cyme, seed shape, size and color, seed weight, and yield per plant in 577 collections represented by 284 accessions of *F. esculentum*, 55 accessions of *F. emarginatum*, 198 accessions of *F. tataricum*, 30 accessions of *F. tataricum* var.

Fig. 7.1 Variations in plant height (**a**), leaf morphology (**b**) and inflorescence color (**c**) in accessions of buckwheat

himalianum, five accessions each of *F. giganteum* and *F. cymosum* from different agroecological regions of the Himalayas. They suggested that leaf width and seed weight in buckwheat were the most potential traits for genetic amelioration. Even though phenotypic plasticity as an important attribute, that enables plants to survive across a range of environments, has been known since the beginning of twentieth century, only in the last 15 years has it been recognized as an important source of diversity (Sultan 2000). Contrary to the neo-Darwinian view which considered phenotypic plasticity as "noise," plasticity is now recognized as a major source of phenotypic variation (Sultan 2003). Phenotypic variation can arise either because of differences in growth rates under environments with contrasting resource availability, i.e., "passive plasticity" (Wright and McConnaughay 2002) or be a consequence "active plasticity," in which fluctuations in growth trajectories are induced by changes in the environment that lead to ontogenetic or true plasticity (Weiner 2004). A logical corollary for modular organization is that the plasticity of the whole plant depends on the component parts or traits that exhibit plasticity as well as the nature of that plastic response.

7.4 Genome Size Variations in *Fagopyrum* and Its Evolutionary Significance

Genome size variations have long been considered as one of the important indicators of genetic complexity. Even though variations in genome size of various organisms have been correlated with physiological and ecological adaptive strategies amongst species (Bennett 1976; Grime 1998; Ohri 1998; Knight and Ackerly 2002) as well as within a species (Laurie and Bennett 1985; Bennett and Leitch 1995; Nevo 2001; Gurushidze et al. 2012), several reports have highlighted the absence of any correlation between the amount of DNA per cell and genetic complexity of the organism (Sparrow et al. 1972; Price et al. 2000). The lack of correspondence between genome size and morphological or physiological complexity of an organism has been historically termed the "C-value paradox" (Thomas 1971). Since the discovery of non-coding DNA and its impact on genome size variation, the term "paradox" has been replaced by "enigma" in an attempt to more appropriately identify the topic as a "perplexing subject" made up of several independent components (Gregory 2005). *Fagopyrum* is no exception to this enigma. In comparison with other taxa, the C-values in *Fagopyrum,* as suggested by Nagano et al. (2000) are distributed rather widely. The values range from 0.55 in *F. tataricum* to 1.92 in *F. urophyllum*, which gives an average variation of about 3.5 fold. Within the two groups of *Fagopyrum*, the values have a 2.53 fold magnitude of variation amongst species within the *cymosum* group and 2.82 fold magnitudes of variation among species within the *urophyllum* group. While the C-value of diploid and tetraploid *F. cymosum* has been reported to be 1.16 and 0.84, respectively, it is 1.12 and 1.3 for the diploid and tetraploid forms of *F. homotropicum,* respectively (Nagano et al. 2000). In an attempt to correlate size of the genome with seed size, Knight and Ackerly (2002) have suggested a triangular relationship between genome size and seed size in plants. They have inferred that while small genomes can be associated with either small or large seeds, plants having large genomes may not have small seeds. However, Beaulieu et al. (2007) could not find any correlation between genome size and seed mass across 1,222 species from 139 families and 48 orders of seed plants. Their observations did, however, indicate that species with very large genome sizes never had small seeds, while species with small genome size had a large range of seed size. Interestingly, the genome size of *F. leptopodum, F. tartaricum* and its subspecies, *F. tartaricum* spp. *Potanini,* all of which produce small seeds, is much smaller than the mean genome size for the entire genus. While Knight and Beaulieu (2008) have highlighted the existence of a positive correlation between genome size variations and morphological characters like cell size and guard cell diameter, they suggested a triangular relationship between genome size and maximum plant height across 324 species of angiosperms. Their observation was that as genome size increases maximum plant height decreases within angiosperm. Contrary to this suggestion, the tallest wild species, *F. urophyllum* has the largest genome size and the smallest species F. *lineare* has the lowest 2C value of 1.08 pg. Contradictions have also been observed in the correlation between reproductive behavior and genome size. While a positive correlation between selfing behavior and low genome size has been reported for many seed plants (Albach and Greilhuber 2004; Wright et al. 2008), observations on the breeding behavior and genome size in buckwheat do not indicate any correlation between the two. Although *F. tartaricum* and *F. tartaricum* spp. *potanini,* which have the smallest genome size within the genus *Fagopyrum,* exhibit self breeding behavior, *F. pleioramosum*, which has second largest genome within the genus, also shows self-breeding habit. On the other hand *F. statice* and *F. leptopodum,* which have comparatively smaller genomes, are out breeders. *F. cymosum* and *F. gracilipes* have exactly same C-value but the former is an out-breeding species, while the latter predominantly displays selfing (Ohnishi 1995).

Finding fossils that provide new insight into the intermediary stages of plant evolution is extremely difficult. However, organisms alive today carry information in their genomes that has been shaped by evolution. Thus, genomes can be viewed as missing links to enable us to connect different species along the evolutionary line. The genome size of an organism is basic information necessary for estimating relationship between genetic distance and physical distance. The underlying causes of genome size variation are not well understood. Much of the data has been interpreted as supporting the idea that the variation has an adaptive basis and is strongly influenced by natural selection (Sparrow et al. 1972; Savolainen et al. 2013). This in turn has led to suggestions regarding genome size variation and its possible relationship to speciation.

The overlapping of C-values among members of the *cymosum* and *urophyllum* groups of the genus *Fagopyrum* has been suggested to imply a dynamic rearrangement of the genome in the genus within a short span of evolution (Nagano et al. 2000). *F. cymosum* had long been suggested to be the wild ancestor of *F. esculentum* and *F. tartaricum* until the discovery of *F. esculentum* spp. *ancestrale* and *F. tartaricum* spp. *potanini* by Ohnishi (1991). The postulates that *F. esculentum* spp. *ancestralis* differentiated from *F. cymosum* much earlier than *F. Tartaricum* spp. *potanini* and that *F. homotropicum* has probably been derived from *F. esculentum* ssp. *ancestralis* by shifting the breeding system from outcrossing to self-fertilization (Ohnishi and Matsuoka 1996; Yasui and Ohnishi 1998) would tend to indicate that genome evolution in *cymosum* group could have progressed in the direction of increase as well as decrease in size. Presumably, the processes leading to evolution of *F. esculentum* from *F. cymosum* was accompanied by decrease of genome size while that leading to evolution of *F. tataricum* progressed with increase in size of the genome. On the other hand, evolutionary processes within the *urophyllum* group involved only decrease in size of the genome. Thus, DNA loss could have been the predominant factor in evolutionary processes within the *urophyllum* group.

Although the coding region of chloroplast genome in buckwheat has been reported to exhibit a highly conserved nature, *rpo*C2, *ycf*3, *acc*D, and *clp*P genes of the LSC region of tartary buckwheat showed a higher evolution rate compared to other genes (Cho et al. 2015). Yamane et al. (2003) has, however, observed a high evolution rate of *acc*D gene in *Fagopyrum* and proposed that this gene was under a weak selection constraint. This is consistent with the high Ks value (2.4538) obtained for *acc*D by Cho et al. (2015). Further, the distribution of Ks values revealed that the LSC region of buckwheat chloroplast genome was under greater selection pressure than the rest, thereby, confirming a positive selection pressure and neutral evolution of the protein coding genes. While the family Polygonaceae, to which the genus *Fagopyrum* belongs, comprises of morphologically diverse group of plants found in a wide range of habitats from the Arctic to the tropics, not much information is available on the plasticity of the genus *Fagopyrum*. While Martynenko et al. (2004) have observed that high level of intrapopulation heterogeny among cultivars of *Fagopyrum* was a consequence of selection methods, Japhet et al. (2009) have ascribed the variations in branch length in buckwheat to the direct effect of size rather than a true plastic investment in branches.

7.5 The Way Forward

Even though buckwheat cultivation has been going on for long time, the crop still has a long way to go in the direction of greater popularity and importance. Some of the inherent limitations in buckwheat like low and unstable yields, indeterminate growth habit, seed shattering, lodging, presence of allergenic compounds, bitter taste, and low shelf life of its flour restrict its popularity. One of the major reasons for low research inputs into the crop has been its low level of plasticity which affects varietal development. However, the low level of plastic responses of buckwheat coupled with self incompability, because of heterostyly, also makes *Fagopyrum* a good model system for correlation of phenotypic variations with genetic plasticity. A major breakthrough in buckwheat breeding was achieved with successful interspecific hybridization between *F. esculentum* and *F. homotropicum* in which the self-compatibility trait was introduced from *F. homotropicum* into *F. esculentum* (Campbell 1995). This successful interspecific cross between *F. esculentum* and *F. homotopicum* at diploid level has opened new possibility toward enhanced plasticity in buckwheat. Unfortunately, *F. homotropicum* has severe seed-shattering, leading to considerable losses in grain yield. The scarcity of photosynthetic products in kernel is also reported to be one of the main factors that cause low yield in buckwheat (Yang et al. 1998). Buckwheat flowering is profuse and the photosynthetic capabilities of the plant do not meet the requirements to properly fill all the seeds. If the yielding ability of common buckwheat were to double, the plants only require 24 % of the flowers now being produced. The nutrients that are now being expended on the production of the 76 % of flowers it does not need, can then be redirected into filling more seeds. Therefore, the present emphasis in the self-pollinating breeding program is a reduction in flower number. The yield potential of the self-pollinating buckwheat would then be determined by the photosynthetic capability of the plants (Woo 2006).

References

Albach DC, Greilhuber J (2004) Genome size variation and evolution in Veronica. Ann Bot 94:897–911

Alpert P, Simms E (2002) The relative advantages of plasticity and fixity in different environments: when is it good for a plant to adjust? Evol Ecol 16:285–297

Beaulieu JM, Moles AT, Leitch IJ, Bennett MD et al (2007) Correlated evolution of genome size and seed mass. New Phytol 173:422–437

Bennett MD, Leitch IJ (1995) Nuclear DNA amounts in angiosperms. Ann Bot 76:113–176

Bennett MD, Smith JB (1976) Nuclear DNA amounts in angiosperms. Philos Trans R Soc Lond B 274:227–274

Campbell CG (1995) Inter-specific hybridization in the genus *Fagopyrum*. In: Proceedings of the 6th international symposium on buckwheat. Shinshu, Japan, pp 255–263

Cho KS, Yun BK, Yoon YH, Hong SY et al (2015) Complete chloroplast genome sequence of tartary buckwheat (*Fagopyrum tataricum*) and comparative analysis with common buckwheat (*F. esculentum*). PLOS ONE. doi:10.1371/journal.pone.0125332

DeWitt TJ, Scheiner SM (2004) Phenotypic plasticity: functional and conceptual approaches. Oxford University Press, Oxford

Eyzaguirre P, Padulosi S, Hodgkin T (1999) IPGRI's strategy for neglected and underutilized species and the human dimension of agrobiodiversity. In: S. Padulosi (ed) Priority setting for underutilized and neglected plant species of the Mediterranean Region. Report of the IPGRI Conference, 9–11 Feb 1998. ICARDA, Aleppo, Syria. International Plant Genetic Resources Institute, Rome, Italy

Gama EEG, Hallauer AR (1980) Stability of hybrids produced from selected and unselected lines of maize. Crop Sci 20:623–626

Garbutt K, Bazzaz FA (1983) Leaf demography, flower production and biomass of diploid and tetraploid populations of *Phlox drummondii* Hook, on a soil moisture gradient. New Phytol 93:129–141

Gregory TR (2005) The C-value enigma in plants and animals: a review of parallels and an appeal for partnership. Ann Bot 95:133–146

Grether GF (2005) Environmental change, phenotypic plasticity, and genetic compensation. Am Nat 166:E115–E123

Grime JP (1998) Benefits of plant diversity to ecosystems: immediate, filter and founder effects. J Ecol 86:902–910

Gurushidze M, Fuchs J, Blattner FR (2012) The evolution of genome size variation in drumstick onions (*Allium* subgenus *Melanocrommyum*). Syst Bot 37:96–104

Jain SK (1978) Inheritance of phenotypic plasticity in soft chess, *Bromus mollis* L. (Gramineae). Experientia 34:835–836

Japhet W, Zhou D, Zhang H, Zhang H et al (2009) Evidence of phenotypic plasticity in the response of Fagopyrum esculentum to population density and sowing date. J Plant Biol 52:303–311

Joshi BD (1999) Status of buckwheat in India. Fagopyrum 16:7–11

Kayashita J, Shimaoka I, Nakajoh M (1995a) Hypocholesterolemic effect of buckwheat protein extract in rats fed cholesterol enriched diets. Nutr Res 15:691–698

Kayashita J, Shimaoka I, Nakajoh M, Arachi Y et al (1995b) Feeding of buckwheat protein extract reduces body fat content in rats. In: Proceedings of the 6th international symposium on buckwheat. Shinshu, Japan, pp 935–940

Kayashita J, Shimaoka I, Nakajoh M, Kato N (1996) Feeding of buckwheat protein extract reduces hepatic triglyceride concentration, adipose tissue weight and hepatic lipogenesis in rats. J Nutr Biochem 7:555–559

Kayashita J, Shimaoka I, Nakajoh M, Kishida N et al (1999) Consumption of buckwheat protein extract retards 7,12-Dimethylbenz[α]anthracene-induced mammary carcinogenesis in Rats. Biosci Biotechnol Biochem 63:1837–1839

Kayashita J, Shimaoka I, Nakajoh M, Yamazaki M et al (1997) Consumption of buckwheat protein lowers plasma cholesterol and raises fecal neutral sterols in cholesterol-fed rats because of its low digestibility. J Nutr 127:1395–1400

Khan MA, Antonovics J, Bradshaw AD (1976) Adaptation to heterogeneous environments. III. The inheritance of response to spacing in flax and linseed (*Linum usitatissimum*). Aust J Agric Res 27:649–659

Knight CA, Ackerly DD (2002) Variation in nuclear DNA content across environmental gradients: a quantile regression analysis. Ecol Lett 5:66–76

Knight CA, Beaulieu J (2008) Genome size scaling in phenotype space. Ann Bot 101:759–766

Laurie DA, Bennett MD (1985) Nuclear DNA content in the genera *Zea* and *Sorghum.* Intergeneric interspecific and intraspecific variation. Heredity 55:307–313

Li S, Zhang QH (2001) Advances in the development of functional foods from buckwheat. Crit Rev Food Sci Nutr 41:451–464

Levin DA (1983) Polyploidy and novelty in flowering plants. Am Nat 122:1–25

Luthar Z (1992) Phenol classification and tanin content of buckwheat seeds. Fagopyrum 12:36–42

Mal B (2007) Neglected and underutilized crop genetic resources for sustainable agriculture. Indian J Plant Genet Resour 20:1–14

Mal B, Joshi V (1991) Underutilized Plant Resources. In: Paroda RS, Arora RK (eds) Plant genetic resources: conservation and management. Malhotra Publishing House, New Delhi, pp 211–229

Martynenko GE, Fesenko NV, Varlakhova LN (2004) Improvement of technologic grain characteristics of determinant buckwheat cultivars in the process of breeding. In: Proceedings of the 9th international symposium on buckwheat, Prague, pp 259–264

Nagano M, All J, Campbell CG, Kawasaki S (2000) Genome size analysis of the genus *Fagopyrum.* Fagopyrum 17:35–39

Nevo E (2001) Evolution of genome–phenome diversity under environmental stress. Proc Natl Acad Sci USA 98:6233–6240

Ogle BM, Grivetti LE (1995) Legacy of the chamaleon: edible wild plants in the Kingdom of Swaziland, Southern Africa. A cultural, ecological, nutritional study. Part II—demographics, species availability and dietary use, analyses by ecological zone. Ecol Food Nutr 17:1–30

Ohnishi O (1991) Discovery of wild ancestor of common buckwheat. Fagopyrum 11:5–10

Ohnishi O (1995) Discovery of new *Fagopyrum* species and its implication for the study of evolution of *Fagopyrum* and the origin of cultivated buckwheat. In: Proceedings of the 6th international symposium on buckwheat. Shinsu, Japan, pp 175–190

Ohnishi O, Matsuoka Y (1996) Search for the wild ancestor of buckwheat II. Taxonomy of *Fagopyrum* (Polygonaceae) species based on morphology, isozymes and cpDNA variability. Genes Genet Syst 71:383–390

Ohri D (1998) Genome size variation and plant systematics. Ann Bot 82(Supplement A):75–83

Ohsako T, Ohnishi O (2000) Intra- and interspecific phylogeny of wild *Fagopyrum* (Polygonaceae) species based on nucleotide sequences of noncoding regions in chloroplast DNA. Am J Bot 87:573–582

Oomah BD, Mazza C (1996) Flavonoids and antioxidative activities in buckwheat. J Agric Food Chem 44:1746–1750

Padulosi S, Mal B, Ravi SB, Gowda J (2009) Food security and climate change: role of plant genetic resources of minor millets. Indian J Plant Genet Resour 22:1–16

Park SS, Ohba H (2004) Suppressive activity of protease inhibitors from buckwheat seeds against human T-acute lymphoblastic leukemia cell lines. Appl Biochem Biotechnol 117:65–74

Pigliucci M, Murren CJ, Schlichting CD (2006) Phenotypic plasticity and evolution by genetic assimilation. J Exp Biol 209:2362–2367

Pistorius R (1997) Scientists, plants and politics–a history of the plant genetic resources movement. In: International Plant Genetic Resources Institute, Rome, Italy, p 134

Pooni HS, Jinks JL (1980) Nonlinear genotype x environment interactions. II. Statistical models and genetical control. Heredity 45:389–400

Prescott-Allen R, Prescott-Allen C (1990) How many plants feed the world? Conservation Biol 4:365–374

Price HJ, Hodnett G, Johnston JS (2000) Sunflower (*Helianthus annuus*) leaves contain compounds that reduce nuclear propridium iodide fluorescence. Ann Bot 86:929–934
Savolainen O, Lascoux M, Merilä J (2013) Ecological genomics of local adaptation. Nat Rev Genet 14:807–820
Scheiner SM (1993) Genetics and evolution of phenotypic plasticity. Ann Rev Ecol Syst 24:35–68
Scheiner SM (2013) The genetics of phenotypic plasticity. XII. Temporal and spatial heterogeneity. Ecol Evol 3:4596–4609
Schlichting CD (1984) Studies on phenotypic plasticity in annual Phlox. Ph.D. thesis, University of Texas, Austin, p 114
Schlichting CD, Levin DA (1986) Effects of inbreeding on phenotypic plasticity in cultivated *Phlox drummondii*. Theor Appl Genet 72:114–119
Sparrow AH, Price HJ, Underbink AG (1972) A survey of DNA content per cell and per chromosome of prokaryotic and eukaryotic organisms: some evolutionary considerations. In: Smith HH (ed) Evolution of genetic systems. Gordon and Breach, New York, pp 451–494
Stibilj V, Kreft I, Smrkolj P, Osvald J (2004) Enhanced selenium content in buckwheat (*Fagopyrum esculentum* Moench) and pumpkin (*Cucurbita pepo* L.) seeds by foliar fertilization. Eur Food Res Technol 219:142–144
Sultan SE (2000) Phenotypic plasticity for plant development, function and life history. Trends Plant Sci 5:537–542
Sultan SE (2003) Phenotypic plasticity in plants: a case study in ecological development. Evol Dev 5:25–33
Thomas CA (1971) The genetic organization of chromosomes. Annu Rev Genet 5:237–256
Tomotake H, Shimaoka I, Kayashita J, Yokoyama F et al (2000) A buckwheat protein product suppresses gallstone formation and plasma cholesterol more strongly than soy protein isolate in hamsters. J Nutr 130:1670–1674
Tomotake H, Yamamoto N, Yanaka N, Ohinata H et al (2006) High protein buckwheat flour suppresses hypercholesterolemia in rats and gallstone formation in mice by hypercholesterolemic diet and body fat in rats because of its low protein digestibility. Nutrition 22:166–173
Watanabe M (1998) Catechins as antioxidants from buckwheat (*Fagopyrum esculentum* Moench) groats. J Agric Food Chem 46:839–845
Wei Y, Hu X, Zhang G, Ouyang S (2003) Studies on the amino acid and mineral content of buckwheat protein fractions. Nahrung/Food 47:114–116
Weiner J (2004) Allocation, plasticity and allometry in plants. Perspect Plant Ecol Evol Syst 6/4:207–215 (Elsevier GmbH)
Woo SH (2006) Breeding improvement of processing buckwheat. J Agric Sci 23:27–34
Wright SI, Rob WN, Foxe JP, Spencer CHB (2008) Genomic consequences of outcrossing and selfing in plants. Int J Plant Sci 169:105–118
Wright SD, McConnaughay KDM (2002) Interpreting phenotypic plasticity: the importance of ontogeny. Plant Species Biol 17:119–131
Yamane K, Yasui Y, Ohnishi O (2003) Intraspecific cpDNA variations of diploid and tetraploid perennial buckwheat, *Fagopyrum cymosum* (Polygonaceae). Am J Bot 90:339–346
Yang W, Hao Y, Li G, Zhou N et al (1998) Relationship between reproductive growth of common buckwheat and light duration. In: Proceedings of the 7th international symposium on buckwheat. Winnipeg, Manitoba, Canada, pp 44–48
Yasui Y, Ohnishi O (1998) Interspecific relationships in *Fagopyrum* (Polygonaceae) revealed by the nucleotide sequences of the *rbc*L and *acc*D genes and their intergenic region. Am J Bot 85:1134–1142
Zeller F (2001) Buckwheat (*Fagopyrum esculentum*Moench): utilization, genetics, breeding. Bodenkultur 52:259–276
Zimmerman CA (1976) Growth characteristics of weediness in *Portulaca oleracea* L. Ecology 57:964–974

Chapter 8
Origin of Genetic Variability and Improvement of Quinoa (*Chenopodium quinoa* Willd.)

Atul Bhargava and Deepak Ohri

Abstract Quinoa is a pseudocereal having a very balanced composition of carbohydrates, fat, and protein. Various studies based on inheritance, molecular cytology, DNA markers, and single locus variability have established it as an allotetraploid (2n = 4x = 36). It has been cultivated for 5 millennia in the Andes where it probably originated from its wild and weedy forms. Domestication process led to loss of many characters disadvantageous to farmers leading to narrowing of the genetic base. However, wide diversity based on plant color, seed color, types of branching and panicles, productivity, abiotic stress tolerance, and disease resistance still exists. This diversity is also reflected at the molecular level and is being used by the plant breeders worldwide to develop improved plant types with respect to uniformity, early maturity, seed yield, protein content, and reduced saponin content in the seeds.

Keywords *Chenopodium quinoa* · Origin · Domestication · Genetic variability · Improvement

8.1 Introduction

Quinoa (*Chenopodium quinoa* Willd.) is a pseudocereal and is one of the 250 species included in the genus *Chenopodium* (Amaranthaceae), commonly known as 'goosefoot' genus (Giusti 1970). The genus comprises herbaceous, suffrutescent, and arborescent perennials, although most species are colonizing annuals (Wilson 1990; Fuentes et al. 2012). Quinoa along with some leafy chenopods (*C. album* and *C. giganteum*) is very important to the food security of marginal farmers as they

A. Bhargava · D. Ohri (✉)
Amity University Uttar Pradesh (Lucknow Campus), Malhaur (Near Railway Station), Gomti Nagar Extension, Lucknow 226028, Uttar Pradesh, India
e-mail: ohri_deepak@rediffmail.com; dohri@lko.amity.edu

A. Bhargava
e-mail: atul_238@rediffmail.com; abhargava@amity.edu

V.R. Rajpal et al. (eds.), *Gene Pool Diversity and Crop Improvement*, Sustainable Development and Biodiversity 10,
DOI 10.1007/978-3-319-27096-8_8

show adaptation to many abiotic and biotic stresses and their ability to grow with minimum inputs (Bhargava et al. 2003, 2006a; Jacobsen et al. 2003; IAEA 2004; Maughan et al. 2009; Bhargava and Srivastava 2013). It is an allotetraploid (2n = 36) annual, self fertile crop as it shows disomic inheritance for most of the traits (Simmonds 1971; Risi and Galway 1984; Ward 2000; Maughan et al. 2004), two different non-transcribed sequence (NTS) classes in 5S rDNA spacer region (Maughan et al. 2006), two homoeologous *SOS1* loci (*cqSOS1A* and *cqSOS1B*) (Maughan et al. 2009), two distinct homoeologs of *GBSS1*gene (*GBSS1A* and *GBSS1B*) (Brown et al. 2014) and by the identification of two distinct subgenomes by FTL intron markers (Storchova et al. 2015).

The small seed of quinoa contains a balanced composition of carbohydrates, fat, and protein (Risi and Galway 1984; Coulter and Lorenz 1990; Chauhan et al. 1992). Moreover, the protein is not only higher (7.5–22.1 %) than major cereals (Tapia et al. 1979) but is also composed of 16 amino acids being rich in lysine, threonine, and methionine (Ruales and Nair 1992; Aubrecht and Biacs 2001; Gorinstein et al. 2002; Wright et al. 2002; Drzewiecki et al. 2003; Vega-Galvez et al. 2010; Stikic et al. 2012; Escuredo et al. 2014). The protein is of high quality containing much higher content of lysine than cereals and even milk and being devoid of gluten makes it suitable for celiac patients (Koziol 1992; Vega-Galvez et al. 2010; Stikic et al. 2012). Quinoa starch is present in the form of small granules of about 1–1.5 μm in diameter and an average molar mass of 11.3×10^6 g/mol (Tang et al. 2002; Wright et al. 2002; Tari et al. 2003; Lindeboom 2005). The small granules and high viscosity of quinoa starch make it useful for specialized industrial applications (Galwey et al. 1990) such as dusting starches in cosmetics and rubber tyre mold release agents (Bhargava et al. 2006a) and as biodegradable fillers in low-density polyethylene (LDPE) films (Ahamed et al. 1996b). Quinoa starch due to its unique mechanical properties can be utilized in the manufacture of carrier bags where tensile strength is important. Because of the freeze–thaw stability and resistance to retrogradation the quinoa starch paste is very suitable in the preparation of frozen and emulsion-type food products (Ahamed et al. 1996a; Bhargava et al. 2006a). It is also rich in essential vitamins and minerals, iron, and calcium (Risi and Galway 1984; Konishi et al. 2004). Quinoa in fact makes a fine example of a functional food being rich in antioxidants (Nisimba et al. 2008; Repo-Carrasco et al. 2003; Koziol 1992). Quinoa seeds contain a number of ecdysteroids beneficial to human health as they reduce glycemia of diabetic patients (Kumpun et al. 2011).

The balanced nutritional superiority of quinoa has been evaluated as a food by the National Research Council and the National Aeronautics and Space Administration (NASA) (Schlick and Bubenheim 1996) and has been recommended as a potential crop for NASA's Controlled Ecological Life Support System (CELSS), which aims to utilize plants to remove carbon dioxide from the atmosphere and generate food, oxygen, and water for the crew of long-term space missions (Schlick and Bubenheim 1996).

8.2 Brief History and Distribution

Quinoa has been recognized for centuries as an important food crop in the high Andes of South America (Tapia 1982). The name quinoa in the Quechua and Aymara languages means 'Mother Grain' and this crop occupied a place of prominence in the Inca Empire next only to maize (Cusack 1984). However, after the conquest of the region by the Spaniards in 1532 A.D. crops such as potato, faba beans, oats, and barley relegated quinoa to the background (Galwey 1995; Bhargava et al. 2006a). During the colonial period the cultivation of quinoa was discouraged, possibly because of its honored position in the Inca society and religion (Risi and Galwey 1989; Ruas et al. 1999). Quinoa's religious significance for the Incas made it a less attractive crop to the Spanish. The status given to it as 'Mother Grain' and the 'Grain of the Gods' put it in direct conflict with the Catholic religion promoted by the Spanish Conquistadors who discouraged its production and consumption in the newly conquered territories. However, the frequent droughts in Andes once again necessitated the cultivation of quinoa as it showed nearly constant yields in severe conditions (Cusack 1984; Bhargava et al. 2006a). During the 1980s, the market for quinoa began expanding in Europe and North America mainly in the health-food sector and the demand was met partly by imports from South America and by the development of quinoa in new regions outside its center of origin.

In its native region, the major areas of current quinoa cultivation appear to extend southward from extreme southern Colombia through Ecuador, Peru, and Bolivia, with extensions into the Chilean altiplano (eastern Tarapaca) and northern Argentina (Jujuy and Salta) (Wilson 1990). According to Rojas (1998) the geographical distribution of quinoa in the region extends from 5°N in Southern Colombia, to 43°S in the Xth Region of Chile. During the last decade, Chile, Ecuador, Argentina, and Colombia have started extensive cultivation and research projects on quinoa, such as SICA (Agricultural Census and Information System) of the Agricultural Ministry of Ecuador; Quinuacoche CANOE program promoted by the Latin American Foundation in Colombia; Provincial Congress for Quinoa promoted by the Deputies Chamber of Salta, Argentina; Program of Encouragement for Business Design; and Innovation promoted by the Euro Chile Foundation (Taboada et al. 2011).

Quinoa was introduced in the 1970s in England and thereafter in Denmark. In 1993, a project titled 'Quinoa-a multipurpose crop' for EC's agricultural diversification was initiated in the European Union (Jacobsen 2003). This led to setting up of laboratories in Scotland and France, and field trials in England, Denmark, the Netherlands, and Italy. Quinoa has been evaluated as a potential crop in Denmark, Poland, Sweden, Italy, and Greece (Iliadis et al. 1997; Gęsiński 2000; Bhargava and Srivastava 2013). It has also been successfully tested in North America and Africa and has been cultivated in the US since the early 1980s and commercially produced since the mid-1980s in the Colorado Rockies, especially in the San Luis Valley (Bhargava and Srivastava 2013). The North American Quinoa Producers Association was formed in 1988 and a small processing plant was started for the crop produced in the area. Production has also been attempted in California, New Mexico, Oregon, and Washington. In Canada, cultivation is done in Saskatchewan and Manitoba most of which is organic (Bhargava and Srivastava 2013).

The Asian experiment on quinoa introduction has been quite impressive with the crop showing good adaptation and abundant yield in the Indian subcontinent. Quinoa was successfully introduced in India in the early 1990s and exhaustive field trials have proved its cultivation as an alternative winter crop for the North Indian Plains (Bhargava et al. 2007a). It was introduced in Pakistan in 2007 in the central Punjab to lessen the dependence of the common people on conventional crops (Munir et al. 2012). Field tests have been done in Japan in the climatic conditions of Southern Kanto District of Japan (Yamashita et al. 2007). Field tests in Kenya have shown seed yield up to 9 t/ha and biomass yield up to 15 t/ha indicating high seed yield comparable to that in the Andean region (Mujica et al. 2001). A partnership between the Danish Company Eghøjgaard and the Egyptian Natural Oil Company (NATOIL) has been constituted since the year 2007 for promoting quinoa in Egypt (Bhargava and Srivastava 2013). Quinoa was formally put in field trials in the Sinai Peninsula with 13 varieties and strains being tested in deserts of South Sinai governorate (near Nuwaiba city) which proved to be a success (Shams 2011). Recent introduction in Morocco has shown a high potential of adaptation in the country (Hirich et al. 2014).

8.3 Domestication

Quinoa is a part of a complex of interfertile New World wild, weedy, and domesticated ecotypes, variously listed as three or four separate taxa (Jellen et al. 2011). Quinoa has been cultivated for more than 5000 years in the Andes but was probably domesticated by ancient civilizations at different times and in different geographic zones, including Peru (5000 BC), Chile (3000 BC), and Bolivia (750 BC) (Tapia 1979; Kadereit et al. 2003). Ancient farmers in the Andean region of South America were the pioneers in domesticating quinoa from its wild or weedy forms until the current known types in a domestication/cultivation period of approximately 5000 years (Bazile et al. 2013). During this period, quinoa was subjected to diverse selection process for desirable traits for its cultivation and consumption by people belonging to different cultures and territories in South America like the Chibchas, Andaki, Inganos in southern Colombia; Aymara and Quechua in areas of Peru, Bolivia, and northern Chile; Diaguitas y Calchaquies in northern Argentina; and Mapuches in southern Chile (Mujica 2004). This process led to the loss of many undesirable characters present in the wild forms such as dehiscent seeds (seed shattering), seed dormancy, and thick seed coats, with simultaneous gain of useful characters such as larger and more starchy seeds, fewer and larger inflorescences, uniform maturity, and environmental adaptations (Galwey 1995; Bazile et al. 2013). This process has been continued by modern breeding techniques for the development of best phenotypes in terms of yield and agronomic performance, which led to the narrowing of the genetic diversity further. However, wide diversity in plant and seed color, types of branching and panicles, as well as grain productivity, abiotic stress tolerance, and disease resistance can be observed in the field (Fuentes and Bhargava 2011; Ruiz-Carrasco et al. 2011).

Quinoa diversity, at a continental scale, has been associated with five main ecotypes viz. Highlands (Peru and Bolivia), Interandean valleys (Colombia, Ecuador and Peru), Salares (Bolivia, Chile and Argentina), Yungas (Bolivia), and Coastal/Lowlands (Chile); each of these is associated with subcentres of diversity that originated around Lake Titicaca (Risi and Galwey 1984). The recent genetic-based analyses have confirmed that quinoa has existed as two distinct germplasm pools: Andean highland quinoa with its associated weedy complex (ajara or ashpa quinoa, *C. quinoa* ssp. *milleanum* Aellen, also referred to as *C. quinoa* var. *melanospermum* Hunziker) and kinwa among the Mapuche people of the central and southern Chilean coastal/lowlands, representing in addition a second center of major quinoa diversity (Christensen et al. 2007; Fuentes et al. 2009; Jellen et al. 2011). The weedy *C. hircinum* from lowland Argentina can be considered as the third distinct germplasm pool, which may represent remnants of archaic quinoa cultivation in that part of South America (Wilson 1990).

Bazile et al. (2013) have recently differentiated 8 types of quinoa growing in the area of its origin. This classification is based on morphology, phenology, cultivation practices, and resistance to biotic and abiotic factors.

8.3.1 Quinoas of the Altiplano (Northern Andean Highlands)

Small plants of different colors adapted to the shore of the Titicaca Lake, having variable saponin content and 6 months of vegetative period with small to medium grain, less resistant to cold and drought, and sometimes adapted to grow in saline soils. Plants in the highlands have few branches and a unique panicle, with abundant foliage. These are moderately resistant to mildew, and are attacked by young plant cutters (*Feltia experta* Walker), Kona Kona (*Eurisacca quinoae* Povolny), and birds (Bazile et al. 2013). They usually require an annual rainfall between 700 and 800 mm and grow at an altitude of 3850 m, for example, Kancolla, Blanca de July, and Chullpi.

8.3.2 Quinoas from the Salars *(Southern Highlands)*

Cultivation is done on the flanks of the volcanos, in the middle of lava blocks and on the slopes where frosts are less frequent. Plants of the southern highlands are large, branched, having different colors, with large grains of 2.2–2.9 mm with high saponin content, drought resistant, adapted to saline and sandy soils of the shore of salt lakes (called *Salares*), and adapted to high, dry, and cold conditions (Bertero et al. 2004; Mujica et al. 2010a; Bazile et al. 2013). The quinoas of this region correspond to a group of landraces from desert areas with low annual rainfall (between 150 and 350 mm) and an altitude of over 3800–4200 m, for example, Pandela, Utusaya, Toledo, and Achachino.

8.3.3 Quinoas from Interandean Valleys

Plants of this type are long, thick stemmed, branched with long vegetative period, have large to small grains of diverse colors, large leaves and inflorescences, susceptible to mildew, and have both high and low contents of saponin (Bazile et al. 2013). These grow between 2500 and 3200 m in areas having annual rainfall of 800–900 mm. These kinds of quinoas are usually called "Quinua" and grouped by their genetic and phenotypic characteristics (Medina et al. 2004), for example, Amarilla de Marangani, Blanca de Junín, Acostambo, Roja Coporaque, Nariño, etc.

8.3.4 Quinoas from Arid Zones and Dry Condition (Eastern Highlands)

Quinoas of these regions are small, with short vegetative period and present morphological, physiological, anatomical, biochemical, and phenological modifications to withstand drought stress (Mujica et al. 2010b). They grow in areas over 3900 m with an annual rainfall of 150–350 mm. Plants have small leaves, deep and highly branched root structure, small to medium sized grains, and high saponin content (Bazile et al. 2013). These quinoas are also called '*quinua,*' for example, Antahuara, Ucha, Ccoyto, and Roja Ayauchana.

8.3.5 Quinoas from High Altitudes and Cool Climate

Plants are small with vivid colors like yellow, reddish, or purple in plants and grains and have small and compact glomerular panicles, bitter grain, high in protein content, cold resistant, with mechanisms of overcooling tolerance, and strong winds (Jacobsen et al. 2007; Bazile et al. 2013). They are resistant to ultraviolet radiation and sown over elevations of 4000 m and in areas with annual rainfall of 800 mm, for example, Huariponcho, Witulla, Kellu, Kancolla, and Roja.

8.3.6 Quinoas from the Coastal Regions and Near the Sea

This group has many types adapted to salty and sandy soils and grows in areas having average annual rainfall of about 500–650 mm (during 4–5 months) and have high evapotranspiration index (Bazile et al. 2010; Núñez et al. 2010). The plants are medium branched, with glomerular panicles and small leaves. All of them have small and hard grain and it is usually protected by perigonium which strongly adheres to the grain (Bazile et al. 2013). Plants are adapted to long days and are resistant to excess moisture with some having the ability to grow in over 2000 mm of annual rainfall, for example, Quinoa Blanca, Kinwamapuche, Lito, Faro and Islunga.

8.3.7 Quinoas from Jungle and Tropical Zones

These are tall quinoas having more branches, long vegetative period, large leaves, bright and intense colors, large and loose panicles, and small grains (Bazile et al. 2013). The category shows resistance to heat and evapotranspiration, and grows in areas having annual rainfall over 1500 mm, i.e., in Tupiza, Sandia, Puno, Ambo-Huánuco, and Lares-Cusco.

8.3.8 Quinoa from High Rainfall and Humidity Zones

These are tall, highly branched, resistant to heavy rainfall and poorly drained soils, high yielding, mildew resistant, and grow in areas having annual rainfall between 2000 and 3000 mm (Bazile et al. 2013), i.e., in Tupiza, Nariño, Sogamoso, Tunkahuan, Sogamoso, Mérida, Tupiza, and Amazonas.

8.4 Cytogenetics

The basic chromosome number in the genus *Chenopodium* is $x = 8$ and $x = 9$ (Kawatani and Ohno 1950, 1956). The number $x = 8$ is restricted to section Ambrina (Uotila 1973) that contains *C. ambrosioides* ($2n = 2x = 16$) as a representative member (Suzuka 1950; Giusti 1970). The number $x = 9$ is found in section Chenopodia, which has been further subdivided into three subsections viz. Leiosperma, Cellulata, and Undata (Risi and Galwey 1984). Cytological studies have established that *C. quinoa* is a tetraploid having a chromosome number $2n = 4x = 36$ (Palomino et al. 1990; Bhargava et al. 2006b). Gandarillas (1979) reported mixoploidy in *C. quinoa* with chromosome numbers of $2n = 18$, $2n = 27$,

2n = 36, and 2n = 45. Wang et al. (1993) studied the somatic chromosomes from root tips of nine taxa of five species of the genus *Chenopodium*. All the three cultivars of *C. quinoa* studied were reported to be tetraploid (2n = 36).

Some workers have described certain conspicuous karyotypic features in some species of the genus (Tanaka and Tanaka 1980; Wang et al. 1993; Kolano et al. 2001). The detailed karyotype of quinoa was described by Catacora (1977) who inferred allopolyploidy based on chromosome arm length ratios, which arranges 36 chromosomes in nine groups. Bhargava et al. (2006b) studied seven accessions of *C. quinoa* and assigned them to two groups based on the ratio between the longest and the shortest chromosomes in the complement which was <2.0 in 1a and >2.0 in 1b types of karyotypes. The symmetry index (TF%) on the basis of arm ratios varied from 43.9 % (PI 584524, most asymmetrical) to 47.4 % (CHEN 58/77, most symmetrical). All taxa were characterized by one satellite pair, the position of which varied according to its comparative size in the complement. The satellite pair was morphologically similar in all the accessions being median (m) or median-submedian (msm), and had the satellite on the short arm. The first chromosome in different complements was either m or msm with arm ratios varying between 1.18 (PI 510537) and 1.56 (CHEN 71/78), while 4th, 9th, and 18th pairs were the most conserved in being median (M or m) in the accessions studied. The greatest variability is observed in 10th and 13th pair with arm ratios ranging between 1.0–1.86 and 1.0–1.78, respectively. *C. berlandieri* subsp. *nuttalliae* also had only one SAT pair with the satellite on the short arm of 3rd pair which was msm. The first pair was msm and the 18th pair was median point (M) and as most of the accessions of *C. quinoa* there was no msm pair in the complement. The symmetry index was 44.1 % and the karyotype belonged to 1a class. The karyotype of *C. berlandieri* subsp. *nuttalliae* did not show any distinct differences and was basically similar to those of different accessions of *C. quinoa*. This was clear from karyotype formula, symmetry index, and one satellite pair of similar morphology as in *C. quinoa*. Similar results were obtained by Palomino et al. (2008) in *C. quinoa* and *C. berlandieri* where all the chromosomes were arranged in nine groups of four each with two pairs of satellite chromosomes in each complement. The accessions of *C. quinoa* studied showed only minor though consistent differences in their karyotypes, which is expected, as *C. quinoa* has a monophyletic origin from Andean crop/weed system (Wilson 1990). These minor differences in karyotypes due to chromosomal alterations are being maintained in quinoa due to predominantly self-pollinating behavior (Risi and Galwey 1984) and are also consistent with some degree of variability in morphological characters (Wilson 1988; Risi and Galwey 1984; Bhargava et al. 2007a), protein profiles (Bhargava et al. 2005), and RAPD profiles (Ruas et al. 1999). The karyotypic analysis has resulted in clearly identifiable 18 pairs thereby indicating allotetraploidy (Bhargava et al. 2006b). This is also supported by duplication of Lap loci (Wilson 1976), disomic inheritance of some characters (Simmonds 1971), and allelic segregation ratios of F_1 and F_2 which indicated disomic-digenic and tetrasomic inheritance in some traits (Ward 2000). The presence of one satellite pair has been corroborated by studies on fluorescent in situ hybridization with 45S rDNA showing two sites of hybridization on two homologous chromosomes (Kolano et al. 2001). Both these rDNA loci are transcriptionally active which means that at least one such locus may have been lost (Kolano et al. 2001). Maughan et al. (2006) also observed one pair of 25S (NOR) rRNA pair and two pairs (terminal and interstitial) of 5S rRNA in *C. quinoa*. However, the number of NOR loci varied in related tetraploids where 2 NOR loci were identified in *C. berlandieri* subsp. *nuttalliae* 'Huauzontle' and only one was observed in *C. berlandieri* subsp. *zschackei* and *C. berlandieri nuttalliae* 'Quelite,' Similarly, *C. berlandieri* subsp. *zschackei* showed two 5S loci and *C. berlandieri* subsp. *nuttalliae* 'Huauzontle' and 'Quelite' showed three 5S loci.

The presence of two subgenomes in quinoa has been demonstrated by FISH using two repetitive sequences, 12-P and 18-24J (Kolano et al. 2011). The specificity of 18-24J to one of the genomes was shown by strong signals on 18 chromosomes in the form of bands of differing intensities on chromosome arms while only minor signals on remaining 18 chromosomes in terminal and centromeric positions. Similar results were obtained with other tetraploids *C. berlandieri* subsp. *nuttalliae* and *C. berlandieri* subsp. *berlandieri*. This sequence also hybridized with genomes of two diploid species *C. album* and *C. pallidicaule*. When simultaneous hybridization was done with 18-24J and rDNA sequences, one 35S rDNA locus and only one of the two 5S rDNA loci in *C. quinoa* and *C. berlandieri* subsp. *berlandieri* were located on chromosome pairs with 18-24J signals, while the second 5S rDNA locus was on chromosome pairs without 18-24J signals. In case of *C. berlandieri* subsp. *nuttalliae* having three 5S rDNA loci, two pairs were located on chromosomes with 18-24J signals. These results, therefore, clearly show that the tetraploid species studied share at least one genome derived from a common ancestor (Kolano et al. 2011). The identification of the two genomes involved in the ancestry of *C. quinoa* has been done by the phylogenetic analysis of two flowering locus T-Like genes CrFTL1 and CrFTL2. One parent was shown to be related to North American *C. standleyanum, C. incanum* or any other related diploid (subgenome 'A') and the other parent belonging to the Eurasian species *C. suecicum, C. ficifolium,* or some related diploid species (subgenome 'B') (Storchova et al. 2015). In this respect, it is interesting to note that *C. quinoa* is inter-crossable with diploid cytotype of *C. album* occurring in North Indian Plains and the resulting triploid shows 18 II and 18I, therefore meaning that one of the genomes of *C. quinoa* is homologous with that of 2x *C. album* (Pal and Ohri, unpublished). The close genetic relationship between *C. quinoa* and 2x *C. album* has also been shown on the basis of RAPD and DAMD studied (Rana et al. 2010).

8.5 Genome Size

Bhargava et al. (2007b) found 4C DNA amounts ranging from 6.34 to 6.47 pg, showing a nonsignificant 1.02-fold variation in 21 accessions of *C. quinoa* by feulgen microdensitometry. Similarly, two accessions of related tetraploid species *C. berlandieri* subsp. *nuttalliae* show 4C DNA amounts of 5.79 and 5.90 pg and their average is 8.31 % less than the mean of 4C DNA values of 21 *C. quinoa* accessions. Palomino et al. (2008) obtained similar results by flow cytometry as *C. quinoa* cv. Barandales showed 2C value of 2.96 pg and six accessions of C. *berlandieri* subsp. *nuttalliae* varied from 2.96 to 3.04 pg. The above results have been supported by Kolano et al. (2012) who showed 2C values ranging from 2.9 to 3.0 pg in 20 accessions of *C. quinoa*. However, significantly lower 2C values of 2.66 pg have been obtained by Bennett and Smith (1991) by microdensitometry and 2.01 pg by flow cytometry (Stevens et al. 2006).

8.6 Interspecific Hybridization

Various attempts have been made to hybridize *C. quinoa* with related wild or cultivated tetraploids. Nelson (1968) produced artificial hybrids between *C. quinoa* and *C. quinoa* var. *melanospermum* and also confirmed the presence of natural hybrids. Heiser and Nelson (1974) produced F_1 hybrids between *C. quinoa* and *C. nuttalliae* or 'huahzontli' which, however, lacked pollen grains as male sterile parent was involved. The F_1 produced seed when backcrossed with the parents showing that two species are closely related. Remarkably, both the parents had light-colored fruits, while the F_1 had black fruit which was interpreted as due to genetic complementation showing thereby that light-colored fruit arose independently in Mexico and S. America. Wilson and Heiser (1979) showed very low pollen fertility (3 %) in hybrids between *C. quinoa* and *C. nuttalliae*, and the hybrids though are self-sterile produce seed when backcrossed with either parent. Similar results were obtained when *C. quinoa* was crossed with its N. American relative *C. berlandieri* (Wilson and Heiser 1979). Wilson (1980) obtained only one hybrid using male sterile *C. quinoa* with S. American diploid *C. petiolare,* out of 17 intersubsectional combinations. The hybrid showed developmental abnormalities and did not reach flowering. Intrasubsectional crosses of *C. quinoa* succeed not only with tetraploids *C. berlandieri* and *C. berlandieri* subsp. *nuttalliae* producing fertile hybrids, but also with diploid *C. neomexicanum* which is sterile due to triploidy. The hybrid between *C. quinoa* and *C. bushianum* is sterile but produced limited back-cross progeny (Wilson 1980).

Natural hybridization has been shown to occur freely between *C. quinoa*, when cultivated in N. America, with related wild species *C. berlandieri* as 30 % of the progeny of the latter was found to be F_1 crop/weed hybrids as confirmed by the presence of polymorphic quinoa isozyme alleles and morphologically intermediate leaves (Wilson and Manhart 1993).

8.7 Genetic Diversity Use of Molecular Markers

Random amplified polymorphic DNA (RAPD) markers were first used in quinoa by Fairbanks et al. (1993) who observed that 26 primers produced polymorphic markers among 16 randomly selected quinoa accessions. The RAPD markers were also used to identify genetic variation among 19 accessions of six species of the genus *Chenopodium* by Ruas et al. (1999). A total of 33 decamer primers generated 399 molecular markers with an average of 12 polymorphisms per primer, which grouped the germplasm collection into five different clusters. The results showed that wild and crop populations of *C. quinoa* shared a low level of molecular variation, without differentiation between sympatric domesticated and weedy populations. RAPD has also been used to study the hierarchical structure among ecotype populations of Highland and Interandean valleys in Bolivia (Del Castillo et al. 2007). The findings reported by scoring 38 selected bands from 10 RAPD primers on eight representative populations (n = 87) directly sampled in farmers' fields, revealed a marked geographical effect on the populations' structure, and explained probably by climatic and orographic barriers present in the studied zone rather than to a distance effect. Thus, the population structure was related to the three major biogeographic zones present in Bolivia: northern and central highland, Interandean valley, and southern Salar. Interestingly, the intrapopulation genetic diversity was higher than expected, due basically to mainly autogamous reproduction, as well as the limited seed exchange among isolated regions considered. The genetic diversity was even higher than that reported in studies based on germplasm collections which suggested that germplasm collection under study may not be representative of the genetic variation of the quinoa complex and that further sampling for ex situ conservation will also have to take into account the hierarchical structure of the genetic variation (Del Castillo et al. 2007).

Rana et al. (2010) assessed the suitability and reliability of RAPD and directed amplification of minisatellite DNA (DAMD) markers to assess molecular diversity in 55 accessions belonging to 14 species of chenopods. A total of 242 polymorphic markers were generated from 12 random primers yielding optimum RAPD profiles, while four DAMD primers resulted in 107 polymorphic bands. The UPGMA tree showed two major clusters: the first cluster grouped all the accessions of *Chenopodium quinoa* and its related species *C. berlandieri* subsp. *nuttalliae*, one *C. album* (4x) from Mexico and three north Indian 2x accessions of *C. album*, while the other cluster comprised mainly 6x accessions of *C. album* and *C. giganteum* in addition to *C. strictum*, *C. bushianum*, *C. opulifolium*, and *C. ficifolium* (Fig. 8.1). No genetic differentiation was observed with regard to the light- and dark-seeded quinoa accessions. The analysis allowed to assess intra- and interspecific variation within cultivated and noncultivated species in this large genus and to solve taxonomic problems either at or below the species level.

Mason et al. (2005) developed first large-scale SSR markers consisting of 208 polymorphic markers, which were validated and characterized in 31 cultivated quinoa accessions representing the main growing areas of South America. A total of

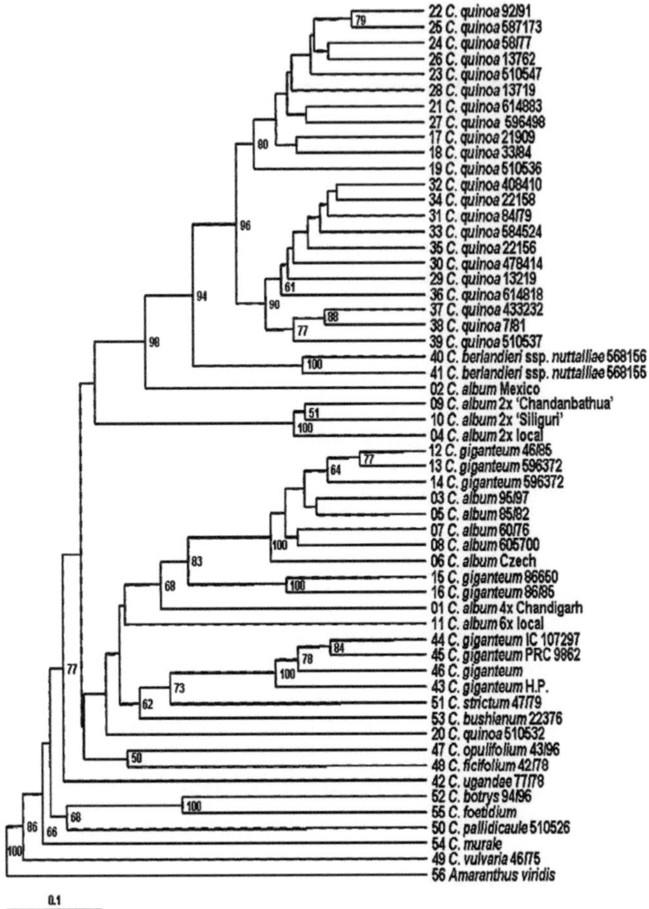

Fig. 8.1 Genetic relationships among *Chenopodium* spp. based on RAPD markers (Rana et al. 2010)

1276 clones were sequenced from three microsatellite-enriched (CA, ATT, ATG) libraries. Four hundred fifty-seven clones (36 %) contained unique microsatellites. The most common repeated motifs, other than CA, AAT, and ATG, were GA and CAA. Flanking primers were designed for 397 microsatellite loci and screened using a panel of diverse quinoa accessions and one accession of *C. berlandieri* Moq. a wild relative of quinoa. Two hundred and eight microsatellite markers (52 %) were polymorphic among the quinoa accessions. An additional 25 of the microsatellite markers (6 %) were polymorphic when the *C. berlandieri* accession was included in the analysis. The genetic analysis performed in the quinoa collection revealed a number of observed alleles ranging from 2 to 13, with an average of four alleles detected per locus. Heterozygosity values ranged from 0.20 to 0.90 with a mean value of 0.57. Sixty-seven markers (32 %) were highly polymorphic

($H \geq 0.70$). This set of SSR markers revealed the potential utility for molecular studies across related species of Chenopodioidae subfamily. The amplification of 202 of the 208 polymorphic SSR markers on a group consisting of two *C. berlandieri* subsp. *nuttalliae* (Huazontle), two *C. pallidicaule* (Canihua), and two accessions of *C. giganteum* (Khan chi) revealed that 67 % of the markers amplified successfully in all groups. The most notable PCR conservation was observed in *C. berlandieri* subsp. *nuttalliae*, with a 99.5 % of reproducible amplifications, of which 81 % were polymorphic.

The level of polymorphism and the genetic relationships were studied by means of molecular markers using the AFLPs and twenty morphological characters (Anabalon-Rodriguez and Isla 2009). Fourteen quinoa landraces from southern Chile, three landraces from highland from northern Chile, and a representation of *C. album* and *C. ambrosioides* were analyzed. The study reported 150 AFLP bands generated by three *Eco*RI-*Mse*I primer combinations of which 130 were polymorphic.

Fuentes et al. (2009) characterized and quantified the genetic diversity within 28 Altiplano and 31 coastal Chilean accessions of quinoa using microsatellite markers. A total of 150 alleles were detected among the quinoa accession, ranging from 2 to 20 alleles per locus and an average 7.5 allele/locus. Both cluster (UPGMA) and principal component analyses separated the accessions into two discrete groups as also shown by isozymes analysis and morphological traits (Wilson 1988), AFLP analysis (Pratt 2003), and microsatellites (Christensen et al. 2007). The first group contained quinoa accessions from the north (Andean highlands) and the second group consisted of accessions from the south (lowland or coastal). The data obtained in the diversity analyses highlights the relationships within and among northern and southern Chilean quinoa accessions and provides a new set of easy to use and highly informative genetic markers.

Fuentes et al. (2012) characterized 20 microsatellite genetic markers in a multi-origin set of 34 quinoa accessions representative of Chile and the South American region to study the impact of farmers' seed exchanges and local production practices on the genetic structure and diversity of quinoa at national scale in Chile. The molecular analysis yielded 118 polymorphic markers for all quinoa accessions assessed, with a mean value of alleles per locus of 5.9. The H for all quinoa accessions ranged between 0.12 (QGA17) and 0.87 (QAAT76) with a mean value of 0.65 which indicated the presence of wide genetic diversity in the quinoa samples and confirmed the highly informative quality of the markers used. The UPGMA analysis using the Jaccard coefficient identified two major groups, which were subdivided into five populations (Fig. 8.2). Population I contained nine accessions representative of the northern zone of Chile; population II seven accessions of the central zone; population III included nine accessions from the southern zone and only one from the central zone (B042); population IV contained six accessions from the highlands of Peru, Bolivia, and Argentina; and population V contained two accessions from Ecuador and Colombia. Thus, the genetic information allowed the detection of variation among and within the populations identified, which matches well with natural geographical–edaphic–climatic

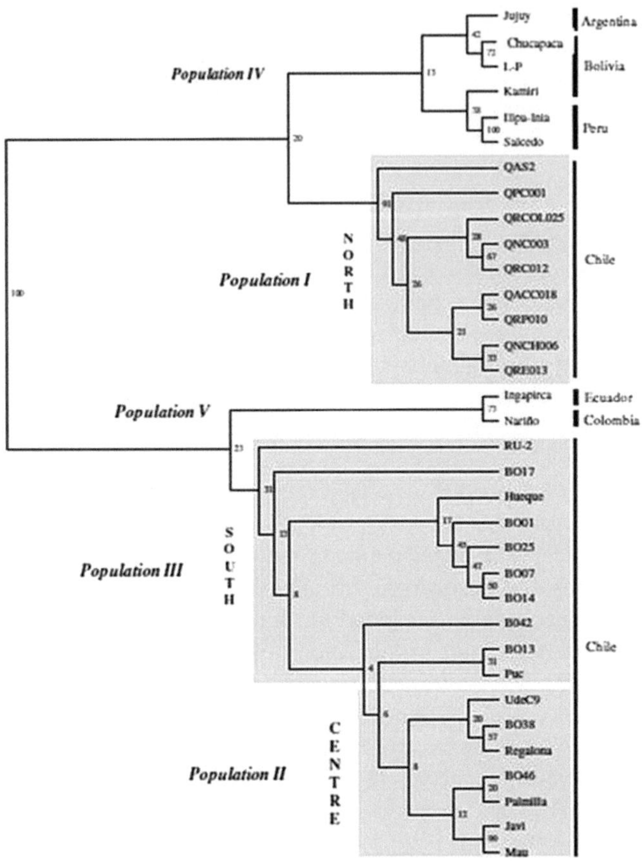

Fig. 8.2 UPGMA cladogram based on Jaccard's similarity coefficient of 34 quinoa accessions performed after 500 replicates for bootstrap test (percentage number between each node) (Fuentes et al. 2012)

constraints to the expansion of quinoa biodiversity. This grouping also correlates well with the social–linguistic context of ancient people inhabiting the Andes region, where agronomic and cultural traditions that have survived until the present time are very different.

In quinoa, the first source of SNP marker was reported by Coles et al. (2005) from an immature seed and floral expressed sequence tags (EST) libraries that were the first EST libraries developed for quinoa. EST sequences are partial sequences from transcribed cDNA sequences that reflect expressed genes in a given tissue type at a specific point of development. A total of 424 ESTs were found that corresponded to 331 sequences from the immature seed cDNA library and 83 sequences from the floral library, with an average length of 581 bp. The SNP studies yielded a total of 51 SNP markers in 20 EST sequences analyzed, consisting in 38 single-base changes and 13 insertions–deletions (Indels), with an average of 1 SNP per 462 base pairs (bp) and 1 Indel per 1812 bp. On inclusion of the *C. berlandieri* subsp. *jonesianum* accession, 81 additional SNPs were identified, bringing the total number of SNPs discovered to 132 (1 per 179 bp).

Maughan et al. (2012) reported a large-scale set of SNP markers and developed functional SNP assays for quinoa. 427 of 511 functional SNP markers were utilized to analyze a set of 113 quinoa accessions which showed MAF (minor allele frequency) values between 0.02 and 0.50. In this study, 90 % of the SNP loci were polymorphic and 46 % were reported to be highly polymorphic. The most frequent point mutation among all SNPs identified corresponded to transitions (A/G or C/T), being 1.6X higher than transversions (A/T, C/A, G/C, G/T). The phenetic analysis separated the accessions into two major groups viz. the Andean and lowland/coastal ecotypes. One interesting observation was the potential transferability of SNP markers to related species including four accessions of *C. berlandieri* (subsp. *nuttalliae*, var. *macrocalycium*, var. *boscianum,* and var. *zschackei*), two accessions of *C. hircinum*, and one accession of both *C. watsonii* and *C. ficifolium*. The two *C. hircinum* accessions and the four *C. berlandieri* accessions presented 81 and 79 % of successful amplification, respectively, which confirmed the close crop–weed sympatric relationship of these two tetraploid species with quinoa.

8.8 Genetic Linkage Maps

Maughan et al. (2004) reported the first quinoa genetic linkage map using AFLP, RAPD, and SSR markers. Selection of the mapping population was based on a preliminary genetic similarity analysis of four potential mapping parents. Breeding lines 'Ku-2' and '0654,' a Chilean lowland type and a Peruvian Altiplano type, respectively, showed a low similarity coefficient of 0.31 and were selected to form an F_2 mapping population. This map consisted of 35 genetic linkage groups containing a total of 230 AFLPs, 19 SSRs, and 6 RAPD markers, spanning 1020 cM with an average marker density of 4.0 cM per marker. This map provided a major breakthrough in the genetic dissection of agronomically important characteristics of quinoa, including seed saponin content, grain yield, maturity, and resistance to disease, frost, and drought.

Jarvis et al. (2008) reported the development of 216 new polymorphic SSR markers from libraries enriched for GA, CAA, and AAT repeats, as well as six SSR markers developed from bacterial artificial chromosome end sequences (BES-SSRs). Heterozygosity (H) values of the SSR markers ranged from 0.12 to 0.90, with an average of 0.57. A linkage map was constructed from a newly developed recombinant inbred line (RIL) population using these SSR markers. The linkage map also contained additional markers, including amplified fragment length polymorphisms (AFLPs), two 11S seed storage protein loci, and the nucleolar organizing region (NOR). The study culminated in the preparation of the first SSR-based map in quinoa and contained 275 markers, including 200 SSR. The map consisted of 38 linkage groups (LGs) covering 913 cM. Segregation distortion was observed in the mapping population for several marker loci, indicating possible chromosomal region associated with selection or gametophytic lethality.

Maughan et al. (2012) prepared first SNP-based integrated linkage map combining data from two mapping populations from a large set of SNP loci. The map consisted of 29 linkage groups with 20 large linkage groups, spanning 1404 cM with a marker density of 3.1 cM per SNP marker. This linkage map was constructed employing a $F_{2:8}$ RIL population from two advanced quinoa mapping populations (Pop1 and Pop39) sharing a common paternal parent (0654, Altiplano type) whose molecular data were combined to construct an integrated linkage map based on 128 individuals. This SNP-based map consisted of approximately twofold number of marker loci as well as spanned a greater genetic coverage than the previous reported maps.

8.9 Bacterial Artificial Chromosome Library

Stevens et al. (2006) constructed two separate quinoa BAC libraries using *BamHI* (26,880 clones) and *EcoRI* (48,000 clones) restriction endonucleases from the "Real" quinoa type. Cloned inserts of *BamHI* and *EcoRI* libraries averaged 113, 130 kb, respectively. The combined quinoa libraries represented about 9.0 di-haploid nuclear genome equivalents. An average of 12.2 positive clones per probe was identified with 13 quinoa single-copy ESTs as probes of the high-density arrayed blots, suggesting that the estimate of 9.0x coverage of the genome is conservative (Stevens et al. 2006). Furthermore, the probing of the library with partial sequence of the 11S globulin seed storage protein gene identified clones that represent two different 11S loci, which suggests the importance of BAC libraries in identifying and cloning the important genes (Stevens et al. 2006).

8.10 Characterization and Expression of Certain Genes

Quinoa is known for its ability to grow under harsh environmental conditions. Maughan et al. (2009) first described the molecular characterization of *Salt Overly Sensitive 1* (*SOS1*) gene while studying the molecular basis of salt tolerance in quinoa. The complete genomic sequence of two homoeologous *SOS1* loci, *cqSOS1A* and *cqSOS1B,* were reported, which spanned 98,357 and 132,770 bp, respectively. The translation of *cqSOS1A* and *cqSOS1B* coding sequences yielded proteins of 1158 and 1161 amino acid, respectively. The comparison of these translated genomic sequences revealed a high degree of similarity with *SOS1* sequence from species belonging to the Caryophyllales order. Under saline conditions (450 mmol/L) relative gene expression of *SOS1* in roots was consistently 3–4 fold higher than in leaf tissue. The *SOS1* expression was more strongly up-regulated by salt stress in leaves as compared to the roots suggesting a constitutive expression of *SOS1* genes in roots and an inducible expression in leaves under stress.

Reynolds (2009) reported the annotation of a large-scale EST collection from maturing quinoa seed tissues expressing saponins, in an attempt to elucidate the genetic components of its biosynthesis. A total of 39,366 unigenes, consisting of 16,728 contigs and 22,638 singletons, were assembled using Sanger and 454 GS-FLX pyrosequencing technologies. The microarray analysis allowed the identification of a set of candidate genes transcriptionally related with saponin biosynthesis that included genes having homology to cytochrome P450s, cytochrome P450 monooxygenases, and glycosyltransferases.

Ruiz-Carrasco et al. (2011) using similar approach reported gene expression analyses for two sodium transporter genes: *CqSOS1* and *CqNHX* genes. Quantitative RT-PCR analyses of these genes revealed that their expression was differentially induced at the shoot and root level, and between genotypes, by 300 mM NaCl.

8.11 Breeding System

Quinoa is gynomonoecious as it has hermaphroditic and female flowers arranged in dichasial cymes. Ten different types of inflorescences were classified in various accessions according to different proportions of hermaphroditic and female flowers and their arrangement on the dichasium (Bhargava et al. 2007c). Although predominantly autogamous, outcrossing occurs in quinoa over considerable distances. Out crossing occurs frequently up to a distance of 1 m and occasionally up to 20 m and its rates at different sowing distances in the Bolivian Altiplano ranging from 0.5 to 9.9 % (Gandarillas 1979). The average rate of outcrossing studied by Lescano (1980) in eight landraces and five varieties of quinoa was 5.8 %. A recent study on *C. quinoa* cv. Sajama gave the outcrossing rates as high as 17.36 % (Silvestri and Gil 2000). This is further corroborated by the absence of any inbreeding depression in characters like weight and height of the above-ground parts of the plant, development of inflorescence and seeds, or on the homogeneity of offsprings in six chenopod species studied (Dostalek 1987). However, some extreme cases of complete self-pollination through cleistogamy (Nelson 1968) and obligate outcrossing by self-incompatibility and male sterility (Nelson 1968; Gandarillas 1969; Simmonds 1971) are also reported. The rate of outcrossing is influenced by wind speed, the proportion of different flower types, and the extent of self-incompatibility prevalent in the plant (Risi and Galwey 1984). As a result of outcrossing, the quinoa landraces have a high level of heterogeneity and heterozygosity and more advanced cultivars are difficult to keep true to type by farmers who propagate their own seeds (Lindhout and Danial 2006).

8.12 Male Sterility

Male sterile lines are necessary for hybrid production (Wilson 1980; Risi and Galwey 1984; Fleming and Galwey 1995; Bhargava et al. 2006a). Both cytoplasmic and genic male sterilities have been reported in quinoa, though the latter is quite less frequent. Gandarillas (1969) reported a recessive nuclear gene controlling male sterility in some Bolivian quinoa lines which when crossed with male fertile line showed Mandelian ratio in F_1 and F_2. However, some other male sterile line from Bolivia showed inconsistent Mandelian ratios which point toward cytoplasmic control (Gandarillas 1969). Rea (1969) also described a male sterile plant having yellowish white or light brown empty anthers but did not study the inheritance of this trait. The genetics of male sterility was studied by Simmonds (1971) in an unnamed quinoa of Bolivian origin and 3 loci: R (red plant) r (green plant); Ax (axil spot) ax (none); and Ms (hermaphrodite) ms (male sterile) were reported. The plants of genotype MsMs and Msms were fertile and showed breeding behavior as expected but that of the recessive msms showed that it carried an erratically expressed or transmitted cytoplasmic factor for male sterility.

A cultivar 'Apelawa' carrying normal and male sterile cytoplasms was isolated by Ward and Johnson (1993). The cross between male sterile and normal male fertile donors consistently produced male sterile offspring, while interspecific crosses between male sterile quinoa plants and *C. berlandieri* produced progenies having partial restoration of male fertility. However, plants of Bolivian cv. 'Amachuma' carrying a single nuclear recessive gene that in homozygous state produced anthers devoid of pollen grains were reported by Ward and Johnson (1994). The heterozygous plants at this locus were indistinguishable from homozygous male fertile ones and further segregation for male sterility followed a normal Mendelian single gene segregation pattern.

Normal hermaphrodite and male sterile quinoa plants were reported in an accession (PI 510536) in the USDA collection of quinoa by Ward (1998), wherein the male sterility was of cytoplasmic nature and was characterized by small shrunken anthers and the absence of pollen. A dominant nuclear allele that interacted with the male sterile cytoplasm to restore male fertility was also present in this accession.

8.13 Breeding and Genetic Improvement

The basic objective of breeding in quinoa is the development of a variety with dwarf, non-branching, and uniformly early maturing plant type to facilitate mechanical harvesting (Jacobsen et al. 1996) and high grain yield with high protein and low saponin content (Bhargava et al. 2006a). Outside the Andean region, breeding research was initiated in Europe in the early 1980s with the objectives of adaptation to local climatic conditions and improvement with respect to uniformity, early maturity, seed yield, protein content, and reduced saponin content in the seeds (Risi and Galwey 1989; Galwey 1993; Jacobsen and Stølen 1993; Jacobsen et al. 1994; Mastebroek and van Soest 1994; Limburg and Mastebroek 1997; Mastebroek and Limburg 1997). Breeding research in Denmark and Sweden was centered around improvement in the fodder quality (Carlsson 1980; Haaber 1991). Breeding research in the Netherlands started in 1986 when trials led to selection of some uniform lines adapted to the Western European climate (Mastebroek and Limburg 1997; Mastebroek et al. 2002). Quinoa was successfully acclimatized and established in India in the early 1990s and thereafter extensive field trials were carried out at the National Botanical Research Institute, India taking into account its importance both as grain and fodder crop (Bhargava et al. 2006a, 2007d).

However, hybridization in quinoa is cumbersome given self-pollinated nature small flowers making emasculation impossible. Inspite of these difficulties, mass selection and hybridization have been practiced in quinoa (Risi and Galwey 1984). A practical approach can be the utilization of morphological markers to distinguish the hybrid from the parents.

8.14 Genetic Diversity

Quinoa displays ample genetic diversity for qualitative as well as quantitative traits, which allows obtaining a wide range of adaptability to agroecological conditions (Rodríguez and Isla 2009). Risi and Galwey (1989) assessed genetic diversity in 294 accessions of quinoa using PCA and canonical analysis. Ortiz et al. (1999) created a phenotypic distance matrix among 76 accessions from a Peruvian quinoa core collection. Rojas (2003) analyzed the genetic diversity in *C. quinoa* using three multivariate methods. Multiple group discriminant function analyses resulted in six statistically significant functions, which separated the different groups. The assessments showed that phenological variables such as initiation of the flowering and mid-bloom date were stronger discriminants as compared to the yield variables. Twenty-nine germplasm lines of *C. quinoa* and two of *C. berlandieri* subsp. *nuttalliae* were evaluated for 19 traits for cluster and PCA in the north Indian conditions (Bhargava et al. 2007e). Multivariate analysis showed that most of the variations were accounted for the first four PCs. Days to maturity, primary branches/plant, chlorophyll content, and seed yield/plant were the main traits that accounted for most variability in both PC1 and PC2. The germplasm lines were grouped into six clusters based on average linkage method (Fig. 8.3). The lines in cluster I were early maturing and high yielding but had low carotenoid content. Cluster II comprised lines having low seed quality but were higher in leaf quality components. Cluster III had highest seed yield and high values for protein and carotenoids. The lines in cluster IV matured earliest and had high seed protein, while cluster V had high seed yield, dry weight/plant, stem diameter, and maximum number of inflorescences. Cluster VI had low values for traits related to seed morphology and quality except for carotenoid content. This investigation also

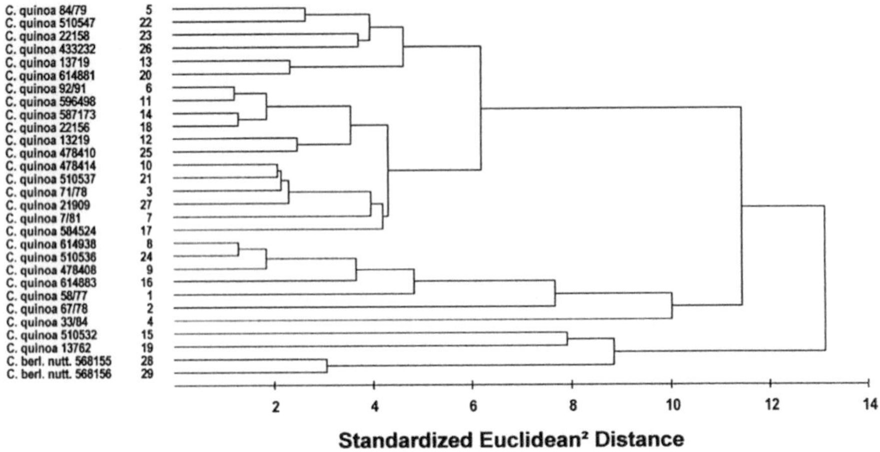

Fig. 8.3 Dendrogram of 29 germplasm lines derived from average linkage method (Bhargava et al. 2007e)

clustered two lines of *C. berlandieri* subsp. *nuttalliae* separately from the quinoa line that is phylogenetically correct. The morphological diversity among 28 quinoa accessions from the Chilean highlands was assessed under desert lowland conditions using multivariate techniques to analyze measurements of 11 morphological descriptors (Fuentes and Bhargava 2011). The first four principal components accounted for 70 % of the total variation among the accessions. A correlation matrix involving the complement of the Pearson coefficient was used to construct a dendrogram using the UPGMA algorithm (Fig. 8.4). Cluster analysis allowed classification of the accessions into six discrete groups.

8.15 Correlation and Path Studies

The study of correlation is regarded as an important step in breeding programs since the information obtained is useful estimating the correlated response to directional selection for the formulation of selection indices. The path coefficient, also known as standardized partial regression coefficient, plays an important role in determining the degree of relationship between yield and yield components since it separates the direct and indirect effects of a correlation coefficient. Espinola and Gandarillas (1985) reported that inflorescence length was the most important component influencing grain yield in *C. quinoa*. Risi and Galwey (1989) reported strong correlation between plant height, stem diameter, inflorescence length, and inflorescence diameter with each other. However, the associations between the durations of the developmental phases were weak suggesting that there is a great scope for manipulation of the pattern of development through breeding. Ortiz et al. (1998) noted high correlation between stem and inflorescence color which

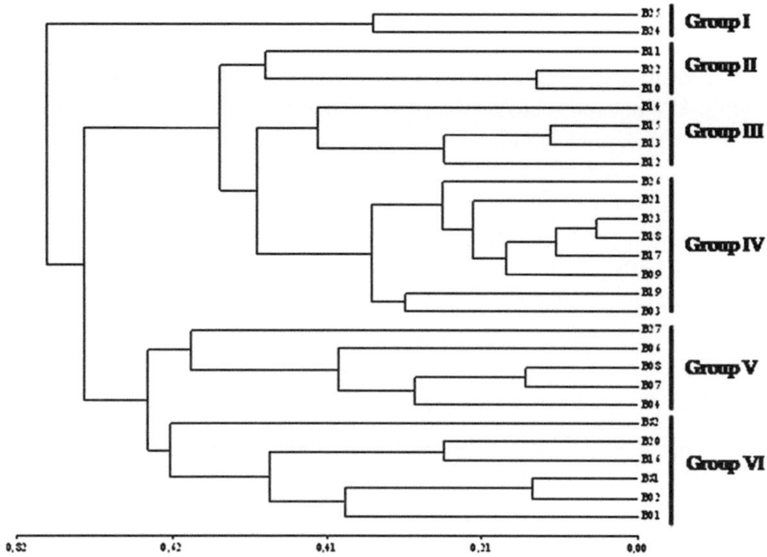

Fig. 8.4 Cluster analysis for characterization of quinoa germplasm grown under lowland desert conditions (Fuentes and Bhargava 2011). Group I: Accessions having large inflorescences and fluctuating grain yields (GYs); group II: medium plant heights and medium GYs; group III: taller plants and low GYs (forage use); group IV: shorter plants with low GYs; group V: medium plant heights and medium-to-low GYs, and group VI: medium plant heights and medium-to-high GYs

confirmed the presence of partial common genetic control for pigmentation in the crop. Bhargava et al. (2003) calculated the correlation coefficients among various traits and their direct and indirect effects on grain yield in quinoa grown on normal and sodic soils. Stem diameter and dry weight/plant were positively correlated with grain yield on both soil types. It was concluded that selection of thick-stemmed plants with more number of inflorescences and high dry weight would be beneficial in breeding for high grain yield in quinoa on sodic soils. Spehar and Santos (2005) reported positive association of inflorescence length and diameter with grain yield, which indicated that the selection for these characters may result in more productive genotypes. Positive correlation was also observed between plant height and inflorescence length which suggested that high grain yield can be attained by selecting for stem/inflorescence ratio. In a more detailed study, the interrelationships among yield and yield components were elucidated in 27 germplasm lines of *C. quinoa* and two lines of *C. berlandieri* subsp. *nuttalliae* by Bhargava et al. (2007e). Significant correlation among branches/plant, inflorescence length, and inflorescence/plant pointed out that plants with good branching habit tend to develop a large number of long inflorescences. Inflorescence length was also positively associated with plant height indicating that lines with greater plant height also developed longer panicles, a fact also reported by Rojas (2003) and Ochoa and Peralta (1988). The nonsignificant correlation between seed yield and seed quality traits and low values of direct path would be beneficial for breeders since it would not hinder attempts to breed lines with both greater grain yield and high seed quality. The path analysis revealed that 1000 seed weight had highest positive direct relationship with seed yield, followed by total chlorophyll and branches/plant. It was concluded that seed yield and seed protein were the only traits exhibiting high positive direct path and significant positive association with harvest index, indicating a true relationship among these traits (Bhargava et al. 2007e). Fuentes and Bhargava (2011) noticed high correlation between stem diameter and plant weight, stem diameter and plant height, plant weight and plant height, plant weight and inflorescence length, plant height and inflorescence length, and leaf length and leaf width. Harvest index showed negative association with stem diameter, plant height, inflorescence length, inflorescence width, and inflorescence branch number.

8.16 Genotype x Environment Interaction (GEI)

The interaction of cultivar with environmental factors is an important consideration for plant breeders. Plant breeders continuously strive to broaden the genetic base of a crop to prevent its vulnerability to the changing environments. A material when planted over different environments exhibits differential responses due to environmental variation (Bhargava et al. 2007d). Consistent performance across different sites and/or years is referred to as stability. A number of statistical models have been put forward for evaluating the yield stability of a genotype in yield trials (Finlay and Wilkinson 1963; Eberhart and Russel 1966; Tai 1971; Shukla 1972; Shafii and Price 1998). The knowledge of GEI and yield stability is important for breeding new cultivars with improved adaptation to the environmental constraints prevailing in the target environment (Bhargava et al. 2007c). Few studies are available on GEI and stability analysis in quinoa (Risi and Galwey 1991; Jacobsen et al. 1996; Jacobsen 1998; Bertero et al. 2004).

A study of the GEI by Risi and Galwey (1991) demonstrated that the GEI differed among the variables measured. Grain yield was strongly dependent on the variety, but micronutrient deficiency and weed competition affected the varieties differently.

The stability of various descriptive characters was studied over a 5-year period in 14 lines of quinoa to determine the most appropriate time in a breeding program when selection could be performed (Jacobsen et al. 1996). GEIs were significant for all characters, but for several traits of interest to the plant breeder, namely earliness, height, and inflorescence size, some of the best lines track the optimum response. The results pointed out that selection for height, inflorescence size, and developmental stage could be satisfactorily performed at an early stage of the breeding program. Potential parents were identified for use in the development of varieties suitable for North European conditions. Jacobsen et al. (1996) studied the stability of quantitative traits in 14 lines of *C. quinoa* and suggested that selection for height, inflorescence size, and developmental stage could be easily performed at an early stage of breeding program. Studies on developmental patterns of quinoa for North

European conditions were carried out in five groups of quinoa lines from different maturity classes in 3 years, and measured on five occasions between bud formation and seeds set (Jacobsen 1997). The seed types originating from Chile were found to be more adapted for growth under North European conditions.

Bertero et al. (2004) examined the size and nature of the GEI effects for grain yield, its physiological determinants, and grain size among 24 cultivars tested in 14 sites in a multi-environment trial across three continents. The G x E interaction to G component of variance ratio was 4:1 and 1:1 for grain yield and grain size, respectively. The clustering separated the cultivars from mid-altitude valleys of the northern Andes, northern Altiplano, southern Altiplano, and sea level. No single genotype group showed consistently superior grain yield across all environment groups. The genotype (G) and GEI effects observed for the duration of the crop cycle had a major influence on the average cultivar performance and on the form of GEIs observed for total above-ground biomass and grain yield. It was observed that good average performance and broad adaptation could come from the combination of medium–late maturity and high harvest index. Correlation analysis revealed no association between the average cultivar responses for grain yield and grain size (Bertero et al. 2004). Both observations indicated that simultaneous selection for grain yield and grain size can be expected from selection.

8.17 Conclusion

Quinoa has been cultivated for more than 5000 years in the Andes and was probably domesticated by ancient civilizations at different times and in different geographic zones. Many wild characters that were disadvantageous to the farmers such as seed shattering, seed dormancy, and thick seed coats were lost, while useful characters such as larger and more starchy seeds, fewer and larger inflorescences, uniform maturity, and environmental adaptations were preserved during domestication. However, wide diversity for plant color, seed color, types of branching and panicles, as well as grain productivity, abiotic stress tolerance, and disease resistance still exists despite narrowing of genetic base during domestication. This diversity is also reflected at the molecular level and is being used by the plant breeders worldwide to develop improved plant types with respect to uniformity, early maturity, seed yield, protein content, and reduced saponin content in the seeds.

References

Ahamed NT, Singhal RS, Kulkarni PR, Pal M (1996a) Physicochemical and functional properties of *Chenopodium quinoa* starch. Carbohydr Polym 31:99–103

Ahamed NT, Singhal RS, Kulkarni PR, Kale DD, Pal M (1996b) Studies on *Chenopodium quinoa* and *Amaranthus paniculatas* starch as biodegradable fillers in LDPE films. Carbohydr Polym 31:157–160

Anabalon-Rodriguez L, Thomet-Isla M (2009) Comparative analysis of genetic and morphologic diversity among quinoa accessions (*Chenopodium quinoa* Willd.) of the south of Chile and highland accessions. J Plant Breed Crop Sci 1:210–216

Aubrecht E, Biacs PA (2001) Characterization of buckwheat grain proteins and its products. Acta Aliment 28:261–268

Bazile D, Olguin Manzano PA, Nunez L, Croce P, Alacron G, Lagos J, Parra F, Peredo P, Negrete Sepulveda J (2010) Differencacion territorial asociada a la quinua en el secano costero dela sexta region, Chile:consideraciones sobre las praticas y representaciones socials para un desarrollo sostenible. In: Anales de la Sociedad Chilena de Ciencias Geograficas, pp 103–109

Bazile D, Fuentes F, Mujika A (2013) Historical perspectives and domestication. In: Bhargava A, Srivastava S (eds) Quinoa botany, production and uses. CAB International, Oxfordshire, pp 16–35

Bennett MD, Smith JB (1991) Nuclear DNA amounts in angiosperms. Phil Trans Roy Soc Ser B Biol Sci 334:309–345

Bertero HD, de la Vega AJ, Correa G, Jacobsen SE, Mujica A (2004) Genotype and genotype-by-environment interaction effects for grain yield and grain size of quinoa (*Chenopodium quinoa* Willd.) as revealed by pattern analysis of international multienvironment trials. Field Crops Res 89:299–318

Bhargava A, Srivastava S (2013) Quinoa botany, production and uses. CAB International, Oxfordshire

Bhargava A, Shukla S, Katiyar RS, Ohri D (2003) Selection parameters for genetic improvement in *Chenopodium* grain on sodic soil. J Appl Hortic 5:45–48

Bhargava A, Rana TS, Shukla S, Ohri D (2005) Seed protein electrophoresis of some cultivated and wild species of *Chenopodium* (Chenopodiaceae). Biol Plant 49:505–511

Bhargava A, Shukla S, Ohri D (2006a) *Chenopodium quinoa*—an Indian perspective. Industrial Cropsand Products 23:73–87

Bhargava A, Shukla S, Ohri D (2006b) Karyotypic studies on some cultivated and wild species of *Chenopodium* (Chenopodiaceae). Genet Res Crop Evol 53:1309–1320

Bhargava A, Shukla S, Ohri D (2007a) Genetic variability and interrelationship among various morphological and quality traits in quinoa (*Chenopodium* Willd.). Field Crops Res 101:104–116

Bhargava A, Shukla S, Ohri D (2007b) Genome size variation in some cultivated and wild species of *Chenopodium* (Chenopodiaceae). Caryologia 60:245–250

Bhargava A, Shukla S, Ohri D (2007c) Gynomonoecy in *Chenopodium quinoa* Willd. (Chenopodiaceae): variation in inflorescence and floral types in some accessions. Biologia 62:19–23

Bhargava A, Shukla S, Ohri D (2007d) Evaluation of foliage yield and leaf quality traits in *Chenopodium* spp. in multiyear trials. Euphytica 153:99–213

Bhargava A, Shukla S, Rajan S, Ohri D (2007e) Genetic diversity for morphological and quality traits in quinoa (*Chenopodium quinoa* Willd.) germplasm. Genet Res Crop Evol 54:167–173

Brown DC, Cepeda-Cornejo V, Maughan PJ, Jellen EN (2014) Characterization of the Granule bound starch synthase I gene in *Chenopodium*. The Plant Genome doi:10.3835/plantgenome2014.09.0051

Carlsson R (1980) Quantity and quality of leaf protein concentrates from *Atriplex hortensis* L. *Chenopodium quinoa* Willd. and *Amaranthus caudatus* L. grown in southern Sweden. Acta Agriculturae Scand 30:418–426

Catacora AG (1977) Determinacion de cariotipoen cinco lineas de quinua (*Chenopodium quinoa* Willd). Ingeniero Agronomo thesis, Universidad Nacionel Technica del Altiplano, Puno, Peru

Chauhan GS, Eskin NAM, Tkachuk R (1992) Nutrients and antinutrients in quinoa seeds. Creal Chem 69:85–88

Christensen SA, Pratt DB, Stevens MR, Jellen EN, Coleman CE, Fairbanks DJ, Bonifacio A, Maughan PJ (2007) Assessment of biodiversity in the USDA and CIP-FAO international nusery collections of quinoa (*Chenopodium quinoa* Willd.) using microsatellite markers. Plant Genet Resour 5:82–95

Coles ND, Coleman CE, Christensen SA, Jellen EN, Stevens MR, Bonifacio A, Rojas-Beltran JA, Fairbanks DJ, Maughan PJ (2005) Development and use of an expressed sequence tag library in quinoa (*Chenopodium quinoa* Willd.) for the discovery of single nucleotide polymorphisms. Plant Sci 168:539–447

Coulter L, Lorenz K (1990) Quinoa-composition, nutritional value, food applications. LWT-Food Sci Technol 23:203–207

Cusack D (1984) Quinoa: grain of the Incas. Ecologist 14:21–31

Del Castillo C,Winkel T, Mahy G, Bizoux JP (2007) Genetic structure of quinoa (*Chenopodium quinoa* Willd.) from the Bolivian altiplano as revealed by RAPD markers. Genet Res Crop Evol 54:897–905

Dostalek J (1987) Influence of the mode of pollination on offsprings of some species of the genus *Chenopodium*. Preslia 59:263–269

Drezewiecki J, Delgado-Licon E, Haruenkit R, Pawelzik E, Martin-Belloso O, Park YS (2003) Identification and differences of total proteins and their soluble fractions in some pseudocereals based on electrophoretic patterns. J Agric Food Chem 51:7798–7804

Eberhart SA, Russel WA (1966) Stability parameters for comparing varieties. Crop Sci 6(1):36–40

Escuredo O, Inmaculada GMM, Moncada GW, Fischer S, Hierro JMH (2014) Amino acid profile of the quinoa (*Chenopodium quinoa* Willd.) using near infrared spectroscopy and chemometric techniques. J. Cereal Sci (in press)

Espinola G, Gandarillas H (1985) Study of correlated characters and their effects on quinoa yield. Bol Genetico 13:47–54

Fairbanks D, Waldrigues A, Ruas CF, Maughan PJ, Robison LR, Adersen WR, Riede CR, Pauley CS, Caeteno LG, Arantes OM, Fungaro MHP, Vidotto MC, Jankevicius SE (1993) Efficient characterization of biological diversity using field DNA extraction and random amplified polymorphic DNA markers. Rev Bras Genet 16:11–22

Finlay KW, Wilkinson GN (1963) The analysis of adaptation in a plant breeding programme. Aust J Agric Res 14:742–754

Fleming JE, Galway NW (1995) Quinoa (*Chenopodium quinoa* Willd.). In: Williams JT (ed) Cereals and Pseudocereals. Chapman and Hall, London, pp 2–83

Fuentes F, Bhargava A (2011) Morphological analysis of quinoa germplasm grown under low land desert conditions. J Agron Crop Sci 197:124–137

Fuentes F, Martinez EA, Hinrichson PV, Jellen EN, Maughan PJ (2009) Assessment of genetic diversity patterns in Chilean quinoa (*Chenopodium quinoa* Willd.) germplas using multiplex fluorescent microsatellite markers. Conserv Genet 10:369–377

Fuentes F, Bazile D, Bhargava A, Martinez EA (2012) Implications of farmers' seed exchanges for on-farm conservation of quinoa, as revealed by its genetic diversity in Chile. J Agric Sci 150:702–716

Galwey NW (1993) The potential of quinoa as a multipurpose crop for agricultural diversification: a review. Ind Crops Prod 1:101–106

Galwey NW (1995) Quinoa and relatives. In: Smartt J, Simmonds NW (eds) Quinoa and relatives. Longman Scientific and Technical, Essex, England

Galwey NW, Leakey CLA, Price KR, Fenwick GR (1990) Chemical composition and nutritional characteristics of quinoa (*Chenopodium quinoa* Willd.). Food Sci Nutr 42F:245–261

Gandarillas H (1969) Esterilidadgenetica y citoplasmica en la quinoa. Turrialba 19:429–430

Gandarillas H (1979) Botanica. Quinua y kaniwa. Cultivos Andinos. In: Tapia ME (ed) Serie Libros y Materiales Educativos. Instituto Interamericano de Ciencias Agricolas, Bogota, Colombia, pp 20–44

Gęsiński K (2000) Potential for *Chenopodium quinoa* Willd acclimatisation in Poland. Crop development of the cool and wet regions of Europe. European Communities, Belgium

Giusti L (1970) El genero *Chenopodium* en Argentina 1: Numeros de cromosomas. Darwiniana 16:98–105

Gorinstein S, Pawelzik E, Gelgado-Licon E, Haruenkit R, Weisz M, Trakhtenberg S (2002) Characterization of pseudocereal and cereal proteins by protein and amino acid analysis. J Sci Food Agric 82:886–891

Haaber J (1991) Chenopodium quinoa Willd. As a green crop for the palleting industry-the effect of heat treatment on the palatability in green pallets made of quinoa. First European symposium on industrial crops and products, Maastricht, The Netherlands

Heiser CB, Nelson DC (1974) On the origin of cultivated chenopods (*Chenopodium*). Genetica 78:503–505

Hirich A, Choukr-Allah R, Jacobsen SE (2014) Deficit irrigation and organic compost improve growth and yield of quinoa and pea. J Agron Crop Sci 200:390–398

IAEA (International Atomic Energy Agency) (2004) Genetic improvement of under-utilized and neglected crops in low income food deficit countries through irradiation and related techniques. In: Proceedings of a final research coordination meeting organized by the joint FAO/IAEA Division of Nuclear Techniques in Food and Agriculture, Pretoria, South Africa, May 19–23, 2003

Iliadis C, Karyotis T, Mitsimponas T (1997) Research on quinoa (*Chenopodium quinoa*) and amaranth (*Amaranthus caudatus*) in Greece. In: Ortiz R, Stolen O (eds) Crop development for the cool and wet regions of Europe. Spelt and Quinoa COST 814, pp 85–91

Jacobsen SE (1997) Adaptation of quinoa (*Chenopodium quinoa*) to northern European agriculture: studies on developmental pattern. Euphytica 96:41–48

Jacobsen SE (1998) Developmental stability of quinoa under European conditions. Ind Crops Prod 7:169–174

Jacobsen SE (2003) The worldwide potential for quinoa (*Chenopodium quinoa* Willd.). Food Rev Int 19:167–177

Jacobsen SE, Stolen O (1993) Quinoa: morphology, phenology and prospects for its production as a new crop in Europe. Eur J Agron 2:19–29

Jacobsen SE, Jorgensen I, Stolen O (1994) Cultivation of quinoa (*Chenopodium quinoa*) under temperate climatic conditions in Denmark. J Agric Sci 122:47–52

Jacobsen SE, Hill J, Stolen O (1996) stability of quantitative traits in quinoa (*Chenopodium quinoa* Willd.). Theor Appl Genet 93:110–116

Jacobsen SE, Mujica A, Jensen CR (2003) The resistance of quinoa (*Chenopodium quinoa* Willd.) to adverse abiotic factors. Food Rev Int 19:99–109

Jacobsen SE, Monteros C, Corcuera LJ, Bravo LA, Christiansen JL, Mujica A (2007) Frost resistance mechanisms in quinoa (*Chenopodium quinoa* Willd.). Eur J Agron 26:471–475

Jarvis DE, Kopp OR, Jellen EN, Mallory MA, Pattee J, Bonifacio A, Coleman CE, Stevens MR, Fairbanks DJ, Maughan PJ (2008) Simple sequence repeat marker development and genetic mapping in quinoa (*Chenopodium quinoa* Willd.). J Genet 87:39–51

Jellen EN, Kolano BA, Sederberg MC, Bonifacio A, Maughan PJ (2011) *Chenopodium*. In: Kole C (ed) Wild crop relatives: genomic and breeding resources, legume crops and forages. Springer, Berlin, pp 35–61

Kadereit G, Borsch T, Weising K, Freitag H (2003) Phylogeny of Amaranthaceae, Chenopodiaceae and the evolution of C4 photosynthesis. Int J Plant Sci 164:959–986

Kawatani K, Ohno T (1950) Chromosome numbers of genus *Chenopodium*, I. Japan J Genet 25:177–180

Kawatani K, Ohno T (1956) Chromosome numbers of genus *Chenopodium*, II. Japan J Genet 31:15–17

Kolano B, Pando LG, Maluszynska J (2001) Molecular cytogenetic studies in *Chenopodium quinoa* and *Amaranthus caudatus*. Acta Soc Botanicorum Pol 70:85–90

Kolano B, Gardunia BW, Michalska M, Bonifacio D, Fairbanks D, Maughan PJ, Coleman CE, Stevens MR, Jellen EN, Maluszynska J (2011) Chromosomal localization of two novel repetitive sequences isolated from the *Chenopodium quinoa* Willd. Genome Genome 54:710–717

Kolano B, Siwinska D, Pando LG, Szymanowska-Pulka J, Maluszynska J (2012) Genome size variation in *Chenopodium quinoa* (Chenopodiaceae). Plant Syst Evol 298:251–255

Konishi Y, Hirano S, Tsuboi H, Wada C (2004) Distribution of minerals in quinoa (*Chenopodium quinoa* Willd) seeds. Biosci Biotech Biochem 68:231–234

Koziol M (1992) Chemical composition and nutritional evaluation of quinoa (*Chenopodium quinoa* Willd.). J Food Compos Anal 5:35–68

Kumpun S, Maria A, Crouzet S, Evrard-Todeschi N, Girault JP, Lafont R (2011) Ecdysteroids from *Chenopodium quinoa* Willd., and ancient Andean crop of high nutritional value. Food Chem 125:1226–1234

Lescano RJL (1980) Avances en la genetic de quinoa. In: Primera Reunion de Genetica y Fitomejoramiento de la quinoa. Universidad Nacional Tecnica del Altiplano, Instituto Boliviano de Tecnologia Agropecuaria, Instituto Interamericano de Ciencias Agricolas, Centro Internacional de Investigaciones para el Desarrollo, Puno,Peru, pp B1–B2

Limburg H, Mastebroek HD (1997) Breeding high yielding lines of *Chenopodium quinoa* Willd. With saponin free seed. In: Stolen O, Bruhn K, Pithan K, Hill J (eds) Small grain Cereals and Pseudocereals. In: Proceedings of the COST 814 workshop, 22–24 Feb 1996, Copenhagen, Denmark, pp 103–114

Lindeboom N (2005) Studies on the characterization, biosynthesis and isolation of starch and protein from quinoa (*Chenopodium quinoa* Willd.), University of Saskatchewan Degree of Doctor

Lindhout P, Danial D (2006) Participatory genomics in quinoa. Tailor Biotechnol 2:31–50

Mason SL, Stevens MR, Jellen EN, Bonifacio A, Fairbanks DJ, McCarty RR, Rasmussen AG, Maughan PJ (2005) Development and use of microsatellite markers for germplasm characterization in quinoa (*Chenopodium quinoa* Willd.). Crop Sci 45:1618–1630

Mastebroek HD, Limburg H (1997) Breedng for harvest security in Chenopodium quinoa. In: Stolen O, Bruhn K, Pithan K, Hill J (eds) Small grain cereals and pseudocereals. In: Proceedings of the COST 814 Workshop. 22–24 February 1996, Copenhagen, Denmark, pp 79–86

Mastebroek HD, van Soest LJM (1994) Gierstmelde blijkt multi-purpose-gewas (*Chenopodium quinoa* proves multipurpose crop). Prophyta 1:15–17

Mastebroek HD, van Loo EN, Dolstra O (2002) Combining ability for seed yield traits of *Chenopodium quinoa* breeding lines. Euphytica 125:427–432

Maughan PJ, Bonofacio A, Jellen EN, Stevens MR, Coleman CE, Ricks M, Mason SL, Jarvis DE, Gardunia BW, Fairbanks DJ (2004) A genetic linkage map of quinoa (*Chenopodium quinoa*) base on AFLP, RAPD and SSR markers. Theor Appl Genet 109:118–1195

Maughan PJ, Kolano BA, Maluszynska J, Coles ND, Bonofacio A, Rojas J, Coleman CE Stevens MR, Fairbanks DJ, Parkinson SE, Jellen EN (2006) Molecular and cytological characterization of ribosomal DNAs in *Chenopodium quinoa* and *Chenopodium berlandieri*. Genome 49:825–839

Maughan PJ, Turner TB, Coleman CE, Elzinga DB, Jellen EN, Morales JA, Udall JA, Fairbanks DJ, Bonofacio A (2009) Characterization of salt overly sensitive (SOS1) gene homoeologs in quinoa (*Chenopodium quinoa* Willd.). Genome 52:647–657

Maughan PJ, Smith S, Rojas-Beltran J, Elzinga D, Raney J, Jellen E, Bonofacio A, Udall J, Fairbanks D (2012) Single nucleotide polymorphisms identifications, characterization and linkage mapping in *Chenopodium quinoa*. Plant Genome 5:1–7

Medina W, Janiak A, Szarejko I, Mujika A, Jacobsen SE (2004) Analisis de relaciones geneticas entre variedads de quinua (*Chenopodium quinoa* Willd.) utilizando la tecnica de AFLP (amplified fragment length polymorphism). In: Libro de Resumenes XI Congresso Internacional de cultivos andinos. Cochabamba, Bolivia

Mujica A (2004) La Quinoa Indigena, caracteristicas e historia. In: Sepulveda J, Thomat MI, Palazuelos F, Mujica A (eds) La kinwa mapuche, Recupracion de un cultivo para la Alimentacion. Fundacion para la Innovacion Agraria, Ministerio de Agricultura Temuco, Chile, pp 22–42, Crop relatives: genomic and breeding resources, legume crops and forages. Springer, Heidelberg, pp 35–61

Mujica A, Jacobsen SE, Ezquierdo J, Marathee JP (2001) Resultados de la Prueba Americana y Europes de la Quinua. CIP, FAO, UNA-Puno, p 51

Mujica A, Chura E, Ruiz E, Martinez R (2010a) Mecanismos de resistencia a sequia de la quina (*Chenopodium quinoa* Willd.). In: Proceedings Primer Congresso Peruano de Majoramiento

Genetico de Plantas y Biotecnologia Agricola, 17–19 May. UNALM, EPG, Lima, Peru, pp 111–114

Mujica A, Chura E, Ruiz E, Rossel J, Pocco M (2010b) Mecanismos de resitencia a sales y seleccion de variedades de quinua (*Chenopodium quinoa* Willd.) resistates a salinidad. In: Anales XII Congraso Nacional de las Ciencias del Suelo y V Congresso Internacional de las Ciencias del Suelo, Arequipa, Peru, 11–15 Oct 2010, pp 187–189

Munir H, Sehar S, Basra SMA, Jacobsen HJ, Rauf S (2012) Growing quinoa in Pakistan as a potential alternative for food security. In: Resilience of agricultural systems against crises, 19–21 Sept 2012, Göttingen-Kassel/Witzenhausen, Germany

Nelson DC (1968) Taxonomy and origins of *Chenopodium quinoa* and *Chenopodium nuttalliae*. Ph.D. thesis, University of Indiana, Bloomington, Indiana

Nisimba RY, Kikuzaki H, Konishi Y (2008) Antioxidant activity of various extracts and fractions of *Chenopodium quinoa* and *Amaranthus* spp. Seeds Food Chem 106:760–766

Nunez Carrasco L, Bazile D, Chia E, Hocde H, Negrete Sapulveda J, Martinez EA (2010) Representaciones socials acerca de la conservacion de la biodiversidad en el caso de peroductores tradicionales de *Chenopodium quinoa* Willd de secano costero en las regionses de O'Higgins y el Maule. Anales de la Sociedad Chilena de Ciencias Geograficas, pp 181–187

Ochoa J, Peralta E (1988) Evaluacion preliminary morfologica y agronomica de 153 entradas de quinua en Santa Catalina. Pichincha. Actas del VI Congreso Internacional sobre Cultivos Andinos. Quito, Ecuador, pp 137–142

Oritz R, Ruiz-Tapia EN, Mujica-Sanchez A (1998) Sampling strategy for a core collection of Peruvian quinoa germplasm. Theor Appl Genet 96:475–483

Oritz R, madsen S, Ruiz-Tapia EN, Jacobsen SE, Mujica-Sanchez A, Christiansen JL, Stolen O (1999) Validating a core collection of Peruvian quinoa germplasm. Genet Res Crop Evol 46:285–290

Palomino GH, Segura MD, Bye RB, Mercado RP (1990) Cytogenetic distinction between *Teloxys* and *Chenopodium* (Chenopodiaceae). Soutwestern Nat 35:351–353

Palomino GH, Hernandez LT, Torres EC (2008) Nuclear genome size and chromosome analysis in *Chenopodium quinoa* and *C. berlandieri* subsp *nuttalliae*. Euphytica 164:221–230

Pratt C (2003) AFLP analysis of genetic diversity in the USDA *Chenopodium quinoa* collection. M.S. Thesis, Brigham Young University, Provo, UT, USA

Rana TS, Narzary D, Ohri D (2010) Genetic diversity and relationships among some wild and cultivated species of *Chenopodium* L. (Amaranthaceae) using RAPD and DAMD methods. Curr Sci 98:840–846

Rea J (1969) Biologia floral de la quinoa (*Chenopodium quinoa*). Turrialba 19:91–96

Rep-Carrasco R, Espinoza C, Jacobsen SE (2003) Nutritional value and use of the Andean crops quinoa (*Chenopodium quinoa*) and kaniwa (*Chenopodium pallidicaule*). Food Rev Int 19: 179–189

Reynolds DJ (2009) Genetic dissection of triterpenoid saponin production in *Chenopodium quinoa* using microarray analysis. M.Sc. thesis, Brigham Young University, Provo, Utah

Risi J, Galway NW (1984) The chenopodium grains of the Andes: Inca crops for modern agriculture. Adv Appl Biol 10:145–216

Risi J, Galwey NW (1989) *Chenopodium* grains of the Andes: a crop for the temperate latitudes. In: Wickens GE, Haq N, Day P (eds) New crops for food and industry. Chapman and Hall, New York

Risi J, Galwey NW (1991) Genotype x environment interaction in the Andean grain crop quinoa (*Chenopodium quinoa*) in temperate environments. Plant Breed 107:141–147

Rodriguez LA, Isla MT (2009) Comparative analysis of genetic and morphologic diversity among quinoa accessions (*Chenopodium quinoa* Willd.) of the south of Chile and highland accessions. J Plant Breed Crop Sci 1:210–216

Rojas W (1998) Análisis de la diversidad genética del germoplasma de quinua (*Chenopodium quinoa* Willd.) de Bolivia, mediante métodos multivariados. Tesis M.Sc., Universidad Austral de Chile, Facultad de Ciencias Agrarias. Valdivia—Chile, 209 p

Rojas W (2003) Multivariate analysis of genetic diversity of Bolivian quinoa germplasm. Food Rev Int 19:9–23

Ruales J, Nair BM (1992) Nutritional quality of protein in quinoa (*Chenopodium quinoa* Willd.) seeds. Plant Foods Hum Nutr 42:1–11

Ruas PM, Bonifacio A, Ruas CF, Fairbanks DJ, Anderson WR (1999) Genetic relationships among 19 accessions of six species of *Chenopodium* L. by random amplified polymorphic DNA fragments (RAPD). Euphytica 105:25–32

Ruiz-Carrasco K, Antognoni F, Coulibali AK, Lizardi S, Covarrubias A, Martinez EA, Molina-Montenegro MA, Biondi S, Zurita-Silva A (2011) Variation in salinity tolerance of four low land genotypes of quinoa (*Chenopodium quinoa* Willd) as assessed by growth, physiological traits, and sodium transporter gene expression. Plant Physiol Biochem 49:1333–1341

Shafii B, Price WJ (1998) Analysis of genotype-by-environment interaction using the additive main effects and multiplicative interaction model and stability estimates. J Agric Biol Environ Stat 3:335–345

Schlick G, Bubenheim DL (1996) Quinoa- candidate crop for NASA's controlled ecological life support systems. In: Janick J (ed) Progress in new crops. ASHS Press, Arlington

Shams A (2011) Combat degradation in rain fed areas by introducing new drought tolerant crops in Egypt. Int J Water Res Arid Environ 1:318–325

Shukla GK (1972) Some statistical aspects of partitioning genotype-environmental components of variability. Heredity 29:237–245

Silvestri V, Gil F (2000) Alogamia en quinoa. Tasa en Mendoza (Argrntina). Revisita de la facultad de Ciencias Agrarias, Universidad Nacional de Cuyo, pp 71–76

Simmonds NW (1971) The breeding system of quinoa. I. Male sterility. Heredity 27:73–82

Spehar CR, Santos RLB (2005) Agronomic performance of quinoa selected in the Brazilian Savannah. Pesquiza Agropecuaria Bras 40:609–612

Stevens MR, Coleman CE, Parkinson SE, Maughan PJ, Zhang HB, Balzotti MR, Kooyman DL, Arumughanathan K, Bonifacio A, Fairbanks DJ, Jellen EN, Stevens JJ (2006) Construction of a quinoa (*Chenopodium quinoa* Willd.) BAC library and its use in identifying genes encoding seed storage proteins. Theor Appl Genet 112:1593–1600

Stikic R, Glamoclija D, Demin M, Vucelic-Radivic B, Jovanovic Z, Milokovic-Opsenica D et al (2012) Agronomical and nutritional evaluation of quinoa seeds (*Chenopodium quinoa* Willd.) as an ingredient in bread formulations. J Cereal Sci 55:132–138

Storchova H, Drabesova J, Chab D, Kolar J, Jellen EN (2015) The introns in Flowering locus T-like (FTL) genes are useful markers for tracing paternity in tetraploid *Chenopodium quinoa*. Genet Res Crop Evol 62:913–925

Suzuka O (1950) Chromosome numbers in pharmaceutical plants I. Seikon Ziho (Rept Kihara Inst Biol Res) 4:57–58

Taboada C, Mamani A, Raes D, Mathijs Erik, García M, Geerts S, Gilles J (2011) Farmers' willingness to adopt irrigation for quinoa in communities of the Central Altiplano of Bolivia. Revista Latinoamericana de Desarrollo Económico 16:7–28

Tai GCC (1971) Genotypic stability analysis and its application to potato regional trials. Crop Sci 11:184–190

Tanaka R, Tanaka A (1980) Karyomorphological studied on halophytic plants. I. Some taxa of *Chenopodium*. Caryologia 45:257–269

Tang H, Watanabe K, Mitsunaga T (2002) Characterization of storage starches from quinoa, barley and adzuki seeds. Carbohydr Polym 49:13–22

Tapia ME (1979) Historia y distribucion geographica. Quinoa y kaniwa. Cultivos Andinos. In: Tapia ME (ed) Serie Libros y Materiales Educativos. Instituto Interamericano de Ciencias Agricolas, Bogota, Colombia, pp 11–15

Tapia M (1982) The environment, crops and agricultural systems in the Andes and Southern Peru, IICA

Tapia M, Gandarillas H, Alandia S, Cardozo A, Mujika A, Oritz R, Otazu V, Rea J, Salas B, Zanabria E (1979) La quinoa y la kaniwa. Centro Internacional de Investigaciones para el Desarollo, Instituto Internacional de Ciencias Agricolas, Bogota, Columbia

Tari T, Annapure U, Singhal R, Kulkarni P (2003) Starch-based spherical aggregates: screening of small granule sized starches for entrapment of a model flavouring compound, vanillin. Carbohydr Polym 53:45–51

Uotila P (1973) Chromosome counts on Chenopodium L. from SE Europe and SW Asia. Ann Botanici Fenn 10:337–340

Vega-Galvez A, Miranda M, Vergara J, Uribe E, Puente L, Martinez E (2010) Nutrition facts and functional potential of quinoa (*Chenopodium quinoa* Willd.) an ancient Andean grain: a review. J Sci Food Agric 90:2541–2726

Wang S, Tsuchiya T, Wilson HD (1993) Chromosome studies in several species of *Chenopodium* from North and South America. J Genet Breed 47:163–170

Ward SM (1998) A new source of restorable cytoplasmic male sterility in quinoa. Euphytica 101:157–163

Ward SM (2000) Allotetraploid segregation for single gene morphological characters in quinoa (*Chenopodium quinoa* Willd.). Euphytica 116:11–16

Ward SM, Johnson DL (1993) Cytoplasmic male sterility in quinoa. Euphytica 66:217–223

Ward SM, Johnson DL (1994) A recessive gene determining male sterility in quinoa. J Hered 85:231–233

Wilson HD (1976) Genetic control and distribution of leucine aminopeptidase in the cultivated chenopods and related weed taxa. Biochem Genet 14:913–919

Wilson HD (1980) Artificial hybridization among species of *Chenopodium* sect. *Chenopodium*. Syst Bot 5:253–263

Wilson HD (1988) Quinoa biosystematics I: domesticated populations. Econ Bot 42:461–477

Wilson HD (1990) Quinua and relatives (*Chenopodium* sect. *Chenopodium* subsect. Cellulata). Econ Bot 44:92–110

Wilson HD, Heiser CB (1979) The origin and evolutionary relationships of 'huauzontle' (*Chenopodium nuttalliae* Safford), domesticated chenopod of Mexico. Amer J Bot 66:198–206

Wilson HD, Manhart J (1993) Crop/weed gene flow: *Chenopodium quinoa* Willd. and *C. berlandieri* Moq. Theor Appl Genet 86:642–648

Wright KH, Pike OA, Fairbanks DJ, Huber SC (2002) Composition of *Atriplex hortensis*, sweet and bitter *Chenopodium quinoa* seeds. Food Chem Toxicol 67:1383–1385

Yamashita A, Isobe K, Ishii R (2007) Agronomic studies on quinoa cultivation in Japan. I. Determination of the proper seeding time in the southern Kanto district for good performance of the grain yield. Japan J Crop Sci 76:59–64

Chapter 9
Emerging Invaders from the Cultivated Croplands: An Invasion Perspective

Neha Goyal and Gyan Prakash Sharma

Abstract Understanding potential sources and pathways of colonization by alien plant propagules in novel environments is crucial for assessing invasion risks posed by aggressive colonizers. With the enormous expansion and intensification of agriculture, cultivated croplands are emerging as potent sources of robust weeds and/or invaders. Ongoing increase in adaptability and evolutionary potential of agricultural systems demands our understanding to better evaluate the invasion risks to heterogeneous environments. The review intends to collate the fine ecological overlap of crops and associated plants with the plant invaders. We begin with an overview of plant invasion process and discuss invasion risks posed by cultivated croplands through putative propagule escape from crops, crop-associated weeds, and feral crop descendants, continuing with a subsequent discussion on their fate. The synthesis concludes with promising prospects for research which may generate better insights on putative invasion risks from croplands.

Keywords Colonizer · Cropland · Crop improvement · Invasion risk · Propagule

9.1 Introduction

The problem of plant invasions is intrinsically associated with the human-mediated dispersal of species to regions beyond their usual range of dispersal (Williamson 1996; Vitousek et al. 1997; Wilson et al. 2009). Human activities significantly contribute to the dispersal of invasive species' propagules, inadvertently promoting plant invasions (Kolar and Lodge 2001; Levine and D'Antonio 2003; Lake and Leishman 2004). The rate and scale of plant invasions has increased at an

N. Goyal · G.P. Sharma (✉)
Department of Environmental Studies, University of Delhi, Delhi 110007, India
e-mail: gyanprakashsharma@gmail.com

N. Goyal
e-mail: nehagoyal.du@gmail.com

V.R. Rajpal et al. (eds.), *Gene Pool Diversity and Crop Improvement*,
Sustainable Development and Biodiversity 10,
DOI 10.1007/978-3-319-27096-8_9

unprecedented level owing to rapid globalization and associated acceleration in trade, travel, and transport (Carlton and Geller 1993; Cohen and Carlton 1998; Ricciardi 2007; Pyšek et al. 2010). Ongoing increase in demand for food in tandem with the increasing trade and development has led escalating economies of the world to focus on agricultural expansion and intensification. Notably, introduction of alien plant species outside the native range is increasing tremendously with the agricultural intensification (Wilson et al. 2009; Ellstrand et al. 2010; Driscoll et al. 2014). Augmenting episodes of dispersal and escape of potential colonizers from the cultivated croplands to heterogeneous lands form an issue of major concern in the current developmental scenario. Evolutionary potential of crops, crop-associated plants, and their descendants into potent weeds and/or invaders adds a serious dimension to the issue.

Intensification of agriculture is important to meet global demand, but it has serious consequences on ecological systems and their functioning (Driscoll et al. 2014). Agriculture sector has significantly been improvised with the advent of modernized present-day technologies and sophisticated practices. In particular, recent advances in molecular approaches have immensely facilitated crop improvement leading to significant alterations in crop gene pool diversity. However, with the ongoing crop improvement, crop plants are increasingly turning weedy and/or invasive which may aggravate plant invasion risks in future (Kowarik 2005; Schierenbeck and Ellstrand 2009; Andersson and de Vicente 2010; Ellstrand et al. 2010). In spite of their utility, numerous crops and their descendants pose significant invasion risks (Ellstrand et al. 2010). Different systems from the cultivated croplands including crops, crop-associated plants, and their descendants behold propensity to emerge as potential invaders. Studies exploring the propagule dispersal from croplands, their establishment as aggressive colonizers, subsequent evolution to potential invaders, and further interplay with the native and invasive species residing in the introduced habitat are scarce. Understanding whether and how inadvertently dispersed propagules from croplands emerge as potential invaders and respond to resident species in heterogeneous environments demands serious research attention to understand their invasion success in all likelihood.

Although a few researches have highlighted the invasion potential of crops and crop-associated plants, understanding the factors that influence their dispersal, establishment, range expansion, and invasiveness remains an understudied domain. Thus, to address an important yet largely ignored facet of agricultural intensification, the chapter attempts to generate an understanding on the invasive propensities of plant systems in cultivated croplands that contribute to the emergence of robust invaders. The main objective of this chapter is to highlight the invasion risks posed by the cultivated croplands. The chapter gives a brief of plant invasion process, invasion risks posed by cultivated croplands including propagule escape from crops, crop-associated weeds, and feral crop descendants, and their fate in heterogeneous environments. We also aim to address the emerging perspectives from the synthesis which could potentially assist in evaluating and managing invasions from croplands.

9.2 Plant Invasions

Growing widespread concern over global biodiversity loss has led to an exponential increase in the study of plant invasions (Gurevitch et al. 2011). Invasive species, second largest threat to the biodiversity (Vitousek et al. 1996; Falk-Peterson et al. 2006; Pyšek et al. 2012; Simberloff et al. 2013), immensely impact ecological, economic, and social systems (Mack 1996; Pimentel et al. 2000; Global Invasive Species Programme (GISP) 2010; Pyšek et al. 2012). Invasion by the alien introduced species may potentially impact different levels of ecological complexity from genes to ecosystems (Mack et al. 2000). Invasive species are the organisms introduced into places out of their natural range of distribution by human intervention, where they establish, disperse, and significantly impact environment, economies and human health (GISP 2010).

Biological invasion is a multi-stage process of species dispersal across various spatial and temporal scales (Richardson et al. 2000a; Sakai et al. 2001; Blackburn et al. 2011). Species introduced to new localities have been varyingly termed as adventive (Mack et al. 2000; Sharma et al. 2005a), alien (Crawley et al. 1996), colonizer (Williamson 1996), exotic (Green 1997), immigrant (Bazzaz 1986), imported (Williamson and Fitter 1996), introduced (Lonsdale 1994), naturalized (Richardson et al. 2000a), non-indigenous (Mack et al. 2000; Kolar and Lodge 2001), and non-native (Davis et al. 2000; Valery et al. 2008). Blackburn et al. (2011) suggested that an introduced species has to overcome a range of barriers, from individual level to population level to advance from one stage of survival to the next; wherein barriers to population survival tend to be more complex than the barriers to individual survival. A species introduced to a novel range has to overcome (i) geographical barrier for movement of species' propagules from current range to the novel range, (ii) barrier of captivity or cultivation to facilitate its introduction, (iii) survival barrier for primary establishment, (iv) reproduction barrier to effectively reproduce and establish to yield self-sustaining populations, (v) dispersal barrier to facilitate further movement of propagules, and lastly (vi) the environmental barrier to emerge as a biotic invader (Blackburn et al. 2011). A species fails to become invasive if it fails to breach some barrier at any stage of the invasion process (Blackburn et al. 2011). A taxon is considered to be successfully naturalized in a novel range after overcoming geographical, environmental, and reproduction barriers; however, to become invasive, the naturalized species has to overcome additional barrier of dispersal to spread into locations away from the point of introduction (Blackburn et al. 2011). Invasion success or failure is determined by the species' ability to invade (*elements of species' invasiveness*; Table 9.1 summarizes important species-level biological attributes associated with invasiveness) in complementarity with features of the novel environment (*components of habitat invasibility*; a few include—absence of predators, gaps generated by disturbance, presence of empty niches, and resource-rich environments) (Mantri et al. 2002; Sharma et al. 2005a; Pyšek et al. 2012).

Table 9.1 Key biological attributes associated with species' invasiveness (Sharma et al. 2005a)

Biological attribute	General concept	Examples
Fitness homeostasis	High ability of an individual or a population to maintain relatively constant fitness over a range of environmental conditions facilitates its invasiveness	Altitudinal range expanse of *Lantana camara* up to 2000 m in the Pulnis hills, Southern India[a]
Propagule number and dimensions	High seed output ensures high propagule pressure, while, high seed size and low seed weight facilitate better seed dispersal, thus promoting species' invasion ability	Lesser seed weight (<50 mg) confers higher invasiveness to woody species in disturbed landscapes[b]
		Production of smaller seeds is associated with greater seed output[c] that facilitates efficient dispersal, thus contributing to invasiveness in *Parthenium hysterophorus*[d], *Echium plantagineum*[e], and *Ricinus communis*[f]
Dispersal ability	Effective dispersal of propagules to different locations enhances prospects for invasion of the recipient habitat	Success of several woody invaders in disturbed as well as undisturbed habitats is owed to vertebrate dispersal[b, g]
		Lantana camara seeds are widely dispersed, predominantly by fruit-eating birds[h], sheep, goats, cattle, foxes, jackals and monkeys, facilitating its wide-range dispersal[i]
		Secondary dispersal of *Ricinus communis* seeds by ants facilitates its spread and enhances seed germination success by elaiosome removal[j, k]
Alternative mode of reproduction	Ability to reproduce vegetatively largely contributes to increased habitat compatibility that further increases the likelihood of successful invasion	Vegetative reproduction is particularly crucial for dispersal in aquatic habitats[l]
		Eichhornia crassipes and *Alternanthera philoxeroides* invasiveness is mainly attributed to its free-floating life form and asexual reproduction by stolons[m, n]
		Lantana camara L. exhibits enormous capability for vegetative spread through layering[i]
Competitive ability	Introduced species belonging to exotic genera possess traits that owe huge competitive ability than those of resident species, hence alien species are more likely to be invasive in comparison to their native congeners	Plant species introduced to novel environment may overcome various biotic and abiotic barriers through non-specific mutualistic associations (root symbionts, pollinators and seed dispersers)[o], and also with competitive superiority due to higher tolerance to low resource conditions[p]
		Short juvenile period (<10 years) of woody invaders owes them a higher competitive ability and thus higher invasion potential in disturbed landscapes[b]
Allelopathy	Ability to produce allelochemicals is one of the key attributes that facilitates a plant species to invade and establish in a new ecosystem	*L. camara* grows in dense thickets by interrupting the regeneration process of native species due to allelopathy[q]
		Allelopathic interactions of *P. hysterophorus* with the resident species inhibit their germination and growth[r]
		Ability of *Eupatorium riparium* and *E. adenophorum* to dominate other plant species in Meghalaya has been ascribed to their allelopathic properties[s, t]
		Prosopis juliflora, highly invasive in India has substantial allelopathic potential[u, v]

(continued)

Table 9.1 (continued)

Biological attribute	General concept	Examples
Phenotypic plasticity	Ability of a genotype to modify its growth and development in response to changes in environment increases the likelihood of invasion	*P. hysterophorus* exhibits substantial plastic responses to soil quality that contributes to its success as an invader[w]
		Ricinus communis shows adaptive modulations in key vegetative and reproductive traits that potentially contributes to its invasiveness in urban habitats[f]
Chromosome number and genome size (ploidy level)	High chromosome number and low genome size are associated with high potential of a plant species to invade a new environment	*Solidago gigantea* occurs as a diploid, tetraploid, and hexaploid in North America (*native range*); however, only tetraploids are invasive in Europe[x] and East Asia (*invasive range*)[y]

[a]Mathews (1972)
[b]Rejmánek and Richardson (1996)
[c]Weiher et al. (1999)
[d]Annapurna and Singh (2003a)
[e]Sharma and Esler (2008)
[f]Goyal et al. (2014)
[g]Binggeli (1996)
[h]Mokotjomela et al. (2013)
[i]Sharma et al. (2005b)
[j]Martins et al. (2006)
[k]Martins et al. (2009b)
[l]Pieterse and Murphy (1990)
[m]Barrett (1989)
[n]Maheshwari (1965)
[o]Richardson et al. (2000b)
[p]Noble and Slatyer (1980)
[q]Gentle and Duggin (1997)
[r]Adkins and Sowerby (1996)
[s]Tripathi et al. (1981)
[t]Rai and Tripathi (1982)
[u]Goel et al. (1989)
[v]Noor et al. (1995)
[w]Annapurna and Singh (2003b)
[x]Schlaepfer et al. (2010)
[y]Schlaepfer et al. (2008)

9.3 Invasion Risks Posed by Cultivated Croplands

Expansion of agriculture has considerably increased the rate and scale of alien species introductions globally (Ellstrand et al. 2010; Driscoll et al. 2014). While attempting to select better crops, it seems that man has inadvertently selected better weeds. Eleven of the world's 18 most aggressive weeds have been reported to be crops (Holm et al. 1997). A seminal work has enlisted instances of key weeds and/or invaders documented to have evolved from crops and/or their wild relatives (Ellstrand et al. 2010). Crops propagated for food, biofuels, and horticultural use pose threat of ecologically damaging invasions (Dehnen-Schmutz et al. 2007a, b; Buddenhagen et al. 2009; Richardson and Blanchard 2011). In general, species into cultivation have characteristics typical of weeds that contribute towards their invasion potential (Raghu et al. 2006; Raghu and Davis 2007; Low et al. 2007; Barney and DiTomaso 2008; Foxcroft et al. 2008). A few of these characteristics include tolerance to a wide range of environmental conditions, high germination and growth rate, and high reproductive potential (Anderson et al. 2006). Further, pronounced phenotypic and genotypic trait diversity in crops concomitantly increases their invasion potential (Ellstrand and Schierenbeck 2000; Anderson et al. 2006; Dehnen-Schmutz et al. 2007b; Buddenhagen et al. 2009).

Species into cultivation typically undergo multiple introductions in large quantities (*high propagule pressure*) that are planted widely in monospecific stands (*dense population*) in resourceful environments (*high resource availability*). Ongoing selection, breeding, and contemporary crop improvement measures to accentuate traits of agroeconomic utility can further complement the traits associated with invasiveness (Richardson 1998; Paynter et al. 2003; Richardson and Rejmánek 2011). High propagule pressure, high growth, and reproductive potential in ample resource conditions, dense population, and high genetic diversity provide increased opportunities for evolution that may enable their higher colonization and establishment rate in all likelihood (Ellstrand and Schierenbeck 2000; Lockwood et al. 2005; Campbell and Snow 2007; Hooftman et al. 2007; Wilson et al. 2009; Hovick et al. 2012). Crop plants' huge resistance to pathogens, biotic and abiotic stresses, and high tolerance to environmental extremes also facilitate their survival and potential to invade anthropic systems, on escape from cultivated croplands (Dehnen-Schmutz et al. 2007b). Other plants in close association with crops including sturdy cropland weeds, crop mimics, and crop feral lineages surviving in the cultivated croplands and/or anthropogenically disturbed lands may possibly pose substantial invasion risks (Ellstrand et al. 2010).

9.3.1 Escapes of Crop Propagules

Agricultural activities have potentially generated new opportunities for alien propagules to disperse and establish (Kalwij et al. 2008). Increasing instances of inadvertent escape of crop propagules from cultivated lands to anthropic habitats, and their subsequent progression to potential weeds or aggressive colonizers has been reported (see Box 9.1) (Kowarik 2005; Raghu et al. 2006, Chapman and Burke 2006; Richardson and Blanchard 2011; Flory et al. 2012; Negussie et al. 2013; Goyal et al. 2014). Recurrent introduction events and crop monocultures contribute to a higher probability of escape from cultivated croplands to adjacent lands.

Box 9.1 *Ricinus communis* L.—an aggressive colonizer

Ricinus communis L. (Castor), native to Ethiopia is a classic case of crop plant that has possibly escaped the cultivation lands and is rapidly expanding its range in India (Goyal et al. 2014). *R. communis* is widely cultivated in India for castor oil which has multifarious industrial applications (Anjani 2012). Goyal et al. (2014) reported that the species is an emerging aggressive colonizer across urban landscapes, where it colonizes heterogeneous habitats *viz.* railway tracks, road verges, garbage dumps, and, wastelands. *R. communis* individuals inhabiting diverse habitat types display immense performance potential through adaptive modulation of traits in spite of considerable environmental stress, enabling its rapid emergence as an aggressive colonizer. High seed production, early maturity, autochorous dispersal, generalist germination behavior, and, rapid growth and recovery after disturbance might ensure high colonization ability of the species in all likelihood (Martins et al. 2009a, b, 2011). Better performance of *R. communis* at elevated levels of carbon dioxide further highlights that it beholds immense invasion potential in future scenarios of climate change (Vanaja et al. 2008).

Crop improvement measures to meet the tremendous global demand for castor oil has led to the development of varieties and hybrids with enhanced yield and oil percentage, disease and insect resistance, and drought tolerance, thus enhancing the existing agronomic diversity (Anjani 2012; Severino et al. 2012). *R. communis* exhibits tremendous diversity in India (Anjani 2012). 17,995 castor germplasm accessions have been reported in the world, of which 4307 castor germplasm accessions exist in India (Anjani 2012). Long history of cultivation coupled with the knowledge on genetic diversity can appropriately be utilized to infer factor(s) behind species' range expansion in India. We anticipate that immense diversity in castor germplasm may enable the species to adapt to diverse ecological niches and expand vigorously. The study calls out for detailed elucidation and categorization of existing variability in nature and in the wild to determine key factors that owe *R. communis*, a high potential of range expansion.

Biofuel crop, *Jatropha curcas* L. is cultivated for its non-edible oil suitable for biodiesel (Achten et al. 2008). Biofuel crops behold immense potential to support the livelihood of rural communities, but their cultivation is highly debatable as they pose serious invasion risks that may aggravate environmental problems (DiTomaso et al. 2010; Gordon et al. 2011). *J. curcas* has been categorized as a high-risk plant due to its invasion potential in many parts of the world (DiTomaso et al. 2007; GISP 2008). Several countries including Australia, South Africa, Hawaii (USA), the Galapagos Islands of Ecuador, Fiji, Guam, New Caledonia, and Samoa have imposed a ban on *J. curcas* cultivation owing to inadvertent escapes from plantations that pose high invasiveness risk in adjacent lands (Negussie et al. 2013). However, Negussie et al. (2015) suggested that consideration of *J. curcas* as an aggressive plant invader is not convincing due to lack of experimental facts. Importantly, in spite of reports and observations regarding the inadvertent propagule escapes from cultivated croplands, experimental findings to substantiate the invasion risks are scanty. There is a need to identify and evaluate the probability of establishment, persistence, and adaptive evolution to predict species' performance and invasive ability, on escape.

9.3.2 Crop-associated Weeds

Certain weeds survive in association with the main crop; their association may be attributed to a number of factors, viz., similarities in life cycle, growth habit, interactions with biotic and abiotic factors, and management practices, etc. During the course of cultivation, crop-associated weeds are confronted with a range of selection pressures resulting from changes in cropping systems and agricultural technology. Such crop-associated weeds also have the potential to escape from the croplands and behave as aggressive colonizers in the disturbed lands. With the current perspective, crop associations and their potential invasion risks can be dealt further in two sub-heads, viz., cropland weeds and crop mimics.

9.3.2.1 Cropland Weeds

Certain weeds grow along with the main crop in the agricultural systems and may evolve into aggressive colonizers during the course of domestication. High colonization ability is due to "*general-purpose genotype*" of cropland weeds (Baker 1965; Clements et al. 2004). Weeds in the cultivated lands often have mixed mating systems with high selfing rates that facilitate colonization (Antonovics 1968; Mulligan and Findlay 1970; Lloyd 1992). Genetic uniformity is more pronounced in weeds that exhibit vegetative reproduction, obligate selfing, and apomixis. Ability of weeds to exhibit plastic responses further adds to the adaptive flexibility in their performance. For example, velvet leaf, *Abutilon theophrasti* populations associated with corn and soybean exhibited adaptive variations in response to light (Weinig 2000). However, cropland weeds are also capable of rapid adaptability and genetic change with the introduction of new crop species and altered agroecological practices (Antonovics 1992; Thompson 1999; Palumbi 2001; Neuhauser et al. 2003).

Shepherd's purse, *Capsella bursa-pastoris* populations in Europe showed greater genetic heterogeneity in cultivated lands than populations growing in ruderal habitats (Bosbach and Hurka 1981). Similarly, groundsel, *Senecio vulgaris* agricultural biotypes were genetically more variable than those inhabiting ruderal habitats (Leiss and Müller-Schärer 2001). Close coupling of ecological and evolutionary changes underwent in agricultural weeds to confront powerful and fluctuating selection pressures in cropping systems facilitates better performance on dispersal to disturbed habitats (Sakai et al. 2001). Evolutionary changes in weeds facilitate their range expansion through new adaptations to local conditions (Gould 1991; Vermeij 1996; Jordan and Jannink 1997; Martínez-Ghersa et al. 2000; Mohler 2001).

9.3.2.2 Crop Mimics

Certain weedy agroecotypes, intimately associated with crops in cultivated lands may mimic a specific crop, whereby the weed resembles the crop at specific stages during its life history (Barrett 1983). Interestingly, the weedy agroecotype may evade eradication as a result of resemblance in phenotypic expressions with the crop. A variety of barnyard grass, *Echinochloa crus-galli* var. *oryzicola* (L.) P. Beauv. represents a classic case of a noxious weed that has evolved to mimic domesticated rice, *Oryza sativa* L. (Barrett 1983). Despite belonging to different genera, rice and its weedy mimic show close morphological and phenological resemblance in their vegetative phase. Certainly, barnyard grass individuals in Japanese rice fields that are morphologically indistinguishable from cultivated rice plants are less likely to be weeded out from rice fields by hand-weeding (Ehara and Abe 1950). Similarly, littleseed canary grass, *Phalaris minor* Retz. mimics wheat, *Triticum* spp. and, resembles the crop in several growth stages, ruling out the probability of its easy eradication (Fig. 9.1). Functional crop mimics such as weedy

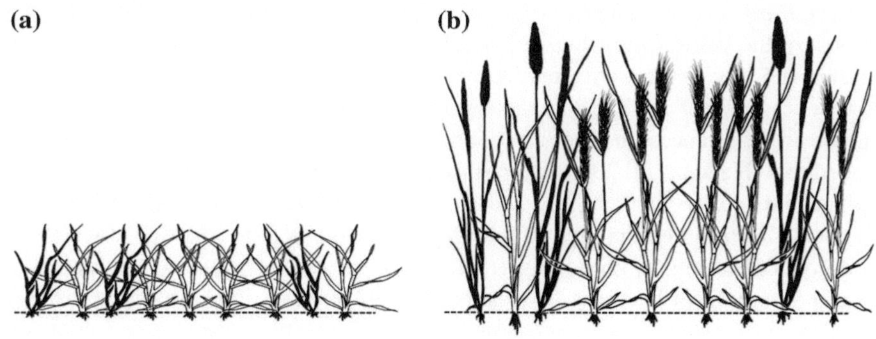

Fig. 9.1 Schematic illustration of cultivated cropland of wheat (*Triticum* spp.) infested with a weed, *Phalaris minor*: a classic example of crop mimicry. **a** *Phalaris minor* (*bold*) very closely resembles wheat (*Triticum* spp.) (*outline*) during early stages of development; **b** *P. minor* is distinguishable from *Triticum* spp. during reproductive phase

rice, weedy beet, weedy rye, and semi-wild wheat are also reported to be morphologically similar to their relatives that reduce their probability of hand-weeding, thereby ensuring enhanced survival (Ellstrand et al. 2010). In fact, morphological crop mimicry is considered an adaptation that has been proposed to be the result of continued selection by human weeding. Hand-weeding triggered adaptation in crop mimics facilitates their enhanced survival in the cropland. Under such scenario, crop mimics may evolve invasive propensities over time.

9.3.3 Feral Lineages Descended from Crop Progenitor(s)

On dispersal of a crop and/or crop descendant to a novel habitat, feral lineages can evolve by two pathways viz. endoferality or exoferality ('endoferal' and 'exoferal' sensu Gressel 2005). Feral lineages that are directly descended from a crop with or without evolutionary change are endoferals, while exoferals are descended from intertaxon hybridization; either between a crop and another taxon or, usually a domesticated crop and its wild relative (Table 9.2). However, feral lineages descended from hybrids between two crops are often termed as exo-endoferals (Table 9.2). Knowledge about crop progenitors is extremely important to deduce the origin of these troublesome feral lineages. For instance, horticulturally valued varieties of *Lantana camara* L. escaped cultivation and hybridized with their wild relatives, leading to the evolution of a species complex, *Lantana camara* L. (sensu lato) (Goyal and Sharma 2015). However, deduction of weedy and/or invasive genets in *Lantana* species complex is a formidable task owing to lack of information about their putative ancestor(s) (Goyal and Sharma 2015).

California wild radish represents a classical example of a hybrid-derived feral lineage descended from crop (cultivated radish, *Raphanus sativus*), and a closely related wild species (wild jointed charlock; sometimes referred as 'wild radish', *R. raphanistrum*) (Hegde et al. 2006; Ridley et al. 2008; Ridley and Ellstrand 2009). The two *Raphanus* taxa, when co-occur, spontaneously hybridize to a limited extent in most of the world (Snow and Campbell 2005); however, extensive hybridization for almost 150 years in California has resulted in a highly feral lineage (Frost 1923; Panetsos and Baker 1967; Chapman and Burke 2006). California wild radish shows significant differences in phenotype and is genetically intermediate to the two progenitors (Hegde et al. 2006). Appropriate amalgamation of traits descended from both parents makes it fit to succeed as a feral lineage in California. Ridley and Ellstrand (2009) noted that hybrid lineage exhibited potential adaptive modulations in reproductive traits by producing greater number of fruits and seeds per plant than either progenitor, in response to contrasting environments in California. Existence of only hybrids in each surveyed population in California generates an impetus to explore the probable differences in hybrid existence, fitness, and adaptive evolution potential elsewhere and regionally adapted hybrid-derived California wild radish (Hegde et al. 2006; Ellstrand et al. 2010). *R. sativus* is an economically important vegetable crop grown and consumed worldwide; however, natural hybridization

Table 9.2 Examples of feral lineages descended from crop progenitor(s) (adapted with permission from Ellstrand et al. 2010)

Weedy lineage (common name)	Progenitor(s)	Study area
Weedy endoferals		
Semi-wild wheat[a, b]	*Triticum aestivum**	Tibet, China
Forrageiro[c]	*Raphanus sativus**	Rio Grande do Sul, Brazil
Weedy rice[d]	*Oryza sativa japonica**	Liaoning, China
'Blackhull' weedy rice[e]	*Oryza sativa indica**	Southeastern USA
Weedy rye, feral rye[f, g]	*Secale cereale**	California and Washington, USA
Weedy exoferals		
Weedy finger millet[h, i, j]	*Eleusine coracana subsp. coracana** × *Eleusine coracana subsp. africana*	Africa
Johnsongrass[k]	*Sorghum bicolor** × *S. halepense*	Nebraska and Texas, USA
Columbus grass[l]	*Sorghum bicolor** × *S. propinquum*	Diverse geographic origins
California wild radish[m, n, o]	*Raphanus sativus** × *R. raphanistrum*	California, USA
'Strawhull' weedy rice[e]	*Oryza sativa indica** × *O. rufipogon*	Southeastern USA
Weed beet[p, q]	*Beta vulgaris subsp. vulgaris** × *B. v. maritima*	France, Germany, Italy
Weedy exo-endoferal		
Weedy rice[r]	*Oryza sativa japonica** × *O.s. indica**	Bhutan

*Domesticated plant
[a]Sun et al. (1998)
[b]Ayal and Levy (2005)
[c]Snow and Campbell (2005)
[d]Cao et al. (2006)
[e]Londo and Schaal (2007)
[f]Suneson et al. (1969)
[g]Burger et al. (2006, 2007)
[h]Hilu et al. (1978)
[i]de Wet et al. (1984)
[j]de Wet (1995)
[k]Morrell et al. (2005)
[l]Paterson et al. (1995)
[m]Hegde et al. (2006)
[n]Ridley et al. (2008)
[o]Ridley and Ellstrand (2009)
[p]Mücher et al. (2000)
[q]van Dijk et al. (2004)
[r]Ishikawa et al. (2005)

events with *R. raphanistrum* may have serious consequences for the crop gene pool diversity. Such interspecific hybridization events may lead to extinction of one or both hybridizing taxa and/or may even lead to de-domestication of certain domesticated cultivars (see Chapman and Burke 2006). Valuable insights into radish genome evolution and the phylogenetic relationships of different radish accessions have been gained through a recent comprehensive analysis of radish expressed sequence tags (ESTs) (Shen et al. 2013). Using insights, evolution of wild radish and its hybrid-derived feral lineages can be explored using the molecular tools. Information of crop feral lineages can be useful to track the ancestry of weedy aggressive colonizers.

9.3.4 Fate of Dispersed Propagules

Not all propagules escaped from the cultivated croplands respond the same to the recipient habitat; some fit and colonize, while others may fail. Colonization refers to the establishment or spread of an organism in a region previously not occupied by the species (Barrett and Husband 1990). Success or failure of colonizing episodes depends on ecological persistence and evolutionary potential of migrant populations. On dispersal, propagules are exposed to a wide variety of environmental stresses existing in recipient environments; wherein, a wide range of human-selected traits of agronomic importance may owe the crops and their descendants, a wider ability to survive in heterogeneous environments (de Wet and Harlan 1975; Mack 2000; Kowarik 2003; Ellstrand et al. 2010). With time, colonizers may undergo significant adaptive changes and evolve into aggressive weeds and/or invaders. Feral lineages of crop progenitor(s) have evolved key traits such as resistance to herbicides, modulations and shifts in plant habit, phenological patterns, mode of reproduction, reproductive output, and, seed shattering and dormancy (Ellstrand et al. 2010). Hybridization of crops with their close wild relatives may contribute to the movement of crop transgenes into descendant feral lineages that may potentially increase their colonization and performance potential in disturbed environments (Ellstrand et al. 1999; Ellstrand and Schierenbeck 2000; Ellstrand 2001; Smith and Barney 2014). Chen et al. (2010) suggested that high invasion potential of crops may be ascribed to lower values of monoploid and basic genome size of crops. Interestingly, significant number of studies in plant invasion ecology suggests that lower genome size correlates with higher plant invasiveness (Rejmànek 1996; Bennett et al. 1998; Knight and Ackerly 2002; Kubešová et al. 2010; Kuester et al. 2014; Pandit et al. 2014; Suda et al. 2015). However, there is a need to examine the suite of factors that facilitate establishment of propagules dispersed from croplands to thoroughly evaluate the pertinent invasion risks.

9.4 Emerging Perspectives

The perspective that agricultural systems may be potential sources and/or progenitors of robust weeds and/or invaders is currently emerging. Although key instances of crops and their derivatives turning weedy are known, the opinion lacks substantial evidences. Crops are broadly the best studied and well-characterized plant systems (Ellstrand et al. 2010). Although many crops are cytogenetically well-characterized, studies examining chromosomal evolution of feral lineages in due course of colonization are lacking. Genomic approaches and advanced molecular tools can be well applied for the identification of traits that may confer high colonization potential to the crops and their descendants. Studies examining adaptive evolution in both weeds and invaders are known (Ehara and Abe 1950; Barrett 1983; Dlugosch and Parker 2008; Ridley and Ellstrand 2009). However, studies pertaining to adaptive evolution of crops and/or their feral descendants compared to their putative source populations are limited (Stewart et al. 2009; Ridley and Ellstrand 2010; Thompson et al. 2012). Primarily, this may be attributed to the lack of information about putative crop progenitors and source populations of the colonizing populations in different habitats. Integrating the knowledge of crop systems with that of plant invasiveness may aid in the prediction of potential invaders from croplands. Prospects of colonization and establishment potential of crop plants in future scenarios of climate change also demands in-depth investigation.

Research attention has primarily been focused on cereal crops and crops cultivated for biofuel and horticulture; however, fruit and vegetable crops are less explored with respect to escape and colonization potential. Wide knowledge gap exists as there are no or limited inventories of potential escapees from croplands, available for many countries. Inventories with continuous updates may expedite early recognition of potential colonizers emerging from croplands, thus enabling informed decision-making for invasion risk monitoring and management. Nevertheless, taking full advantage of the '*knowing*' requires needful '*doing*' (Esler et al. 2010). In conclusion, hitchhiking propagules from croplands may lead us to substantiate key questions on plant invasiveness and invasion risks associated with the current agricultural intensification.

Acknowledgments Authors thank Dr. V.R. Rajpal and Prof. S.N. Raina for the invitation to write this chapter. GPS acknowledges funding from University of Delhi and Department of Science and Technology, India. NG acknowledges Senior Research Fellowship (SRF) support from University Grants Commission, India. Constructive suggestions by Prof. A.S. Raghubanshi, Banaras Hindu University are duly acknowledged.

References

Achten W, Verchot L, Franken Y (2008) *Jatropha* bio-diesel production and use. Biomass Bioenergy 32:1063–1084

Adkins SW, Sowerby MS (1996) Allelopathic potential of the weed *Parthenium hysterophorus* L. in Australia. Plant Prot Q 11:20–23

Anderson NO, Galatowitsch SM, Gomez N (2006) Selection strategies to reduce invasive po-tential in introduced plants. Euphytica 148:203–216

Andersson MS, de Vicente MC (2010) Gene flow between crops and their wild relatives. Johns Hopkins University Press, Baltimore

Anjani K (2012) Castor genetic resources: a primary gene pool for exploitation. Ind Crops Prod 35:1–14

Annapurna C, Singh JS (2003a) Variation of *Parthenium hysterophorus* in response to soil quality: implications for invasiveness. Weed Res 43:190–198

Annapurna C, Singh JS (2003b) Phenotypic plasticity and plant invasiveness: case study of congress grass. Curr Sci 85:197–201

Antonovics J (1968) Evolution in closely adjacent plant populations V. Evolution of self-fertility. Heredity 23:219–238

Antonovics J (1992) Toward community genetics. In: Fritz RS, Simms EL (eds) Plant resistance to herbivores and pathogens: ecology, evolution, and genetics. University of Chicago Press, Chicago, pp 426–449

Ayal S, Levy AA (2005) Wheat domestication and de-domestication–What are the odds? In: Gressel J (ed) Crop ferality and volunteerism. CRC Press, Boca Raton, pp 167–173

Baker HG (1965) Characteristics and modes of origin of weeds. In: Baker HG, Stebbins GL (eds) The Genetics of Colonizing Species. Academic Press, New York, pp 147–168

Barney JN, DiTomaso JM (2008) Nonnative species and bioenergy: are we cultivating the next invader? Bioscience 58:64–70

Barrett SH (1983) Crop mimicry in weeds. Econ Bot 37:255–282

Barrett SCH (1989) Waterweed invasions. Sci Am 260:90–97

Barrett SCH, Husband BC (1990) Genetics of plant migration and colonization. In: Brown AHD, Clegg MT, Kahler AL, Weir BS (eds) Plant Population Genetics. Breeding and Genetic Resources, Sinauer Associates, Massachusetts, pp 254–277

Bazzaz FA (1986) Life history of colonizing plants: some demographic, genetic and physiological features. In: Mooney HA, Drake JS (eds) Ecology of biological invasions of North America and Hawaii. Springer, New York, pp 96–110

Bennett MD, Leitch IJ, Hanson L (1998) DNA amounts in two samples of angiosperm weeds. Ann Bot 82:121–134

Binggeli P (1996) A taxonomic, biogeographical and ecological overview of invasive woody plants. J Veg Sci 7:121–124

Blackburn TM, Pyšek P, Bacher S et al (2011) A proposed unified framework for biological invasions. Trends Ecol Evol 26:333–339

Bosbach K, Hurka H (1981) Biosystematic studies on *Capsella bursa-pastoris* (Brassicaceae): enzyme polymorphism in natural populations. Plant Syst Evol 137:73–94

Buddenhagen CE, Chimera C, Clifford P (2009) Assessing biofuel crop invasiveness: a case study. PLoS ONE 4(4):e5261

Burger JC, Lee S, Ellstrand NC (2006) Origin and genetic structure of feral rye in the western United States. Mol Ecol 15:2527–2539

Burger JC, Holt JS, Ellstrand NC (2007) Rapid phenotypic divergence of feral rye from domesticated cereal rye. Weed Sci 55:204–271

Campbell LG, Snow AA (2007) Competition alters life history and increases the relative fecundity of crop-wild radish hybrids (*Raphanus* spp.). New Phytol 173:648–660

Cao Q, Lu BR, Xia H et al (2006) Genetic diversity and origin of weedy rice (*Oryza sativa f. spontanea*) populations found in North-eastern China revealed by simple sequence repeat (SSR) markers. Ann Bot 98:1241–1252
Carlton JT, Geller JB (1993) Ecological roulette: the global transport of nonindigenous marine organisms. Science 261:78–82
Chapman MA, Burke JM (2006) Radishes gone wild. Heredity 97:379–380
Chen GQ, Guo SL, Yin LP (2010) Applying DNA C-values to evaluate invasiveness of angiosperms: validity and limitation. Biol Invasions 12:1335–1348
Clements DR, DiTommaso A, Jordan N et al (2004) Adaptability of plants invading North American cropland. Agric Ecosyst Environ 104(3):379–398
Cohen AN, Carlton JT (1998) Accelerating invasion rate in a highly invaded estuary. Science 279:555–558
Crawley MJ, Harvey PH, Purvis A (1996) Comparative ecology of the native and alien floras of the British Isles. Philos Trans R Soc Lond B Biol Sci 351:1251–1259
Davis MA, Grime JP, Thompson K (2000) Fluctuating resources in plant communities: a general theory of invasibility. J Ecol 88:528–534
de Wet JMJ (1995) Finger millet. In: Smartt J, Simmonds NW (eds) Evolution of crop plants, 2nd edn. Longman, Harlow, pp 137–140
de Wet JMJ, Harlan JR (1975) Weeds and domesticates: evolution in the man-made habitat. Econ Bot 29:99–107
de Wet JMJ, Prasada Rao KE, Brink DE et al (1984) Systematics and evolution of *Eleusine coracana* (Gramineae). Am J Bot 71:550–557
Dehnen-Schmutz K, Touza J, Perrings C et al (2007a) The horticultural trade and ornamental plant invasions in Britain. Conserv Biol 21:224–231
Dehnen-Schmutz K, Touza J, Perrings C et al (2007b) A century of the ornamental plant trade and its impacts on invasion success. Divers Distrib 13:527–534
DiTomaso JM, Barney JN, Fox A (2007) Biofuel feedstocks: the risk of future invasions. Council for Agricultural Science and Technology Commentary, QTA2007-1
DiTomaso JM, Reaser JK, Dionigi CP et al (2010) Biofuel vs bioinvasion: seeding policy priorities. Environ Sci Technol 44:6906–6910
Dlugosch KM, Parker IM (2008) Invading populations of an ornamental shrub show rapid life history evolution despite genetic bottlenecks. Ecol Lett 11(7):701–709
Driscoll DA, Catford JA, Barney JN et al (2014) New pasture plants intensify invasive species risk. Proc Natl Acad Sci USA 111:16622–16627
Ehara K, Abe S (1950) Classification of the forms of Japanese barnyard millet. Proc Crop Sci Soc Jpn 20:245–246
Ellstrand NC (2001) When transgenes wander, should we worry? Plant Physiol 125:1543–1545
Ellstrand NC, Schierenbeck KA (2000) Hybridization as a stimulus for the evolution of invasiveness in plants? Proc Natl Acad Sci USA 97:7043–7050
Ellstrand NC, Prentice HC, Hancock JF (1999) Gene flow and introgression from domesticated plants into their wild relatives. Annu Rev Ecol Evol Syst 30:536–539
Ellstrand NC, Heredia SM, Leak-Garcia JA et al (2010) Crops gone wild: evolution of weeds and invasives from domesticated ancestors. Evol Appl 3:494–504
Esler KJ, Prozesky H, Sharma GP et al (2010) How wide is the "knowing-doing" gap in invasion biology? Biol Invasions 12:4065–4075
Falk-Peterson J, Bøhn T, Sandlund OD (2006) On the numerous concepts in invasion biology. Biol Invasions 8:1409–1424
Flory SL, Lorentz KA, Gordon DR et al (2012) Experimental approaches for evaluating the invasion risk of biofuel crops. Environ Res Lett 7(4):045904
Foxcroft LC, Richardson DM, Wilson JRU (2008) Ornamental plants as invasive aliens: prob-lems and solutions in Kruger National Park, South Africa. Environ Manage 41:32–51
Frost HG (1923) Heterosis and dominance of size factors in *Raphanus*. Genetics 8:116–153
Gentle CB, Duggin JA (1997) Allelopathy as a competitive strategy in persistent thickets of *Lantana camara* L. in three Australian forest communities. Plant Ecol 132:85–95

GISP (2008) Biofuels run the risk of becoming invasive species. The Global Invasive Species Programme. http://www.issg.org/pdf/publications/GISP/Resources/BiofuelsReport.pdf
GISP (2010) The IAS problem. The Global Invasive Species Programme. http://www.gisp.org/about/IAS
Goel U, Saxena DB, Kumar B (1989) Comparative study of allelopathy as exhibited by *Prosopis juliflora* swartz and *Prosopis cineraria* (L) Druce. J Chem Ecol 15:591–600
Gordon DR, Tancig KJ, Onderdonk DA et al (2011) Assessing the invasive potential of biofuel species proposed for Florida and the United States using the Australian Weed Risk Assessment. Biomass Bioenerg 35:74–79
Gould F (1991) The evolutionary potential of crop pests. Am Sci 79:496–507
Goyal N, Sharma GP (2015) *Lantana camara* L. (*sensu lato*): an enigmatic complex. NeoBiota 25:15–26
Goyal N, Pardha-Saradhi P, Sharma GP (2014) Can adaptive modulation of traits to urban environments facilitate *Ricinus communis* L. invasiveness? Environ Monit Assess 186: 7941–7948
Green RE (1997) The influence of numbers released on the outcome of attempts to introduce exotic bird species to New Zealand. J Anim Ecol 66:25–35
Gressel J (2005) Introduction—the challenges of ferality. In: Gressel J (ed) Crop ferality and volunteerism. CRC Press, Boca Raton, pp 1–7
Gurevitch J, Fox GA, Wardle GM et al (2011) Emergent insights from the synthesis of conceptual frameworks for biological invasions. Ecol Lett 14(4):407–418
Hegde SG, Nason JD, Clegg J et al (2006) The evolution of California's wild radish has resulted in the extinction of its progenitors. Evolution 60:1187–1197
Hilu K, de Wet JMJ, Seigler D (1978) Flavonoid patterns and systematics in *Eleusine*. Biochem Syst Ecol 6:247–249
Holm L, Doll J, Holm E, Pancho J, Herberger J (1997) World weeds: natural histories and distribution. Wiley, New York
Hooftman DAP, De Jong MJ, Oostermeijer JGB et al (2007) Modelling the long-term consequences of crop-wild relative hybridization: a case study using four generations of hybrids. J Appl Ecol 44:1035–1045
Hovick SM, Campbell LG, Snow AA et al (2012) Hybridization alters early life-history traits and increases plant colonization success in a novel region. Am Nat 179:192–203
Ishikawa R, Toki N, Imai K et al (2005) Origin of weedy rice grown in Bhutan and the force of genetic diversity. Genet Resour Crop Ev 52(4):395–403
Jordan NR, Jannink JL (1997) Assessing the practical importance of weed evolution: a research agenda. Weed Res 37:237–246
Kalwij JM, Robertson MP, van Rensburg BJ (2008) Human activity facilitates altitudinal expansion of exotic plants along a road in montane grassland, South Africa. Appl Veg Sci 11:491–498
Knight CA, Ackerly DD (2002) Variation in nuclear DNA content across environmental gradients: a quantile regression analysis. Ecol Lett 5:66–76
Kolar CS, Lodge DM (2001) Progress in invasion biology: predicting invaders. Trends Ecol Evol 16:199–204
Kowarik I (2003) Human agency in biological invasions: secondary releases foster naturalisation and population expansion of alien plant species. Biol Invasions 5:293–312
Kowarik I (2005) Urban ornamentals escaped from cultivation. In: Gressel J (ed) Crop ferality and volunteerism. CRC Press, Boca Raton, pp 97–121
Kubešová M, Moravcová L, Suda J et al (2010) Naturalized plants have smaller genomes than their non-invading relatives: a flow cytometric analysis of the Czech alien flora. Preslia 82:81–96
Kuester A, Conner JK, Culley T et al (2014) How weeds emerge: a taxonomic and trait-based examination using United States data. New Phytol 202:1055–1068
Lake JC, Leishman MR (2004) Invasion success of exotic plants in natural ecosystems: the role of disturbance, plant attributes and freedom from herbivores. Biol Cons 117:215–226

Leiss KA, Müller-Schärer H (2001) Adaptation of *Senecio vulgaris* (Asteraceae) to ruderal and agricultural habitats. Am J Bot 88:1593–1599

Levine JM, D'Antonio CM (2003) Forecasting biological invasions with increasing international trade. Conserv Biol 17:322–326

Lloyd DG (1992) Self- and cross-fertilization in plants. II. The selection of self-fertilization. Int J Plant Sci 153:370–380

Lockwood JL, Cassey P, Blackburn T (2005) The role of propagule pressure in explaining species invasions. Trends Ecol Evol 20:223–228

Londo JP, Schaal BA (2007) Origins and population genetics of weedy red rice in the USA. Mol Ecol 16:4523–4535

Lonsdale WM (1994) Inviting trouble: introduced pasture species in Northern Australia. Aust J Ecol 19:345–354

Low T, Booth C, Council I (2007) The weedy truth about biofuels. In: Invasive Species Council, Melbourne, Australia, p 46

Mack RN (1996) Predicting the identity and fate of plant invaders: emergent and emerging approaches. Biol Conserv 78:107–121

Mack RN (2000) Cultivation fosters plant naturalization by reducing environmental stochasticity. Biol Invasions 2:111–122

Mack RN, Simberloff D, Lonsdale WM et al (2000) Biotic invasions: causes, epidemology, global consequences and control. Ecol Appl 10:689–710

Maheshwari JK (1965) Alligator weed in Indian lakes. Nature 205:1270

Mantri A, Annapurna C, Singh JS (2002) Terrestrial plant invasion and global change. In: Tripathi G, Tripathi YC (eds) Bioresource and Environment Campus Book International, New Delhi, p 25–44

Martínez-Ghersa MA, Ghersa CM, Satorre EH (2000) Co-evolution of agricultural systems and their weed companions: implications for research. Field Crop Res 67:181–190

Martins VF, Guimarães PR, Silva RR (2006) Secondary seed dispersal by ants of *Ricinus communis* (Euphorbiaceae) in the Atlantic forest in Southeastern Brazil: influence on seed germination. Sociobiology 47:265–274

Martins VF, Guimarães Jr PR, Haddad CRB et al (2009a) The effect of ants on the seed dispersal cycle of the typical myrmecochorous *Ricinus communis*. Plant Ecol 205:213–222

Martins VF, Haddad CRB, Semir J (2009b) Seed germination of *Ricinus communis* in predicted settings after autochorous and myrmecochorous dispersal. J Torrey Bot Soc 136:84–90

Martins VF, Haddad CRB, Semir J (2011) Responses of the invasive *Ricinus communis* seedlings to competition and light. N Z J Bot 49:263–279

Mathews KM (1972) The high altitude ecology of *Lantana*. Indian For 97:170–171

Mohler CL (2001) Weed evolution and community structure. In: Liebman M, Mohler C, Staver CP (eds) Ecological management of agricultural weeds. Cambridge University Press, Cambridge, pp 444–493

Mokotjomela TM, Musil CF, Esler KJ (2013) Potential seed dispersal distances of native and non-native fleshy fruiting shrubs in the South African Mediterranean climate region. Plant Ecol 214(9):1127–1137

Morrell PL, Williams-Coplin TD, Lattu AL et al (2005) Crop-to-weed introgression has impacted allelic composition of johnsongrass populations with and without recent exposure to cultivated sorghum. Mol Ecol 14:2143–2154

Mücher T, Hesse P, Pohl-Orf M et al (2000) Characterization of weed beet in Germany and Italy. J Sugar Beet Res 37:19–38

Mulligan GA, Findlay JN (1970) Reproductive systems and colonization in Canadian weeds. Can J Bot 48:859–860

Negussie A, Achten WM, Aerts R et al (2013) Invasiveness risk of the tropical biofuel crop *Jatropha curcas* L. into adjacent land use systems: from the rumors to the experimental facts. Glob Change Biol Bioenergy 5(4):419–430

Negussie A, Nacro S, Achten WM et al (2015) Insufficient evidence of *Jatropha curcas* L. invasiveness: experimental observations in Burkina Faso, West Africa. Bioenergy Res 8(2):570 580

Neuhauser C, Andow DA, Heimpel GE et al (2003) Community genetics: Expanding the synthesis of ecology and genetics. Ecology 84:545–558

Noble IR, Slatyer RO (1980) The use of vital attributes to predict successional changes in plant communities subject to recurrent disturbances. succession. Springer, Netherlands, pp 5–21

Noor M, Salam U, Khan MA (1995) Allelopathic effects of *Prosopis juliflora* Swartz. J Arid Environ 31:83–90

Palumbi SR (2001) Humans as the world's greatest evolutionary force. Science 293:1786–1790

Pandit MK, White SM, Pocock MJO (2014) The contrasting effects of genome size, chromosome number and ploidy level on plant invasiveness: a global analysis. New Phytol 203:697–703

Panetsos CA, Baker HG (1967) The origin of variation in "wild" *Raphanus sativus* (Cruciferae) in California. Genetica 38:243–274

Paterson AH, Schertz KF, Lin YR et al (1995) The weediness of wild plants: molecular analysis of genes influencing dispersal and persistence of johnsongrass, *Sorghum halepense* (L.). Pers. Proc Natl Acad Sci USA 92:6127–6131

Paynter Q, Csurhes SM, Heard TA et al (2003) Worth the risk? Introduction of legumes can cause more harm than good: an Australian perspective. Aust Syst Bot 16:81–88

Pieterse AH, Murphy KJ (eds) (1990) Aquatic Weeds. Oxford University Press, Oxford

Pimentel D, Lach L, Zuniga R et al (2000) Environmental and economic costs of non-indigenous species in the United States. Bioscience 50(1):53–65

Pyšek P, Jarošík V, Hulme PE et al (2010) Disentangling the role of environmental and human pressures on biological invasions across Europe. Proc Natl Acad Sci USA 107:12157–12162

Pyšek P, Jarošík V, Hulme PE et al (2012) A global assessment of invasive plant impacts on resident species, communities and ecosystems: the interaction of impact measures, invading species' traits and environment. Glob Change Biol 18:1725–1737

Raghu S, Davis A (2007) Exotic grasses as biofuels the concerns. Ill Nat Steward 16:30–32

Raghu S, Anderson RC, Daehler CC et al (2006) Adding biofuels to the invasive species fire? Science 313:1742

Rai JPN, Tripathi RS (1982) Allelopathy as a factor contributing to dominance of *Eupatorium riparium* Regel. Indian J Ecol 9:14–20

Rejmánek M (1996) A theory of seed plant invasiveness: the first sketch. Biol Cons 78:171–181

Rejmánek M, Richardson DM (1996) What attributes make some plant species more invasive? Ecology 77:1655–1662

Ricciardi A (2007) Are modern biological invasions an unprecedented form of global change? Conserv Biol 21:329–336

Richardson DM (1998) Forestry trees as invasive aliens. Conserv Biol 12:18–26

Richardson DM, Blanchard R (2011) Learning from our mistakes: minimizing problems with invasive biofuel plants. Curr Opin Environ Sustain 3:36–42

Richardson DM, Rejmánek M (2011) Trees and shrubs as invasive alien species—a global review. Divers Distrib 17:788–809

Richardson DM, Pyšek P, Rejmánek M et al (2000a) Naturalization and invasion of alien plants: concept and definitions. Divers Distrib 6:93–107

Richardson DM, Allsopp N, D'Antonio C et al (2000b) Plant invasions—the role of mutualisms. Biol Rev 75:65–93

Ridley CE, Ellstrand NC (2009) Evolution of enhanced reproduction in the hybrid-derived invasive, California wild radish (*Raphanus sativus*). Biol Invasions 11:2251–2264

Ridley CE, Ellstrand NC (2010) Rapid evolution of morphology and adaptive life history in the invasive California wild radish (*Raphanus sativus*) and the implications for management. Evol Appl 3(1):64–76

Ridley CE, Kim SC, Ellstrand NC (2008) Bidirectional history of hybridization in California wild radish, *Raphanus sativus* (Brassicaceae), as revealed by chloroplast DNA. Am J Bot 95: 1437–1442

Sakai AK, Allendorf FW, Holt JS et al (2001) The population biology of invasive species. Annu Rev Ecol Evol Syst 32:305–332
Schierenbeck KA, Ellstrand NC (2009) Hybridization and the evolution of invasiveness in plants and other organisms. Biol Invasions 11:1093–1105
Schlaepfer DR, Edwards PJ, Semple JC et al (2008) Cytogeography of *Solidago gigantea* (Asteraceae) and its invasive ploidy level. J Biogeogr 35(11):2119–2127
Schlaepfer DR, Edwards PJ, Billeter R (2010) Why only tetraploid *Solidago gigantea* (Asteraceae) became invasive: a common garden comparison of ploidy levels. Oecologia 163(3):661–673
Severino LS, Auld DL, Baldanzi M et al (2012) A review on the challenges for increased production of castor. Agron J 104:853–880
Sharma GP, Esler KJ (2008) Phenotypic plasticity among *Echium plantagineum* populations in different habitats of Western Cape, South Africa. S Afr J Bot 74:746–749
Sharma GP, Singh JS, Raghubanshi AS (2005a) Plant invasions: Emerging trends and future implications. Curr Sci 88:726–734
Sharma GP, Raghubanshi AS, Singh JS (2005b) *Lantana* invasion: an overview. Weed Biol Manag 5:157–165
Shen D, Sun H, Huang M et al (2013) Comprehensive analysis of expressed sequence tags from cultivated and wild radish (*Raphanus* spp.). BMC Genom 14(1):721
Simberloff D, Martin JL, Genovesi P et al (2013) Impacts of biological invasions: what's what and the way forward. Trends Ecol Evol 28:58–66
Smith LL, Barney JN (2014) The relative risk of invasion: evaluation of *Miscanthus* × *giganteus* seed establishment. Invasive Plant Sci Manag 7:93–106
Snow AA, Campbell LG (2005) Can feral radishes become weeds? In: Gressel J (ed) Crop Ferality and Volunteerism. CRC Press, Boca Raton, pp 193–208
Stewart CN Jr, Tranel PJ, Horvath DP et al (2009) Evolution of weediness and invasiveness: charting the course for weed genomics. Weed Sci 57:451–462
Suda J, Meyerson LA, Leitch IJ et al (2015) The hidden side of plant invasions: the role of genome size. New Phytol 205:994–1007
Sun Q, Ni Z, Liu Z et al (1998) Genetic relationships and diversity among Tibetan wheat, common wheat and European spelt wheat revealed by RAPD markers. Euphytica 99:205–211
Suneson CA, Rachie KO, Khush GS (1969) A dynamic population of weedy rye. Crop Sci 9:121–124
Thompson JN (1999) The evolution of species interactions. Science 284:2116–2118
Thompson GD, Bellstedt DU, Byrne M et al (2012) Cultivation shapes genetic novelty in a globally important invader. Mol Ecol 21:3187–3199
Tripathi RS, Singh RS, Rai JPN (1981) Allelopathic potential of *Eupatorium adenophorum*—a dominant ruderal weed of Meghalaya. Proc Indian Natl Sci Acad Part B 47:458–465
Valery L, Fritz H, Lefeuvre JC et al (2008) In search of a real definition of the biological invasion phenomenon itself. Biol Invasions 10:1345–1351
van Dijk H, den Nijs HCM, Bartsch D, Sweet J (2004) Gene exchange between wild and crop in *Beta vulgaris*: how easy is hybridization and what will happen in later generations? In: den Nijs HCM, Bartsch D, Sweet J (eds) Introgression from Genetically Modified Plants into Wild Relatives. CABI Publishing, Wallingford, pp 53–61
Vanaja M, Jyothi M, Ratnakumar P et al (2008). Growth and yield responses of castor bean (*Ricinus communis* L.) at two enhanced CO_2 levels. Plant Soil Environ UZPI (Czech Republic)
Vermeij GJ (1996) An agenda for invasion biology. Biol Conserv 78:3–9
Vitousek PM, D'Antonio CM, Loope LL et al (1996) Biological invasions as global environmental change. Am Sci 84:468–478
Vitousek PM, Mooney HA, Lubchenco J et al (1997) Human domination of earth's ecosystems. Science 277:494–499

Weiher EV, Werf A, Thompson K et al (1999) Challenging theophrastus: a common core list of plant traits for functional ecology. J Veg Sci 10:609–620
Weinig C (2000) Plasticity versus canalization: population differences in the timing of shade avoidance responses. Evolution 54:441–445
Williamson M (1996) Biological invasions. Chapman and Hall, London, p 244
Williamson M, Fitter A (1996) The varying success of invaders. Ecology 77:1661–1666
Wilson JRU, Dormontt EE, Prentis PJ et al (2009) Something in the way you move: dispersal pathways affect invasion success. Trends Ecol Evol 24:136–144

Chapter 10
Chromosome Engineering for High Precision Crop Improvement

Harinder Kumar Chaudhary, Vineeta Kaila, Shoukat Ahmad Rather, Navdeep Singh Jamwal and Anila Badiyal

Abstract Logarithmically increasing population and steadily changing climatic conditions have created a threatening situation of food insecurity worldwide and pose a challenge to breeders. In view of the narrow genetic background of the cultivated crop species, it has become imperative to broaden their genetic base by introgressing alien genes. However, monitoring the introgression(s) is indispensable for accelerated and high precision crop improvement. This chapter reveals the application of various innovative approaches like haploid inducer genes and chromosome elimination-mediated doubled haploidy breeding in barley, maize, wheat and potato required for the acceleration of breeding endeavours. It also covers the strategic chromosome engineering techniques needed for the alien chromatin introgression in wheat and further monitoring by use of novel molecular cytogenetic tools, including GISH and FISH for the targeted genetic upgradation with high precision.

Keywords Wheat · Barley · Wheat × maize · Wheat × *Imperata cylindrica* · Haploid inducer genes · GISH · FISH

H.K. Chaudhary (✉) · V. Kaila · S.A. Rather · N.S. Jamwal · A. Badiyal
Molecular Cytogenetics and Tissue Culture Lab, Department of Crop Improvement, CSK Himachal Pradesh Agricultural University, Palampur 176062, India
e-mail: ctlhkc@rediffmail.com; mctlhkc@gmail.com

V. Kaila
e-mail: vineetakaila@yahoo.com

S.A. Rather
e-mail: shoukat.pbg@gmail.com

N.S. Jamwal
e-mail: navdeeppau@gmail.com

A. Badiyal
e-mail: sharma.anila2013@gmail.com

V.R. Rajpal et al. (eds.), *Gene Pool Diversity and Crop Improvement*, Sustainable Development and Biodiversity 10,
DOI 10.1007/978-3-319-27096-8_10

10.1 Introduction

The world population is estimated to reach 9.1 billion by the middle of this century, 34 % higher than today, which will mainly comprise of an urban population (nearly 70 %) compared to 49 % today. Thus, the consumers will be far more than the food growers. Use of modern agricultural practices and development of elite wheat and rice cultivars revolutionized food production in the 1960s, resulting in a hike in crop yields. This period was commonly known as the 'Green revolution', but since the past few decades a plateau has been observed. Limitations of land resources, steadily changing climatic conditions and continuously evolving pathogens have further complicated the situation. To feed the logarithmically growing richer population, the scientific community across the world will have to join hands to produce elite crop cultivars with better adaptability in diverse environmental conditions. The focus of breeders is the sustainable genetic enrichment of crops in an accelerated and précised manner to address the upcoming issues of endlessly growing demands for food, feed, fodder, fuel, fibre and pharmaceuticals.

Traditionally, improvement in crop yield was achieved through hybridization among the related species and selection thereafter, however, the introduction of novel biotechnological tools has overcome crossing barriers and has led to the stable introduction of newer genes into the crops. Manipulation at the DNA level, commonly called 'genetic engineering', leading to the creation of improved plants with wider adaptability in diverse circumstances has instigated a new era in crop improvement. Introduction of a whole new genome (synthetic polyploids), addition or replacement of a complete chromosome (addition or substitution lines), or a small segment of alien chromatin (introgression) to the chromatin of crop species is collectively known as 'chromosome engineering', which creates variation at the chromosome level. Segregation in further generations can lead to loss of the newly introduced variations, so this has to be fixed immediately. Doubled haploidy breeding not only results in the attainment of a homozygous population in just one step but the manipulation is also integrated stably in the genetic complement of the crop. DH breeding accompanied with marker assisted selection can result in upgradation of elite cultivars with high precision in a very short time span.

This chapter summarizes haploid induction technology and chromosome engineering in a comprehensive manner, which may be of prospective use for future breeders.

10.2 Development of Haploid Plants via Chromosome Engineering

The sporophytes that contain gametic chromosome numbers are generally referred to as haploids, while the doubled haploids are the genotypes produced as a result of doubling of chromosomes in a haploid organism. Haploids originate either spontaneously or can be induced artificially. The former case is a rare occurrence and is of less practical utility. Since the discovery of spontaneous haploid plants in *Datura* by Belling and Blakeslee in 1922 and induction of haploid plantlets via anther culture by Guha and Maheshwari in 1964, a lot of efforts have been invested to understand the phenomenon deeply and establish protocols for haploid induction in different crop species. Development of haploids and doubled haploids has various applications in crop improvement, the most important being the recovery of complete homozygous lines in just 2 years, thereby saving time and energy. Although haploids were first reported in the 1920s, they were not put to use till the 1950s due to unavailability of the chromosome doubling mechanism. After the discovery of colchicine as chromosome doubling agent the practical utility of haploids in crop improvement increased manifold.

10.2.1 Haploid Inducer Genes

Apart from spontaneous development of haploids as a result of meiotic irregularity, there are certain genetic factors identified that lead to development of haploids of higher frequency than expected via natural induction. Such genetic factors are commonly known as 'haploid induction genes' and the special genetic stocks or plants are referred to as 'inducers'. The haploid induction genes are commonly found and used in barley and maize.

10.2.1.1 Haploid Initiator Gene (*hap*) in Barley

The haploid inducing phenomenon in barley was described by Hagberg and Hagberg in 1980 after isolating a haploid inducing mutant from a mutant stock developed by Gustafsson and his associates (1960). The original mutant was developed in 'Bonus' variety of barley by giving 0.3 % ethyl methane sulphonate (EMS) treatment. The resulting mutant was a *tigrina* type chlorophyll mutant. The chlorophyll mutation was controlled by *tigrina* locus (*tig*) which revealed a balanced inheritance of 1:1 upon selfing of green plants due to the presence of another genetic factor responsible for male sterility (*let*). The cytological analysis of *tigrina* and green plants revealed that some of the plants were haploids and led to the isolation of haploid mutants having only the haploid induction factor (*hap*) and not the chlorophyll deficiency (*tig*) or male sterility (*let*) controlling factors.

The *hap* initiator gene is a mutation that leads to egg parthenogenesis accompanied by normal endosperm development producing normal appearing seed. After pollination, the male sperm nuclei take about 1 h to reach the synergid cells; other nuclei reach polar nuclei and form triploid endosperm. In *hap/hap* and *hap/+* genotypes, the egg cell is not always reached by the other sperm nucleus. Thus, in a *hap/hap* plant, evidently about half of the eggs stay unfertilized, and some of these develop into haploid embryos. The *hap* initiator gene controls survival or abortion of abnormal embryos and endosperms.

The haploid initiator gene is active in both heterozygous and homozygous conditions, though the frequency of haploid plant production can be four times greater in homozygous *hap/hap* plants. Hagberg and Hagberg (1980) also reported that the *hap* mutation was an incomplete dominant mutation since the recovery of haploids in heterozygous (*hap*/+) mutants is nearly four times less than those of homozygous (*hap/hap*) plants. The *hap* gene was reported to influence double fertilization in 10–14 % (haploid induction frequency) of the selfed seeds (Hagberg and Hagberg 1981). The frequency of haploid occurrence is highly influenced by the genotype and also by the environment. Using the *hap* system with marker genes, breeders need not use the embryo culture technique. However, they have to make a large number of crosses—a greater number than is needed using the *Hordeum bulbosum* technique (Hagberg and Hagberg 1987).

A unique advantage of the *hap* method is that the haploid embryos do not require in vitro culture as seed development is relatively normal. However, the disadvantages of the technique are that (a) spontaneously doubled haploids cannot be easily distinguished from hybrid embryos and (b) the technique is limited to genotypes having the *hap* gene.

The technique of doubling the barley haploids is well developed and it should be a fairly simple procedure to isolate series of haploid plants in the F_2 generation and double their chromosome number. The question is whether the frequency of haploids will be sufficiently high to yield enough homozygous breeding lines for the selection programme. However, a further advantage of the *hap/hap* method over the *H. bulbosum* technique lies in the segregation of *hap/hap* genotypes in the F_3 and subsequent generations, with further possibilities for crossing over to occur between linked loci that are still heterozygous. This will allow for new recombinations to occur in each generation instead of just the F_1 generation, as is the case for the *H. bulbosum* technique. This is also under test at present.

10.2.1.2 Maize Haploid *Inducer* Line

Haploid induction in maize, particularly using in vivo mode of haploid induction, became a widely used tool in maize research and breeding during the past decade. The technology is used to develop doubled haploid (DH) mapping populations, and analyse linkage disequilibrium and haplotype/trait associations. The first haploid maize plant was reported by Randolph as early as 1932; however, it was only more than a decade later that Chase (1947, 1951) reported the spontaneous haploid induction rate in maize of 0.1 % and suggested that haploids could be used for line development in hybrid breeding. During initial works on induction of haploids in maize, the low recovery of spontaneous haploid induction posed a major hurdle in utilization of the available technique in development of maize haploids and their utilization in maize improvement programmes. However, looking into the potential of haploid plants in genetic studies as well as maize improvement, a number of researchers worked on identification of efficient haploid inducing stocks. With more divergent germplasm involved and more refinement in the haploid generation protocols, the haploid recovery has been gradually yet significantly enhanced to the tune of 10 % (Chalyk 1999; Sarkar et al. 1994; Shatskaya et al. 1994; Zhang et al. 2008).

The in vivo haploid induction in maize can be categorized into two classes, maternal and paternal haploids. The genomes of maternal haploids originate exclusively from the seed parent plant where the pollen parent is the haploid inducer (Coe 1959). In the other case where paternal haploids are developed, the pollinator serves as the genome donor and the female parent acts as the inducer (Kermicle 1969). In the former case, i.e. induction of maternal haploids a male inducer line derived from Stock 6 line of maize is used, whereas for the latter, i.e. paternal haploids, the inducer line having *ig1* mutant is used as the female inducer.

10.2.2 Haploid Induction Using Indeterminate Gametophyte Mutant

Utilization of *ig1* mutant gene gives rise to indeterminate gametophyte as well as increases the frequency of haploid progenies when the line having this gene is crossed with any other normal maize line. Homozygous *ig1* mutants show several embryological abnormalities including egg cells without a nucleus. Post double fertilization of such eggs with normal male gametes, the cells give rise to haploid embryos. As only the pollen parent contributes towards the genetic composition of the haploid plants, the haploids are termed as paternal haploids. This system also provides a reliable system for conversion of an inbred line to its cytoplasmic male sterile form using *ig1/ig1* genetic stock; however, the low frequency of haploid recovery, as well as changes in the constitution of cytoplasm from the donor genotype, renders this system not so attractive as in vivo haploid generation in maize (Schneerman et al. 2000).

10.2.2.1 Induction of Maternal Haploids

Maternal haploids carry both the cytoplasmic and nuclear genome of the maternal (donor) parent, whereas the nuclear genome of the inducer line is gradually eliminated. Over a period of time the haploid induction efficiency of maternal haploids has been enhanced by identifying more efficient inducers from Stock 6, MHI, WS14, HZI1, etc. (Coe 1959; Lashermes and Beckert 1988). For easy identification of maternal haploids from diploids, some colour marker genes have been incorporated into male inducer lines for purple leaf, sheath and plants or purple endosperm crown and purple plumule colour to facilitate identification of haploid seeds at the ear level.

10.3 Mechanism of In Vivo Haploid Induction

Although the mechanism of induction of maternal haploids is poorly understood, some hypotheses have been put forward. Wedzony et al. (2004) proposed that one of the two sperm cells provided by the inducer is defective but still fuses with the egg cell. During subsequent cell divisions, the inducer chromosomes get gradually isolated from the maternal chromosomes and get gradually degenerated from the primordial cells. The other well functional sperm cells fuse with the polar nuclei and give rise to regular endosperm. Another hypothesis supported by Chaylk et al. (2003) suggests that one of the two sperm cells is not able to fuse with the egg cell but instead triggers haploid embryogenesis while the secondary fertilization goes normal. If the functional sperm fuses with the egg cell and the defective sperm fuses with the secondary nuclei, this will result in kernel abortion.

10.4 Chromosome Elimination Approaches of Haploid Induction

Elimination of chromosomes is a common phenomenon in wide hybrids, ranging from loss of one or two alien chromosomes to elimination of whole chromosome complement from one parent. Such elimination leads to the development of haploids which are then given chemical treatment to double the chromosome number. The doubled haploidy (DH) breeding following chromosome elimination approach has been used in crops such as barley, wheat, oats, triticale, rye and potato, where the other techniques of haploid induction like anther, pollen/microspore culture and ovule culture were not so efficient. In order to apply the DH systems successfully to a breeding programme, a technique should fulfill the following three criteria: (1) DH line(s) should be produced efficiently from all the genotypes, (2) DH should represent a random sample of the parental gametes and (3) DH should be genetically normal and stable (Snape et al. 1986).

Several explanations have been proposed to account for uniparental chromosome elimination, viz. difference in timing of essential mitotic processes attributable to asynchronous cell cycling (Gupta 1969) and asynchrony in nucleoprotein synthesis leading to loss of the most retarded chromosomes (Bennett et al. 1976; Laurie and Bennett 1989). Other hypotheses put forward are the formation of multipolar spindles (Subrahmanyam and Kasha 1973), spatial separation of genomes during interphase (Finch 1983; Linde-Laursen and von Bothmer 1999) and genome elimination by nuclear extrusions (Gernand et al. 2005, 2006). In addition, degradation of alien chromosomes by host-specific nucleases (Davies 1974), uniparental nondisjunction of anaphase chromosomes (Ishii et al. 2010) and parent-specific inactivation of centromeres (Finch 1983; Jin et al. 2004; Mochida et al. 2004) have been suggested. The actual cellular mechanism involved in the process of uniparental chromosome elimination remains poorly understood.

Various chromosome elimination-mediated approaches of doubled haploid production in various crops are given below.

10.4.1 Haploid Induction in Barley (*Hordeum vulgare*)

The first method in cereals based on wide crossing following chromosome elimination was *H. vulgare* × *H. bulbosum*, commonly known as 'bulbosum method' (Stephan 1969; Kasha and Kao 1970; Lange 1971). During early embryogenesis, chromosomes of the wild relative are preferentially eliminated from the cells of developing embryos leading to the formation of haploid embryos. The endosperm is frequently formed, but its development is usually disturbed, hence at 12–14 days of pollination, the embryos are excised from developing caryopsis and are cultured in vitro. The bulbosum method was the first haploid induction method to produce a large number of haploids across most genotypes and this method quickly entered into breeding programmes. Kasha and Kao (1970) presented evidence to show that these haploids are not caused by parthenogenesis but by the elimination of *H. bulbosum* chromosomes. This elimination is under genetic control (Ho and Kasha 1975). Haploids of *H. vulgare* are also obtained when it is used as a male parent in the wide hybridization programme. This method represents a considerably advanced approach in the production of barley haploids and it has a number of advantages over the anther culture. In particular, haploids can be produced from any cultivar of barley, whereas in the anther culture, success is dependent on the genotype. The parent-specific inactivation of centromeres during the mitosis-dependent process of chromosome elimination in *H. vulgare* × *H. bulbosum* hybrids was confirmed by Sanie et al. (2011). They reported that the loss of centromeric histone H3 (CENH3) from centromeres precedes uniparental chromosome elimination in interspecific barley hybrids. Gernand et al. (2006) studied the mechanism underlying selective elimination of the paternal chromosomes during the development of *H. vulgare* × *H. bulbosum* hybrid embryos that is restricted to an early stage of development. In almost all embryos, most of the *H. bulbosum* chromatin undergoes a fast rate of elimination within 9 days after pollination.

10.4.2 Haploid Induction in Wheat (*Triticum aestivum*)

10.4.2.1 Wheat × *Hordeum bulbosum*

Haploid wheat plantlets were obtained when 'Chinese Spring' variety of *Triticum aestivum* ($2n = 6x = 42$) was crossed with *H. bulbosum* ($2n = 2x = 14$ or $4x = 28$). This happened as a result of elimination of *H. bulbosum* chromosomes from the interspecific hybrid during its early embryogenesis (Barclay 1975; Zenketler and Straub 1979). However, this method was not successful with other wheat varieties as in the anther culture due to the effect of dominant crossability inhibitor alleles *Kr1, Kr2, Kr3* and *Kr4* located on 5B, 5A, 5D and 1A chromosome arms (Riley and Chapman 1958; Krolow 1970; Sitch et al. 1985; Zheng et al. 1992), which prevent the entry of *H. bulbosum* pollen tube into the ovary of wheat. The 'Chinese Spring' variety of bread wheat possesses recessive crossability alleles, that is, *Kr1* and *Kr2*. Jalani and Moss (1980) reported that crossability genes have little effect on pollen germination and on the time taken for the pollen tubes to reach the micropyle. The number of pollen tubes reaching the micropyle is, however, affected by the *Kr* genes, as high crossable genotypes have more pollen tubes than low crossable ones. Factors affecting crossability between 'Chinese Spring' wheat and *H. bulbosum* were also found on chromosomes 3A, 3B and 3D (Miller et al. 1983). This system was hence useful to a limited extent due to the sensitivity of the *H. bulbosum* pollen to the crossability inhibitor genes.

10.4.2.2 Wheat × Maize

Zenkteler and Nitzsche (1984) reported for the first time that embryos were frequently formed when hexaploid wheat was pollinated with maize. Later, their results were confirmed by Laurie and Bennett (1986). They cytologically demonstrated that the maize pollen normally germinated and grew into the wheat embryo sac where the wheat egg was fertilized by the maize pollen. A hybrid zygote with 21 wheat chromosomes and 10 maize chromosomes was produced (Laurie and Bennett 1988). The hybrid zygotes were karyotypically unstable and the maize chromosomes failed to move towards the spindle poles during cell division. Possibly, their centromeres failed to attach to the spindle microtubules due to progressive loss of centromere activity. Resultantly, maize chromosomes were eliminated after three to four mitotic cell divisions forming wheat haploid embryo with $n = 21$ chromosomes (Laurie and Bennett 1989). Some earlier studies show that the wheat × maize system is more efficient in embryo formation compared to other techniques. For haploid embryo production a system of wheat × maize crossing is widely used due to higher production of haploid embryos compared to other grass species pollination systems (Inagaki and Tahir 1991; Kisana et al. 1993; Inagaki and Mujeeb-Kazi 1995). This system is fast, economically viable, easy for application and more efficient than others due to its low level of genotype specificity (Cherkaoui et al. 2000). The maize chromosome elimination system in wheat is insensitive to crossability inhibitor genes (Laurie and Bennett 1989) and enables the production of a large number of haploids from any genotype including those recalcitrant to androgenesis (Inagaki et al. 1998; David et al. 1999; Cherkaoui et al. 2000; Chaudhary et al. 2002; Singh et al. 2004; Pratap et al. 2006). Several other investigations of haploid wheat production through wide crossing have since been reported (Laurie and Bennett 1989; Laurie and Reymondie 1991; Matzk and Mahn 1994; Suenaga 1994; Morshedi and Darvey 1995). It appears that a wide range of wheat and maize genotypes can be used to produce haploid wheats, although there is evidence to suggest that the efficiency of production is variable (Suenaga 1994).

Haploid production efficiency is affected by the proportion of pollinated florets that develop haploid embryos. Yields of haploid embryos have been reported to be as high as 53 % (Morshedi and Darvey 1995) and as low as 1 % (Suenaga and Nakajima 1989), depending on a wide range of variables. Factors that affect the yield of haploid embryos include genotypic differences between individual wheat and maize lines (lnagaki and Tahir 1990; Suenaga 1994; Chaudhary et al. 2002; Sharma et al. 2005; Pratap and Chaudhary 2007; Dhiman et al. 2012), the timing and use of exogenous growth substances to stimulate ovule development (Suenaga and Nakajima 1989) and environmental factors (especially temperature) during and after pollination. Laurie and Bennett (1989) reported that all maize chromosomes were lost during the first three cell division cycles in most embryos. All embryos with four or more cells had micronuclei, showing that embryo development was dependent on fertilization. The only primary endosperm metaphase obtained in the experiment had 42 wheat and 10 maize chromosomes, and the presence of micronuclei in most developing endosperms showed that at least 85 % were of hybrid origin.

10.4.2.3 Wheat × *Imperata cylindrica*

Considering the above chromosome elimination-mediated haploid induction systems, no alternative pollen source was reported to overcome the problems of wheat × maize system, viz. non-synchronization of flowering with wheat naturally and poor performance in producing haploids from triticale × wheat and wheat × rye derivatives. These constraints made it imperative to search for some other pollen source. Professor H.K. Chaudhary and his associates (2005) pioneered in inventing wheat × *Imperata cylindrica*, a highly efficient doubled haploidy breeding system in wheat of the existing systems.

Among all the Gramineae genera, viz. *Zea mays, Sorghum bicolor, Pennisetum americanum, Setaria italica, Festuca arundinacea, Imperata cylindrica, Cynodon dactylon, Lolium temulentum* and *Phalaris minor* tested for haploid plant production, and *I. cylindrica* produced more embryos and haploids over others (Chaudhary et al. 2005). Cogon grass (*I. cylindrica*) (Fig. 10.1) is a wild weedy perennial grass ($2n = 2x = 20$) that does not require repeated sowing and its flowering coincides well with that of wheat and triticale under natural conditions. Furthermore, it is available under natural conditions in almost all parts of the world wherever wheat is cultivated. The *I. cylindrica*-mediated chromosome elimination approach (Fig. 10.2) of doubled-haploidy breeding is a non-specific genotype for hybridization with any variety of wheat (Chaudhary et al. 2005), triticale or their derivatives (Pratap et al. 2005).

Imperata cylindrica has been reported to perform significantly better than maize for all the haploid induction parameters in wheat and triticale and their derivatives (Chaudhary 2008a, b, 2012, 2013; Jeberson et al. 2012; Kishore et al. 2011; Chaudhary et al. 2013a, b; Badiyal et al. 2014). Cytological investigation of the wheat × *I. cylindrica* chromosome elimination system has shown that there is no

Fig. 10.1 Spike of a wildly growing plant of *Imperata cylindrica*, an efficient pollen source for haploid induction in wheat

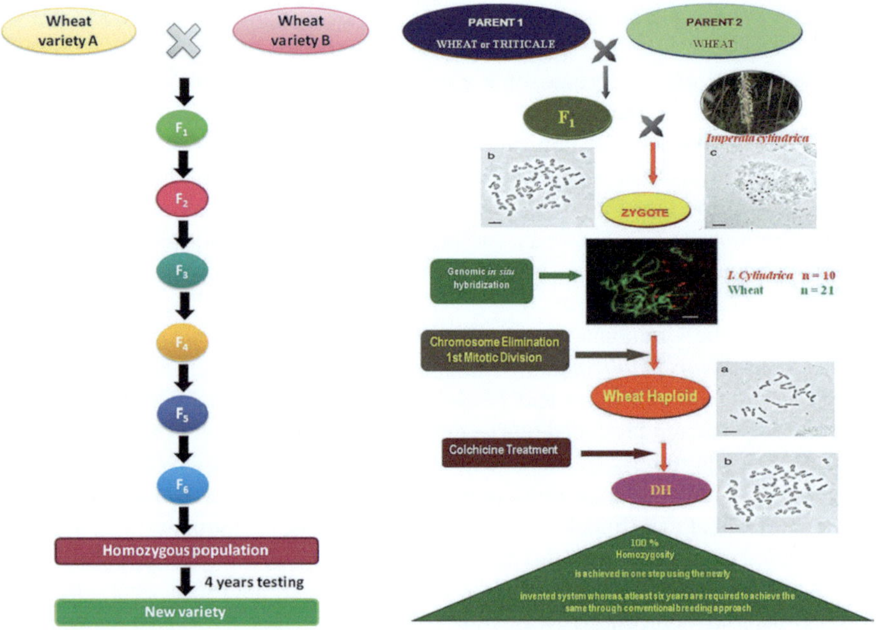

Fig. 10.2 Flowchart exhibiting comparison of the conventional (*left flank*) and DH breeding approaches (*right flank*) (Chaudhary et al. 2005; Komeda et al. 2007)

endosperm formation and the elimination of chromosomes of *I. cylindrica* takes place in the first zygotic division during the process of seed development, thus allowing the production of embryo-carrying pseudoseeds (Komeda et al. 2007;

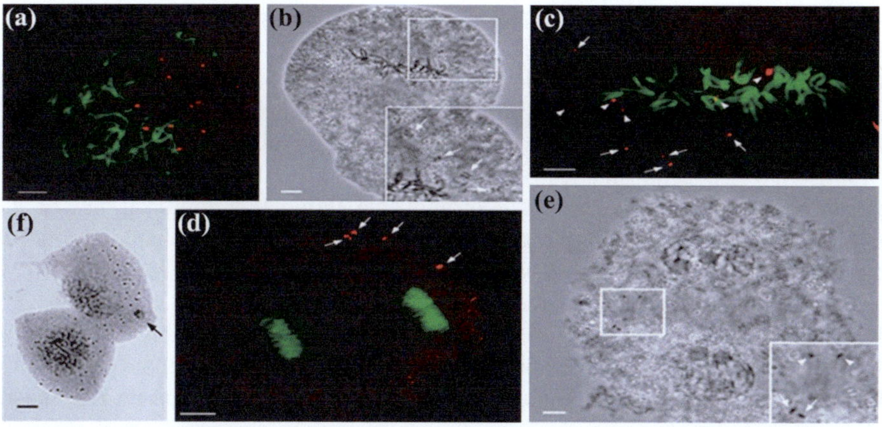

Fig. 10.3 Molecular cytological evidence of sequential elimination of alien chromosomes in wheat × *I. cylindrica* hybrids during first zygotic mitosis (*Green* Wheat chromosomes, *Red I. cylindrica* chromosomes). **a** Interspecific hybrid. **b**, **c** Aberrant movement of *I. cylindrica* chromosomes. **d** Anaphase cell showing elimination of *I. cylindrica* chromosomes. **e** Unsquashed zygote at telophase showing extruding *I. cylindrica* chromosomes. **f** Interphase daughter cells after 3–4 days of pollination showing extruded *I. cylindrica* micronucleus (Komeda et al. 2007)

Fig. 10.3). Recently, Tayeng et al. (2012) reported that in vivo application of colchicine (2,000 ppm) enhances the doubled-haploid production efficiency in wheat × *I. cylindrica*-mediated chromosome elimination approach of doubled haploidy breeding. The haploid chromosome set of wheat (n = 21) is obtained after wheat × *I. cylindrica* hybridization. Similar to wheat × maize system, the mean response of wheat and *I. cylindrica* to haploid induction varies from genotype to genotype (Rather et al. 2014). The morphological marker, that is, absence of endosperm in haploid embryo-carrying seeds developed from wheat × *I. cylindrica* hybridization, can be used efficiently to exploit the asynchronous behaviour of anthesis within wheat spikes (Chaudhary et al. 2013a, b) for undertaking this wide hybridization without emasculation. This endeavour has saved considerable time and energy required otherwise for emasculation in wheat × *I. cylindrica* hybridization.

10.4.3 *Solanum tuberosum* × *S. phureja*

Doubled haploids can be produced from tetraploid genotypes of *S. tuberosum* (cultivated potato) by pollination with the diploid potato species, *S. phureja* (Mendiburu et al. 1974; De Maine 2003). In about 0.5 % of pollinated ovules, both male sperm cells of *S. phureja* take part in the formation of functional endosperm. The best pollinator lines of *S. phureja* were bred for a dominant purple spot embryo marker; thus, seeds containing haploid embryos can be easily distinguished from hybrid *S. tuberosum* × *S. phureja* seeds. Methods of more effective chromosome number duplication were developed more recently and production of potato can now be obtained by androgenetic methods with better efficiency (Jacobsen and Ramanna 1994; Rokka et al. 1996; Rokka 2003). Moreover, androgenesis is applicable to a wider range of *Solanum* species in comparison to crosses with *S. phureja* (Jacobsen and Ramanna 1994; Aziz et al. 1999; Rokka 2003). Montelongo-Escobedo and Rowe (1969) reported that the superior pollinator in potato haploidy breeding following chromosome elimination approach may be the one that produces a high frequency of restitution sperm nuclei. Dihaploid potatoes can be used for breeding purposes, including alien germplasm introgression or selection at the diploid level, but such plants are not homozygous. Haploids have a significant role in potato breeding programmes, as they enable interspecific hybridization which would not be otherwise possible due to differences in ploidy level and endosperm balance numbers. The gene pool of potato can be broadened, and certain valuable traits, such as disease resistance characters from the wild solanaceous species, can be more efficiently introgressed into cultivated potato (Rokka 2009).

10.5 Centromere Sequences

Centromeres are the chromosomal loci that attach to spindle microtubules to mediate faithful inheritance of the genome during cell division. These are epigenetically specified by incorporation of CENH3, a histone H3 variant that replaces conventional H3 in centromeric nucleosomes16. Sanei et al. (2011) isolated CENH3-1, an embryo-lethal null mutant in *Arabidopsis thaliana* that allows us to completely replace native CENH3 with modified variants. CENH3-1 plants complemented by transgenic green fluorescent protein-tagged CENH3 (GFP–CENH3) have a wild-type phenotype. Komeda et al. (2007) reported that the loss of *I. cylindrica* chromosomes is due to a lack of kinetochore activity of the chromosomes as a result of which the chromosomes become laggards and get eliminated from the hybrid cell leading to formation of wheat haploids. In oat × maize hybrids, maize centromeres were reported to have lost the transcriptional activity in genetic background of oat. Synthesis of oat CENH3 proteins and their incorporation in one or a few maize kinetochores led to the normal segregation of maize chromosomes. Whereas the chromosomes without CEN proteins incorporated in their kinetochores fail to attach with the spindle fibres and hence became laggards and eventually got eliminated (Jin et al. 2004). Maruthachalam and Chan (2010) reported that CENH3 null mutants expressing altered CENH3 proteins can be successfully exploited for induction of haploids in *A. thaliana*. Similarly, the loss of centromeric histone H3 (CENH3) from centromeres of *H. bulbosum* was revealed in elimination of its chromosomes in *H. vulgare* × *H. bulbosum* hybrids. They revealed that centromere inactivity of *H. bulbosum* chromosomes was due to centromeric loss of CENH3 protein rather than uniparental silencing of CENH3 genes to trigger the mitosis-dependent process of uniparental chromosome elimination. They also revealed that the diploid barley species encode two CENH3 variants, the proteins of which are intermingled within centromeres throughout mitosis and meiosis (Sanei et al. 2011).

10.6 Development of Recombinant, Addition and Substitution Lines

Wild relatives and related species are important resources for broadening the genetic variability of crop plants. The transfer of such desirable traits from wild species is possible at three levels:

(i) At the level of whole genome for the production of amphidiploids (*Raphanobrassica* and *Triticale*)
(ii) At the level of individual whole chromosome for the production of alien addition and substitution lines
(iii) At the level of chromosome segment as done in the production of intercalary and terminal translocations

As crop plants are often bred for specific quality attributes, such as high yield potential and plant type suited to specific agronomic practices, the amount of alien genetic material introduced into an elite cultivar has to be carefully controlled. The commercially viable and useful modes of enhancing genetic diversity using manipulations at whole chromosome and chromosome segment level are discussed under the following heads:

10.6.1 Alien addition and substitution lines
10.6.2 Translocation genetic stocks/lines.

10.6.1 Production of Alien Addition or Substitution Lines

The transfer of a single or pair of chromosome(s) from one species to another can be useful for the introduction of desirable traits such as resistance against biotic and abiotic stresses. For this, aneuploid plants are produced that contain an extra single chromosome or an extra chromosome pair from a donor plant, called monosomic or disomic addition lines, respectively. Such addition lines can be created by hybridization between donor and recipient plant lines followed by repeated back-crossing with the recipient plant line. Donor and recipient parents are usually from different species.

After achieving the objective of producing alien addition lines in different crops, their evaluation suggested that they were invariably unstable and at meiosis they exhibited a higher frequency of univalents than normal. In view of this problem with alien addition lines and due to their undesirable effects on the phenotype, the addition lines could not be considered suitable for commercial use. It was argued that if chromosome number is maintained at normal euploid level by substituting a pair of alien chromosomes for a pair of normal recipient crop chromosomes, the product may be more desirable and, therefore, acceptable for cultivation.

One of the major achievements of alien addition and substitution lines has been the wheat–rye introgression in which individual whole chromosome alien addition lines were introduced into wheat, utilizing *Secale cereale* as donor species. An amphidiploid ($2n = 56$) between wheat and rye is first produced following the normal method of crossing the two species followed by doubling the chromosome number in the F_1 hybrid. The amphidiploid is backcrossed to wheat giving a heptaploid with $21^{II} + 7^{I}$, where the bivalents belong to wheat and univalents belong to rye. On selfing of these heptaploids ($2n = 7x$), monosomic ($21^{II} + 1^{I}$ rye) and disomic ($21^{II} + 1^{II}$ rye) are obtained. Whenever monosomic additions are available, these may be selfed to get the disomic addition lines.

The wheat–rye substitution lines are developed by crossing the alien addition lines as male parent with the monosomic lines of wheat. F_1 plants with $20^{II} + 2^{I}$ were selected, which either on selfing or backcrossing to the same addition line gave rise to monosomic or disomic substitution lines.

Other alien addition lines utilizing whole chromosomes from *Aegilops* or barley to wheat were developed by Islam et al. 1981. The success in wheat–barley cross was achieved for the first time by Kruse (1973) followed by Islam et al. (1975). Viable hybrid plants could be easily produced when barley was used as a female parent and when the developing embryos were treated with gibberallic acid and transferred to an artificial culture medium. The success was 5.8 % over all the crosses made by Islam et al. (1975), although 15.9 % success could be achieved when Chinese Spring wheat was crossed with Betzes barley. Substitution of one chromosome or a complete pair can also be useful, e.g. the introduction of disease resistance by the exchange of chromosome 1B of wheat for 1R of rye (Khush 1973). As discussed earlier, addition lines often show reproductive instability and are mostly not of direct use in breeding. However, they can be used for the localization of genes for particular traits to specific chromosomes. For this purpose, other variants can also be used, namely monosomic or trisomic lines within crop species. Such lines can be made by applying the mitotic inhibitor colchicine, radiation or selection in the progeny of triploid plants (Khush 1973). Sets of such lines have been compiled that together represent the complete haploid chromosome complement of a crop species. Such sets are for instance known for oat, barley, wheat, rice, sorghum, cotton, asparagus, pepper, tomato and tobacco.

Resistance to common root rot and black point, caused by *Cochliobolus sativus*, was evaluated in alien chromosome substitution and addition lines of the cultivars ‘Cadet’ and ‘Rescue’ (Conner et al. 1993). Substitution of chromosome 5B in ‘Rescue’ with 5Ag from *Agropyron elongatum* decreased root rot susceptibility to a level intermediate between that in the susceptible ‘Rescue’ and the resistant ‘Cadet’. The substitution of ‘Rescue’ chromosome 5A or 5D with 5Ag, or the addition of 5Ag to ‘Rescue’ complement had no consistent effect on root rot susceptibility. The root rot resistance of ‘Cadet’ was unaffected by substitution of chromosomes 5A, 5B, or 5D with 5Ag, or the addition of 5Ag.

Solanum brevidens is a wild diploid potato species possessing high level of resistance to several major potato diseases. Dong et al. (2005) developed fertile somatic hybrids between *S. brevidens* and the cultivated potato (*S. tuberosum*) in order to introgress disease resistance from this wild species into potato. A series of backcross progenies were developed from a hexaploid somatic hybrid A 206. Using a combination of *S. brevidens*-specific randomly amplified polymorphic DNA (RAPD) markers and a sequential genomic in situ hybridization (GISH) and fluorescence in situ hybridization (FISH) technique, they identified all 12 *S. brevidens* chromosomes in the backcross progenies. Seven potato *S. brevidens* monosomic chromosome addition lines (chromosomes 1, 3, 4, 5, 8, 9 and 10) and one monosomic substitution line (chromosome 6) were identified, and the remaining four *S. brevidens* chromosomes (2, 7, 11, and 12) were included in two other lines. These chromosomal addition/substitution stocks provide valuable tools for potato cytogenetic research, and can be used to introgress disease resistance from *S. brevidens* into potato.

The identification of addition or substitution lines for different alien chromosomes is achieved by any one of the following techniques:

(i) Morphology: Addition of different chromosomes sometimes leads to modification of morphology in different directions so that the different addition lines can be distinguished from each other. These addition lines also provide an opportunity to study the effect of individual alien chromosomes in the uniform wheat background.
(ii) Karyotype: The alien addition lines can also be identified on the basis of the chromosome size and morphology.
(iii) Intercrossing: Intercrossing between alien addition lines may help to find out if two plants have addition for the same or different chromosomes through the study of meiosis in F_1 hybrid, which should exhibit $21^{II} + 2^{I}$ if they differ.

10.7 Translocation Breeding

Reduced chromosome pairing and lack of recombination is an important problem in the production of interspecific hybrids. Pairing of chromosomes that normally show a low extent of pairing, that is, between the homoeologous chromosomes derived from the different parental species, can be induced by specific mutations. Recombination frequencies can also be increased artificially by use of various chemical agents, physical stress such as a temperature shock or by UV irradiation. In breeding with complex allopolyploids, there is also the possibility of suppression of genes favouring homologous pairing in order to induce exchange of segments between homoeologous chromosomes. While irradiation produces random interchanges even between non-homoeologous chromosomes, recombination produces interchanges between homeologous chromosome segments. On the other hand, formation of multivalents during meiosis, particularly in newly created autopolyploids, is preferably avoided in normal propagation because of the consequent reproductive instabilities. This may be achieved by selection for improved fertility of progeny and/or induction of chromosomal changes leading to a 'diploid' (or 'allopolyploid') behaviour of the polyploid (diploidization), so that, in practice, mainly bivalents are formed during meiosis. Sometimes, supernumerary chromosomes (so-called B chromosomes, versus the normal A chromosomes complement) can also have an effect of favouring recombination between homologous chromosomes over that between closely related homeologous ones.

10.7.1 Translocations Using Irradiation

The first useful transfer by irradiation involved a segment from *Aegilops umbellulata* chromosome 6U carrying resistance to wheat leaf rust (*Lr 9*) to chromosome arm 6BL of wheat (Sears 1956). To achieve this goal, as cross of hexaploid wheat and *Ae. umbellulata* was not successful, an amphidiploid (*T. Dicoccoides* × *Ae. umbellulata*) with $2n = 21^{II}$ was used as a bridge. The amphidiploid was crossed with 'Chinese Spring' variety of wheat followed by two backcrosses with Chinese Spring and selection for leaf rust. This gave a plant with $2n = 43$, in which extra chromosome was from *Ae. umbellulata* and carried the gene for rust resistance associated with some undesirable characters. This plant was exposed to a heavy dose of X-rays before meiosis. Pollen from this irradiated plant was used for pollinating Chinese Spring wheat. Only balanced pollen with $n = 21$ would function and may or may not carry an intercalary translocation. Among 6091 offsprings, 132 were resistant, of which 40 had translocations and only one of them was intercalary. Plants for this intercalary translocation resembled Chinese Spring, except for rust resistance and slightly late maturity. The intercalary translocation was later found to be on a chromosome that was homeologous to the alien chromosome involved.

Following irradiation of seeds and plants of an alien substitution line carrying *Agropyron elongatum* chromosome, a segment carrying leaf rust resistance was transferred to the bread wheat variety 'Thatcher'. The stock was called *Translocation-4* (Sharma and Knott 1966).

In general, translocations can also be induced by irradiation of a trisomic addition line (with one extra homoeologous chromosome from another species) and by subsequent recovery of plants with chromosome segments of the added chromosome incorporated into their genome.

10.7.2 Translocations Through Recombination

10.7.2.1 Recombination via Manipulation of Diploidizing System

As mentioned above, translocations can also be obtained through recombination among the homeologous chromosomes which is facilitated by the suppression of genes that otherwise facilitate bivalent formation instead of multivalent formation. This type of recombination can be best explained by illustrating wheat genome. In wheat, there are various approaches as given below for manipulating the regulatory mechanism of homeologous recombination:

- Use of high pairing *Ph1* deficient stocks (*ph1b* in bread wheat and *ph1c* in durum wheat);
- Use of 5B-deficient stocks such as 5D(5B) substitution lines;
- Hybridization with alien species, which epistatically control the activity of *Ph* locus.

10.7.2.2 Use of High Pairing *Ph1* Deficient Stocks

Sears (1982) developed first mutant *Ph1b* at *Ph1* locus and later it was found to be deficient for dominant *Ph1* locus at intercalary position. Similarly, a mutant *Ph2b* was developed which was known to have *Ph1b*-like effect in wheat-alien bridge species (Wall et al. 1971; Sears 1977). The mutation in the *Ph1* locus strongly promotes homeologous chromosome pairing within wheat and between wheat and its relatives (Yu et al. 1995). Genetic transfers of alien genes from *Agropyron* and *Aegilops* spp. into wheat were achieved through use of *ph1b* stocks (Yu et al. 1995). Dundas et al. (2007) produced bread wheat introgression lines containing *SrR* (*Secale cereale*), *Sr26* (*Thinopyrum ponticum*), *Sr32* and *Sr39* (*Ae. speltoides*) and *Sr40* (*Triticum timopheevii*) using *ph1b* mutants. Similarly, *ph1c* mutant stocks in durum were used to introgress genes from *Aegilops* species such as *Sr39* from *Ae. speltoides* (Mago et al. 2009) as well as distant relatives such as *Secale cereale* (Giorgi and Barbera 1981) and *Th. bessarabicum* (King and Laurie 1993). Several other works also show the effectiveness of the recessive gene *ph1b* in inducing homeologous pairing of wheat with *Secale*, *Agropyron* and *Agrotricum* hybrids (Dhaliwal et al. 1977; Naranjo et al. 1988; Wu et al. 1989; Ahmad and Comeau 1991).

Mutants at *Ph* locus on 5BL chromosome arm of wheat were also isolated and used for facilitating homeologous recombination. To achieve this, special stocks (triple monosomics) with single doses each of 5B, the alien chromosome and a wheat homoeologue were prepared. This can be done by crossing monosomic 5B with an alien substitution line (20^{II} + 1^{II} alien). This method was successfully utilized for homeologous recombination among wheat—*Agropyron* and wheat—rye chromosomes (Islam and Shepherd 1991).

10.7.2.3 Use of 5B-Deficient Stocks

As the dominant *Ph1* locus is situated at the long arm of 5B chromosome of wheat, 5B deficient stocks such as Langdon 5D (5B) disomic substitution lines can be utilized to promote intergenomic chromosome pairing leading to effective interspecific and intergeneric gene transfers (Sears 1981; Jauhar et al. 2009). Using this method, the genes for scab resistance were transferred successfully from diploid wheat grass *Lophopyrum elongatum* to durum wheat (Jauhar et al. 2009). The same strategy was employed for transferring genes from *Th. bessarabicum* and *Th. curvifolium* into durum.

Nullisomy for 5B was first used by Riley (1966) for transfer of segment from *Ae. bicornis,* but a more efficient method was suggested by Sears (1972), where a nulli-5B-tetra-5D line was crossed with an alien substitution line to produce plants which were nulli-5B-tri-5D and double monosomic for alien and wheat homeologues ($18^{I} + 1^{III} + 2^{I}$). The absence of 5B allowed the alien chromosome to pair with its wheat homeologue and the progeny obtained by pollinating these plants by euploid wheat was tested for rust resistance. Using this method, Sears (1972, 1973) was successful in transferring segments from 3Ag (carrying *Lr 24* gene) and 7Ag (carrying *Lr 19* gene) to 3D and 7D, respectively.

10.7.2.4 Suppression of *Ph1* Activity Through Epistasis

Homoeologous pairing can also be enhanced among wheat and *Ae. speltoides* by suppressing the action of the *Ph1* gene due to the presence of the *Ph1* suppressor genes *Su1-Ph1* and *Su2-Ph1* (Dvorak et al. 2006). Leaf rust and stripe rust resistance genes from *Ae. umbellulata* (Chhuneja et al. 2008) and from *Ae. triuncialis* and *Ae. geniculata* (Aghaee-Sarbarzeh et al. 2002) were successfully transferred to wheat following this approach. A number of genes for disease resistance were transferred from wild relatives such as *Ae. ovata*, *Dasypyrum villosum* (Blanco et al. 1988), and *Thinopyrum curvifolium* (Jauhar and Almouslem 1998) to durum wheat utilizing this strategy. However, the existence of such genes is limited to some genotypes and the efficiency in inducing pairing between homeologous does not appear to be comparative to 5B deficient or *ph1b/ph1c* mutants (Jauhar and Chhibbar 1999). Thus this approach is the least exploited for alien gene transfers into wheat.

For a stable introgression, mere identification of the transferred gene/genes is not enough. The transferred genetic content should be inherited stably in the progeny to be manifested as the desired trait. Hence, the progeny is critically screened for the trait governed by the candidate gene. To identify the extent of homeologous recombination leading to successful alien introgressions in wheat, several techniques have been utilized. Earlier, chiasmata measurement and C-banding techniques were used by (Sears 1952, 1973; Gill and Chen 1987) to measure the homeologous recombination. But nowadays, the use of modern cytological procedures such as GISH and FISH (Schwarzacher et al. 1989; Yamamoto and Mukai 1989; Jenczewski et al. 2003; Ji and Chetelat 2003; Khrustaleva et al. 2005;

Schwarzacher et al. 2011) and novel biotechnological tools such as molecular markers have revolutionized the assessment of introgression breeding (Rogowsky et al. 1993; Lukaszewski et al. 2004). Works have identified wheat–rye translocations such as 1BL.1RS (Fig. 10.4) after successful wide hybridization among triticale and wheat genotypes (Chaudhary et al. 2004; Jeberson 2010). Simultaneously, GISH can also be used to identify the chromosomes of different species in an intergeneric cross such as wheat × *Imperata cylindrica* the wheat genetic background (Chaudhary et al. 2013a, b). The crop genome-specific DNA markers available today can easily detect the transferred genes into the crop background and, thus, they offer highly amenable and cost-effective tools for such assessments (Peng and Lapitan 2005).

Homoeologous recombination for transfer of alien segments can also be achieved by using certain strains of *Ae. speltoides* and *Ae. mutica* to suppress the effect of *Ph* in the appropriate hybrids. If the genes are to be transferred from the same species as that used for suppression, the transfer can be achieved by making and backcrossing the hybrids with wheat. In other cases, crosses are first made with the alien species and suppression of 5B system is achieved later by incorporating suitable genotype of *Ae. speltoides* or *Ae. mutica.* Riley et al. (1968) used this method for transfer of stripe rust resistance from *Ae. comosa* to wheat.

10.7.2.5 Recombination Without Manipulation of Diploidizing System

In some crops such as groundnut, a diploidizing genetic system is either absent or weak. Thus, in hybrids with related diploid species, occasional allo-syndetic pairing occurs and recombination is achieved through production of F_1 hybrids followed by backcrossing accompanied with selection for desirable traits. Alternatively, F_1 hybrids may be selfed to allow further recombination and stable tetraploids may be reconstituted through backcrossing. Using this approach, alien gene introgressions for different traits have been carried out in different crops such as groundnut (Singh et al. 1991), sugar beet (Nakamura and Tsuchiya 1982), etc.

10.8 Homologous Recombination for Chromosome Engineering

The recombinant lines developed after the successful introgression of alien chromatin into the wheat background, through homoeologous recombinations, are mainly agronomically inferior to elite cultivars. Hence, there is a need for transferring the desired genes (without linkage drag) into the elite background through homologous recombination using recurrent backcrossing (Valkoun 2001). Homology among recombinants can also be utilized by intercrossing them to generate secondary and tertiary recombinants, which are further screened for the introgressed trait. This approach was followed by Lukaszewski and Xu (1995) to engineer 1BL.1RS translocation. Genes for leaf and stripe rust resistance were transferred from *Ae. umbellulata* accn. 3237 to durum wheat through the production of amphiploids (AABBUU) between *Ae. umbellulata* and T. *durum* cv. WH890 and were further crossed to CS *PhI* to promote homeologous pairing. The F_1s (AABBDU) thus generated were backcrossed to a susceptible hexaploid wheat cultivar WL711 and introgression lines carrying resistance to leaf and stripe rust were screened in the backcross progenies (Chhuneja et al. 2007). Genes conferring resistance against powdery mildew (*Pm13)* from the diploid grass *Aegilops longissima*, for leaf rust and yellow pigmentation (*Lr19* + *Yp)* from the decaploid *Thinopyrum ponticum*, and the genes controlling gluten and gliadin (*Glu-D1* or *Gli-D1*/*Glu-D3)* from *T. aestivum* were separately introduced into durum wheat. Different strategies were employed for such transfers. Transfer of the *Pm13* gene was first obtained at the 6x level (Ceoloni et al. 1992) by use of the *ph1b* Chinese Spring mutant (Sears 1977) and then moved into a 4x background by homologous recombination (Ceoloni et al. 1996). For other transfers, wheat-alien chromosome recombination was induced using *ph1c* durum wheat mutants (Ceoloni et al. 2005). Several genes for disease resistance and quality attributes were transferred and stably inherited as reduced segments in the elite wheat background in backcrossed progenies. Such introgressions could be detected using GISH/FISH and molecular markers. The pattern of inheritance of the traits can be studied using F_2 population as the mapping population.

10.9 Reverse Breeding

In plant breeding, the importance of heterosis is well known where the hybrid (F_1) progeny is typically superior in comparison to its homozygous parents. The unpredictable nature of this phenomenon poses many problems for breeders (Fernandez-Silva et al. 2009). The performance of crop varieties cannot be optimized with the unknown constituents. Hence breeders evaluate heterosis by controlled crosses of inbred lines (i.e. by a prior selection and combination of unknown alleles). This hit-or-miss nature approach makes it difficult to optimize the effects of heterosis. Another barrier in plant breeding programmes is the reproduction of desirable heterozygotes by seeds. Favourable allele combinations of the elite heterozygote are lost in the next generation due to the segregation of traits.

Reverse breeding can be utilized as an alternative strategy to meet the challenge of fixing complex heterozygous genomes by constructing complementing homozygous lines (Dirks et al. 2009). In this method the order of events leading to the production of a hybrid plant variety is reversed (Fig. 10.4). It facilitates the production of homozygous parental lines that, once hybridized, reconstitute the genetic composition of an elite heterozygous plant, without the need for backcrossing and selection.

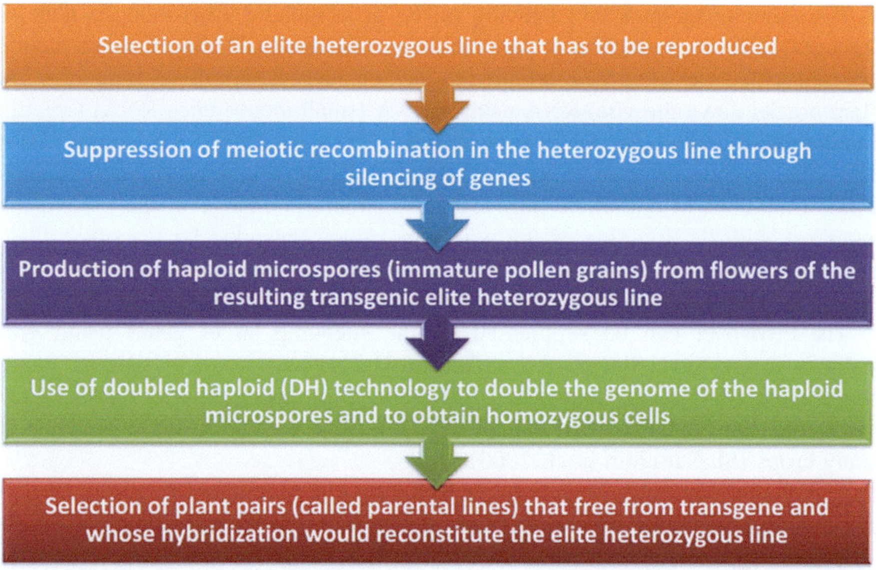

Fig. 10.4 Steps involved in reverse breeding

10.10 Crossover Suppression

The first step in reverse breeding is to produce gametes from the desired heterozygote without crossover recombination. To achieve this aim, genes controlling recombination are to be suppressed to ensure that gametes are formed without crossing over among the homologous chromosomes. There are a number of homologous chemicals available to suppress such genes, i.e. *MutL* homolog1 (suppresses *MutL*, which encodes a mismatch repair protein), *MutS* homolog2 suppresses *MutS* that encode mismatch repair proteins and promote homeologous recombination, RAD51 homolog (competes with RAD51, a gene involved in DNA repair) and homologs for proteins controlling crossing over such as SPORULATION-DEFICIENT11 (SPO11) a key protein for crossover initiation, univalent-producing mutants (desynaptic1 [*dsy1*], and meiotic prophase amonipeptidase1 [*mpa1*]) required for meiotic recombination. One of the successful examples of genome silencing is in the model plant *Arabidopsis* by Wijnker et al. (2012), who used RNA interference (RNAi) to knockdown the function of the *RecA* homolog DMC1, a meiosis-specific recombinase essential for the formation of crossovers. As RNAi is genetically dominant, it is easy to obtain progeny devoid of the RNAi cassette that would otherwise cause sterility in phenotypes among reverse-breeding offspring (Wijnker et al. 2012). RNAi silencing is easy to implement in many crops, and a single cassette targeting a well-conserved meiotic gene can be used across multiple crop species.

The other techniques for silencing of the gene include virus-induced gene silencing (VIGS) for induction of post-transcriptional gene silencing where a plant is infected with a virus that was modified to include a target gene RNA sequence. The plant breaks down the viral RNA using siRNA (small interference RNA) targeting simultaneously the plants' endogenous mRNA (Ruiz et al. 1998; Baulcombe 2004). The other technique used in reverse breeding is dominant negative mutations, where mutations in one allele may lead to a structural change in the protein that interferes with the function of the wild-type protein encoded by the other allele. These mutations are characterized by a dominant or semi-dominant phenotype (Rine and Herskowitz 1987).

The crossover can be suppressed by the silencing target genes using graft transmission method. Shoots of the plant in which genes are to be silenced would be grafted on transgenic rootstocks. The advantage of this method is that only a few transgenic rootstocks would be required to routinely apply reverse breeding in many crops (Shaharuddin et al. 2006).

The other methods include complete knockout of a gene by a recessive mutation, but this is not suitable for reverse breeding as it would reintroduce the same mutation into the offspring. There is a recent approach based on use of chemicals, i.e. 'mirin', an inhibitor for the Mre11-Rad50-Nbs1 (MRN) protein complex which plays a significant role in the processing of double strand breaks (Dupre et al. 2008). Other chemicals may be used, i.e. DIDS (Ishida et al. 2009), PSF (Morozumi et al. 2009), Halenaquinone (Takaku et al. 2011), Ri-1 (Budke et al. 2012) to inhibit RAD51. Exogenous application of compounds that cause inhibition or omission of recombination during meiosis would speed-up the application of reverse breeding enormously. A major advantage of using chemicals in reverse breeding is that their effect is not heritable hence the resultant reverse breeding products (DHs) are free of transgenes. This is important because the reverse breeding products are destined to be used in further breeding schemes, and should not have an achiasmatic phenotype. Perhaps contrary to intuition, DHs produced by transgene-mediated methods can be transgene-free. If a dominant knockdown construct is present in the hemizygous state, half of the gametes that are formed will not carry the transgene and, hence, are non-transgenic. Multiple transgenic lines with knockdown constructs on different chromosomes can be used to generate a full array of complementary DHs that do not carry transgenes (Wijnker and de Jong 2008). After the transformation non-recombinant the plant is subjected to the next step for production of doubled haploid.

10.11 Doubled Haploid Production

The second step in reverse breeding is to convert haploid gametes, carrying non-recombined chromosomes, into homozygous diploid adults. Efficient production of haploid plants that inherit a balanced number of chromosomes from gametes can greatly accelerate reverse breeding. The probability of desired haploids in which at least one copy of all chromosomes present is quite low because achiasmatic chromosomes have high chances of making unbalanced univalents. However, these unbalanced chromosomes containing haploids cannot pass through all the developmental stages, from cell division and embryogenesis to plant regeneration. Haploids generated from non-recombinant individuals are converted to diploid instant homozygous lines, bypassing the generations of inbreeding. This can be achieved through different methods, depending on the plant species, as described earlier in the section on haploid induction. As reverse breeding is still in its infancy, most of the research is restricted to the model plant *A. thaliana.* The haploid plants were produced by microspore culture (Wijnker et al. 2012, 2014). Another approach to haploid production is through centromere-mediated genome elimination in *A. thaliana*. In this approach, manipulation of a single centromere protein, the centromere-specific histone CENH3, has been used. The mutant was crossed to wild type and the chromosomes from mutant were eliminated and haploids were generated. The success of reverse breeding is limited to those crops where DH technology is the common practice. For the great majority of crop species this technology is well established and professional breeding companies routinely use such techniques in their breeding programmes (Forster et al. 2007).

10.12 Applications

This technique can be used for preserving elite genotypes. Through reverse breeding homozygous parental lines can be produced from a heterozygous plant, which shows the potential of an elite variety. These parental lines can then be crossed to achieve hybrids which reconstruct the heterozygous genotype of the elite plant. With conventional methods it would not be possible to produce a variety that maintains the genotype of such an elite plant.

10.12.1 Production for Substitution Lines for Chromosomal Studies

Reverse breeding provides plant breeders with new possibilities for further applications in breeding. The chromosome substitution lines can be obtained when the F_1 hybrid is subjected to reverse breeding. As a resultant homozygous chromosome, substitution lines for single and multiple chromosomes are generated. Additionally, for generating heterozygous chromosome substitution lines, homozygous chromosome substitution lines are further backcrossed with one of the original parents. In *Arabidopsis*, a traditionally complete set of chromosome substitution lines have been generated using more generations of crossing and extensive genotyping.

Whereas with reverse breeding Wijnker et al. (2012) obtained a complete set of Landsberg erecta (*Ler*) chromosome substitutions in the Columbia (Col-0) background, as well as two substitutions of a Col-0 chromosome into a *Ler* background from the population of 36 *Arabidopsis* doubled-haploids just in two generation.

These chromosome substitutions have various potential applications, as in the generation of near-isogenic lines by recurrent backcrosses. Such lines are extremely valuable for mapping quantitative trait loci (QTL) and for advanced forms of marker-assisted breeding. It can be also applied to improve parental lines or for genetic studies (Driks et al. 2009). The quantitative traits are always a matter of interest in crops, as most agronomically important traits are polygenic in nature and are located on different chromosomes. The study of epistatic interactions between the background and genes contributed by the substitution chromosome are facilitated by reverse breeding. Offspring of plants in which just one chromosome is heterozygous will segregate for traits present on that chromosome only. Selfing plants that carry a substituted chromosome (or using recurrent backcrosses) allow breeders to fine-tune interesting characteristics on a single chromosome scale. This could bring forth improved breeding lines carrying introgressed traits. Reverse breeding provides the opportunity to have control over homozygosity or heterozygosity at the single chromosome level.

10.12.2 Production of Genetic Resources

The reverse breeding process provides a set of germplasm along with complementary parents to the hybrid. That germplasm may contain euploids, and can serve as potential germplasm for various agronomically important traits and their genetic interactions. Less number of doubled haploid plants is required, which is necessary to reconstruct the starting plant at different levels of probability. In maize ($x = 10$) just 98 DHs are expected to contain a set of two reciprocal genotypes ($P = 99\ \%$); the rest of doubled haploids can be used in crop improvement programmes.

10.12.3 Creation of Additional Hybrid Parental Lines

In vegetables crops, i.e. cauliflower, seed production problem hinders the commercialization of hybrid varieties. While applying reverse breeding to these heterozygous hybrids it is possible to produce the same variety with two other parental lines, with potentially better reproducibility. The additional parental line can provide better scope for adaptability of hybrids (Lussern et al. 2011).

10.12.4 Marker-Assisted Reverse Breeding (MARB)

Reverse breeding in combination with molecular markers can act as a more effective tool for plant breeding. Use of high throughput markers, i.e. SSR (simple sequence repeats) and SNPs (single nucleotide polymorphism) can speed-up the process of identification of complementary parents in populations of DHs in the early stages. Marker-assisted reverse breeding has far more importance in the study of gene interactions of the heterozygous inbred families (HIFs) produced from reverse breeding. It can be used in the screening of population segregate for the quantitative traits. Reverse breeding can be used to provide highly valuable insights into the nature of heterotic effects using HIFs for the generation of chromosome-specific linkage maps and the fine mapping of genes and alleles. In the case of *Arabidopsis*, Wijnker et al. (2012) use evenly spaced SNPs markers at approximately 4-Mb intervals to identify the reverse breeding F_1s from wild-type F_1s haploid lines.

10.13 Limitations

The major limitation of this method is that it cannot be applied to crops with large genome size because most of the resulting spores are unbalanced, containing either none, one or two copies of a given chromosome. However, balanced spores, containing one copy of each chromosome, will be formed at a probability of (1/2) x, where x equals the basic chromosome number. Consequently, the chance of obtaining balanced spores decreases exponentially with the chromosome number and seems feasible for species in which the chromosome number equals 12 or less. Unintended effects include the silencing of other homologous sequences in the genome as a result of the presence of the RNAi construct. This would not induce genomic changes, but could affect expression levels. Another unintended effect of the technique could be an incomplete suppression of meiosis. This would lead to some degree of meiosis and recombination, which are natural processes in plants.

10.14 Conclusion

In a world with a population of almost one billion endlessly struggling to fight hunger, the need for enhanced crop yields is a must within a short span of time. The integrated approach using haploid technology with the conventional breeding methods and molecular techniques has revolutionized crop improvement programmes across the world, though the technology is not available for all crops. During introgression breeding, the first and foremost requirement in the development of widely adaptable cultivars is to confirm the alien chromatin. In recent years, it has been established that genetic and physical maps are not directly comparable to the chiasmata owing to their unequal distribution over the chromosomes. Molecular cytogenetic tools like genomic in situ hybridization (GISH) and fluorescence in situ hybridization (FISH) have been greatly used for the proper identification and localization of alien chromosomes or chromosome segments in the crop background. These techniques have undoubtedly revolutionized introgression breeding, but to get most of these tools, proper understanding of the gene interaction among themselves and with the environment is required.

In conclusion, haploid technology holds great promise for cost-effective highly efficient crop improvement, in a sustainable way. Use of this dynamic technology in alliance with other novel tools in chromosome engineering can boost the targeted farm productivity and food quality in a sustainable and eco-friendly way. Thus, greater emphasis and significant investment must be focused on the use of such innovative techniques for accelerated and high precision crop improvement.

Acknowledgments The authors are highly obliged to Prof. Yasuhiko Mukai, Osaka Kyoiku University, Japan and Dr. Trude Schwarzacher, Department of Biology, University of Leicester, UK, for extending their expertise in getting the resolution of certain results mentioned in this article.

References

Aghaee-Sarbarzeh M, Ferrahi M, Singh S, Singh H, Friebe B, Gill BS, Dhaliwal HS (2002) *Ph I* - induced transfer of leaf and stripe rust-resistance genes from *Aegilops triuncialis* and *Ae. geniculata* to bread wheat. Euphytica 127:377–382

Ahmad F, Comeau A (1991) A new intergeneric hybrid between *Triticum aestivum* L. and *Agropyron fragile* (Roth) Candargy: variation in *A. fragile* for suppression of the wheat *Ph*-Locus activity. Plant Breed 106:275–283

Aziz AN, Seabrook JEA, Tai GCC, de Jong H (1999) Screening diploid *Solanum* genotypes responsive to different anther culture conditions and ploidy assessment of anther-derived roots and plants. Am J Potato Res 76:9–16

Badiyal A, Chaudhary HK, Jamwal NS, Hussain W, Mahato A, Bhatt AK (2014) Interactive genotypic influence of triticale and wheat on their crossability and haploid induction under varied agroclimatic regimes. Cereal Res Commun 42(1):1–10

Barclay IR (1975) High frequencies of haploid production in wheat (*Triticum aestivum*) by chromosome elimination. Nature 256:410–411

Baulcombe D (2004) RNA silencing in plants. Nature 431:356–363

Belling J, Blakeslee AF (1922) The assortment of chromosomes in triploid *Daturas*. Am Nat 56:339–346

Bennett MD, Finch RA, Barclay IR (1976) The time rate and mechanism of chromosome elimination in *Hordeum* hybrids. Chromosoma 54:175–200

Blanco A, Perrone V, Simeone R (1988) Chromosome pairing variation in *Triticum turgidum* L. × *Dasypyrum villosum* (L.) Candargy hybrids and genome affinities. In: Proceedings of 7th international wheat genetics symposium, Cambridge, pp 63–67

Budke B, Logan HL, Kalin JH, Zelivianskaia AS, Cameron McGuire W, Miller LL, Stark JM, Kozikowski AP, Bishop DK, Connell PP (2012) RI-1: a chemical inhibitor of *RAD51* that disrupts homologous recombination in human cells. Nucl Acids Res 40(15):7347–7357

Ceoloni C, Signore G, Ercoli L, Donini P (1992) Locating the alien chromatin segment in common wheat–*Aegilops longissima* mildew resistance transfers. Hereditas 116:239–245

Ceoloni C, Biagetti M, Ciaffi M, Forte P, Pasquini M (1996) Wheat chromosome engineering at the 4′ level: the potential of different alien gene transfers into *durum* wheat. Euphytica 89:87–97

Ceoloni C, Forte P, Gennaro A, Micali S, Carozza R, Bitti A (2005) Recent developments in durum wheat chromosome engineering. Cytogenet Genome Res 109:328–334

Chalyk ST (1999) Creating new haploid-inducing lines of maize. Maize Genet Coop Newslett 73:53–54

Chalyk S, Bauman A, Daniel G, Eder J (2003) Aneuploidy as a possible cause of haploid induction in maize. Maize Genetic Coop Newslett 77:29–30

Chase SS (1947) Techniques for isolating monoploid maize plants. J Bot 34:582

Chase SS (1951) The monoploid method of developing inbred lines. In: Proceedings of 6th annual hybrid corn industry research conference. Illinois, Chicago, pp 29–30

Chaudhary HK (2008a) Dynamics of wheat × *Imperata cylindrica*—a new chromosome elimination mediated system for efficient haploid induction in wheat. In: Appels R et al (eds) Proceedings of the 11th international wheat genetics symposium. University of Sydney Press, Sydney, pp 647–650

Chaudhary HK (2008b) Dynamics of doubled haploidy breeding and molecular cytogenetic approaches in bread wheat. In: Taniguchi K, Zhang X (eds) Focus on north–west Himalayan regions. Advances in chromosome science, vol 3(2). The Society of Chromosome Research, Hiroshima, pp 67–69

Chaudhary HK (2012) New frontiers in chromosome engineering for enhanced and high precision crop improvement. In: Proceedings of national seminar on plant cytogenetics: new approaches, Department of Botany, Punjabi University, Patiala, 23–24, pp 35–36

Chaudhary HK (2013) New frontiers in chromosome elimination-mediated doubled haploidy breeding for accelerated and high precision genetic upgradation in wheat. In: Proceedings of international Triticeae mapping initiative and plant & animal genome XXI Conference, San Diego, pp 12–16, p 26

Chaudhary HK, Schwarzacher T, Heslop-Horrison JS (2004) Detection and characterization of rye (Secale cereale) chromatin introgression into wheat (Triticum aestivum) through fluorescence in situ hybridization. Report submitted to Commonwealth Commission, London. p 1–6

Chaudhary HK, Singh S, Sethi GS (2002) Interactive influence of wheat and maize genotypes on haploid induction in winter × spring wheat hybrids. J Genet Breed 56:259–266

Chaudhary HK, Sethi GS, Singh S, Pratap A, Sharma S (2005) Efficient haploid induction in wheat by using pollen of *Imperata cylindrica*. Plant Breed 124:96–98

Chaudhary HK, Tayeng T, Kaila V, Rather SA (2013a) Use of asynchrony in flowering for easy and economical polyhaploid induction in wheat following *Imperata cylindrica*- mediated chromosome elimination approach. Plant Breed 132(2):155–158

Chaudhary HK, Tayeng T, Kaila V, Rather SA (2013b) Enhancing the efficiency of wide hybridization mediated chromosome engineering for high precision crop improvement with special reference to wheat × *Imperata cylindrica* system. The Nucleus 56:7–14

Cherkaoui S, Lamsaouri O, Chlyah A, Chlyah H (2000) Durum wheat × maize crosses for haploid wheat production: influence of parental genotypes and various experimental factors. Plant Breed 119:31–36

Chhuneja P, Kaur S, Goel RK, Aghaee-Sarbarzeh M, Dhaliwal HS (2007) Introgression of leaf rust and stripe rust resistance genes from *Aegilops umbellulata* to hexaploid wheat through induced homoeologous pairing. Wheat Product Stress Environ Dev Plant Breed 12:83–90

Chhuneja P, Kaur S, Goel RK, Aghaee-Sarbarzeh M, Prashar M, Dhaliwal HS (2008) Transfer of leaf rust and stripe rust resistance from *Aegilops umbellulata* Zhuk. to bread wheat (*Triticum aestivum* L.). Genet Res Crop Evol 55:849–859

Coe EH (1959) A line of maize with high haploid frequency. Am Nat 93:381–382

Conner RL, Whelan ED, Laroche A, Thomas JB (1993) Reaction of alien chromosome substitution and addition lines of hard red spring wheat to common root rot and black point. Genome 36(1):173–180

David JL, Dusautoir JC, Raynauld C, Roumet P (1999) Heritable variation in the ability to produce haploid embryos *via* pollination with maize and embryo rescue in durum wheat. Genome 42:338–342

Davies DR (1974) Chromosome elimination in inter-specific hybrids. Heredity 32:267–270

De Maine MJ (2003) Potato haploid technologies. In: Maluszynski M et al (eds) Doubled haploid production in crop plants: a manual. Kluwer, Dordrecht, pp 241–247

Dhaliwal HS, Gill BS, Waines JG (1977) Analysis of induced homoeologous pairing in a *ph* mutant wheat × rye hybrid. J Hered 68:206–209

Dhiman R, Rana V, Chaudhary HK (2012) Himalayan maize—potential pollen source for maize mediated system of chromosome elimination approach in DH breeding of bread wheat. Cereal Res Commun 40:246–255

Dirks R, van Dun K, de Snoo CB, van den Berg M, Lelivelt CL, Voermans W, Woudenberg L, de Wit JP, Reinink K, Schut JW, van der Zeeuw E, Vogelaar A, Freymark G, Gutteling EW, Keppel MN, van Drongelen P, Kieny M, Ellul P, Touraev A, Ma H, de Jong H, Wijnker E (2009) Reverse breeding: a novel breeding approach based on engineered meiosis. Plant Biotechnol J 7:837–845

Dong F, Tek AL, Frasca ABL, McGrath JM, Wielgus SM, Helgeson JP, Jiang J (2005) Development and characterization of potato- *Solanum brevidens* chromosomal addition/substitution lines. Cytogenet Genome Res 109:368–372

Dundas IS, Anugrahwati DR, Verlin DC, Park RF, Bariana HS (2007) New sources of rust resistance from alien species: meliorating linked defects and discovery. Aust J Agric Res 58:545–549

Dupre A, Boyer-Chatenet L, Sattler RM, Modi AP, Lee JH, Nicolette ML, Kopelovich L, Jasin M, Raer R, Paull TT, Gautier J (2008) A forward chemical genetic screen reveals an inhibitor of the *mre11-rad50-nbs1* complex. Nat Chem Biol 4:119–125

Dvorak J, Deal KR, Luo MC (2006) Discovery and mapping of wheat *Ph1* suppressors. Genetics 174:17–27

Fernandez-Silva I, Moreno E, Eduardo I, Arus P, Alvarez JM, Monforte AJ (2009) On the genetic control of heterosis for fruit shape in melon (*Cucumis melo* L.). J Hered 100:229–235

Finch RA (1983) Tissue-specific elimination of alternative whole parental genomes in one barley hybrid. Chromosoma 88:386–393

Forster BP, Heberle-Bors E, Kasha KJ, Touraev A (2007) The resurgence of haploids in higher plants. Trends Plant Sci 12:368–375

Gernand D, Rutten T, Varshney A, Rubtsova M, Prodanovic S, Brub C, Kumlehn J, Matzk F, Houben A (2005) Uniparental chromosome elimination at mitosis and interphase in wheat and pearl millet crosses involves micronucleus formation, progressive heterochromatinization, and DNA fragmentation. Plant Cell 17:2431–2438

Gernand D, Rutten T, Pickering R, Houben A (2006) Elimination of chromosomes in *Hordeum vulgare* × *H. bulbosum* crosses at mitosis and interphase involves micronucleus formation and progressive heterochromatinization. Cytogenet Genome Res 114:169–174

Gill BS, Chen PO (1987) Role of cytoplasm specific introregression in the evolution of polyploidy wheats. Proc Natl Acad Sci USA 84:6800–6804

Giorgi B, Barbera F (1981) Increase of homoeologous pairing in hybrids between a *ph* mutant of *T. turgidum* L. var. *durum* and two tetraploid species of *Aegilops kotschyi* and *Ae. cylindrica*. Cereal Res Commun 9:205–211

Guha S, Maheshwari SC (1964) *In vitro* production of embryos from anthers of *Datura*. Nature 204:497

Gupta SB (1969) Duration of mitotic cycle and regulation of DNA replication in *Nicotiana plumbaginifolia* and a hybrid derivative of *N. tabacum* showing chromosome instability. Can J Genet Cytol 11:133–142

Gustafsson A, Hagberg A, Lundqvist U (1960) The induction of early mutants in Bonus barley. Hereditas 46:675–699

Hagberg A, Hagberg G (1980) High frequency of spontaneous haploids in the progeny of an induced mutation barley. Hereditas 93:341–343

Hagberg G, Hagberg A (1981) Haploidy initiater gene in barley. In: Barley Genetics IV proceedings of 4th international barley genetics symposium. Edinburg, pp 686–689

Hagberg A, Hagberg G (1987) Production of spontaneously doubled haploids in barley using a breeding system with marker genes and the "hap"-gene. Biologisches Zentralblatt 106:53–58

Ho KM, Kasha KJ (1975) Genetic control of chromosome elimination during haploid formation in barley. Genetics 81:263–275

Inagaki MN, Mujeeb-Kazi A (1995) Comparison of polyhaploid production frequencies in crosses of hexaploid wheat with maize, pearl millet and sorghum. Breed Sci 45:157–161

Inagaki MN, Tahir M (1991) Efficient production of wheat haploids through intergeneric crosses. TARC Newslett 2:4

Inagaki MN, Varughese G, Rajaram S, van Ginkel M, Mujeeb-Kazi A (1998) Comparison of bread wheat lines selected by doubled haploid, single-seed descent and pedigree selection methods. Theor Appl Genet 97:550–556

Inagkai MN, Tahir M (1990) Comparison of haploid production frequencies in wheat varieties crossed with *Hordeum bulbosum* L. and maize. Jpn J Breed 40:209–216

Ishida T, Takizawa Y, Kainuma T, Inoue J, Mikawa T, Shibata T, Suzuki H, Tashiro S, Kurumizaka H (2009) DIDS, a chemical compound that inhibits *RAD51*-mediated homologous pairing and strand exchange. Nucleic Acids Res 37:3367

Ishii T, Ueda T, Tanaka H, Tsujimoto H (2010) Chromosome elimination by wide hybridization between Triticeae or oat plant and pearl millet: pearl millet chromosome dynamics in hybrid embryo cells. Chromosome Res 18:821–831

Islam AKMR, Shepherd KW (1991) Alien genetic variation in wheat. In: Gupta PK, Tsuchiya T (eds) Chromosome engineering in plants: genetics, breeding, evolution, Part A. Elsevier, Amsterdam, pp 291–319

Islam AKMR, Shehpherd KW, Sparrow DHB (1975) Addition of individual barley chromosomes to wheat. In: Gaul H (ed) *Barley Genetics III.* Proceedings of 3rd international barley genetics symposium, Garching, pp 260–270

Islam AKMR, Shepherd KW, Sparrow DHB (1981) Isolation and characterization of euplasmic wheat-barley chromosome addition lines. Heredity 46:161

Jacobsen E, Ramanna MS (1994) Production of monohaploids of *Solanum tuberosum* L. and their use in genetics, molecular biology and breeding. In: Breadshaw JE, Mckay GR (eds) Potato genetics. CAB International, Wallingford, pp 155–170

Jalani BS, Moss JP (1980) The site of action of the crossability genes (*Kr1, Kr2*) between *Triticum* and *Secale* in pollen germination, pollen tube growth, and number of pollen tubes. Euphytica 29:571–579

Jauhar PP, Almouslem AB (1998) Production and analyses of some intergeneric hybrids between durum wheat and *Thinopyrum* species. In: Jaradat AA (ed) Triticeae III. Scientific Publishers Inc., New Hampshire, pp 119–126

Jauhar PP, Chibbar RN (1999) Chromosome-mediated and direct gene transfers in wheat. Genome 42:570–583

Jauhar PP, Peterson TS, Xu SS (2009) Cytogenetic and molecular characterization of a durum alien disomic addition line with enhanced tolerance to *Fusarium* head blight. Genome 52:467–483

Jeberson MS (2010) Physical mapping of some triticale × wheat derived rye chromatin introgressed wheat recombinants through fluorescent in situ hybridization. Ph. D. thesis. CSK HP Agricultural University, Palampur, Himachal Pradesh, p 127

Jeberson MS, Chaudhary HK, Kishore N (2012) Molecular cytogenetic studies for detection and characterization of alien chromosome/chromatin introgressions in triticale × wheat derived wheat stable lines. In: Proceedings of national symposium on plant cytogenetics: new approaches, Punjabi University, Patiala, p 99

Jenczewski E, Eber F, Grimaud A, Huet S, Lucas MO, Monod H, Chevre AM (2003) *Pr Bn*, a major gene controlling homeologous pairing in oilseed rape (*Brassica napus*) haploids. Genetics 164:645–653

Ji Y, Chetelat RT (2003) Homoeologous pairing and recombination in *Solanum lycopersicoides* monosomic addition and substitution lines of tomato. Theor Appl Genet 106:979–989

Jin W, Melo JR, Nagaki K, Talbert PB, Henikoff S, Dawe RK, Jiang J (2004) Maize centromeres: organization and functional adaptation in the genetic background of oat. Plant Cell 16:571–581

Kasha KJ, Kao KN (1970) High frequency haploid production in barley (*H. Vulgare* L.). Nature 225:874–876

Kermicle JL (1969) Androgenesis conditioned by a mutation in maize. Science 166:1422–1424

Khrustaleva LI, de Melo PE, van Heusden AW, Kik C (2005) The integration of recombination and physical maps in a large-genome monocots using haploid genome analysis in a trihybrid *Allium* population. Genetics 169:1673–1685

Khush GS (1973) Cytogentics of aneuploids. Academic Press, New York, pp 178–208

King IP, Laurie DA (1993) Chromosome damage in early embryo and endosperm development in crosses involving the preferentially transmitted *4Sl* chromosome of *Aegilops sharonensis*. Heredity 70:52–59

Kisana NS, Nkongolo KK, Quick JS, Johnson DL (1993) Production of doubled haploids by anther culture and wheat × maize method in a wheat breeding programme. Plant Breed 110:96–102

Kishore N, Chaudhary HK, Chahota RK, Kumar V, Sood SP, Jeberson S, Tayeng T (2011) Relative efficiency of the maize and *Imperata cylindrica* -mediated chromosome elimination approaches for induction of haploids of wheat-rye derivatives. Plant Breed 130:192–194

Komeda N, Chaudhary HK, Mukai Y (2007) Cytological evidence for chromosome elimination in wheat × *Imperata cylindrica* hybrids. Genes Genetic Syst 82:241–248

Krolow KD (1970) Investigations on compatibility between wheat and rye. Z. Pflanzenzuchtung 64:44–72

Kruse A (1973) *Hordeum* × *Triticum* hybrids. Hereditas 73:157–161

Lange W (1971) Crosses between *Hordeum vulgare* L. and *H. bulbosum* L. 1.Production, morphology and meiosis of hybrids, haploids and dihaploids. Euphytica 20:14–29

Lashermes P, Beckert M (1988) A genetic control of maternal haploidy in maize (*Zea mays* L.) and selection of haploid inducing lines. Theor Appl Genet 76:405–410

Laurie DA, Bennett MD (1986) Wheat × maize hybridization. Can J Genet Cytol 28:313–316

Laurie DA, Bennett MD (1988) The production of haploid wheat plants from wheat × maize crosses. Theor Appl Genet 76:393–397

Laurie DA, Bennett MD (1989) The timing of chromosome elimination in hexaploid wheat × maize crosses. Genome 32:953–961

Laurie DA, Reymondie S (1991) High frequencies of fertilization and haploid seedling production in crosses between commercial hexaploid wheat varieties and maize. Plant Breed 106:182–189

Linde-Laursen I, von Bothmer R (1999) Orderly arrangement of the chromosomes within barley genomes of chromosome-eliminating *Hordeum lechleri* × barley hybrids. Genome 42:225–236

Lukaszewski AJ, Xu X (1995) Screening large populations of wheat hybrids by C-banding. Cereal Res Commun 23:9–13

Lukaszewski AJ, Rybka K, Korzun V, Malyshev SY, Lapinski B, Whitkus R (2004) Genetic and physical mapping of homoeologous recombination points involving wheat chromosome 2B and rye chromosome 2R. Genome 47:36–45

Lussern M, Parisi C, Plan D, Rodríguez-Cerezo E (2011) Drivers & constraints. New plant breeding techniques State-of-the-art and prospects for commercial development. Luxembourg publication, European Union, pp 45–46

Mago R, Zhang P, Bariana HS, Verlin DC, Bansal UK (2009) Development of wheat lines carrying stem rust resistance gene *Sr39* with reduced *Aegilops speltoides* chromatin and simple PCR markers for marker-assisted selection. Theor Appl Genet 119:1441–1450

Maruthachalam R, Chan SWL (2010) Haploid plants produced by centromere-mediated genome elimination. Nature 464:615

Matzk F, Mahn A (1994) Improved techniques for haploid production in wheat using chromosome elimination. Plant Breed 113:125–129

Mendiburu AO, Peloquin SJ, Mok DWS (1974) Potato breeding with haploids and 2n gametes. In: Kasha KJ (ed) Haploids in higher plants. Guelph University Press, Guelph, ON, pp 249–259

Miller TE, Reader SM, Gale MD (1983) The effect of homoeologous group 3 chromosomes on chromosome pairing and crossability in *Triticum aestivum*. Can J Genet Cytol 25:634–641

Mochida K, Tsujimoto H, Sasakuma T (2004) Confocal analysis of chromosome behaviour in wheat × maize zygotes. Genome 47:199–205

Montelongo-Escobedo H, Rowe PR (1969) Haploid induction in potato: cytological basis for the pollinator effect. Euphytica 18:116–123

Morozumi Y, Takizawa Y, Takaku M, Kurumizaka H (2009) Human PSF binds to *RAD51* and modulates its homologous-pairing and strand-exchange activities. Nucleic Acids Res 37:4296–4307

Morshedi AR, Darvey NL (1995) High frequency of embryos in wheat × maize crosses. SABRAO J 27:17–22

Nakamura C, Tsuchiya T (1982) Cytogentics of alien addition trisomics in sugar beets. I. Meiotic chromosome behaviour in nematode resistant trisomics. Biologisches Zentralblatt 101:227–240

Naranjo T, Roca A, Giráldez R, Goicoechea PG (1988) Chromosome pairing in hybrids of *ph1b* mutant wheat with rye. Genome 30:639–646

Peng JH, Lapitan NLV (2005) Characterization of EST-derived microsatellites in the wheat genome and development of eSSR markers. Funct Integr Genomics 5:80–96

Pratap A, Chaudhary HK (2007) Genetic studies on the effect of triticale × wheat F_1s and maize genotypes on haploid induction following wheat × maize system. J Genet Breed 60:119–124

Pratap A, Sethi GS, Chaudhary HK (2005) Relative efficiency of different Gramineae genera for haploid induction in triticale and triticale × wheat hybrids through chromosome elimination technique. Plant Breed 124:147–153

Pratap A, Sethi GS, Chaudhary HK (2006) Relative efficiency of anther culture and chromosome elimination technique for haploid induction in triticale × wheat and triticale × triticale hybrids. Euphytica 150:339–345

Randolph LF (1932) Some effects of high temperature on polyploidy and other variations in maize. Genetics 18:222–229

Rather SA, Chaudhary HK, Kaila V (2014) Proportional contribution and potential of maternal and paternal genotypes for polyhaploid induction in wheat × *Imperata cylindrica* chromosome elimination approach. Cereal Res Commun 42(1):19–26

Riley R (1966) Cytogenetics and wheat breeding. Contemporary. Agriculture 11–12:107–117

Riley R, Chapman V (1958) Genetic control of the cytologically diploid behaviour of hexaploid wheat. Nature 182:713–715

Riley R, Chapman V (1967) The inheritance in wheat of crossability with rye. Genet Res 9:259–267

Riley R, Chapman V, Johnson R (1968) Introduction of yellow rust resistance of *Aegilops comosa* into wheat by genetically induced homoeologous recombination. Nature 217:383–384

Rine J, Herskowitz I (1987) Four genes responsible for a position effect on expression from HML and HMR in *Saccharomyces cerevisiae*. Genetics 116:9–22

Rogowsky PM, Sorrells ME, Shepherd KW, Langridge P (1993) Characterization of wheat-rye recombinants with RFLP and PCR probes. Theor Appl Genet 85:1023–1028

Rokka VM (2003) Anther culture through direct embryogenesis in a genetically diverse range of potato (*Solanum*) species and their interspecic and intergeneric hybrids. In: Maluszynski M et al (eds) Doubled haploid production in crop plants. A manual. Kluwer, Dordrecht, pp 235–245

Rokka VM (2009) Potato haploids and breeding. In: Touraev A et al (eds) Advances in haploid production in higher plants. Springer Science + Business Media B.V, Germany, pp 199–208

Rokka VM, Pietila L, Pehu E (1996) Enhanced production of dihaploid lines *via* anther culture of tetraploid potato (*Solanum tuberosum* L. ssp. *tuberosum*) clones. Am J Potato Res 73:1–12

Ruiz MT, Voinnet O, Baulcombe DC (1998) Initiation and maintenance of virus-induced gene silencing. Plant Cell 10:937–946
Sanei M, Pickering R, Kumke K, Nasuda S, Houben A (2011) Loss of centromeric histone H3 (CENH3) from centromeres precedes uniparental chromosome elimination in interspecific barley hybrids. Proc Natl Acad Sci USA 108:498–505
Sarkar KR, Pandey A, Gayen P, Mandan JK, Kumar R, Sachan JKS (1994) Stabilization of high haploid inducer lines. Maize Genet Coop Newslett 68:64–65
Schneerman MC, Charbonneau M, Weber DF (2000) A survey of ig containing materials. Maize Genet Coop Newslett 74:92–93
Schwarzacher T, Leitch AR, Bennett MD, Heslop-Harrison JS (1989) In situ localization of parental genomes in a wide hybrid. Ann Bot 64:315–324
Schwarzacher T, Ali N, Chaudhary HK, Graybosch R, Kapalande HV, Kinski E, Heslop-Harrison JS (2011) Fluorescent in situ hybridization as a genetic technology to analyzing chromosomal organization of alien wheat recombinant lines. IAEA- TECDOC-1664: physical mapping technologies for the identification and characterization of mutated genes to crop quality. IAEA TECDOC 1664 Subject Classification: 0203-Mutation plant breeding. IAEA, Vienna. ISBN 978-92-0- 119610-1 ISSN 1011-4298
Sears ER (1952) Misdivision of univalents in common wheat. Chromosoma 4:535–550
Sears ER (1956) The transfer of leaf rust resistance from *Aegilops umbellulata* to wheat. Brookhaven Sympos Biol 9:1–22
Sears ER (1972) Chromosome engineering in wheat. In: Stadler genetics symposium 4, University of Missouri, Columbia, pp 23–38
Sears ER (1973) *Agropyron*- wheat transfers induced by homoeologous pairing. In: Sears ER, Sears LMS (eds) Proceedings of 4th international wheat genetics symposium. Columbia, Missouri, pp 191–199
Sears ER (1977) An induced mutant with homoeologous pairing in common wheat. Can J Genet Cytol 19:585–593
Sears ER (1981) Transfer of alien genetic material to wheat. In: Evans LT, Peacock WJ (eds) Wheat science-today and tomorrow. Cambridge University Press, Cambridge, pp 75–89
Sears ER (1982) A wheat mutation conditioning an intermediate level of homoeologous pairing. Can J Genet Cytol 24:715–719
Shaharuddin NA, Yuanhuai H, Hongying L, Grierson D (2006) The mechanism of graft transmission of sense and antisense gene silencing in tomato plants. FEBS Lett 580:6579–6586
Sharma D, Knott DR (1966) The transfer of leaf rust resistance from *Agropyron* to *Triticum* by irradiation. Can J Genet Cytol 8:137–143
Sharma S, Sethi GS, Chaudhary HK (2005) Influence of winter and spring wheat genetic backgrounds on haploid induction parameters and trait correlations in the wheat × maize system. Euphytica 144:199–205
Shatskaya OA, Zabirova ER, Shcherbak VS, Chumak MV (1994) Mass induction of maternal haploids in corn. Maize Genet Coop Newslett 68:51
Singh AK, Stalker HT, Moss JP (1991) Cytogenetics and use of alien genetic variation in groundnut improvement. In: Tsuchiya T, Gupta PK (eds) Chromosome engineering in plants: genetics, breeding, evolution, Part B. Elsevier, Amsterdam, pp 65–78
Singh S, Sethi GS, Chaudhary HK (2004) Differential responsiveness of winter and spring wheat genotypes to maize-mediated production of haploids. Cereal Res Commun 32:201–207
Sitch LA, Snape JW, Firman SJ (1985) Intra chromosomal mapping of crossability genes in wheat (*Triticum aestivum*). Theor Appl Genet 70:309–314
Snape JW, Simpson E, Parker BB (1986) Criteria for the selection and use of doubled haploid systems in cereal breeding programs. In: Horn W et al (eds) Genetic manipulation in plant breeding. Walter de Gruiter, New York, pp 217–229
Stephan S (1969) Haploid barley from crosses of *Hordeum bulbosum* (2x) × *Hordeum vulgare* (2x). Can J Genet Cytol 11:602–608
Subrahmanyam NC, Kasha KJ (1973) Selective chromosomal elimination during haploid formation in barley following interspecific hybridization. Chromosoma 42:111–125

Suenaga K (1994) Doubled haploid system using the intergeneric crosses between wheat (*Triticum aestivum*) and maize (*Zea mays*). Bull Natl Inst Agrobiol Res 9:83–139

Suenaga K, Nakajima K (1989) Efficient production of haploid wheat (*Triticum aestivum*) through crosses between Japanese wheat and maize (*Zea mays*). Plant Cell Rep 8:263–266

Takaku M, Kainuma T, Ishida-Takaku T, Ishigami S, Suzuki H, Tashiro S, van Soest RW, Nakao Y, Kurumizaka H (2011) Halenaquinone, a chemical compound that specifically inhibits the secondary DNA binding of *RAD51*. Genes Cells 16(4):427–436

Tayeng T, Chaudhary HK, Kishore N (2012) Enhancing doubled haploid production effi ciency in wheat (*Triticum aestivum* L. em. Thell) by *in vivo* colchicine manipulation in *Imperata cylindrica* - mediated chromosome elimination approach. Plant Breed 131:574–578

Valkoun JJ (2001) Wheat pre-breeding using progenitors. Euphytica 119:17–23

Wijnker E, Dun K, de Snoo CB, Lelivelt CL, Keurentjes JJ, Naharudin NS, Ravi M, Chan SW, de Jong H, Dirks R (2012) Reverse breeding in *Arabidopsis thaliana* generates homozygous parental lines from a heterozygous plant. Nat Genet 44:467–470

Wall AM, Riley R, Chapman V (1971) Wheat mutants permitting homoeologous meiotic chromosomes pairing. Genetics Res 18:311–328

Wedzony M, Röber F, Geiger HH (2004) Chromosome elimination observed in selfed progenies of maize inducer line RWS. In: VII. Intern. Congress on Sexual Plant Reproduction. Maria Curie-Sklodowska University Press, Lublin, p 173

Wijnker E, de Jong H (2008) Managing meiotic recombination in plant breeding. Trends Plant Sci 13:640–646

Wijnker E, Deurhof L, Jose B, de Snoo CB, Blankestijn H, Becker F, Ravi M, Chan SWL, van Dun K, Lelivelt CL, de Jong H, Dirks H, Keurentjes JJ (2014) Hybrid recreation by reverse breeding in *Arabidopsis thaliana*. Nat Protoc 9:761–772

Wu L, Zheng C, Jia Z, Yuan J (1989) Chromosome pairing in hybrids of *ph1b* and nulli5B-tetra5D wheat with rye and *Agrotricum*. Plant Breed 102:281–285

Yamamoto M, Mukai Y (1989) Application of fluorescence in situ hybridization to molecular cytogenetics of wheat. Wheat Inf Serv 69:30–32

Yu MQ, Jaheir J, Person-Dedryver F (1995) Studies on the effect of *Ph1b* gene on the F_1, BC_1, BC_2, BC_3 crosses between wheat and *Ae. variabilis* and the transfer of cereal root knot nematode (*M. naasi*) resistance. Acta Agronomica Sinica 21(2):136–144

Zenketler M, Straub J (1979) Cyto-embryological study on the process of fertilization and the development of haploid embryo of *Triticum aestivum* (2n = 42) after crossing with *Hordeum bulbosum* (2n = 14). Z Pflanzenzuchtung 82:36–44

Zenkteler M, Nitzsche W (1984) Wide hybridization experiments in cereals. Theor Appl Genet 68:311–316

Zhang Z, Qiu F, Liu Y, Ma K, Li K, Xu S (2008) Chromosome elimination and *in vivo* haploid production induced by *Stock 6*-derived inducer line in maize (*Zea mays* L.). Plant Cell Rep 27:1851–1860

Zheng YL, Luo MC, Yen C, Yang JL (1992) Chromosome location of a new crossability gene in common wheat. Wheat Inf Serv 75:36–40

Chapter 11
Overview of Progress and Potentials of Improving Commonly Used *Allium* species in India

R.N. Gohil and Veenu Kaul

Abstract The genus *Allium* L. is an assemblage of about 780 species distributed all over the world. Two species, *Allium cepa* L. and *Allium sativum* L., are valued throughout the world not only as spice or food but also as medicinal herbs. Both are important and nearly indispensable seasoning agents in most of the kitchens. More than 40 species of this genus are widely distributed across the temperate and alpine regions of the Indian subcontinent. Nearly all these are edible and used as vegetables, spices and condiments. While most of these are used in folk medicine some are repositories of important genes. Although no significant pest or insect problems have been recorded in the wild taxa, cultivated species are susceptible to many diseases that drastically affect their productivity. Besides diseases, physiological features like asynchronous seed maturation and shattering of seeds from the mature capsule are major problems that require remediation. For a commodity worth millions in world trade the limitations acquire great significance and need immediate attention. To mitigate these common problems, attempts have been made, within and outside the Indian subcontinent, to tailor genetic makeup of cultivated taxa by introducing useful genes from wild species. Intervarietal and interspecific hybridizations have been effective in producing new races in short period of time with desirable traits transferred from one species to another, directly or through bridge species. F_1 hybrids obtained in many of these crosses are superior to regular cultivars in vigour, flavour, productivity and insect, pest and disease resistance. Techniques like protoplast fusion coupled with GISH, mutagenesis, marker assisted breeding and in vitro and/or biotechnological interventions, can help Allium breeders in making important breakthroughs through rapid multiplication and disease eradication in these taxa of high economic value. Further with the increasing awareness and availability of the databases of traditional knowledge many lesser known and underutilized species are, now, gaining attention for their potential in food, pharmaceutical and horticultural industry. The chapter, while providing a

R.N. Gohil (✉) · V. Kaul
Department of Botany, University of Jammu, Jammu 180006, India
e-mail: rng04@rediffmail.com

V. Kaul
e-mail: veenukaul@yahoo.co.in

V.R. Rajpal et al. (eds.), *Gene Pool Diversity and Crop Improvement*,
Sustainable Development and Biodiversity 10,
DOI 10.1007/978-3-319-27096-8_11

general overview of the species of Allium available in India, focuses mainly on the work done on the cultivated species.

Keywords Alliaceae · Alliaceous · Polymorphism · Relationships · Phylogeny · Origin · Onion · Garlic · Shallot

11.1 Introduction

Used in folk medicine since prehistoric times, *Allium cepa* L. and *Allium sativum* L. are valued throughout the world not only as spice or food (Afzal et al. 2000; Kakar et al. 2002; Mahmood et al. 2002; Bhuiya et al. 2003; Rahman et al. 2003; Islam et al. 2004) but also as medicinals (Pooler and Simon 1994; Afzal et al. 2000; Sata et al. 2001; Yamashita et al. 2002; Haque et al. 2003; Rahman et al. 2004; Islam et al. 2007; Khan and Hanif 2006). Onion is the second most valuable vegetable in the world preceded only by tomato (FAOSTAT 2011). Its importance can be gauged from the fact that Germans call it the 'queen of the kitchen' and people all over the world, with minor exceptions, consume it as a substantial dietary ingredient or as a flavouring agent. It is favoured as a spice, vegetable or accompaniment for its crisp, juicy, mild and sweet nature. Because of varying tastes, preferences, wide choices of cultivation practices under different environmental conditions the onion breeders have the onerous task not only to develop cultivars to meet these specific goals but also to ensure good storage quality without compromising on its flavouring and eating delights.

The genus on account of its size (ca.780 species), is one of the largest repositories of gene pool amongst all crops. This characteristic is of immense importance considering that the present-day plant breeders are equipped with a variety of modern, sophisticated and precise techniques to overcome even the crossability barriers.

The genus *Allium* L. belongs to family Alliaceae Batsch ex Borkh. subfamily Alloideae (Takhtajan 1987, 1997; Fay and Chase 1996; APG II 2003; Chuda and Adamus 2009). The family comprises of biennial or perennial bulbous or rhizomatous herbs usually with alliaceous (onion-like) odour, characterized by plants bearing narrow leaves with sheathing bases, an umbellate inflorescence enclosed in a spathe when young, and a superior ovary (Nasir 1975; Simpson 2010). An assemblage of about 780 species distributed all over the world, *Allium* is the second largest genus of monocotyledons in general and probably the largest of petaloid monocots in particular (Stearn 1992; Friesen et al. 2006; Stavelikova 2008). According to FAO, it represents one of the world's most economically and ecologically important crops because of the various uses its species are put to. While cultivated species of *Allium* form important and essential components of cooking, the wild taxa are also widely used, sometimes even preferred for culinary purposes wherever these are available in plenty. Till date no wild species of this genus has been found or reported to be poisonous or harmful. Because of the usefulness of both the cultivated and wild species, the genus has a unique place insofar as its indispensability as a culinary agent and importance in medicine is concerned.

Taxonomic placement of *Allium,* sharing characteristics of two established families Liliaceae and Amaryllidaceae (Nasir 1975) has been quite difficult. This is further complicated on account of 'no single uniting morphological trait (Stearn 1946a), assimilation of complex chromosome aberrations (Ved Brat 1965), accumulation of unequal amounts of heterochromatin or spacer DNA (Jones and Rees 1968) and the probable loss of many species during the previous ice age (Traub 1968)' (see Havey 1992). While Bentham and Hooker (1883) placed the genus in Liliaceae because of its superior ovary; Hutchinson (1934) shifted it to Amaryllidaceae due to its scapigerous inflorescence and presence of spathe. As at present, the taxonomists favour its inclusion in a distinct family Alliaceae (Purseglove 1972; Hara et al. 1978–1982; Polunin and Stainton 1984).

Large size coupled with extreme polymorphism prevalent in the genus *Allium* necessitated its subdivisions into various subgenera, sections and subsections. This infrageneric classification has been the subject of great debate depending on the parameters used by various workers. Earliest classifications of the genus date back to Regel (1875) who identified 6 sections [informal groupings of Don (1832)] and 262 species. Others classified it into 9 sections and 228 species (Vvedensky 1935); 3 subgenera, 36 sections and subsections and ca. 600 species (Traub 1968); 6 subgenera, 44 sections and subsections (Kamelin 1973); 6 subgenera, 50 sections and subsections for 600–700 species (Hanelt et al. 1992). Classification of *Allium by* Hanelt et al. (1992) into six subgenera and 43 sections (excluding subsections) was based mainly on the presence or absence of bulbs or rhizomes, leaf and flower morphologies, length of leaf sheaths and structure of bulb tunics. The six subgenera as per Hanelt et al. (1992) are *Rhizirideum, Amerallium, Bromatorrhiza, Allium, Melanocrommyum* and *Caloscordum.* Two subgenera, *Bromatorrhiza* (Hanelt and Fritsch 1994) and *Melanocrommyum* (Khassanov and Fritsch 1994) were later subjected to further refinement (see Klaas 1998). While subgenera *Amerallium, Allium, Melanocrommyum* and two species of *Caloscordum* bear true bulbs on a scarcely developed rhizome and are adapted to arid conditions, those of *Bromatorrhiza* are adapted to strongly humid conditions and consist of very specialized and heterogeneous taxa with thick fleshy roots, without well-developed bulbs or rhizomes. Morphological, karyological and molecular characteristics make subgenus *Rhizirideum* a highly heterogeneous group (Pich et al. 1996). Its $\sim$200 species in Eurasia and 2 in North America bearing rhizome, a primitive feature, are characterized by less specialized traits. These are divided into sections on the basis of leaf, scape and floral morphologies. This assemblage, including crop species like *A. cepa* L. (Common onion and shallot), *A. fistulosum* L. (Welsh or Japanese bunching onion), *A. schoenoprasum* L. (chive), *A. chinense* G. Don (rakkyo) and *A. tuberosum* Rottl. Ex Spreng. (Chinese chive), is usually considered to be the basic stock from which other species of the genus are supposed to have evolved (Klaas 1998).

Studies using molecular markers suggest further changes in the structure of the genus. Minor to major modifications of the earlier grouping have been possible depending upon the type of molecular marker used and interpretations made thereupon. Integrating these earlier types of studies with their own ITS data on 221 accessions (196 of Allium and 5 outgroup species), Friesen et al. (2006) recognized 15 subgenera and 56 sections. In accordance with their taxonomic conspectus different species of the genus have diversified along three major lines of evolution; first includes 3 subgenera; second 5 and third 7. Subgenus *Allium* with 300 species is the largest and falls in the third group.

11.2 Origin

Allium is believed to have originated along the Irano–Turanian biogeographical region with the Mediterranean basin and Western North America as the secondary centres of origin (Vavilov 1926; Hanelt 1990; Fritsch and Friesen 2002; Friesen et al. 2006). The main centre of diversity is in South west and Central Asia (Friesen et al. 2006). The genus is, however, widely distributed mainly in the seasonally dry regions of the northern hemisphere (Kamenetsky and Rabinowitch 2006; Friesen et al. 2006). Domestication of its wild, edible and medicinal species is still continuing (Fritsch and Friesen 2002). Despite such a long history of cultivation, not much is known about inter- and intraspecific relationships within this genus. Adaptation of the species to a wide array of geoclimatic conditions and the consequent polymorphism has added to the complexity (Kamenetsky and Rabinowitch 2006; Singh et al. 2014).

While the few cultivated species including onion and garlic are grown worldwide, the rest grow wild mainly in the Northern Hemisphere. Though some species are found in places like North Africa and Ceylon, majority are temperate in distribution, growing successfully in Central Asia, Persia and California (Stearn 1946a). The chive (*A. schoenoprasum* L.) is of special interest as it extends up to the arctic region and is found growing wild both in the New and the Old Worlds. In India while *A. cepa* and *A. sativum* are very widely cultivated, cultivation of *A. ascalonicum* L. and *A. ampeloprasum* L. is confined to some specific regions and communities. With over 40 wild species the Indian subcontinent, especially the Himalayan belt, represents a major gene pool for the genus. *Allium tuberosum* Rott. ex Spreng. and *A. rubellum* M Bieb. are also found growing wild in the plains besides the cultivated species (Hooker 1892; Stearn 1945; Gohil 1973). Wild populations of the former perhaps are escapes.

Garlic probably originated in the mountainous regions of the Central Asia/Southern Siberia from where it is thought to have spread first to the Old world and then to the New world (Bozzini 1991; Shemesh et al. 2008). European traders facilitated its further distribution, making it an important flavouring commodity throughout the world (Kamenetsky 2007).

The place and time of origin of the cultivated species of *Allium* is still a matter of speculation. Among the earliest records are the Egyptian pyramids dating back to 3200–2700 B.C. where workers are depicted eating onions and garlic (Fakhry 1961). Engeland (1991) is of the opinion that about 10000 years ago wild garlic was not only widely dispersed in Central Asia but also cultivated by semi-nomadic tribes who might have been instrumental in introducing garlic to the Mediterranean basin, India and China. Its popularity can be judged by the umpteen references to it in the ancient Indian and Chinese scriptures. The oldest record of garlic cultivation found in Sumerian literature dates back to 2600–2100 B.C. Garlic was found in the tomb of Tutankhamen besides probably used by Israelis during their exodus from Egypt as mentioned in the Holy Bible (Harris et al. 2001). Introduction to Egypt must have occurred much earlier. Ancient Egyptians inscribed its uses in cooking, embalming and to mend broken bones on the walls of places of worship and pyramids (Kumar and Berwal 1998). The Holy Quran also mentions garlic and onions (Part 1: 61) (Jones and Mann 1963). Its strong flavour is reported to have been disliked by the Romans who usually gave it to their labourers and soldiers to increase their stamina (Kamenetsky 2007). Although the foregoing account does not confirm the place and origin of cultivated species of *Allium*, it certainly points towards their importance even in prehistoric times.

People in India seem to have been growing onion even before the sixth century B.C. A reference to it has been made by Charaka in his compilation titled Charaka Samhita written in the first century A.D. (based on earlier records dating back to the sixth century B.C.). He described the antirheumatic and diuretic properties of onions and their usefulness in the ailments of eyes and digestive tract and in stimulating the heart. On the other extreme, the use of onions was abhorred by some who believed that it caused headache, hurt the eyes, rendered one dim sighted and caused flatulence and over sleepiness, especially when taken raw (Gerard 1597; Arber 1938). It is also mentioned as an aphrodisiac in ancient Hindu scripture Garuda Purana (Basak 1987). Some sects/communities in India do not use onion or garlic at all.

Besides the cultivated forms, many wild taxa are also reported to be consumed by the natives for different purposes in various forms. Fernald and Kinsey (1943) have pointed out that although all *Allium* species of North America appear to be edible and are, in fact, collected and eaten, none has been domesticated as a food plant. That is precisely true for the Indian species also. During many plant exploration trips to various parts of the country, it was found that wild *Alliums* are being consumed in large quantities by the locals, though none so far has been tried for cultivation (Gohil 1973, 1992; Negi 2006). Species like *A. paradoxum* (Bieb.) G. Don, *A. triquetrum* L., *A. ursinum* L., *A. victorialis* L., *A. atropurpureum* Waldst. Et Kit., *A. narcissiflorum* Vill. and *A. rubellum* M. Bieb. possess dense flower heads of attractive colours on account of which they are prized as popular garden ornamentals.

11.3 Diversity, Domestication, Varietal Groups and Relationships

Not known in the wild form, domestication of *Allium cepa* L. the common onion, for instance, is thought to have started more than 4000 years ago (Puizina 2013) probably in the Middle East (Hanelt et al. 1992). For a long time,

A. oschaninii O. Fedtsch (*A. cepa* var. *silvestra* Regel) from Central Asia was believed to represent the wild ancestor of *A. cepa* (Stearn 1980; Hanelt et al. 1992). Studies on karyotype banding patterns, crossing barriers and chloroplast phylogeny, however, do not support this hypothesis (Vosa 1976; Raamsdonk et al. 1992; Havey 1992). Recent RFLP analyses of nuclear DNA, three chloroplast loci and ITS of nuclear rDNA strongly favour *A. vavilovii* M. Popov et Vved. as the possible progenitor of *A. cepa* (Bradeen and Havey 1995a, b; Raamsdonk et al. 2000; Gurushidze et al. 2007).

Normally, three major varietal groups have been recognized within *Allium cepa*. These are the common onion group (var. *cepa*), the aggregatum group (var. *aggregatum*) and the proliferum group (var. *proliferum*) (Jones and Mann 1963). The first group comprises strains exhibiting maximum diversity in photoperiod, storage life, flavour, skin colour and pungency. All the commercially important bulb onions belong to the common onion group. The group is characterized by plants having large, usually single bulbs with inflorescences devoid of bulbils and reproducing usually by sexual means. Aggregatum group resembles common onion group in bearing inflorescences devoid of bulbils and could be seed producing but differs in having clusters of bulbs and shoots produced from a single planted bulb. Occasionally, however, these are sterile dependent on almost exclusive vegetative propagation. It is further distinguished into three forms namely; the potato or multiplier onion, the ever-ready onion and the commercial shallot of US and Europe (Jones and Mann 1963). The potato or multiplier onion group stands out in bearing fairly large, oblate and broader than long bulbs with numerous laterals enclosed in outer scales. Laterals produce separate tops and bulbs in the second year of growth; flower heads either do not appear or do so sporadically. Ever-ready or the Welsh onion 'resembles the biennial, large bulbed forms of *A. cepa* in floral characteristics and leaf morphology but is perennial with prolific growth, much narrower bulb and narrower leaves, a shorter flower stalk and smaller umbel' (Stearn 1943). Plants of this group rarely flower and propagation is vegetative (Jones and Mann 1963).

Shallots differ from the common onion group as these produce a cluster of bulbs from a single planted bulb. Though smaller than *A. cepa*, the flowers and inflorescences are typical of common onion. Some shallots do not flower at all, others do so rarely while still others flower freely and set abundant seed. Shallots differ from ever-ready onion in that they produce well-defined bulbs, the tops die and the plants enter a state of rest as in the common onion group. They differ from the potato and multiplier onion mainly in bulb size. Their laterals do not aggregate but separate as individual bulbs. Earlier named as *A. ascalonicum* L., shallots are now grouped under *A. cepa* var. *aggregatum* (Rabinowitch and Kamenetsky 2002). Similarly, a number of hybrids are cultivated that have *A. cepa* as one of the parents. These include the top onion, the tree onion, the Catawirsa onion, the Egyptian onion (*A.* × *proliferum*), Wakegi onion (*A.* × *wakegi*) and the triploid onion (*A.* × *cornutum*) which fall under the *proliferum* group (Jones and Mann 1963). Plants of this group bear poorly developed bulbs and inflorescences with preponderance of bulbils and no true seed formation. Propagation is, thus, vegetative by bulbils. It includes two botanical varieties; viviparous *A. cepa* var. *viviparum* (Metz.) Alef. and proliferous *A. cepa* var. *proliferum* Targioni-Tozzetti. While some workers do not differentiate between these two, others consider them as distinct taxa (see Jones and Mann 1963). Further, the proliferous races of *A. cepa* have often been confused with *A. canadense* L, a North American species with races having bulbiferous umbels.

Shallots, a substitute for bulb onion, are preferred in south East Asia and some African countries because of the difficulty to grow and produce true seeds of common onion. Morphological and RAPD analyses have revealed that sexually propagated shallots are closer to common onion than vegetatively propagated ones (Le Thierry D' Ennequin et al. 1997). This is because these have both sexual and asexual forms and are more closer to European than tropical onions (Le Thierry D' Ennequin et al. 1997). France and Italy have long been cultivating a specific form of shallots called grey shallot. A comparative analysis of the isozyme profile of 30 accessions of the French grey shallot, 466 accessions of bulb onions and other shallots, 15 accessions of *A. oschaninii* and 22 of *A. vavilovii* revealed a greater and close relationship of grey shallot with *A. oschaninii* and *A. vavilovii* than to *A. cepa* (Maass 1996). Interestingly, the scape and flowers of its bolting mutant 'Grisombelle' strongly resemble Central Asian *A. oschaninii*. RAPD and GISH results confirmed that grey shallots belong to *A. oschaninii* (Friesen and Klaas 1998).

Triploid onion, i.e. *Allium* × *cornutum* (Clementi ex Visiani 1842) with $2n = 3x = 24$, grown as a minor garden crop, is mostly consumed in south eastern Asia and Europe because of its uniquely tasting bulbs and leaves. Known earlier as *A.cepa* var *viviparum* (Metzg.) Alef. (Jones and Mann 1963; Langer and Kaul 1983; Puizina and Papes 1996, 1997; Puizina et al. 1999) its nomenclature was later changed to *A.* × *cornutum* because of its hybrid origin. A Dalmatian bulbiferous taxon *A. cornutum* Clementi ex Visiani described first from a rocky shore at Budva by Visiani (1842) was affiliated to *A. cepa* by Stearn (1980). Since its description is in agreement with that for triploid viviparous onions therefore as per Stearn (1980) these plants may have escaped from cultivation and the name could be used as *A.* x *cornutum* Clementi ex Visiani for triploid onion unambiguously. While confirming its highly heterozygous karyotype, cytogenetic analyses also pointed towards its complex triparental genomic origin. Based on the detailed comparison of morphology, pmc meiosis and karyotypes of *A. cepa, A. fistulosum* and Pran (i.e. *A.cepa* var *viviparum*) it was thought to have originated from a cross between *A. cepa* and *A. fistulosum* (Singh et al. 1967; Koul and Gohil 1971). While Singh and his colleagues (1967) on the basis of karyotypic studies considered it a genomic allotriploid of AAB types, Koul and Gohil (1971) combining karyotypic studies and pairing behaviour during meiosis favoured segmental allotriploid status for it with the genomic formula A A′ A″. Subsequently, *A. cepa* L. and *A. roylei* Stearn. were considered as the two putative parents, whereas the third remained unknown for a pretty long time. While *A. fistulosum* was excluded as a progenitor by Puizina and Papes (1996), Maass (1997a, b) removed it along with *A. roylei* from the list of putative ancestors of triploid onion. They opined and confirmed that one of the three species of section Cepa (*A. cepa* L., *A. oschaninii* O. Fedtsch and *A. vavilovii* M. Popov et Vved.) to be the parent of triploid onion. In 2014, Fredotovic's group worked out the sequences of ITS1-5.8S—ITS2 of 35S rDNA and the NTS of 5S rDNA of this triploid and its relatives within section *cepa.* Analysis of the sequence data revealed intra-individual variation in triploid onion which clustered into three main clades each with high-sequence homology to one of the three other species of section *cepa*; *A. cepa, A. roylei* Stearn. and unexpectedly the wild Asian species, *A. pskemense* B. Fedtsch. Inferences drawn on the basis of these studies assign the third hitherto unknown parental status to *A. pskemense*. Subsequently using double FISH in *Allium* x *cornutum* chromosomes 35S and 5S rRNA genes were localized to their respective places in the three putative parents; *A. cepa, A. roylei* and *A. pskemense* (Lepen and Puizina 2011). GISH corroborated these results (Fredotovic et al. 2014). It is now confirmed that *A.* x *cornutum* has triparental genomic origin with *A. cepa, A. roylei* and *A. pskemense* as its parents. The triparental origin is in consonance with the conclusions drawn by Koul and Gohil (1971) on the basis of the cytological studies.

Top onion or tree onion or Egyptian onion, are the common names of diploid viviparous onion. Classified earlier as a variety of common onion, it was named as *A. cepa* var. *proliferum* (Targioni-Tozzetti) or *A. cepa* var. *viviparum* (Metzg.) Alef. (Helm 1956; Jones and Mann 1963; Mc Callum 1974; Puizina and Papes 1996; Puizina 2013). Now it has been established that these plants are the spontaneous hybrids between *A. cepa* (common onion) and *A. fistulosum* (Japanese bunching onion) and should be named *A.* × *proliferum* (Moench) Schrad. (Schubert et al. 1983; Puizina 2013). It had been described first as *Cepa prolifera* from the Botanical Garden of University of Marburg (Moench 1794) and is widely cultivated for pickling its bulbils (Ker Gawler 1812). It differs from *A. cepa* var. *viviparum* Metzger in frequent absence of flowers. The plants are somewhat gigantic with poorly developed bulbs and inflorescences enveloped by a long spathe. Karyotypically it is an allodiploid with $2n = 16$ and regarded a hybrid with one genome contributed by *A. fistulosum* and other by *A. cepa* (Havey 1991a, b). Propagated vegetatively in Europe, North East Asia and North America, its hybrid origin has been proved by C-banding pattern of karyotype, meiosis (Puizina and Papes 1995, 1997, 1999; Puizina 2013); isozyme analysis (Maass 1997a); RAPD markers (Friesen and Klaas 1998), GISH experiments (Friesen and Klaas 1998; Puizina and Papes 1999) and localization of 5S rRNA loci (Hizume 1994). Another hybrid with a similar genomic constitution is Wakegi onion, i.e. *A.* x *wakegi* Araki (Hizume 1994). Cultivated widely in Asia, China and Japan, the *A. cepa* parent of wakegi onion is a shallot and that of top onion a common onion (Maass 1997b). To assess the origin of *A.* x *wakegi*, RAPD and PCR-RFLP analyses were conducted among collected accessions of shallot and *Allium* x *wakegi.* Together the results indicated that *A.* x *wakegi* originated from an interspecific hybridization between shallot as the maternal and Welsh onion as the paternal parent as well as from reciprocal crosses (Arifin et al. 2000). As mentioned earlier shallots, either *A. cepa* var. *ascalonicum* or *A. cepa* Aggregatum group, are more popular and important than common onion in Southeast Asian countries. In some places of Indonesia cultivation of shallot and wakegi onion is mixed probably due to limited morphological and physiological divergence between the two species, and that this place must be one of the germplasm centres of *A.* x *wakegi.*

In spite of *A. cepa* differing from *A. fistulosum* in the colour and shape of perianth, leaf anatomy, degrees of bulbing and scape swelling a close phylogenetic relationship is thought to exist between the two. The view is further supported by identical number ($2n = 2x = 16$) of similar chromosome types, cellular DNA amounts; 'cladistic analysis of RFLPs in chloroplast genome and naturally occurring, sterile hybrids (*A.* × *proliferum*) that reproduce asexually by vivipary' (Bark et al. 1994).

Relationships between 6 species of *Allium,* namely *A. cepa, A. galanthum* Kar. Et Kir., *A. vavilovii, A. oschaninii, A. fistulosum* and *A. altaicum* Pall., are rather complex. That between *A. altaicum* and *A. fistulosum* and, *A. cepa* and *A. vavilovii* have been proved by a high level of crossability/fertility; rest exhibit extremely low cross-fertilities among them. Results indicated that *A. oschaninii* was closely related to *A. galanthum* than to *A. cepa* and *A. vavilovii* with the latter an intermediate progenitor of *A. cepa*. While *A. altaicum* grows wild in Northern parts of Mongolia and Transbaikala, *A. fistulosum* is thought to exist only in cultivation and is morphologically similar to *A. altaicum*. Both are interfertile leading Bradeen and Havey (1995a, b) to propose that *A. altaicum* may be the progenitor of or conspecific with *A. fistulosum*. Raamsdonk et al. (1997) on the basis of the morphological and RAPD data supported this view; both have weak sister group relation with *A. galanthum* (see also Dubouzet et al. 1997). Freisen and his group (1999) confirmed the single origin of *A. fistulosum* from *A. altaicum* using data on RAPD and non-coding chloroplast DNA.

Cytoplasmic male sterility (CMS) was induced in *A. fistulosum* by cytoplasmic substitution with the cytoplasm of a wild species *A. galanthum* Kar. et. Kir. (Yamashita et al. 1999a). Further studies revealed the presence of a single dominant fertility restoring gene (R_f) whose origin was from the nuclear genome of *A. galanthum.* Isozyme and RAPD markers completely linked to the R_f gene for a cytoplasmic male sterile *A. fistulosum* with the cytoplasm of *A. galanthum* were developed. Results exhibited interspecific polymorphism between the two species. All male fertile plants in B_3, B_4 and B_5 generation had both alleles of *A. galanthum* and *A. fistulosum* at *Pgi-1* locus, while male sterile had only the allele of *A. fistulosum* indicating that *Pgi-1* of A. *galanthum* is linked to the R_f locus. Similarly, all male fertile plants had *A. galanthum* specific band (OPEO 3700) which was lacking in all male sterile ones.

Earlier using alien monosomic addition lines of *A. fistulosum* with extra chromosome from shallot (*A. cepa* Aggregatum group) led to the location of *Pgi-1* on chromosome 5A of shallot (Shigyo et al. 1995c). Since chromosomes of *A. fistulosum* are homeologous to those of *A. galanthum,* it was assumed that *Pgi-1* is located on chromosome 5G of *A. galanthum* implying thereby that R_f gene is also present on the same chromosome (Yamashita et al. 1999b).

In garlic Helm (1956) recognized three botanical varieties, namely var. *sativum*, var. *ophioscorodon* (Link) Doll and var. *pekinense* (Prokh.) Maekawa. Later in

1978, Kazakova proposed a scheme with two subspecies, ssp. *sativum* and ssp. *asiae-mediae* Kaz., with two forms each to classify a collection from former Soviet Union. However, there was no delimitation against *A. longicuspis* Regel. Some authors believe that *A. longicuspis* Regel. is the wild progenitor of cultivated garlic (Figliuolo et al. 2001). This is supported by the formation of a fertile hybrid in the crosses made between *A. longicuspis* and some lines of seed bearing garlic (Etoh 1985).

11.4 Importance Vis-a-Vis Chemical Composition

Compared to other fresh vegetables, onions are relatively high in food energy, intermediate in protein content and rich in calcium and riboflavin (Jones and Mann 1963). Fresh bulbs also have vitamin A, thiamine, niacin and ascorbic acid with carbohydrates as the major component of the bulb dry matter.

A. sativum, a natural storehouse of innumerable chemicals, also has abundant carbohydrates (77 % of their dry weight) in its bulbs. These include non-structural as well as structural carbohydrates like glucose, fructose, sucrose and a series of oligosaccharides, the fructans. The adult plant is equally rich in proteins, pectins, minerals, polyamines, saponins and selenium (Kamenetsky 2007). Contrary to this, carbohydrates in onion are devoid of fructose polysaccharides and fructosans.

The characteristic odours of garlic and onion are caused by the volatile oils comprising dimethyl disulphide, diallyl di- and tri-sulphides, methyl-*n*-propyl disulphide, methyl allyl di- and tri-sulphides, di-*n*-propyl disulphide, *n*- propyl allyl disulphide and diallyl disulphide (Bernhard 1970). It may, however, be pointed out that such a smell is not exclusive to *Allium* for some species of other genera like *Ipheion, Androstephium, Hesperocallis, Tulbhagia, Nectaroscordum* and *Milula* too possess similar odour (Jones and Mann 1963). Many of these organo-sulphur compounds like diallyl di- and, tri-sulphides, methyl allyl trisulfide and propyl allyl disulfide are responsible for the characteristic flavour and potent biological health benefits of garlic (Edris and Fadel 2002). It contains 3–100 times higher levels of sulphur compounds in comparison to other plant species (Kamenetsky 2007). Onion, on the other hand, has methyl and propyl derivatives of alliin (Takada et al. 1994) but no true alliin. That is why garlic oil largely yields allyl compounds and onion oil methyl and propyl compounds (Jones and Mann 1963). Methyl homologue of alliin is found in *Ipheion uniflorum* (Alliaceae) and four dicot families but none in any other *Allium* species.

Onion germplasm is rich in the variety of colours that its bulbs are known to exist in. Bulb colour, an important trait in onion, ranges from different shades of white to yellow to chartreuse to gold to red and many others. According to Slimestad et al. (2007) 54 kinds of flavonoids, present in onion, are responsible for imparting colour to its bulbs. The most abundant flavonoid in bulbs and flowers is quercetin along with pigments like anthocyanins. Because of this, the bulbs and flowers of onion have been used in German households for dyeing Easter eggs, cotton, woollen and linen fabrics. The colours imparted thereof vary from yellow and its different shades through brown to grey. Tannin-like substances are thought to contribute to brown and deep brown colour formations. Flavonoids besides their involvement in UV protection, fertility and pigmentation in plants (Shirley 1996) also act as antioxidants (Lotito and Frei 2006).

Early uses of onion as medicinal ranged from a cure for small pox, to hysteria, pulmonary conditions, lowering blood pressure, as anthelmintic, intestinal antispasmodic and rubefacients or irritants (Jones and Mann 1963). It is also known to be protective against stomach cancer and certain infections, and is known to be beneficial to asthmatics (Sharma 2014). Medicinal importance of onion has been attributed to the presence of quercetin acting as a blood thinner to prevent clot formation; lowers blood pressure, triglycerides, total body cholesterol and boosts HDL; fights asthma, chronic bronchitis, hay fever and so on. Similarly, garlic is prized as a potent antioxidative (Kim et al. 1997; Dhawan and Jain 2005; Park et al. 2009), antithrombotic (Sata et al. 2001), antituberculosis (Pooler and Simon 1994), antidiabetic (Modak et al. 2007), antibronchial (Afzal et al. 1985), antigenotoxic (Park et al. 2009) and anti-inflammatory activities. When raw garlic bulb is crushed, alliin (*S*-allyl-*L*-cystein sulfoxide) is converted by alliinase into allicin (diallyl disulphide-*S*-oxide), the flavour producing compound. It is also known to enhance the effects of monosodium glutamate and disodium inosinate which are added in Chinese and Japanese soups and dishes. These, thereby, improve the taste of the food (Ueda et al. 1990). Spraying with garlic powder increases the shelf life and prevents spoilage in processed foods (Kumar and Berwal 1998). In fact, allicin is responsible for imparting antibacterial and antifungal properties to garlic (Hayashi et al. 1993). However, because of its instability it is rapidly converted to diallyl disulfide (DADS) and other sulphur compounds. One of the derivatives of allicin is Ajoene. Chemically (E, Z)-4, 5, 9-trithiadodela-1, 6, 11-triene-9-oxide, Ajoene is a major sulphur-containing compound in oil-macerated extract and reported to have antimicrobial, antiviral, antiprotozoal, antimutagenic and anticancer (Nishikawa et al. 2002) activities. It is also antithrombotic preventing platelet aggregation and exhibits strong hepatoprotective effects (Hattori et al. 2001).

Numerous studies have demonstrated that major S compounds, namely *S*-allylcysteine, diallyl disulfide and allicin in garlic extracts inhibit the activation of nuclear factor kappa B (NF-k B), a central transcription factor. A rise in the activity of NF-k B is known to be associated with an enhanced risk of arthritis, atherosclerosis, cardiovascular diseases, Alzheimer's disease and cancer (Youn et al. 2008 and references therein). Similarly many bioactive compounds are known to control atherosclerosis by reducing cholesterol level in the body (Afzal et al. 1985) or by acting as a vasodilator (Ganado et al. 2004). Abdo and Al-Kafawi (1969) have recommended garlic juice for treating various forms of ulcers, in regulating menstrual cycle, as a rubefacient in skin diseases, as an antipyretic and lotion for washing the wounds and ulcers.

11.5 Chromosome Biology

With large chromosomes amenable to all types of observations, *A. cepa* and other species of the genus have been the favourite of everybody—students to hardcore cytologists as well as those in search of good systems for their experimental work. On account of these qualities studies on *Allium* chromosomes were initiated as early as 1898 when Nemec and Schaffner individually described the cytology of *A. cepa*. Voluminous literature is now available on the cytology of genus *Allium*. In view of the limited coverage envisaged, that too on the species available in India, only the most important and significant works are referred to.

Three basic chromosome numbers $x = 7$, 8 and 9 have been suggested for *Allium* (Stearn 1946a, b; Darlington 1937, 1940; Eid 1963; Fedorov 1969) with $x = 8$ as primitive and $x = 7$, 9 as derived. All the studies conducted and conclusions drawn point towards $x = 8$ as the primary base number from which the other numbers are derived (Mensinkai 1939; Ved Brat 1965; Traub 1968; Kollmann 1973). Exceptions like *A. kujukense* Vved., *A. decipens* Fisch. ($x = 10$) and, *A. hookeri* Thwaites and *A. chamaemoly* L. ($x = 11$) (Vakhtina 1964; Pedersen and Wendelbo 1966; Tzanoudakis 1992; Sharma et al. 2011c) also appear to have derived base numbers. Majority (~65 %) of the species of the genus are diploid (Fedorov 1969) with polyploidy having been reported in about 35 % of the species (Hamoud et al. 1990).

Allium cepa, the most favoured and well-worked species is uniformly diploid with $2n = 16$. Anomalies like structural heterozygosity etc. have, however, been reported in some stray collections (Koul 1963; Sharma and Gohil 2011b). As mentioned earlier, this most widely cultivated and consumed species is a favoured material for various cytological and experimental studies. It has long chromosomes that stain well. Two chromosomes of the complement bear distinct satellites. Chromosome complements of its various groups like cepa, shallot, proliferum, etc., differ from each other with respect to various chromosome metrics. While cepa and shallot exist only in diploid form ($2n = 16$), the proliferum group has variants with $2n = 16$ and 24. Both the cytotypes of the proliferum group are of hybrid origin as is evident even from their karyological parameters. Interestingly both the diploid and triploid cytotypes have only a single chromosome with NOR. On account of their hybrid nature these cytotypes undergo abnormal meiosis that renders them sterile. Compared to these the cepa and shallot group strains are mostly fertile.

Insofar as the chromosome number of garlic is concerned it is uniformly diploid with $2n = 2x = 16$ (Mensinkai 1939; Verma and Mittal 1978; Roy 1978; Etoh et al. 1992; Eom and Lee 1999; Yuzbasioglu and Unal 2004; Ipek et al. 2005; Chuda and Adamus 2009), excepting two reports of $2n = 12$ (Banerjee 1980) and $2n = 18$ (Sharma and Bal 1959; Etoh 1986). Apart from these stray reports of numerical variation, considerable heteromorphism at the structural level in its chromosomes is very well documented. Variation in arm ratio and total chromosome length among the European clones was documented by Battaglia (1963) despite strict uniformity in the total number of chromosomes ($2n = 16$) and also number of nucleolar chromosomes (4) per complement. Subsequent to Battaglia's (1963) work,

Konvicka and Levan (1972) reported 2–4 nucleolar chromosomes in the complements studied by them. Similar observations were recorded by Verma and Mittal (1978), Koul et al. (1979), Langer (1980), Sato et al. (1980), Sharma et al. (1988), Wajahatullah and Vahidy (1990), Sharma et al. (1992) and Yuzbasioglu and Unal (2004). Based on both conventional and more specific silver staining schedules, these workers confirmed the earlier observations on the variable number and metrics of nucleolar chromosomes. On the basis of all these reports it can be safely presumed that the most common number of secondarily constricted chromosomes per complement in *A. sativum* is 4 followed by 3 and 2.

Since 1970, when Gohil and Koul first reported meiosis and subsequently desynapsis (Gohil and Koul 1971) in totally male sterile *A. sativum* (Koul and Gohil 1970b), a number of new observations have been made on the reduction division in the species. While Etoh and Ogura (1978) and Etoh (1979) reported multivalents in 43 clones of *A. sativum*, Pooler and Simon (1994) found 8 regular bivalents at metaphase I with normal segregation at anaphase I in all the 29 clones studied by them. Other anomalies like desynapsis etc. have also been reported in this species. Irrespective of meiosis being normal or abnormal, the strains studied by all these workers are sexually defunct. In the light of this, the prospects of improving garlic for the production of variants and elite genotypes through conventional hybridization techniques have not been possible for breeders. Koul et al. (1979) in view of garlic's strong apomictic nature advocated that induced somatic mutations are likely to widen the spectrum of variability leading to identifying, isolating and maintaining the more promised lines. Treatment of garlic with diethyl sulphate and gamma rays resulted in some clastogenic changes as manifested by many anomalies including the formation of centric and acentric ring chromosomes (Gohil and Koul 1983). Besides the cytological changes, morphological variants were also observed in M_1 progeny.

Besides *A. cepa* group, the commonly cultivated *A. ampeloprasum* (leek), is a tetraploid ($2n = 4x = 32$) with centric localization of chiasmata (Koul and Gohil 1970a). While centric localization has sexually stabilized the seemingly segmental allotetraploid species by ensuring normal meiosis, possibility of increasing spectrum of variability has got limited. Most of the variability through sexual means is perhaps generated because of independent assortment and not recombination (Gohil and Koul 1977).

Other less grown species *A. tuberosum* has also attracted the attention of many cytologists on account of its tetraploid nature ($2n = 4x = 32$) coupled with abnormal meiosis but followed by abundant seed set through apomictic development of the ovule (Gohil and Kaul 1979, 1980, 1981, 1983; Sharma and Gohil 2011a, 2013).

Polyploidy in both wild and cultivated species of the genus is accompanied by some genetic evolutionary process which have compensated for the initial adaptive disadvantage, attendant in raw polyploids. These processes or mechanisms include centric chiasma localization, imposition of genetic changes, bulbifery and vegetative reproduction (Gohil 1973, 1979, 1998; Gohil and Koul 1971, 1978, 1981a, b).

Some species of the genus, both cultivated and wild have also been assessed for their recombination potential in the two sexes. These works have clearly shown that while in some species there is no difference between the two sexes, in others some compensatory mechanisms are at work (Kaul 1980; Gohil and Kaul 1980, 1981; Sharma and Gohil 2011a). Mechanism of apomixis has also been worked out in *A. tuberosum* (Kaul 1980; Gohil and Kaul 1979, 1981, 1983; Geeta 2005; Sharma and Gohil 2013). Such works can help breeders in planning future strategies.

11.6 Molecular Studies

When compared to other major crops, molecular studies on species of *Allium* commenced rather late (Rabinowitch 1988). Acquisition of knowledge at the molecular level and application of modern tools and techniques in this otherwise well-studied genus are still lagging. Work on RAPD, AFLP and DNA fingerprinting to study diversity at intra- and interspecific levels (Kamenetsky and Rabinowitch 2006), and vector-mediated and direct gene transfer systems for *Allium* species have, however, been initiated (Havey 2002, 2013; Havey and Leite 1999; Havey et al. 1996, 2001).

Determination of the sources of CMS genes and the corresponding restorers, a very important trait for breeding programme; inheritance patterns thereof and those of other economically important traits have also been achieved to some extent through the use of molecular markers (Havey 2002, 2013; Havey and Leite 1999; Havey et al. 1996, 2001). For instance, the most widely used source of CMS, the S cytoplasm (Jones and Clarke 1943) is in all likelihood an alien cytoplasm transferred to onion via *Allium* × *proliferum* (Moench.) Schrad. (Havey 1993) and differs from N cytoplasm (Courcel et al. 1989; Havey 1993; Sato 1998). Another one is that of T cytoplasm (Berninger 1965; Schweisguth 1973). To distinguish them from S cytoplasm, N and T were assigned to M group (Courcel et al. 1989) for their most likely origin from M cytoplasm which is thought to have arisen from the wild species, *Allium vavilovii* M. Pop.et Vved. (Havey 1997). S cytoplasmic sources have been independently obtained from Italian Red (Jones and Emsweller 1936) and Nasik White Globe (Pathak and Gowda 1993). de Vries and Wietsma (1992) screened out a male fertile hybrid in the progeny of a cross between Hygro F_1 as the male sterile female parent and *A. roylei* Stearn. According to them *A. roylei* possessed nuclear factors which could restore male fertility for onion male sterile cytoplasm. Later Havey (2000) demonstrated that Hygro F_1 possessed a T-cytoplasmic like CMS system and also that male restorers in *A. roylei* may not be the same as the Ms locus for S cytoplasm. Together these results indicate that 'identical or very similar, male sterile cytoplasms have been independently isolated and exploited for hybrid seed production'.

Onion bulb colours too have been the subject of interest for geneticists on account of the tremendous diversity found worldwide. Inherited in a complex manner and involving epistatic interaction, onion bulb colours are probably under the control of loci that might code for enzymes required for anthocyanin synthesis (El-Shafie and Davis 1967; Koops et al. 1991). Kim et al. (2004, 2005a, b, 2007, 2009) have traced the colour variation in onion bulbs to mutations at several steps in the anthocyanin biosynthesis pathway.

'In *Allium*, the RAPD analyses have been performed to assist breeding for resistance (de Vries et al. 1992a, b, c), examine variability among species (Wilkie et al. 1993) and cultivars (Roxas and Peffley 1992), assess integrities of inbred (Bradeen and Havey 1995b) and double haploid lines (Campion et al. 1995). This technique has also helped to confirm hybridities (Friesen et al. 1997), study phylogenetic relationships' (Maab and Klass 1995; Hong et al. 1996; Arifin et al. 2000), assign chromosomal locations to some markers (Shigyo et al. 1997) and confirm the monophyletic origin of *A. fistulosum* from *A. altaicum* (Friesen et al. 1999) (see Shigyo et al. 1997).

Analysis of RAPD markers from 07 cultivars of *A. cepa* and one each of Japanese bunching onion, chive, leek and *A. roylei,* indicate that *A. roylei* is the closest relative of *A. cepa* (Susan et al. 1993). Based on these results, Susan et al. (1993) questioned the current placement of *A. cepa* in the section cepa of subgenus Rhizirideum. Using 90 primers Sangeetha et al. (2006) classified 24 cultivars of onion into those belonging to Northern and Southern regions of India while 10 varieties were distinguished by Maniruzzaman et al. (2010) into genetically similar and dissimilar groups. These were Faridpuri and Bhati, and Bermi and India-2 respectively. Besides RAPD, ISSR, SSR and EST-SSR markers have been employed in studying various parameters of *A. cepa.* Using ISSR, 32 cultivars were grouped into 05 groups, namely Yellow Sweet Spanish, Bejo Daytona, Yellow Globe, Yellow Globe Danvers and Yellow Danvers (Qijiang and Jia 2007). On the basis of 398 SNPs, indels and SSRs Jakse et al. (2005) distinguished 35 elite onion populations. In India, Mahajan and his colleagues (2009) determined the extent of differences between 14 short-day and 2 long-day cultivars of exotic and native origin. The exotic cultivars (Alisa Craig and Brigham Yellow Globe) differed from the Indian ones. Two of the latter (Nashik Red and Poona Red) were indistinguishable and other two (N-53 and Bombay Red) quite close to each other. This work was subsequently followed by the diversity assessment of Tropical Indian Onion by cross-amplification of genomic and EST-SSR markers (Khar et al. 2010).

Isozymes (or isoenzymes), the powerful molecular tools for examining gene variability within and between populations of plants and animals (Zeidler 2000), were the first molecular markers to have been applied to *Allium* species to study the intraspecific diversity in *A. cepa* (Makinen 1968; Loaiza-Figueroa and Weeden 1990). Many workers used isozymes as markers to study variation and compared the results so obtained with the ones obtained from RAPD. Maab and Klaas (1995) subjected 300 accessions of *A. sativum* to isozyme analysis. Of these, 48 were taken up for RAPD and the two marker systems compared. Their collection had many accessions from areas close to the centre of origin, i.e. Central Asia and, hence seemed suitable for investigating the genetic relationship between cultivated clones with primitive features, derived strains and an uncultivated accession of *A. longicuspis*. Although both the markers gave a good general delimitation of varieties *sativum* and *ophioscorodon*, RAPD markers allowed a more detailed distinction; even bolting and non-bolting types could be separated. Isozyme patterns classified 300 clones into 16 types which could further be divided into 4 groups. These are the *ophioscorodon*, the *subtropical*, the *longicuspis* and the *sativum* groups. The *ophioscorodon* group is distributed from Central and Eastern Europe to the Caucasus, the *sativum* in the Mediterranean and the *subtropical* seems to have originated a long time ago from the *longicuspis* group, perhaps in northern India. On the basis of this work it was proposed that the *longicuspis* group was the oldest and the original one. Similar work on 20 Australian accessions was carried out by Bradley et al. (1996) with five RAPD primers. While investigating 27 garlic cultivars Al-Zahim et al. (1997) obtained results which differed in some interesting aspects. Sixty three polymorphic bands were generated from 26 primers. Based on these results, 11 accessions each were assigned to variety *ophioscorodon* and variety *sativum*, and 5 to *A. longicuspis*. In agreement with Maab and Klaas (1995), the accessions of variety *sativum* grouped together. Since only its non-bolting accessions were included, the conclusions thus drawn have their limitations.

Although RAPD markers in garlic have been extensively employed to study the phylogenetic relationships, Hong et al. (1997) tried to find the ones related to the pollen fertility traits. Twelve pollen fertile and sterile clones were screened using 60 RAPD primers to identify the DNA fragment related to pollen fertility. Of the 60, two viz. $OPJ12_{1300}$ and $OPJ12_{1700}$ amplified only in the pollen fertile clones with the primer OPJ12 (5'-GTC CCG TGG T-3'). The selected primer was then tested on all the 60 clones. Two markers were present in all of the 31 pollen fertile clones and absent in the 29 sterile ones with the exception of one ($OPJ12_{1300}$) which amplified in three sterile clones. These markers could, therefore, provide substantial evidence in favour of the genetic control of fertility.

Similar studies were carried out on varying number of cultivars by different groups from different countries (Eom and Lee 1999; Xu et al. 2001; Paredes et al. 2008). Eom and Lee (1999) used 15, 18–24 nucleotide long primers, 12 of which produced polymorphic bands. The remaining three which yielded monomorphic bands were found useful in detecting genetic variation among four species of *Allium,* namely *A. fistulosum*, *A. cepa*, *A. victorialis* var. *platyphyllum* and local garlic (*A. sativum*) cultivar.

Paredes et al. (2008) applied RAPD to 65 garlic clones collected from Chile but introduced from different countries and maintained ex situ in the Germplasm Bank at INIA (Instituto de Investigaciones Agropecuarias). Forty primers generated a total of 398 bands, 87 % of which were polymorphic. Despite this magnitude of polymorphism, the clones analysed had little genetic diversity. The dendrogram grouped the accessions into two major clusters. The first group including 44 clones (68 %) is homogeneous, with 100 % genetic similarity. The second group clustered 21 clones (16.9 %), with a relatively high genetic diversity. Clones of different origins and with variable phenotypic characteristics were found clustered together in both these groups. For instance clones with and without floral stalks got distributed in both the clusters. Previously, however, Al-Zahim et al. (1997) narrowed down to a group of accessions that differentiated floral stalk from those that did not. The explanation Paredes et al. (2008) gave for such a condition was that genotypes and RAPD primers used by them were different. This observation of a paradoxical relation between RAPD and geographical origin has also been made by Abdoli and his co-workers (2009). Al-Zahim et al. (1999) also applied this technique to quantify somaclonal variations in plants regenerated from long term callus cultures of garlic. They observed a novel band in two regenerants that was having the same size as an original band of another parental clone. It suggested that a specific region in the garlic genome may be extremely susceptible to mutation generated by in vitro culture conditions and in all likelihood resemble somatic mutations occurring in nature and being the source of forming new clones in an otherwise sterile species (Al-Zahim et al. 1999).

Using AFLP (Amplified Fragment Length Polymorphism) on Argentinean clones, Lampasona et al. (2003) could establish an appreciating association between AFLP and geographical origin of the clones. This is in agreement with Lallemand et al. (1997) who could associate geographical origins of some cultivars with their isozyme patterns. Similarly, clustering was also in accordance to the physiological groups and bulb colour. Ipek et al. (2003) developed several locus specific RAPD and AFLP markers which could be used as a 'tool for rapid characterization of garlic germplasm collections' (see also Chinnappareddy et al. 2013). Recently, a distinct correlation was found to exist between the genetic basis of 135 garlic accessions and cysteine sulphoxide content. Two hundred eighty six AFLP fragments grouped 135 accessions in 08 clusters; two contained only *A. sativum* L. var. *ophioscorodon* clones, four grouped together both non-bolting and semi-bolting types and the remaining two only non-bolting ones. An efficient distinction could be achieved between var. *ophioscorodon* from var. *sativum* using these markers (Ovesna et al. 2011).

Using seven polymorphic ISSR markers developed from 31 Tunisian and 4 French classified genotypes of garlic, Jabbes et al. (2011) reported higher polymorphism in Tunisian germplasm than that generated by RAPD for Brazilian and Chilean germplasm (Buso et al. 2008; Paredes et al. 2008). On the other hand, Gantait et al. (2010) using 10 ISSR primers tested the genetic fidelity of in vitro regenerated clones of *A. ampeloprasum* and *A. sativum* refuting the view that 'long term multiplication of micropropagated plantlets might accelerate the rise of somaclonal and epigenetic variation'.

Peroxidase isozymes of 93 clones of *A. sativum* resolved into 16 types of zymograms which helped Etoh and Ogura (1981) to differentiate between clones from subtropical regions and those which did not develop any flower buds. They could also distinguish between the early and the late or medium cultivars but found no correlation between the chromosome configuration at meiosis and the patterns of peroxidase isozymes. Taking into consideration these profiles Etoh et al. (1992) compared various cultivars or strains of leek (*A. porrum*) and kurrat (*A. kurrat*) with fertile and sterile clones of *A. sativum*. They concluded that garlic and leek may have a common ancestor, and kurrat with 5 isozyme bands may have been derived from leek. Working on similar lines, Pooler and Simon (1993a) tried to draw a correlation between variation in the isozyme pattern and morphological traits in 110 diverse *A. sativum* and *A. longicuspis* clones. All these segregated into 17 distinct isozyme groups. Floral traits/characters correlated well with isozyme data, bulb related characters or geographical origin had little predictive value for the genetic relation of accessions. Unlike floral traits, bulb related traits were not associated with isozyme phenotypes. Clones of *A. longicuspis* did not display unique morphological or isozyme characters. Although, *A. longicuspis* is considered an ancestor of *A. sativum* by many workers, Pooler and Simon (1993a, b)'s study does not warrant the former as a separate species. These workers extended this technique to seedlings of *A. sativum* and confirmed their sexual and/or syngamic origin (Pooler and Simon 1994).

Shemesh et al. (2008) used NBS (Nucleotide Binding Site) profiling of the randomly selected seedlings raised from the seeds of a single mother plant. The results showed that though cross-pollination was dominant, self-pollination was also common in the fertile garlic—the first molecular evidence for the mixed pollination mode in the species. While 25 plants had a certain fraction of foreign DNA which is not present in the mother plant, two seedlings had DNA profile similar to the mother plant indicating their origin from possible selfings. Knowledge of this property of garlic could be handy in its future breeding programmes. However, this is subject to the availability of fertile clones of the species that can be multiplied and used.

Nearly 24 different isozymes have been characterized in onion seeds and roots (Rabinowitch 1988). These molecular markers have been primarily used to compare other *Allium* species with *A. cepa* and to identify the origins of chromosomal regions in interspecific hybrids (Peffley et al. 1985; Havey et al. 1996). Polymorphism in these markers is on record for different species of *Allium* including *A. cepa*, *A. fistulosum*, etc. (Loaiza-Figueroa and Weeden 1990; Magnum and Peffley 1994, 2005). The genetic variability in enzymatic patterns using six enzymes (G-6-PDH, AAT, PGM, EST, ACP, PGI) among populations of *A. carinatum* L. subsp. *carinatum* (wild garlic) were studied from the erstwhile Czechoslovakia. Coupled with mixed phenotype analysis, low among—population variability was reported on the basis of dendrogram of dissimilarity. Getting low values is attributed to clonality, geographic shaping, distance, weak dispersion, fixed genotype and action of somatic mutations (Zeidler 1999).

*A.*x *wakegi, A. cepa, A. fistulosum* and *A. altaicum* and 14 artificial diploid hybrids between the latter three species were analysed for 6 enzyme (GPI, IDH, PDG, PGM, SKD and TPI) systems and compared with top onion *(A. × proliferum)*. Results while confirming the hybrid origin of top onion, and *A.*

altaicum as the wild progenitor of *A. fistulosum*, could not distinguish between cepa and aggregatum groups (Maass 1997a, b).

Further, through isozyme analysis on 16 triploid clones and various accessions of *A. altaicum, A. cepa (cepa* group, *aggregatum* group, clone *perutile), A. fistulosum, A. galanthum, A. oschaninii, A.* x *proliferum, A. pskemense, A. roylei, A. schoenoprasum* and *A. vavilovii,* Puizina and Papes (1996) excluded *A. fistulosum* from the ancestry of triploid onions and narrowed down to either of the 3 species (*A. cepa, A. oschaninii* and *A. vavilovii*) being the parent of triploid onions. While agreeing with these results, Maaβ (1997b) also excluded A. *roylei* from the ancestry of triploid onions. Results of GISH confirmed that DNA of grey shallot is closely related to *A. oschaninii* than to *A. cepa* or *A. vavilovii* and originated parallel to common shallot (*A. cepa* var. *aggregatum*).

Many other species of *Allium* (*A. fistulosum* and *A. cepa*) have been subjected to isozyme analyses resulting in the determination of chromosomal locations of some genes (Peffley and Currah 1988; Haishima et al. 1993; Shigyo et al. 1994). To establish the genetic system of Glutamate dehydrogenase in section *Cepa*, Shigyo et al. (1995b) worked along similar lines and found predominant interspecific polymorphism. Similar studies have been carried out in section *Cepa* of *Allium* using phosphoglucoisomerase (PGI) (Shigyo et al. 1996). Polymorphisms in different PGM and ADH loci in the native varieties of *A. fistulosum* (Haishima et al. 1993) were used to assign few isozyme genes to specific locations on chromosomes (Peffley and Currah 1988; Shigyo et al. 1994). Similarly location of five isozyme gene loci (LAP-1, GOT-1, 6-PGDH-2, ADH-1 and GDH-1) vis-a-vis chromosomes has been accomplished in shallot, i.e. *Allium cepa* L. Aggregatum group (Shigyo et al. 1995a). Krzyminska et al. (2008) applied the technique to Alliums to verify the similarity of 13 species and 5 ornamental allium cultivars and classify them into particular groups based on morphological and isozyme variation. The data showed enormous inter- as well as intraspecific variability despite the mode of propagation being vegetative.

Studies on response of isozymes in *Allium* to thermal and aerobic stress to isolate stress tolerant variants have also been carried out. For this two species of *Allium, A. cepa* cv 'Temprana' and *A. fistulosum* var. 'Heshiko' were tested using 18 enzyme systems. Of the 18, only alcohol dehydrogenase exhibited differential expression in different stress conditions but the inducibility was inconsistent (Peffley et al. 2001). Employing isoelectric focusing in polyacrylamide gel with Ampholine pH 3.5–10.0, better resolution was obtained in MDH zymograms of 10 *Allium* species and 6 cultivars of *A. cepa* which could be used for their characterization (Hadacova et al. 1983). Similarly esterase isozymes were used to determine the degree of apomixis in *A. tuberosum* Rottl. ex. Spreng (Kojima et al. 1991, 1992; Kojima and Nagato 1992). While comparing parent and progeny plants of *A. tuberosum* for GOT, MDH and esterase enzymes, Geeta (2005) found similarity in the number and position of bands. This led her to conclude that the parent and progeny plants are similar in their enzymatic composition. She, however, could not conclusively establish its apomictic nature due to her inability to perform any controlled crossing. This was because of the non-availability of variants in the basic stock.

Cryder et al. (1991) carried out isozyme analysis of progeny derived from (*Allium fistulosum* × *Allium cepa*) × *Allium cepa*. Two backcross populations, BC 1034 and BC 1040, distinguished by different parents, were categorized for *Idh-1*, *Adh-1* and *Pgi-1*. Then, cell probabilities for a mixed population of F_2 and BC_1 progeny were calculated using statistical methods. The isozyme loci, *Idh-1* and *Pgi-1* appear to be linked with a map distance estimated at 33 Centimorgans (cM) in BC1034 and 42 cM in BC1040. The distorted segregation ratios for these pairs of loci are on account of genetic incompatibility between the two species. Studies have been conducted to determine DNA polymorphism in cytoplasm of *A. cepa* highlighting its implications regarding the origin of onions using RAPD and GISH (Courcel et al. 1989; Friesen and Klaas 1998). Besides, alien addition, cytoplasmic male sterile and alloplasmic male sterile lines were also obtained in some species of *Allium* (Peffley et al. 1985; Yamashita et al. 2002, 2005; Vu et al. 2012).

11.7 Agronomy and Improvement

11.7.1 Traditional

Genetic variation available in cultivated taxa or wild relatives has been used time and again by plant breeders to improve the commonly cultivated strains of various crops. Similar attempts have also been made at combining and recombining the germplasm over time to change/alter/modify the genetic structure of onions.

On account of their importance in the day-to-day life and consequently in the economy of both the farmers and the common people and the countries, agronomical aspects of *A. cepa* and *A. sativum* have been studied throughout the globe, especially in the prime onion and garlic producing countries (Ahmad and Iqbal 2002; Kakar et al. 2002; Mahmood et al. 2002; Bhuiya et al. 2003; Haque et al. 2003a, b; Rahman et al. 2003, 2004; Islam et al. 2004, 2007). Because of the varied nature of propagation available in different cultivated taxa of this genus, both agronomical practices and prospects of improvement vary from species to species.

According to the global review, onion ranks second in area and third in production in the world. FAO reports of 2005 place world production of onion at 57.91 million tonnes from 3.18 million hectare area. According to National Horticulture Research and Development Foundation (NHRDF) estimates of 2005, China tops the list both in area and production followed by India. In India, Maharashtra ranks first in onion production followed closely by Gujarat, Orissa, Karnataka, Tamil Nadu, Madhya Pradesh, Uttar Pradesh, Rajasthan and Andhra Pradesh in that order. Onion is the only commodity amongst fruits and vegetables where Indian figures are prominent than the world's production and export (Sharma 2014). Despite second ranking, production of onion in India is far too low, a mere 10.38 million tonnes per hectare as against 55.92 million tonnes per hectare in Korean Republic. Main reasons for low productivity have been attributed to traditional cultivation methods, use of local varieties, manual packaging practices and so on. In India, losses in onion yields are to the tune of 10–15 % on an annual basis which has been attributed to two fungal diseases and one pest. The three namely stemphylium blight, purple blotch diseases and onion thrip infestation have been considered of national concern while *Colletotrichum* blight was restricted to Maharashtra (Gupta et al. 1993).

As of now, a number of different *A. cepa* forms and varieties available have been produced by hybridization and selection which can be grown in diverse climates. In India, many types of onion are grown; big, small and multiplier. The former with bulb colours ranging from light red to dark red are grown in most parts of the country. Small onions fall into rose and Krishnapuram categories which are cultivated largely in Kolar (Karnataka) and Cuddapah (Andhra Pradesh) districts. The third or multiplier type called Podisu and shallots are restricted in cultivation to Tamil Nadu, Puducherry and Andhra Pradesh. NHRDF has improved all the three types and released six self raised varieties along with one [Arad (H)] from Hazera Seed Co. of Israel as shown in the table below.

Type	Variety	Bulb characters				Yield q/ha	Storage ability
		Colour	Shape	Maturity (Days)	Diameter (cm)		
Big	(i) Agrifound dark red	Dark red	Global round medium to big	90–100	–	300–400	Medium
	(ii) Agrifound light red	Light red	Global round, big, compact	120	–	300–325	Good
Small	Agrifound rose	Scarlet red	Flattish round	95–110	2.5–3.5	190–200	Good
Multiplier	Agrifound red	Brick red	5–6 bulblets per clump	66–67	2–2.5	180–200	Good
	Arad (H)	Yellow	Big sized, global round	90–100	6–8	500–800	Poor
	Pusa Red	Red	Flattish round	–	–	–	Very good
	Agrifound white	Whitish	–	–	–	–	Suitable for dehydration; long term storage

Sources http://www.nhrdf.com/htmlfiles/onion/onionexport.htm. Indian Horticulture Database (2011)

Varying in colour, shape, maturity, productivity and storage potential all of them are important from trading point of view as preferences vary from place to place both within and outside country. While 4–6 cm light to dark red onions are preferred by people in Middle East and Far East, 6–7 cm yellowish/brown with mild pungency by European countries and Japan, 3–4 cm sized onions are preferred in Bangladesh.

Except for a few stray reports (Etoh 1979, 1983a, b; Etoh and Ogura 1981; Etoh et al. 1992; Etoh and Simon 2002), *A. sativum* is completely sterile and is, therefore, propagated only vegetatively (Koul and Gohil 1970b; Verma and Mittal 1978; Nagakubo et al. 1993; Pooler and Simon 1993b, 1994; Maab and Klaas 1995; Barandiaran et al. 1999; Eom and Lee 1999; Haque et al. 1999, 2003a, b; Figliuolo et al. 2001; Goleniowski et al. 2001; Kamenetsky and Rabinowitch 2001; Sata et al. 2001; Kim et al. 2003; Kamenetsky et al. 2004; Paredes et al. 2008; Shemesh et al. 2008). Increase in genetic variation in this taxon, therefore, is possible only through random or induced mutations and/or somaclonal variation with the new cultivars so obtained bred by clonal selection (Kamenetsky and Rabinowitch 2001). In the absence of sexual reproduction variations can be fixed easily through asexual mode of propagation without the risk of their getting flushed out at any point of time and in any geographical location. This imparts an added advantage to the species by helping it tide over diverse ecological conditions encountered over its large geographical range (Singh et al. 2014). Strains of *A. sativum* face ever-changing environments be they climate changes, pollution or those brought about by diseases, pests or parasites. To cope with these changes, Panes et al. (2007) opined that garlic must either evolve or become extinct. As per Singh et al. (2014) *A. sativum* has chosen to remain in evolution. Since evolution is dependent on variation, mostly generated through sexual mode of propagation, garlic being an asexually propagating system had to find some other avenues to overcome this bottleneck. From the literature available it is quite apparent that this taxon has taken recourse to generate variations at the cytological/molecular level to meet this goal. Vast information generated on cytological variability in various collections of this species corroborates this view (Battaglia 1963; Konvicka and Levan 1972; Verma and Mittal 1978; Koul et al. 1979; Singh 2013).

Although garlic is regarded a representative of the obligate apomicts by Fryxell (1957), some fertile garlic clones have been reported from Soviet Central Asia (Etoh 1983a, b; 1985, 1986) and Campania region of Italy (Bozzini 1991). These findings are likely to provide a window to the garlic breeders for increasing its spectrum of variability, thereby improving this so far asexually propagating individuals through conventional breeding techniques. These reports have also initiated discussions as to whether garlic acquired sterility after it was brought under cultivation or sterile variants were already available in the wild form too. Sterility coupled with the ease of vegetative propagation might have been instrumental for its selection. As mentioned above this sterility has not hampered the ability of this species to acquire morphological, cytological and molecular variability (Singh 2013; Singh et al. 2009, 2014). As such local genotypes in Central Asia and adjoining regions are highly variable in economically important traits like disease/pest resistance and better adaptation to abiotic stresses. In order to broaden the variability range in the cultivated garlic, large collection of landraces and wild populations, some producing flowers and setting seeds, have been identified, characterized and maintained (Shemesh et al. 2008).

Despite reports of fertile lines of garlic, in all the cultivated stock sterility is widespread and its cause has been discussed several times. Khoshoo et al. (1960) attributed the failure of sexual reproduction to the degeneration of sporogenous tissue both within the anther and the ovule. However, Koul and Gohil (1970b) while describing normal pairing of chromosomes in its PMCs proposed that the sterility was caused by nutritional competition between flowers and bulbils. Since continuous removal of developing bulbils enhanced development of floral buds, Koul and Gohil (1970b) opined that the competition for the nutrients between flowers and bulbils is the apparent cause of the sterility in the species. On the other hand, Pooler and Simon (1994) are of the opinion that since garlic has been propagated asexually for many generations, accumulation of chromosome aberrations could have reduced its potential to form balanced gametes.

As per Etoh (1985) garlic clones might have acquired the non-bolting habit by artificial selection. According to him, since the bulbil formation on the top of the flower stalk frequently causes decrease of bulb mass, and the bulbils themselves are too small to be used economically, non-bolting condition with very small bulbils might have been preferred to complete bolting by the growers. This conclusion is in line with the observation that the farmers remove the scapes from the bolting types to divert the entire photosynthetic products towards the bulbs. The selective pressure by human beings may, therefore, have resulted in the evolution and mass cultivation of non-bolting from the complete bolting habit. According to Bozzini (1991), sexual reproduction in garlic is blocked in different clones at different stages of the floral organogenesis, i.e. at sporogenesis, gametogenesis, fertilization and finally during seed development.

Important garlic varieties released by NHRDF have been developed by mass selection from collections obtained from local farms and/or markets of different regions of India as summarized in the following table.

Population used from	Year	Name of the Variety	Features of the				Yield q/Ha	Storage Quality	Others	Recommendation w.r.t cultivation etc.
			Bulb		Clove					
			Diameter (cm)	Colour, etc.	No./bulb	Size, shape				
Bihar	1989	Agrifound white	3.5–4.5	Compact, silvery white with creamy flesh	20–25	Bigger elongated	130	Good	Susceptible to purple blotch (P. B) and stem phylium blight (S. P.B)	Areas with low incidence of diseases
Delhi	1989	Yamuna safed (G-1)	4.0–4.5	-do-	25–30	Sickle shaped	150–175	Good	tolerant to P.B, S.P.B and insect pests and onion thrips	All over the country
Haryana	1996	Yamuna safed 2 (G-50)	3.5–4.0	-do-	35–40	Sickle shaped	150–200	Good	-do-	Northern India
Tamil Nadu	1999	Yamuna safed 3 (G-282)	5–6	Creamy white	15–16	Sickle shaped	175–200	Good	-do-	North and central parts of India suitable for export
Hongkong	1992	Agrifound parvati	5–6.5	Creamy white with pinkish tinge	10–16	Sickle shaped	175–225	Medium	Tolerant	-do-
UP	2006	Yamuna safed 4 (G-323)	4–5	creamy white with creamy flesh	18–23 (Cloves mature in 165–175 days)	Sickle shaped	200–250	Better than Y. S.3	Tolerant	-do-

Agriclr.in/tnauEAgri/eagri50/HORT281/pdf/lec16.pdf and Indian Horticulture Database

To store variability for use in future, National Bureau of Plant Genetic Resources (NBPGR) is maintaining over 2200 accessions of onion and garlic including land races, farmers' traditional cultivars and wild relatives namely *A. ampeloprasum, A. auriculatum* Kunth., *A. ascalonicum, A. carolinianum* DC, *A. chinense* G. Don, *A. wallachi* Kunth., *A. tuberosum* and *A. rubellum.* About 1100 accessions introduced from over 40 countries fall into different categories of improved cultivars, germplasm collections and related species. Accessions tolerant/resistant to diseases such as purple blotch, stem phylium blight and garlic mosaic virus have been identified. Those resistant to neck rot, powdery mildew, black rot and black mould, however, are yet to be screened/narrowed down (Singh and Rana 1993).

Attempts have been on to expand the genetic base of onion since 1940s and the initial crosses led to the production of hybrids (Emsweller and Jones 1935; Jones and Emsweller 1936). Currently onion breeding programmes focus mainly on the improvement of existing cultivars. Cultivated and wild *Allium* species possess many desirable traits like disease and pest resistance, CMS, bulb quality and so on which might be useful for genetic improvement of the bulb onion (Yamashita et al. 1999; Kik 2002; Kielkowska and Adamus 2010). Of all the species belonging to genus *Allium,* closely allied *A. cepa, A. vavilovii, A galanthum. A. roylei, A. fistulosum, A. altaicum, A. pskemense and A. oschaninii* have been recognized as the most important gene pools of onion (Shigyo and Kik 2008). For example, *A. fistulosum,* commonly called Japanese bunching onion or Welsh onion, the main garden onion of China and Japan, possesses many traits agronomically useful for onion (Peffley and Hou 2002; Kik 2002; Yamashita et al. 2005; Chuda and Adamus 2012). These include resistance to onion leaf blight, pink root, anthracnose and onion fly; earliness, high dry matter content and winter hardiness. Due to these desirable traits, Japanese bunching onion has been used to expand the genetic variation of onion since last century.

Interspecific hybridizations between *A. cepa* and *A. fistulosum,* initiated in 1935 by Emsweller and Jones, have been repeatedly attempted but backcrosses of F_1 hybrids to *A. cepa* have proved quite problematic on account of stylar incongruity, nuclear–cytoplasmic or central cell nuclear–cytoplasmic incompatibilities (see Chuda and Adamus 2012 and references therein). On account of these bottlenecks, only 4 new varieties of *A. fistulosum* (Beltsville Bunching, Delta Giant, Top onion and Wakegi onion) have been developed so far (Kik 2002).

A. roylei, a wild species from the Indian subcontinent, has been considered as one of the most promising species for onion breeders (de Vries et al. 1992b). Interspecific hybrids of *A. fistulosum* × *A. roylei* (Mc Collum 1982; Bark et al. 1994) and *A. roylei* × *A. cepa* (Van deer Meer and de Vries 1990) are on record. While confirming the extent of proximity between these species, crosses enabled breeders to modify the genetic composition of some cultivated taxa. Genes imparting resistance against downy mildew and leaf blight were successfully transferred from *A. roylei* to *A. cepa* (de Vries et al. 1992a; Scholten et al. 2007). *A. roylei* also harbours a single and dominantly inherited anthracnose resistant gene (Galvan et al. 1997). Of the 906 plants recovered through embryo rescue technique by Chuda and Adamus (2012), 884 true F_1 interspecific *A. cepa* × *A. roylei* hybrids were identified. The hybridity of all the regenerants was verified with three molecular markers namely SIR (sulphite reductase), ACS (anthocyanidin synthase) and A (*A. cepa* allinase) by using a specific GISH protocol. AFLP analysis of hybrids of *A. cepa* × *A. roylei*, have resulted in the identification of 186 markers in *A. cepa* with 10 different primer combinations. Of these, 51 absent in *A. roylei* could be used as markers in mapping eight *A. cepa* linkage groups (Van Heusden et al. 2000a). Coupled with GISH analysis, these studies have revealed that the *A. roylei* fragment harbouring the downy mildew resistance probably is at the distal end of the long arm of chromosome 3 of the BC progenies (Van Heusden et al. 2000b, Scholten et al. 2007).

A. roylei has also been utilized as a bridge species for successful introgression of some agronomical traits from *A. fistulosum* into *A. cepa* (Khrustaleva and Kik 1998, 2000). Alan et al. (2003) raised gynogenic plants from hybrids of *A. cepa* and *A. roylei* by culturing flower buds from two *A. cepa* lines and several generations of plants derived from crosses of *A. cepa* and *A. roylei*. An analysis was also done using integrated mapping in the genotypes of trihybrid [*A. cepa* × (*A. roylei* × *A. fistulosum*)].

In vitro techniques have helped in producing interspecific hybrids within edible Alliaceae (Gonzalez and Ford-Lloyd 1987), e.g. *A.cepa* × *A. fistulosum* (Dolezel et al. 1980), onion and fertile garlic (Ohsumi et al. 1993), onion and leek (Peterla et al. 1997) and *A. ampeloprasum* × *A. sativum* (Yanagino et al. 2003). Ohsumi's group (1992, 1993) not only succeeded in obtaining interspecific hybrid between onion and fertile garlic but established its hybridity by morphological, chromosome, isozyme and rDNA analyses. The hybrid plants reportedly have flavours and contents of some compounds intermediate between that of onion and garlic (Ohsumi and Hayashi 1994). Earlier, Sugimoto et al. (1991) attempted hybridization between 7 cultivars of leek and 7 lines of fertile garlic and obtained triploid and aneuploid plants by embryo culture. Yanagino et al. (2003), however, not only succeeded in producing interspecific hybrids between garlic and leek but also confirmed their hybridity at the morphological, cytological, molecular and biochemical levels. The hybrids had intermediate characteristics, were vigorous but sterile, formed bulbs and propagated vegetatively. Similarly somatic hybridization was successful between leek and onion (Buiteveld et al. 1998).

11.7.2 Heterotic Breeding

Discovery and characterization of CMS, an important landmark in improving onion, in 'Italian Red 15-53' not only helped in the transfer of this trait to most onion hybrids grown today (Jones and Clarke 1943; Simon et al. 1991; Ryder 2003; Simon 2005) but also gave the required boost to onion breeding. Exploitation of CMS, widely existing in most natural populations of onion, for the production of hybrid seeds can be fruitful provided male sterile lines and their maintainers are identified. This becomes all the more essential since conventional methods need large scale emasculations which are impractical and genetic analysis takes

4–8 years to identify CMS in onion (Havey 1995, 2000; Cho et al. 2006). Two well-characterized sources of CMS used widely in hybrid onion breeding programmes are CMS-S and CMS-T systems. While the former results from an interaction between a cytoplasmic factor(s) and a single nuclear restoring gene (Jones and Clarke 1943), the latter is complex and under the control of three independently segregating loci in the nuclear genome (Berninger 1965; Schweisguth 1973). Of the two, CMS-S system has been used on a greater scale because of its stability over many environments (Havey 1995, 2000). Using techniques like conventional RFLP, PCR-RFLP and SNPs, DNA markers related to CMS have been reported in onion chloroplast and mitochondrial genome sequences (Havey 1995, 2000; Sato 1998; Cho et al. 2006). Many such RFLPs distinguish between the normal N male fertile and male sterile S cytoplasms. Havey (2000) also identified RFLPs in the organellar genomes from the commercially utilized male sterile lines from Holland, Japan and India and compared them with the usual S and T cytoplasms. Of the 13 (10 from India) Asian populations, 12 did not possess S cytoplasm and the 13th (Indian Poona Red) had a mixture of N and S cytoplasms with N predominating (Havey 1993). Five putative CMS lines (OM 113, 7, 5, 8 and M1111) from another Indian population Nasik White Globe were identical or very similar to S cytoplasm (Havey 2000). Earlier, this cultivar was identified and extracted from a population at the Indian Institute of Horticultural Research, Bangalore (Pathak and Gowda 1993). Their objective was to develop a male sterile line for commercial exploitation of heterosis in onion which would satisfy the short photoperiodic environments of India. This trait with a strong cytoplasmic basis has been successfully transferred to six different genotypes which are now being used for exploiting heterosis. Two of the 75 test crosses namely Hybrid 1 and Hybrid 5 yield 45–50 tonnes of good quality bulbs per hectare (Pathak and Gowda 1993). More work along similar lines is required to build a strong base for the development of male sterile cultivars.

Meanwhile using probes for mitochondrial genes like *atp A*, *atp 6*, *atp 9*, *cox 1*, *cob*, *nad 3*, *nad 4* and *nad 6* (Szklarczyk et al. 2002), these markers were located in a chloroplast *psb A* gene amplicon which can distinguish male fertile (N) and male sterile (S) cytoplasm in onions (Cho et al. 2006). Evaluation of nuclear and mitochondrial genomic diversities by employing RAPD, SSR and RFLP markers (Chaurasia et al. 2010) led to the identification of at least 03 genes required for restoration of fertility in CMS-T and single for that in CMS-S male sterile plants. Gai and Meng (2010) identified one marker each of SCAR and RAPD which could distinguish between N and S cytoplasm in Welsh onions.

To conclude it is apparent that there have been many efforts to improve the cultivated species of this economically important genus through controlled crossings, exploiting hybrid vigour, mutagenesis and now biotechnological interventions. Introgression of genes from wild relatives, directly or through bridge species, has also been achieved. Continuous efforts are required for improving onion and evolve strains which are better yielding, resistant to diseases during cultivation and post harvest and have good storage potential.

While there have been successes, a lot more needs to be done especially in the vegetatively propagated taxa. Isolation and identification of some fertile collections of garlic has provided the breeders a window to improve this species.

There is also need to explore the possibility of bringing under cultivation some of the promising wild taxa of the genus that are used by tribals, etc., on a large scale wherever these are found in abundance. It is pertinent to mention that these wild species have already passed the normal hit and trial process of acceptability—the first stage in domestication and evolution of crop plants.

Acknowledgments The authors are thankful to Drs. Geeta Sharma, Lakhvinder Singh and Beetika Kohli for sharing literature on Alliums and, Mr. Sonam Tamchos, Ms. Shveta Saroop and Ms. Madhu Raina for help in typing the manuscript.

References

Abdo MS, Al-Kafawi AA (1969) Biological activities of *Allium sativum*. Jpn J Pharma Chem 19:1–4

Abdoli M, Habibi-Khaniani B, Baghalian K, Shahnaz S, Rassouli H (2009) Classification of Iranian garlic (*A. sativum L.*) ecotypes using RAPD markers. J Med Plant 8:45–51

Afzal M, Hassan RAH, El-Kazimi AA, Fattah RMA (1985) *Allium sativum* in the control of atherosclerosis. Agric Biol Chem 49(4):1187–1188

Afzal M, Ali M, Thomson M, Armstrong D (2000) Garlic and its medicinal potential. Inflammopharmacology 8(2):123–148

Agriclr.in/tnauEAgri/eagri50/HORT281/pdf/lec16.pdf; Indian Horticulture Database

Ahmad S, Iqbal J (2002) Optimizing plant density cum weed control method for *Puccinia* rust management and yield in garlic. Asian J Plant Sci 1(2):197–198

Alan AR, Mutschler MA, Brants A, Cobb Ed, Earle ED (2003) Production of gynogenic plants from hybrids of *Allium cepa* L. and *A. roylei* Stearn. Plant Sci 165:1201–1211

Al-Zahim M, Newbury HJ, Ford-Lloyd BV (1997) Classification of genetic variation in garlic (*Allium sativum* L.) revealed by RAPD. Hortsci 32(6):1102–1104

Al-Zahim MA, Ford-Lloyd BV, Newbury HJ (1999) Detection of somaclonal variation in garlic (*Allium sativum* L.) using RAPD and cytological analysis. Plant Cell Rep 18:473–477

APG II (2003) An update of the Angiosperm Phylogeny Group classification for the orders and families of flowering plants: APG II. Bot J Linn Soc 141:399–436

Arber A (1938) Herbals-their origin and evolution. A chapter in the history of Botany, 2nd ed. Cambridge University Press, London

Arifin NS, Ozaki Y, Okubo H (2000) Genetic diversity in Indonesian shallot (*Allium cepa* var. *ascalonicum*) and *Allium* X *wakegi* revealed by RAPD markers and origin of *A.* x *wakegi* identified by RFLP analyses of amplified chloroplast genes. Euphytica 111:23–31

Banerjee N (1980) Chromosome studies in some species of *Allium*. Proc Ind Sci Congr Assoc 67 (IVA):35

Barandiaran X, Martin N, Alba C, Rodriguez-Conde MF, Pietro AD, Martin J (1999) An efficient method for the in vitro management of multiple garlic accessions. In Vitro Cell Dev Biol Plant 35:466–469

Bark OH, Havey MJ, Corgan JN (1994) Restriction fragment length polymorphism (RFLP) analysis of progeny from an *Allium fistulosum* x *A. cepa* hybrid. J Am Soc Hort Sci 119 (5):1046–1049

Basak R (1987) The Hindu concept of the natural world. In: Morgan K (ed) The religion of the Hindus. Motilal Banarsidass, Delhi, pp 111–112

Battaglia E (1963) Mutazione chromosomica E cariotipo fondamentale in *Allium sativum* L. Caryo 16:1–46
Bentham G, Hooker JD (1883) Genera Plantarum, vol 3, Part 2. London
Bernhard RA (1970) Chemotaxonomy: Distribution studies of sulfur compounds in *Allium*. Phytochem 9:2019–2027
Berninger ET (1965) Contribution à l'étudede la sterile male de ioignon (*Allium cepa* L.) Ann Amelior Plant 15:183–199
Bhuiya MAK, Rahim MA, Chowdhury MNA (2003) Effect of planting time, mulch and irrigation on the growth and yield of garlic. Asian J Plant Sci 2(8):639–643
Bozzini A (1991) Discovery of an Italian fertile tetraploid line of garlic. Eco Bot 45(3):436–438
Bradeen JM, Havey MJ (1995a) Restriction fragment length polymorphisms reveal considerable nuclear divergence within a well-supported maternal clade in *Allium* section *Cepa* (Alliaceae). Am J Bot 82(11):1455–1462
Bradeen JM, Havey MJ (1995b) Randomly amplified polymorphic DNA in bulb onion and its use to assess inbred integrity. J Am Soc Hort Sci 120:752–758
Bradley KF, Rieger MA, Collins GG (1996) Classification of Australian garlic cultivars by DNA fingerprinting. Aus J Exp Agric 36:613–618
Buiteveld J, Suo Y, Van Lookeren Campagne MM, Creemers-Molenaar J (1998) Production and characterization of somatic hybrid plants between leek (*Allium ampeloprasum* L.) and onion (*Allium cepa* L.). TAG 96:765–775
Buso GSC, Paiva MR, Torres AC, Resende FV, Ferreira MA, Buso JA, Dusi AN (2008) Genetic diversity studies in Brazilian garlic cultivars and quality control of garlic-clover production. Genet Mol Res 7(2):534–541
Campion B, Bohanec B, Javornik B (1995) Gynogenic lines of onion (*Allium cepa* L.): evidence of their homozygosity. TAG 91:598–602
Chaurasia AK, Adsul GG, Nair D, Subramaniam VR, Krishna B, Sane PV (2010) Diversity in Indian and some exotic onion cultivars as revealed by genomic and mitochondrial DNA. Acta Hort (ISHS) 859:207–220
Chinnappareddy LRD, Khandagale K, Chennareddy A, Ramappa VG (2013) Molecular markers in the improvement of *Allium* crops. Czech J Genet Plant Breed 49(3):131–139
Cho KS, Yang TJ, Hong SY, Kwon YS, Woo JG, Park HG (2006) Determination of cytoplasmic male sterile factors in onion plants (*Allium cepa* L.) using PCR-RFLP and SNP markers. Mol Cells 21(3):411–417
Chuda A, Adamus A (2009) Aspects of interspecific hybridization within edible Alliaceae. Acta Physiol Planta 31:223–227
Chuda A, Adamus A (2012) Hybridization and molecular characterization of F_1 *Allium cepa* x *Allium roylei* plants. Acta Biol Cracoviensia Series Bot 54(2):25–31
Courcel de, Vedel AF, Boussac J (1989) DNA polymorphism in *Allium cepa* cytoplasm and its implications concerning the origin of onions. TAG 77:793–798
Cryder CM, Corgan JN, Urquhart NS, Clason D (1991) Isozyme analysis of progeny derived from (*Allium fistulosum* x *Allium cepa*) x *Allium cepa*. TAG 82:337–345
Currah L (1993) Network approach to onion varietal evaluation in the tropics. Acta Hort (ISHS) 358, International Symposium on Alliums for the Tropics; www.actahort.org/books/358/358_31.html
Darlington CD (1937) What is a hybrid? J Hered 28:308
Darlington CD (1940) Taxonomic species and Genetic systems. In: Huxley J (ed) The new systematics. Clarendon Press, Oxford, pp 322–323
de Visiani R (1842) Flora Dalmatica, vol 1. Lipsiae Hofmeister
de Vries JN, Wietsma WA (1992) *Allium roylei* Stearn. restores cytoplasmic male sterility of Rijnsburger onion (*A. cepa* L.). J Genet Breed 46:379–382
de Vries JN, Wietsma WA, de Vries T (1992a) Introgression of leaf blight resistance from *Allium roylei* Stearn into onion (*A. cepa* L.) Euphytica 62(2):127–133

de Vries JN, Wietsma WA, Jongerius MC (1992b) Introgression of characters from *Allium roylei* Stearn. into *A. cepa* L. In: Hanelt P, Hammer K, Knüpffer H (eds) The genus *Allium*—taxonomic problems and genetic resources. Gatersleben, Germany, pp 321–325

de Vries JN, Jongerius R, Sandbrink H, Lindhout P (1992c) RAPD markers assist in resistance breeding. Prophyta 46:50–51

Dhawan V, Jain S (2005) Garlic supplementation prevents oxidative DNA damage in essential hypertension. Mol Cell Biochem 275:85–94

Doležel J, Novâk FJ, Lužney J (1980) Embryo development and in vitro culture of *Allium cepa* and its interspecific hybrid. Z Planzenzűcht 85:177–184

Don GA (1832) A monograph on the genus *Allium*. Mem Nera Nat Hist Soc Edinburgh 6:1–102

Dubouzet JG, Shinoda K, Murata N (1997) Phylogeny of *Allium* L. sub genus *Rhizirideum* (G. Don ex Koch) Wendelbo according to dot blot hybridization with randomly amplified DNA probes. TAG 95:1223–1228

Edris AE, Fadel HM (2002) Investigation of the volatile aroma components of garlic leaves essential oil—possibility of utilization to enrich garlic bulb oil. Euro Food Res Techn 214:105–107

Eid SE (1963) Cytological studies in section *Molium* of the genus *Allium*. Genet Today 1: 134 (Proc XIth Internat Cong Genet)

El Shafie MW, Davis GN (1967) Inheritance of bulb colour in the onion (*A. cepa* L.). Hilgar 38:607–622

Emsweller S, Jones HA (1935) An interspecific hybrid in Allium. Hilgar 9:265–273

Engeland RI (1991) Growing great headed garlic. Filaree Prod, Okanogan, pp 11–30

Eom E, Lee D (1999) Characterization of chromosomal DNA and DNA polymorphism in Korean cultivars of *Allium sativum* L. J Plant Biol 42(2):159–167

Etoh T (1979) Variation of chromosome pairings in various clones of garlic, *Allium sativum* L. Mem Fac Agri, Kagoshima University 15:63–72

Etoh T (1983a) Accomplishment of microsporogenesis in a garlic clone. Mem Fac Agri, Kagoshima University 19:55–63

Etoh T (1983b) Germination of seeds obtained from a clone of garlic, *Allium sativum* L. Proc Jpn Acad 59(Ser. B (4)):83–87

Etoh T (1985) Studies on the sterility in garlic, *Allium sativum* L. Mem Fac Agri, Kagoshima University 21:77–132

Etoh T (1986) Fertility of the garlic clones collected in Soviet Central Asia. J Jpn Soc Hort Sci 55 (3):312–319

Etoh T, Ogura H (1978) Multivalent chromosomes in garlic, *Allium sativum* L. Mem Fac Agri, Kagoshima University 14:53–59

Etoh T, Ogura H (1981) Peroxidase isozyme in the leaves of various clones of garlic, *Allium sativum* L. Mem Fac Agri, Kagoshima University 17:71–77

Etoh T, Simon PW (2002) Diversity, fertility and seed production of garlic. In: Rabinowitch HD, Currah L (eds) Allium crop science: recent advances. CABI publishing, Wallingford, UK, pp 101–117

Etoh T, Sakai Y, Johjima T (1992) Peroxidase isozyme in various cultivars of Leek and Kurrat. Mem Fac Agri, Kagoshima University 28:75–82

Fakhry A (1961) The pyramids. University of Chicago Press, Chicago

FAOSTAT (2011) Food and Agricultural Organisation of the United Nations Agricultural data. http://faostat.foa.org/site291/default.aspx

Fay MF, Chase MW (1996) Resurrection of Themidaceae for the *Brodiaea* alliance, and recircumscription of Alliaceae, Amaryllidaceae and Agapanthoideae. Taxon 45:441–451

Fedorov AA (1969) Chromosome numbers of flowering plants. Leningrad

Fernald ML, Kinsey AC (1943) Edible wild plants of eastern North America. Idle wild press, New York

Figliuolo G, Candido V, Logozzo G, Miccolis V, Zeuli PLS (2001) Genetic evaluation of cultivated garlic germplasm (*Allium sativum* L. and *A. ampeloprasum* L.). Euphytica 121:325–334

Fredotovic Z, Samanic I, Weiss—Schnee Weiss Kamnejarin J, Jang TS, Puizina J (2014) Triparental origin of triploid onion *Allium* x *cornutum* (Clement ex Visiani, 1842), as evidenced by molecular phylogenetic and cytogenetic analyses. BMC Plant Biol 14:24
Friesen N, Klass M (1998) Origin of some minor vegetatively propagated *Allium* crops studied with RAPD and GISH. GRACE 45:511–523
Friesen N, Fritsch R, Bachmann K (1997) Hybrid origin of some ornamentals of *Allium* subgenus *Melanocrommyum* verified with GISH and RAPD. TAG 95:1229–1238
Friesen N, Pollner S, Bachmann K, Blattner FR (1999) RAPDs and non coding chloroplast DNA reveal a single origin of the cultivated *Allium fistulosum* from *A. altaicum* (Alliaceae). Am J Bot 86(4):554–562
Friesen N, Fritsch RM, Blattner FR (2006) Phylogeny and new infrageneric classification of *Allium* (Alliaceae) based on nuclear ribosomal DNA ITS sequences. Aliso 22:372–395
Fritsch RM, Friesen N (2002) Evolution, domestication and taxonomy. In: Rabinowitch HD, Currah L (eds) *Allium* crop science: recent advances. CABI publishing, Wallingford, pp 5–30
Fritsch R, Khassanov FO, Friesen N (1998) New taxa, new combinations and taxonomic remarks on *Allium* L. from Fergan depression, Middle Asia. Linz Biol Beitr 30:281–292
Fryxell PA (1957) Mode of reproduction in higher plants. Bot Rev 23:135–233
Gai S, Meng X (2010) Application of molecular markers linking to cytoplasmic male sterile loci to assist maintainer line selection and their selection efficiency in Welsh onion (*A. fistulosum* L.). Agric Sci China 9:1571–1576
Galvan GA, Wietsma WA, Putrasemedja S, Permadi AH, Kik C (1997) Screening for resistance to anthracnose (*Colletotrichum gloeosporioides* Penz.) in *Allium cepa* and its relatives. Euphytica 95(2):173–178
Ganado P, Sanz M, Padilla E, Tejerina T (2004) An in vitro study of different extracts and fractions of *Allium sativum* (Garlic): vascular reactivity. J Pharmacol Sci 94:434–442
Gantait S, Mandal N, Bhattacharya S, Das PK (2010) Determination of genetic integrity in long term micro propagated plants of *A. ampeloprasum* L. using ISSR markers. Biotech 9:218–223
Geeta (2005) Cytogenetic studies in *Allium tuberosum* Rottl. Ex. Spreng with emphasis on the phenomenon of apomixis. Ph.D Thesis, University of Jammu
Gerard J (1597) The herbal or general historie of plantes. London
Gohil RN (1973) Cytogenetic studies on some Indian Alliums. Ph.D. Thesis, University of Kashmir, Srinagar
Gohil RN (1979) Structural heterozygosity in tetraploid *Allium platyspathum*. Bot Not 132:325–328
Gohil RN (1992) Himalayan representatives of *Allium*. In: Hanelt P, Hammer K, Knüpffer H (eds) The genus *Allium*—taxonomic problems and genetic resources. Gatersleben, Germany, pp 335–340
Gohil RN (1998) Cytology of Indian Alliums—A brief overview. Prog Cytogen:95–115
Gohil RN, Kaul R (1979) Seed progeny studies in Alliums I. Numerical variants in the progeny of tetraploid *Allium tuberosum* Rottl. ex. Spreng. Beit Zur Biol Pflan 54:305–309
Gohil RN, Kaul R (1980) Studies on male and female meiosis in Indian *Allium* I. Four diploid species. Chromo 77:123–127
Gohil RN, Kaul R (1981) Studies on male and female meiosis in Indian Allium II. Autotetraploid *Allium tuberosum*. Chromo (Berl) 82:735–739
Gohil RN, Kaul R (1983) Seed progeny studies in Alliums II. Male meiosis in the progeny plants of tetraploid *Allium tuberosum* Rottl. ex. Spreng. Cytologia 48:109–118
Gohil RN, Koul AK (1971) Desynapsis in some diploid and polyploid species of *Allium*. Can J Genet Cytol 13:723–728
Gohil RN, Koul AK (1977) The cause of multivalent suppression in *Allium ampeloprasum* L. Beitr zur Biol Pflan 53:473–478
Gohil RN, Koul AK (1978) Structural hybridity in *Allium consanguineum*. Cytologia 43:243–247
Gohil RN, Koul AK (1981a) Cytology of tetraploid desynaptic *Allium porrum*. The Nucleus 24:79–83

Gohil RN, Koul AK (1981b) Cytology of the tetraploid *Allium chinense* G. Don. Caryologia 34:73–81
Gohil RN, Koul AK (1983) Formation of ring chromosomes by diethyl sulphate and gamma rays. Experientia 39:1150–1153
Goleniowski M, Del longo O, De Forchetti SM, Arguello JA (2001) Relationships between peroxidases and in vitro bulbification in garlic (*Allium sativum* L.). In Vitro Cell Dev Biol-Plant 37:683–686
Gonzalez LG, Ford-Lloyd BV (1987) Facilitation of wide-crossing through embryo rescue and pollen storage in inter-specific hybridization of cultivated *Allium* species. Plant Breed 98:318–322
Gupta RP, Srivastava KJ, Pandey UB (1993) Diseases and insect pests of Onion in India. In: Acta Hort (ISHS) 358, International Symposium on Alliums for the Tropics; www.actahort.org/books/358/358_43.html
Gurshidze M, Mashayeki S, Blattner FR, Friesen N, Fritsch RM (2007) Phylogenetic relationship of wild and cultivated species of *Allium* section *cepa* inferred by nuclear rDNA ITS sequence analysis. Plant Syst Evol 369:259–269
Hadacova V, Svachulova J, Klozova E, Hadac E, Pitterova K (1983) Use of esterase isoenzymes revealed by gel isoelectric focusing as an aid in chemotaxonomical study of the genus *Allium*. Biol Plant Prague 259:36–42
Haishima M, Kato J, Ikehashi H (1993) Isozyme polymorphism in native varieties of Japanese Bunching Onion (*Allium fistulosum* L.). Jpn J Breed 43:537–547
Hamoud MA, Badr A, Turki Z (1990) Cytotaxonomic relationships of some taxa of Egyptian *Allium* L. Cytologia 55:161–167
Hanelt P (1990) Taxonomy, evolution and history. In: Rabinowitch HD, Brewster JL (eds) Onions and allied crops, vol 1. CRC Press, Boca Raton, pp 1–26
Hanelt P, Fritsch R (1994) Notes on some infrageneric taxa in *Allium* L. Kew Bull 49:559–564
Hanelt P, Schultz-Motel J, Fritsch R, Kruse J, Maab HI, Ohle H, Pistrick K (1992) Infrageneric grouping of *Allium*—the Gatersleben approach. In: Hanelt P, Hammer K, Knüpffer H (eds) The Genus *Allium*- taxonomic problems and genetic resources. Gatersleben, Germany, pp 107–123
Haque MS, Wada T, Hattori K (1999) Anatomical changes during in vitro direct formation of shoot bud from root tips in garlic (*Allium sativum* L.). Plant Prod Sci 2(2):146–153
Haque MS, Wada T, Hattori K (2003a) Shoot regeneration and bulblet formation from shoot and root meristem of garlic cv Bangladesh local. Asian J Plant Sci 2(1):23–27
Haque MdS, Islam MdR, Karim MA, Khan MAH (2003b) Biomass production and growth rates at different phenophases of garlic as influenced by natural and synthetic mulches. Asian J Plant Sci 2(1):90–96
Hara H, Stearn WT, Williams LHJ (1978-82) An enumeration of the flowering plants of Nepal, vol 1–3. Trustee of the British Museum, London
Harris JC, Cottrell SL, Plummer S, Lloyd D (2001) Antimicrobial properties of *Allium sativum* (garlic). Appl Microbiol Biotech 57:282–286
Hattori A, Yamada N, Nishikawa T, Fukuda H, Fujino T (2001) Protective effect of ajoene on acetaminophen-induced hepatic injury in mice. Biosci Biotechnol Biochem 65(11):2555–2557
Havey MJ (1991a) Phylogenetic relationships between cultivated *Allium* species from restriction enzyme analysis of the chloroplast genome. TAG 81:752–757
Havey MJ (1991b) Molecular characterization of the interspecific origin of viviparous onion. J Hered 82:501–503
Havey MJ (1992) Molecular phylogenetics of cultivated and wild *Alliums*. In: Hanelt P, Hammer K, Knupffer H (eds) The Genus Allium- taxonomic problems and genetic resources. Gatersleben, Germany, pp 125–127
Havey MJ (1993) A putative donor of S-cytoplasm and its distribution among open pollinated populations of onion. TAG 86:128–134
Havey MJ (1995) Cytoplasmic determinations using the polymerase chain reaction to aid in the extraction of maintainer lines from open- pollinated populations of onion. TAG 90:263–268

Havey MJ (1997) On the origin and distribution of normal cytoplasm of onion. GRACE 44:307–313
Havey MJ (2000) Diversity among male sterility inducing and male fertile cytoplasms of onion. TAG 101:778–782
Havey MJ (2002) Genome organization in *Allium*. In: Rabinowitch HD, Currah L (eds) *Allium* crop science: recent advances. CABI t, Wallingford, pp 59–79
Havey MJ (2013) Single nucleotide polymorphisms in linkage disequilibrium with the male fertility restoration (Ms) locus in open pollinated and inbred populations of onion. J Am Soc Hort Sci 138:306–309
Havey MJ, Leite DL (1999) Towards the identification of cytoplasmic male sterility in leek: evaluation of organellar DNA diversity among cultivated accessions of *Allium ampeloprasum*. J Am Soc Hort Sci 124(2):163–165
Havey MJ, King JJ, Bradeen JM, Bark O (1996) Molecular markers and mapping in bulb onion, a forgotten monocot. Hort Sci 31:1116–1118
Havey MJ, King JJ, Bradeen JM, Bark O, Sato Y, Gokce AF (2001) A low density genetic map of Onion (*A. cepa* L.) and its use for marker facilitated selection of maintainer lines. Acta Hort (ISHS) 555:87–89
Hayashi T, Sano K, Ohsumi C (1993) Gas chromatographic analysis of alliin in the callus tissues of *Allium sativum*. Biosci Biotechnol Biochem 57(1):162–163
Helm J (1956) Die Zu Wűrz- und Speisez wecken Kultivierten Arten der Gattung *Allium* L. – Kultur pflanze 4:130–180
Hizume M (1994) Allodiploid nature of *Allium wakegi* Araki revealed by genomic in-situ hybridization and localization of 5S and 18S rDNAs. Jpn J Genet 69:407–415
Hong C, Arisumi K, Etoh T (1996) RAPD analysis of ornamental *Alliums* for phylogenetic relationship. Mem Fac Agri, Kagoshima University 32:51–58
Hong C, Etoh T, Landry B, Matsuzoe N (1997) RAPD markers related to pollen fertility in garlic (*Allium sativum* L.). Breed Sci 47:359–362
Hooker JD (1892) The flora of British India, vol VI. L. Reeve & Co., London
http://www.nhrdf.com/htm/files/onion/oni_export.htm 2011. Indian Horticulture Database
Hutchinson J (1934) The families of flowering plants, vol II. Macmillan Co., London
Ipek M, Ipek A, Simon PW (2003) Comparison of AFLPs, RAPD markers and isozymes for diversity assessment of garlic and detection of putative duplicates in germplasm collection. J Am Soc Hort Sci 128(2):246–252
Ipek M, Ipek A, Almquist SG, Simon PW (2005) Demonstration of linkage and development of the first low-density genetic map of garlic, based on AFLP markers. TAG 110:228–236
Islam MJ, Islam MA, Tania SA, Saha SR, Alam MS, Hasan MK (2004) Performance evaluation of some garlic genotypes in Bangladesh. Asian J Plant Sci 3(1):14–16
Islam MJ, Hossain AKMM, Khanam F, Majumder UK, Rahman MM, Rahman MS (2007) Effect of mulching and fertilization on growth and yield of garlic at Dinajpur in Bangladesh. Asian J Plant Sci 6(1):98–101
Jabbes N, Clerc V, Dridi B, Hannechi C (2011) Inter simple sequence repeat finger prints for assessing genetic diversity of Tunisian garlic populations. J Agric Sci 3:77–85
Jakse J, Martin W, McCallum J, Havey MJ (2005) Single nucleotide polymorphisms, indels, and simple sequence repeats for onion cultivar identification. J Am Soc Hort Sci 130:912–917
Jones HA, Clarke AE (1943) Inheritance of male sterility in the onion and the production of hybrid seed. Proc Am Soc Hort Sci 43:189–194
Jones H, Emsweller S (1936) A male sterile onion. Proc Am Soc Hort Sci 34:582–585
Jones HA, Mann LK (1963) Onions and their allies. Leonard Hill Books, London
Jones R, Rees H (1968) Nuclear DNA variation in Allium. Heredity 23:591–605
Kakar AA, Abdullahzai AK, Saleem M, Shah SAQ (2002) Effects of nitrogenous fertilizer on growth and yield of garlic. Asian J Plant Sci 1(5):544–545
Kamelin RV (1973) Florogenetichesky analiz estestvennoj flory gernoj Sredney Azii. Nauka, Leningrad Russia 354
Kamenetsky R (2007) Garlic: botany and horticulture. Hort Rev 33:123–172

Kamenetsky R, Rabinowitch HD (2001) Floral development in bolting garlic. Sex Plant Reprod 13:235–241
Kamenetsky R, Rabinowitch HD (2006) The genus *Allium*: A developmental and horticultural analysis. Hort Rev 32:329–337
Kamenetsky R, Shafir IL, Zemah H, Barzilay A, Rabinowitch HD (2004) Environmental control of garlic growth and florogenesis. J Am Soc Hort Sci 129(2):144–151
Kaul R (1980) Studies on the female meiosis of some Indian *Alliums*. Ph. D thesis, University of Kashmir, Srinagar
Kazakova AA (1978) Luk tartarka (*Allium fistulosum* L.). In: Brezhnev DD (ed) Kul'turnaja flora SSSR, vol 10. Leningrad Koloz, pp 54–61
Kergawler G (1812) *Allium cepa* (the bulb bearing variety). Bot Mag 36(1469)
Khan MI, Ch MA, Hanif W (2006) Ethno veterinary medicinal uses of plants from Samahni valley Dist. Bhimber, (Azad Kashmir) Pakistan. Asian J Plant Sci 5(2):390–396
Khar A, Lawande KE, Negi KS (2010) Microsatellite marker based analysis of genetic diversity in short day tropical Indian Onion and cross amplification in related *Allium*. GRACE 58:741–752
Khassanova FO, Fritsch RM (1994) New taxa in *Allium* L. subg. *Melanocrommyum* (webb et. Berth.) Rouy. Linz Biol Beitr 26:965–990
Khoshoo TN, Atal CK, Sharma VB (1960) Cytotaxonomical and chemical investigations on the North West Indian garlics. Res Bull Punj Univ 11:37–47
Khrustaleva LI, Kik C (1998) Cytogenetical studies in the bridge cross *Allium cepa* (*A. fistulosum* x *A. roylei*). TAG 96:8–14
Khrustaleva LI, Kik C (2000) Introgression of *A. fistulosum* into *A. cepa* mediated by *A. roylei*. TAG 100:17–26
Kielkowska A, Adamus A (2010) Morphological, cytological and molecular evaluation of interspecific F_1 (*A. galanthum* x *A. cepa*) hybrids. Biotechnology 2:146–155
Kik C (2002) Exploitation of wild relatives for the breeding of cultivated *Allium* species In: Rabinowitch HD, Currah L (eds) Advance in *Allium* science. CAB International, New York, pp 81–100
Kim SM, Kubota K, Kobayashi A (1997) Antioxidative activity of sulfur-containing flavor compounds in garlic. Biosci Biotechnol Biochem 61(9):1482–1485
Kim EK, Hahn EJ, Murthy HN, Paek KY (2003) High frequency of shoot multiplication and bulblet formation of garlic in liquid cultures. Plant Cell Tissue Organ Cult 73:231–236
Kim S, Jones R, Yoo K, Pike LM (2004) Gold colour in onions (*A. cepa* L.): a natural mutation of the *chalcone isomerase* gene resulting in a premature termination codon. Mol Genet Genom 272:411–419
Kim S, Rick J, Kil-Sun Y, Leonard MP (2005a) Development of a co-dominant, PCR-based marker for allelic selection of the pink trait in onions (*A. cepa* L.), based on the insertion mutation in the promoter of the anthocyanidin synthase gene. TAG 110:628–633
Kim S, Rick J, Kil-Sun Y, Leonard MP (2005b) The L-locus, one of complementary genes required for anthocyanin production in onion (*A. cepa* L.), encodes anthocyanidin synthase. TAG 111:120–127
Kim S, Haejeen B, Kil-Sun Y, Leonard MP (2007) Marker assisted genotype analysis of bulb colours in segregating populations of Onion (*A. cepa* L.). Mol Cell 23:192–197
Kim S, Doohyun B, Dong YC, Eut-lai L, Moo-Ky-oung Y (2009) Identification of two novel inactive *DFR-A* alleles responsible for failure to produce anthocyanin and development of a simple PCR-base molecular marker for bulb colour selection in Onion (*A. cepa* L.). TAG 118:1391–1399
Klaas M (1998) Application and impact of molecular markers on evolutionary and diversity studies in the genus *Allium*. Plant Breed 117:297–308
Kojima A, Nagato Y (1992) Pseudogamous embryogenesis and the degree of parthenogenesis in *Allium tuberosum*. Sex Plant Reprod 5:79–85
Kojima A, Nagato Y, Hinata K (1991) Degree of apomixis in Chinese chive (*Allium tuberosum*) established by esterase isozyme analysis. Jpn J Breed 41:73–83

Kojima A, Nagato Y, Hinata K (1992) Apomixis in *Allium tuberosum* and *A. ramosum*. In: Hanelt P, Hammer K, Knupffer H (eds) The Genus *Allium*—taxonomic problems and genetic resources, Gatersleben pp 161–165
Kollmann F (1973) Karyology of some species of *Allium* section Molium in Israel. Israel J Bot 22:92–112
Konvicka O, Levan A (1972) Chromosome studies in *Allium sativum*. Hereditas 72:129–148
Koops AJ, Hall RD, Kik C (1991) Bolkleur vorming bij de ui:een biochemisch genetisch model. Prophyta 6:20–22
Koul AK (1963) A spontaneously occurring translocation heterozygote of *Allium cepa*. J Ind Bot Soc XLII 3:416–418
Koul AK, Gohil RN (1970a) Cytology of the tetraploid *Allium ampeloprasum* with chiasma localization. Chromosoma 29:12–19
Koul AK, Gohil RN (1970b) Causes averting sexual reproduction in *Allium sativum* Linn. Cytologia 35(2):197–202
Koul AK, Gohil RN (1971) Further studies on natural triploidy in viviparous onion. Cytologia 36 (2):253–261
Koul AK, Gohil RN (1972) Causes and consequences of seedlessness in garlic. Proc Symp Biol Land Plant 266–270
Koul AK, Gohil RN, Langer A (1979) Prospects of breeding improved garlic in the light of its genetic and breeding system. Euphytica 28:457–464
Krzyminska A, Gawlowska M, Wolko B, Bocianowski J (2008) Genetic diversity of ornamental *Allium* species and cultivars assessed with isozymes. J Appl Genet 49(3):213–220
Kumar M, Berwal JS (1998) Sensitivity of food pathogens to garlic (*Allium sativum*). J Appl Microbiol 84:213–215
Lallemand J, Messian CM, Briand F, Etoh E (1997) Delimitation of varietal groups in garlic (*Allium sativum* L.) by morphological, physiological and biochemical characters. Acta Hort 433:123–132
Lampasona SG, Martinez L, Burba JL (2003) Genetic diversity among selected Argentinean garlic clones (*Allium sativum* L.) using AFLP (Amplified Fragment Length Polymorphism). Euphytica 132:115–119
Langer A (1980) Studies on nucleolar organization in some angiosperms. Ph.D. thesis, University of Jammu
Langer A, Koul AK (1983) Studies on nucleolar chromosomes in angiosperms VII. Nature of nucleolar chromosome polymorphism in *Allium cepa* var. *viviparum* (Metzg.) Alef. Cytologia 48:323–332
Lepen I, Puizina J (2011) Fish mapping of 18S- 5.8S- 26S rRNA genes and fluorochrome banding in the triploid viviparous onion *Allium* x *cornutum* (Clementi ex Visiani 1842). Acta Biol Cracoviensia series Bot 53(1):111–116
Le Thierry D', Ennequin M, Panaud D, Robert, T, Ricroch A (1997) Assessment of genetic relationships among sexual and asexual forms of *Allium cepa* using morphological traits and RAPD markers. Heredity 78:403–409
Loaiza-Figueroa F, Weeden NF (1990) Effects of seed increase procedures on isozyme polymorphism in *Allium*. FAO/IBPGR Plant Genet Res Newslett 83(84):1–3
Lotito SB, Frei B (2006) Consumption of flavonoids rich foods and increased plasma antioxidant capacity in humans: cause, consequence or epiphenomenon? Free Rad Biol Med 41:1727–1746
Maab HI (1997) Studies on triploid viviparous onions and their origin. GRACE 44:95–99
Maab HI, Klaas M (1995) Infraspecific differentiation of garlic (*Allium sativum* L.) by isozyme and RAPD markers. TAG 91:89–97
Maass HI (1996) About the origin of grey shallot. GRACE 43:291–292
Maass HI (1997) Genetic diversity in top onion, *Allium* x *proliferum* analysed by isozymes. Plant Syst Evol 208:35–44
Mahajan V, Jakse J, Havey MJ, Lawande KE (2009) Genetic finger printing of Onion cultivars using SSR markers. Ind J Hort 66:62–68

Mahmood T, Hussain SI, Khokhar KM, Bhatti MH, Laghari H (2002) Comparative performance of garlic cultivars. Asian J Plant Sci 1(2):160–161

Makinen Y (1968) Isozyme polymorphism in flowering plants VI. Variation of isoenzyme patterns in onion seedlings. Physiol Planta 1(4):858–865

Magnum PD, Peffley EB (1994) Inheritance of ADH, 6-PGDH, PGM, and SKDH in *Allium fistulosum* L. J Am Soc Hort Sci 119(2):335–338

Magnum PD, Peffley EB (2005) Central cell nuclear cytoplasmic incongruity: a mechanism for segregation distortion in advanced backcross and selfed generations of [(*A. cepa* L. x *A. fistulosum* L.) x *A. cepa* L.] interspecific hybrid derivatives. Cytogen Geno Res 109:400–407

Maniruzzaman ME, Haque MM, Haque MA (2010) Molecular characterization of onion (*Allium cepa* L.) using RAPD markers. Bangl J Agric Res 35:313–322

McCallum G (1974) Hybrid origin of top onion, *Allium cepa* var *viviparum*. Z Pflanzenz licht 71:222–232

McCallum J, Clarke A, Pither-Joyce M, Shaw M, Butler R, Brash D, Scheffer J, Sims I, Heusden S, Shigyo M, Havey M (2006) Genetic mapping of a major gene affecting onion bulb fructan content. TAG 112:958–967

McCollum DG (1982) Experimental hybrids between *Allium fistulosum* and *A. roylei*. Bot Gaz 143 (2):238–242

Mensinkai SW (1939) Cytogenetic studies in the genus *Allium*. J Genet 39:1–45

Modak M, Dixit P, Londhe J, Ghaskadbi S, Devasagayam TPA (2007) Indian herbs and herbal drugs used for the treatment of diabetes. J Clinical Biochem Nutr 40:163–173

Moench C (1794) Methodus plantas horti botanici et agri marbur gensis Marburg

Nagakubo T, Nagasawa A, Ohkawa H (1993) Micropropagation of garlic through in vitro bulblet formation. Plant Cell Tissue Organ Cult 32:175–183

Nasir E(1975) Flora of West Pakistan. No. 83 Alliaceae Stewart Herbarium, Gordon College, Rawalpindi

Negi KS (2006) *Allium* species in Himalayas and their uses, with special reference to Uttranchal. Ethnobot 18:53–66

Nemec B (1898) Uber abnorme Kernteilungen in der Wurzelspitze von *Allium cepa*. Sitzungsber Bohm. Ges. Wiss Prague, Math Nat Kl 4:1–10

Nishikawa T, Yamada N, Hattori A, Fukuda H, Fujino T (2002) Inhibition by Ajoene of skin-tumor promotion in mice. Biosci Biotechnol Biochem 66(10):2221–2223

Ohsumi C, Hayashi T (1994) Carbohydrate analysis of an interspecific hybrid between onion and garlic. Biosci Biotechnol Biochem 58(5):959–960

Ohsumi C, Kojima A, Hinata K, Etoh T, Hayashi T (1992) Interspecific hybrids between onion anmd garlic. In: P. Hanelt, K. Hammer and H. Knupffer (eds) The Genus *Allium*- taxonomic problems and genetic resources, Gatersleben pp 233–239

Ohsumi C, Kojima A, Hinata K, Etoh T, Hayashi T (1993) Interspecific hybrid between *Allium cepa* and *Allium sativum*. TAG 85:969–975

Ovesna J, Kucera L, Hornickova J, Svobodova L, Stavelikoa H, Velisek J, Milella L (2011) Diversity of S-alk(en)yl cysteine sulphoxide content within a collection of garlic (*Allium sativum* L.) and its association with the morphological and genetic background assessed by AFLP. Sci Hort 129:541–547

Panes VA, Lasala RAS, Tolentino VS, Chan MA (2007) Genetic diversity of Philippine *Allium sativum* L. (Alliaceae) using random amplified polymorphic DNA (RAPD) analysis. Loy Sch Rev 6:43–63

Paredes CM, Becerra VV, Gonzalez AMI (2008) Low genetic diversity among garlic (*Allium sativum* L.) accessions detected using Random Amplified Polymorphic DNA (RAPD). Chil J Agric Res 68(1):3–12

Park J, Park YK, Park E (2009) Antioxidative and antigenotoxic effects of garlic (*Allium sativum* L.) prepared by different processing methods. Plant Foods Human Nutr 64:244–249

Pathak CS, Gowda RV (1993) Breeding for the development of onion hybrids in India: problems and prospects. Acta Hort (ISHS) 358:239–242

Pedersen K, Wendelbo P (1966) Chromosome numbers of some SW Asian *Allium* species. Blyttia 24:307–313

Peffley EB, Currah L (1988) The chromosomal locations of enzyme-coding genes in *Adh-1* and *Pgm-1* in *Allium fistulosum* L. TAG 75:945–949

Peffley EB, Hou A (2002) Bulb type onion introgressants possessing *Allium fistulosum* L. genes recovered from interspecific hybrid back crosses between *A. cepa* L. and *A. fistulosum* L. TAG 100:528–534

Peffley EB, Corgan JN, Horak KE, Tanksley SD (1985) Electrophoretic analysis of *Allium* alien addition lines. TAG 71:176–184

Peffley EB, Magnum PD, Burke JD (2001) Response of isozymes in *Allium* to thermal and aerobic stress. Texas J Agric Nat Resour 14:91–95

Peterla H, Budahn H, Schrader O (1997) Interspecific hybrids between Onion (*Allium cepa* L.) with S-cytoplam and leek (*Allium ampeloprasum* L.) TAG 94:383–389

Pich U, Fritsch R, Schubert I (1996) Closely related *Allium* species (Alliaceae) share a very similar satellite sequence. Plant Syst Evol 202:255–264

Polunin O, Stainton A (1984) Flowers of the Himalaya. Oxford University Press, New Delhi pp 413–416

Pooler MR, Simon PW (1993a) Characterization and classification of isozyme and morphological variation in a diverse collection of garlic clones. Euphytica 68:121–130

Pooler MR, Simon PW (1993b) Garlic flowering in response to clone, photoperiod, growth temperature and cold storage. Hortscience 28(11):1085–1086

Pooler MR, Simon PW (1994) True seed production in garlic. Sex Plant Reprod 7:282–286

Puizina, J, Papes D (1995) Highly conservative karyotype of triploid shallot (*Allium cepa* var. *viviparum*) from Croatia. Chrom Res 3(Suppl 1):58

Puizina J (2013) Shallots in Croatia—genetics, morphology and nomenclature. Acta Bot Croat 72 (2):387–398

Puizina J, Papes D (1996) Cytogenetical evidences for hybrid structure and origin of diploid and triploid shallots (*Allium cepa* var. *viviparum,* Liliaceae) from Dalmatia (Croatia). Plant Syst Evol 199:203–215

Puizina J, Papes D (1997) Further cytogenetic analyses of the Croation triploid shallot "Ljutika" (*Allium cepa* var. *viviparum,* Alliaceae) and its comparison with the Indian triploid "Pran". Plant Syst Evol 208:11–23

Puizina J, Papes D (1999) Classical and molecular cytogenetic studies of top onion, *Allium* x *proliferum* (Moench) Schrader Acta Bot Croat 58:65–77

Puizina J, Javornik B, Bohanec B, Schweizer D, Maluszynska J, Papes D (1999) Random amplified polymorphic DNA analysis, genome size, and genomic in situ hybridization of triploid viviparous onion. Genome 42:1208–1216

Purseglove JW (1972) Alliaceae. In: Longman Group Ltd. Tropical crop (Monocotyledons), pp 37–57

Qijiang CC, Jia T (2007) Genetic diversity analysis of onion (*Allium cepa* L.) germplasm by ISSR markers. Chin Agric Sci Bull 23:126–130

Raamsdonk LWD, van Wietsma WA, de Vries JN (1992) Crossing experiments in *Allium* L. section *Cepa*. Bot J Linn Soc 109:293–303

Raamsdonk LWD, van Smiech MP, Sanbrink JM (1997) Introgression explains incongruence between nuclear and chloroplast DNA based phylogenies in *Allium* section *Cepa*. Bot J Linn Soc 123:91–108

Raamsdonk LWD, van Ginkel MV, Kick C (2000) Phylogeny reconstruction and hybrid analysis in *Allium* subgenus *Rhizirideum*. TAG 100:1000–1009

Rabinowitch HD (1988) Genetics and breeding: state of the art, or: too slow but not too late. In: Riggs TJ, Astley D, Brewster JL, Dawson PR, Maude, RB (eds) Proceedings of the 4th Eucarpia Allium symposium Wellesbourne UK, pp 57–70

Rabinowitch HD, Kamenetsky R (2002) Shallots (*Allium cepa* Aggregatum group). In Rabinowitch, H.D. and Currah L (eds) *Allium* crop science: recent advances. CABI, Wallington UY, pp 409–430

Rahman Md H, Haque MS Ahmed M (2003) Pre-planting temperature treatments for breaking dormancy of garlic cloves. Asian J Plant Sci 2 (1):123–126

Rahman Md S, Islam Md A, Haque Md S, Karim Md A (2004) Effects of planting date and gibberellic acid on the growth and yield of garlic (*Allium sativum* L.). Asian J Plant Sci 3 (3):344–352

Regel E (1875) Alliorum adhuc cognitorum monographia. Trudy Imper S Peterburgsk Bot Soda 3:1–266

Roxas VP, Peffley EB (1992) Short-day onion varietal identification using molecular (RAPD) markers. Allium Improv Newslett 2:15–17

Roy SC (1978) Polymorphism in giemsa banding patterns in *Allium sativum*. Cytologia 43:97–100

Ryder EJ (2003) Perspectives on germplasm. HortScience 38:922–927

Sangeeta KM, Gowda RV, Lalitha A (2006) Analysis of genetic diversity among Indian short day onion (*A. cepa* L.) cultivars using RAPD markers. J Hort Sci Biotechnol 81:774–777

Sata SJ, Bagatharia SB, Thakar VS (2001) Induction of direct somatic embryogenesis in garlic (*Allium sativum*). Meth Cell Sci 22:299–304

Sato Y (1998) PCR amplification of CMS- specific mitochondrial nucleotide sequence to identify cytoplasmic genotypes of onion (*Allium cepa*). TAG 96:367–370

Sato S, Hizume M, Kawamura S (1980) Relationship between secondary constriction and nucleolus organizing region in *Allium sativum* chromosomes. Protoplasma 105:77–85

Schaffner JH (1898) Karyokinesis in the root tips of *Allium cepa*. Bot Gaz 26:225–238

Scholten OE, Heusden AW, Khrustaleva LI, Burger-Meijer K, Mank RA, Antonise RGC, Harrewijn JL, Van haecke W, Oost EH, Peters RJ, Kik C (2007) The long and winding road leading to the successful introgression of downy mildew resistance into onion. Euphytica 156 (3):345–353

Schubert I, Ohle H, Hanelt P (1983) Phylogenetic conclusions from Giemsa banding and NOR staining in top Onions (Liliaceae). Plant Syst Evol 143:245–256

Schweisguth B (1973) Etude d' un nouvea type de sterilite male che zl orignon, *Allium cepa* L. Ann Amelior Pl 23:221–223

Sharma A (2014) Nutritional benefits of onion. Facts for You, pp 27–30. www.ffymag.com

Sharma AK, Bal AK (1959) A study of spontaneous fragmentation in two varieties of *Allium sativum* and interpretation of their karyotypes. In: Proceedings of the 46th session of Indian science congress Part III, p 352

Sharma G, Gohil RN (2008) Intrapopulation Karyotypic variability in *Allium roylei* Stearn- a threatened species. Bot J Linn Soc 158:242–248

Sharma G, Gohil RN (2011a) Occurrence of differential meiotic associations and additional chromosomes in the embryosac mother cells of *Allium roylei* Stearn. J Genet 90:45–49

Sharma G, Gohil RN (2011b) Occurrence of multivalents and additional chromosomes in the pollen mother cells of *Allium cepa* L. The Nucleus 54(3):137–140

Sharma G, Gohil RN (2013) Double hypoploid of *Allium tuberosum* Rottl. ex Spreng. ($2n = 4x = 30$): its origin and cytology. GRACE 60:2283–2292

Sharma N, Koul AK, Langer A (1988) Karyotype polymorphism in *Allium sativum* L. The Nucleus 31(1, 2):31–36

Sharma N, Dhar MK, Langer A, Koul AK (1992) Studies on nucleolus and nucleolar chromosomes of some Indian clones of *Allium sativum* L. through conventional and silver staining techniques. The Nucleus 35(1):44–49

Sharma G, Gohil RN, Kaul V (2011c) Cytological status of *Allium hookeri* Thwaites ($2n = 22$). GRACE 58:1041–1050

Shemesh E, Scholten O, Rabinowitch HD, Kamenetsky R (2008) Unlocking variability: inherent variation and developmental traits of garlic plants originated from sexual reproduction. Planta 227:1013–1024

Shigyo M, Kik C (2008) Onion. In: Prohens J, Nuez F (eds) Vegetables II: Fabaceae, Liliaceae, Solanaceae and Umbelliferae. Springer, New York, pp 121–159

Shigyo M, Tashiro Y, Miyazaki S (1994) Chromosomal locations of glutamate oxaloacetate transaminase gene loci in Japanese bunching onion (*Allium fistulosum* L.) and shallot (*A. cepa* L. *Aggregatum* group). Jpn J Genet 69:417–424

Shigyo M, Tashiro Y, Isshiki S, Miyazaki S (1995a) Chromosomal locations of five isozyme gene loci (*Lap-1, Got-1, 6Pgdh-2, Adh-1* and *Gdh-1*) in shallot (*A. cepa* L. *Aggregatum* group). Jpn J Genet 70:339–407

Shigyo M, Tashiro Y, Isshiki S, Miyazaki S (1995b) Chromosomal locations of isocitrate dehydrogenase and phoshphoglucoisomerase gene loci in shallot (*A. cepa* L. *Aggregatum* group). Jpn J Genet 70:627–632

Shigyo M, Tashiro Y, Isshiki S, Miyazaki S (1995c) Genetic analyses of Glutamate Dehydrogenase isozymes in cultivated and wild species of section *Cepa* in *Allium*. J Jpn Soc Hort Sci 64(1):121–127

Shigyo M, Tashiro Y Isshiki S, Miyazaki S (1996) Establishment of a series of alien monosomic addition lines of Japanese bunching onion *Allium fistulosum* L. with extra chromosomes from shallot (*A. cepa* L. *Aggregatum* Group). Genes Genet Syst 71:363–371

Shigyo M, Miyazaki T, Isshiki S, Tashiro Y (1997) Assignment of randomly amplified polymorphic DNA markers to all chromosomes of shallot (*Allium cepa* L. Aggregatum group). Genes Genet Syst 72:249–252

Shirley BW (1996) Flavonoid biosynthesis. 'new' functions for an 'old' pathway. Trends Plant Sci 1:377–382

Simon PW (2005) Collections. HortScience 40(2):309–310

Simon PW, Gabelman WH, Franklon DF (1991) Henry A. Jones. Hort Sci 26:1115–1118

Simpson MG (2010) Plant systematics 2nd edn. Academic Press, London

Singh L (2013) Analysis of cytogenetic and molecular variability in *Allium sativum* Linn. Ph. D. Dissertation, University of Jammu, Jammu

Singh BP, Rana RS (1993) Collection and Conservation of *Allium* genetic resources: an Indian perspective. Acta Hort (ISHS) 358. In: International Symposium on Alliums for the Tropics. www.actahort.org/books/358/358_31.html

Singh FS, Ved Brat S, Khoshoo TN (1967) Natural triploidy in viviparous onion. Cytologia 32:403–407

Singh, L., Kaul, V. and Gohil, R.N. (2009). Optimization of DNA isolation and PCR protocols for RAPD analysis of *Allium sativum* L. (Garlic). Ind J Bot Res 5 (3&4):289–294

Singh L, Kaul V, Gohil RN (2014) Analysis of morphological variability in the Indian germplasm of *Allium sativum* L. Plant Syst Evol 300:245–254

Slimestad R, Fossen T, Vagen IM (2007) Onions: a source of unique dietary flavonoids J Agri. Food Chem 55:10067–10080

Stavelikova H (2008) Morphological characteristics of garlic (*Allium sativum* L.) genetic resources collection-Information. HortScience (Prague) 35(3):130–135

Stearn WT (1943) The Welsh onion and the ever ready onion. The Gard Chron III 114:86–88

Stearn WT (1945) The Alliums of British India- Sir Joseph. D. Hooker (revised and supplemented by William. T. Stearn). Herb 12:73–84

Stearn WT (1946a) Notes on the genus *Allium* in the old world. Herb 11:11–34

Stearn WT (1946b) The floristic regions of the USSR with reference to the genus *Allium*. Herb 11:45–63

Stearn WT (1980) *Allium* L., *Nectaroscordum, Nothoscordum.* In: Tutin TG et al (eds) Fl Europ 5:49–70 Cambridge University Press, Cambridge

Stearn WT (1992) How many species of *Allium* are known? Kew Mag 9:180–182

Sugimoto H, Tsuneyoshi T, Tsukamoto M, Uragami Y, Etoh T (1991) Embryo cultured hybrids between garlic and leek. Allium Improv Newslett 1:67–68

Susan EW, Peter GI, Robert JS (1993) Random amplified polymorphic DNA (RAPD) markers for genetic analysis in *Allium*. TAG 86:497–504

Szklarczyk M, Simlat M, Jagosz B, BaG (2002) The use of cytoplasmic markers in onion hybrid breeding. Cell Mol Biol Lett 7:625–634

Takada N, Matsuda T, Otoshi T, Yano Y, Otani S, Hasegawa T, Nakae D, Konishi Y, Fukushima S (1994) Enhancement by organo sulphur compounds from garlic and onions of diethyl nitrosamine induced glutathione-S-transferase positive foci in the rat liver. Cancer Res 54(11):2895–2899

Takhtajan A (1987) Systema magnoliophytorum. Nauka Leningrad Russia 438 pp

Takhtajan A (1997) Diversity and classification of flowering plants. Columbia University Press, New York

Traub HP (1968) The subgenera sections and subsections of *Allium* L. Plant Life 24:147–163

Tzanoudakis D (1992) Karyotype variation and evolution in the Greek *Allium*. In: Hanelt P, Hammer K, Knüpffer H (eds) The Genus Allium—Taxonomic problems and genetic resources. Gatersleben Germany, pp 305–320

Ueda Y, Sakaguchi M, Hirayama K, Miyajima R, Kimizuka A (1990) Characteristic flavor constituents in water extract of garlic. Agric Biol Chem 54(1):163–169

Vakhtina LI (1964) Kariotipy 20-Khromosomaykh vidvo Luka-*Allium kujukense* Vved. : *A. decipiens* Fisch. Tsitologiya 6:577–582

van der Meer QP, de Vries JN (1990) An interspecific cross between *Allium roylei* Stearn and *Allium cepa* L. and its backcross to *A. cepa*. Euphytica 47:29–31

van Heusden AW, van Ooijen JW, Vrielink vGR, Verbeek WHJ, Wietsma WA, Kik C (2000a) A genetic map of an interspecific cross in *Allium* based on amplified fragment length polymorphism (AFLP) markers. TAG 100:118–126

van Heusden AW, Shigyo M, Tashiro Y, Vrielink-van Ginkel R, Kik C (2000b) AFLP linkage group assignment to the chromosomes of *Allium cepa* L. via monosomic addition lines. TAG 100:480–486

Vavilov NI (1926) Studies on the origin of cultivated plants. Trudy Byuro Prikalodnoy Botaniki 16:1–139

Ved Brat S (1965) Genetic systems in *Allium*. I. Chromosome variation. Chromosoma 16:486–499

Verma SC, Mittal RK (1978) Chromosome variation in the common garlic, *Allium sativum* L. Cytologia 43:383–396

Vosa CG (1976) Heterochromatic patterns in *Allium*. I The relationship between the species of the *Cepa* group and its allies. Heredity 36:383–392

Vu HQ, Yoshimatsu Y, Khrustaleva LI, Yamauchi N, Shigyo M (2012) Alien genes introgression and development of alien monosomic addition lines from a threatened species, *Allium roylei* Stearn to *Allium cepa* L. TAG 124(7):1241–1257

Vvedensky AI (1935) Rod 267. Luk—*Allium* L. In: Komarov VL (eds) Flora URSS, Vol. 4, Izd. Akad Nauk SSSR Leningrad, Russia, pp 112–280

Wajahatullah MK, Vahidy AA (1990) Karyotyping and localization of nucleolar organizer regions in garlic, *Allium sativum* L. Cytologia 55:501–504

Wilkie SE, Isaac PG, Slater RJ (1993) Random Amplified Polymorphic DNA (RAPD) markers for genetic analysis in Allium. TAG 86:497–504

Xu P, Liu H, Gao Z, Srinivas P, Yang CY (2001) Genetic identification of garlic cultivars and lines by using RAPD assay. Acta Hort (ISHS) 555:213–220

Yamashita K, Arita H, Tashiro Y (1999a) Cytoplasm of a wild species, *A. galanthum* Kar. et Kir is useful for developing the male sterile line of *A. fistulosum* L. J Japan Soc Hort Sci 68:788–797

Yamashita KI, Oyama T, Noda R, Miyazaki T, Tashiro Y (1999b) Comparative study on methods for identification of chloroplast DNA of cultivated and wild species in section *Cepa* of *Allium*. Bull Fac Agric Saga University 83:111–120

Yamashita K, Arita H, Tashiro Y (1999b) Isozyme and RAPD markers linked to fertility restoring gene for cytoplasimic male sterile *Allium fistulosum* L. with cytoplasm of *Allium galanthum* Kar. et Kir. J Japan Soc Hort Sci 68(5):954–959

Yamashita K, Hisatsune Y, Sakamoto T, Ishizuka K, Tashiro Y (2002) Chromosome and cytoplasm analyses of some hybrids between onion (*Allium cepa* L.) and garlic (*A. sativum* L.). Euphytica 125:163–167

Yamashita K, Takatori Y, Tashiro Y (2005) Chromosomal location of a pollen fertility—restoring gene Rf, for CMS in Japanese bunching onion (*Allium fistulosum* L.) possessing the cytoplasm of *A. galanthum* Kar et Kir revealed by genomic in situ hybridization. TAG 111:15–22

Yanagino T, Sugawara E, Watanabe M, Takahata Y (2003) Production and characterization of an interspecific hybrid between leek and garlic. TAG 107:1–5

Youn H, Lim HJ, Lee HJ, Hwang D, Yang M, Jeon R, Ryu (2008) Garlic (*Allium sativum*) extract inhibits lipopolysaccharide-induced toll-like receptor for dimerization. Biosci Biotechnol Biochem 72 (2):368–375

Yuzbasioglu D, Unal F (2004) Karyotyping, C- and NOR banding of *Allium sativum* L. (Liliaceae) cultivated in Turkey. Pak J Bot 36 (2):343–349

Zeidler M (1999) Genetic variability among populations of *Allium carinatum* subsp. *carinatum*. Thaiszia—J Bot Košice 9:81–90

Zeidler M (2000) Electrophoretic analysis of plant isozymes. Acta Univ Palack Olomuc Facult Rerum Nat Biol 38:7–16

Chapter 12
Utilization of Germplasm for the Genetic Improvement of Mung bean [*Vigna radiata* (L.) Wilczek]: The Constraints and the Opportunities

Ruchi Vir, Suman Lakhanpaul, Sonal Malik, Sooraj Umdale and Kangila Venkataramana Bhat

Abstract Pulses are rich in proteins and serve as a main source of this essential component of nutrition particularly for the predominantly vegetarian population of India and adjacent countries. Mung bean [*Vigna radiata* (L.) Wilczek] also known as green gram is an important pulse crop due to its widespread consumption throughout the Indian subcontinent. It is increasingly becoming popular in other parts of the world in recent years due to its value added products that are rich in several nutrients. However, unlike cereal crops, mung bean yields have not been able to meet the demands of the consumers leading to its import from other countries thereby resulting in steep rise in prices. Low productivity in mung bean is pushing it to the marginal lands and further decreasing its competitiveness in comparison to other crops. Despite developing several cultivars suitable for specific agro-climatic zones, mung bean crop is affected by a wide range of biotic and abiotic stresses. Further, some quality traits of mung bean also need to be improved for enhancing its nutritional value. Large germplasm collections are available in national and international gene banks; however, their vast potential is yet to be exploited. Effective utilization of these genetic resources requires their trait-based evaluations for identification of the elite genotypes and core sets. Conventional breeding approaches will get strong impetus by the identification of primary, secondary and tertiary genepools in order to select the donor and design the judicious approaches for the transfer of useful genes. Detailed molecular characterization of genetic diversity of the available germplasm and assessment of phylogenetic relationships among the related taxa done so far can provide useful leads in this regard.

R. Vir · S. Lakhanpaul · S. Malik
Department of Botany, University of Delhi, Delhi 110007, India

S. Umdale · K.V. Bhat (✉)
National Bureau of Plant Genetic Resources, Pusa Campus, New Delhi 110012, India
e-mail: kvbhat2001@yahoo.com

V.R. Rajpal et al. (eds.), *Gene Pool Diversity and Crop Improvement*,
Sustainable Development and Biodiversity 10,
DOI 10.1007/978-3-319-27096-8_12

Recent developments in the large-scale genomic tools have resulted in the availability of whole genome sequence of mung bean which is a significant boost for the exploitation of biotechnological advancements for its improvement. However, lack of efficient protocols for transformation and regeneration of mung bean pose the important challenges that need to be addressed at the earliest. Appropriate combination of conventional and molecular approaches aimed at exploitation of the available germplasm is the need of the hour for successful development of high yielding cultivars in this crop.

Keywords Pulses · *Vigna radiata* · Green gram · Mung bean · Wild relatives · Germplasm · Genomic resources

12.1 Introduction

Pulses or food legumes including beans, peas and lentils are the most versatile and nutritious food available. They are the major source of protein for the predominantly vegetarian population of Indian subcontinent and other adjacent countries. Pulses provide significant nutritional and health benefits as they are low in fats contain no cholesterol and are high in proteins, folate, potassium, iron and magnesium. They also have phytochemicals; a group of compounds that are known for preventing chronic ailments such as cardiovascular diseases and cancers. In addition, pulses are a good source of fibre that can reduce the risk of developing diabetes and help in lowering of blood cholesterol levels, thereby also help in reducing the risk of heart diseases. Due to their ability to fix atmospheric nitrogen, they play a key role in maintaining soil fertility and ensuring sustainability of production system, particularly in low input, small-scale agriculture.

India is the largest producer, importer and consumer of the pulses in the world, accounting for 25 % of the global production, 15 % of the trade and 27 % consumption, as sizeable population of the country still depends on the pulses to meet its protein requirement. Cereals are generally deficient in the lysine which is an essential amino acid and can be provided by legumes. On the other hand, legumes are low in sulphur containing amino acids, which cereals can provide (Thirumaran and Seralathan 1988). Therefore, a combination of cereals and pulses is the most prevalent meal in common households in many countries including India.

The country produces variety of pulses including chickpea, pigeon pea, urdbean, mung bean, lentil, field pea and others to the tune of 13–15 million tonnes from an area of 22–33 million hectares with an average yield of 600–650 kg/h. In India, Madhya Pradesh, Maharashtra, Uttar Pradesh, Andhra Pradesh, Karnataka and Rajasthan are the major states which grow pulses and contribute to the 80 % of the total production of pulses.

Domestic production of pulses after its peak of 14.94 million tonnes in 2003–04 had declined to 13.38 million tonnes in 2004–05 and to 13.11 in 2005–06, due to adverse climatic conditions prevalent in the major production zones. In the year 2007–08, the production has increased marginally to 14.9 million tones. This still fell short of the domestic requirement, which is increasing consistently with growing population, rising income, value addition and market opportunities. The widening gap in demand and supply has continued and led to soaring prices of pulses during past 2 years. Also, inclusion of pulses in future trading and limited availability in the international market has further fueled the prices.

12.2 Constrains that Results in Low Productivity of Pulses

The stagnant production of pulses in the past decade is raising a concern among the stakeholders and thus demands technological breakthrough immediately for improving productivity to balance the demand and supply of pulses. The major factor responsible for poor growth in the production of food legumes is their low productivity that in turn makes pulses economically less competitive in comparison to other crops. In fact, there has been only a marginal annual increase (less than 1 %) over the past more than five decades in case of pulse crops. They are therefore increasingly getting pushed to less endowed and marginal lands (areas), particularly in developing countries. The greatest challenge for legume researchers is to enhance the economic competitiveness of these legumes by improving their niches available in various cropping systems, enhancing their end use quality for diversified uses, and reducing their susceptibility to a host of biotic stresses (diseases, insect pests, parasitic and other weeds, etc.) and abiotic stresses (drought, extremes of temperature, salinity, nutrient deficiencies and toxicities, etc.). Therefore, conventional as well as biotechnological interventions that could either enhance their productivity or improve their resistance against biotic stresses and abiotic stresses are required.

Further in India, the national agriculture policies stress on promotion of cereal crops as pulses are generally considered neither as staple food nor as cash crops. They occupy low priority in the farming system because of lack of stability and high risk involved in their production. Also, pulses are prone to high losses during storage. Further, they have so far proved to be less responsive to inputs including irrigation (Singh and Kochhar 2005). To add to it, there are extremely fluctuating markets for the pulses. As productivity of pulses is low, they are generally grown in rain-fed area with poor soil fertility and low moisture retention capacity. Thus, these crop face moisture stress at various growth stages. Nonavailability of quality seed is another important constrains that limits pulse production. Seeds of newly developed varieties are either not available to the farmers or are not available in required quantities. Thus, varieties with better yield advantages and desired characteristics to suit various agro- climatic conditions need to be developed.

12.3 Mung Bean

12.3.1 Taxonomy and Distribution

The genus *Vigna* named after Domenico Vigna, Professor of Botany in Pisa (Savi 1826; Baudoin and Maréchal 1988) of the family Fabaceae includes many pulse-yielding species. It is an important legume taxon widely distributed in tropical and subtropical regions of both hemispheres and contains about 150 species, divided into seven subgenera. Of which, subgenus *Ceratotropis* mainly comprising species found in Asia, has been differentiated into morphologically homogenous group with specialized and complex floral morphology (Maréchal et al. 1978). *Vigna mungo* (L.) Hepper (urdbean) and *V. radiata* (L.) Wilczek (mung bean) are the two most important pulse-yielding taxa belonging to subgenus *Ceratotropis* of genus *Vigna*. Indian centre is extremely rich and possesses considerable landrace diversity of these species and several allied ancestral wild taxa. *V. radiata* and *V. mungo* are morphologically similar; however, minor differences exist in morphology and few other important plant characters. Variations have also been reported in karyotype, DNA content, proteins, amino acid composition and species hybridity.

12.3.2 Cultivation and Economic Importance of Mung bean

In general, pulses are grown mostly in two seasons; warmer, rainy season or kharif (June-Oct) and cool, dry season or rabi (Oct-April). Mung bean is grown on approximately six million hectares, mainly in Asia; however, cultivation has been extended to parts of Australia, USA, Canada and Ethiopia (Schafleitner et al. 2015). However, 90 and 54 % of world production of mung bean comes from Asia and India, respectively. Thus, 65 % of the world acreage under mung bean is accounted by India alone. In India, mung bean is grown on 34.4 lakh hectares that accounts for 18.07 % of the land under pulse production, whereas the production is approximately 14 lakh tonnes accounting for 11.48 % of the total pulse production. With a productivity of 406.98 kg/ha, mung bean is primarily grown in seven states in India, namely Rajasthan, Maharashtra, Andhra Pradesh, Karnataka, Orissa, Tamil Nadu and Uttar Pradesh. However, yield of mung bean has not shown any dramatic increase during the past several years (Fig. 12.1).

12.3.2.1 Export–Import Status and World Trade

India imports substantial amount of pulses from the world market for its domestic consumption. Import of mung bean and urdbean has been increased to 624.12 thousand tonnes in the year 2013–2014. These pulses are mainly imported from Myanmar (82.83 %), Tanzania (4.23 %), Kenya (3.55 %), Australia (3.05 %) and

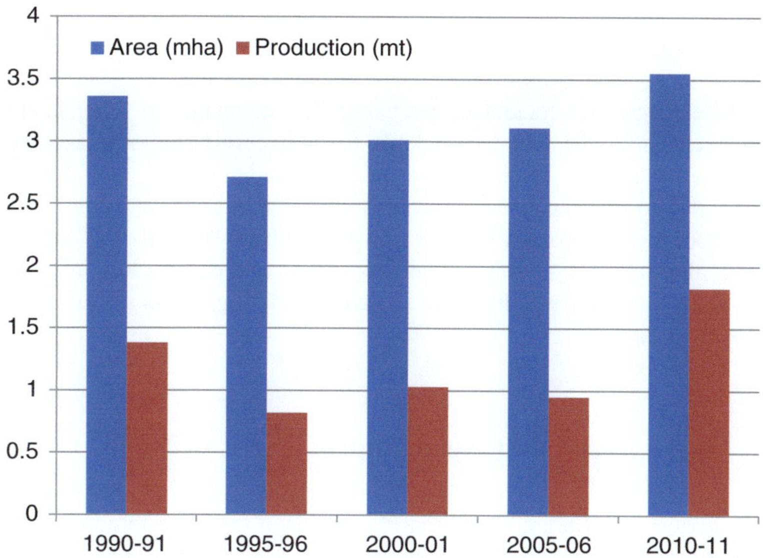

Fig. 12.1 Graphical representation showing area and production of *Vigna* sp. across India during 1990–2010

Mozambique (1.61 %). Interestingly, India exports also a small quantity of urad and mung bean (approximately 1.16 % of the total pulses export) to USA, Sri Lanka and Kenya.

12.3.3 Nutrients

Mung bean along with other pulses constitutes the major source of protein in Asia and constitutes an important supplement to the predominantly cereal-based diet. Seed of mung bean contain 26 % protein, 62.5 % carbohydrates, 1–2 % fats and 4.2 % fibres. Mung bean is also used for bean sprouts, starch noodles, green pods as peas in cooking and mung bean soup. Seeds and sprouts of mung bean with high levels of total phenols and flavonoids, DPPH (2,2-diphenyl-1-picrylhydrazyl), tyrosinase inhibition, anti-proliferative and ADH (Alcohol Dehydrogenase) activities could be recommended as preventative or/and therapeutic agents for human diseases in addition to normal prescription drugs. Yao et al. (2008) discussed the antidiabetic properties of extracts from mung bean sprouts and mung bean seed coat in their study of type 2 diabetic mice. Thus, selection of varieties with high levels of polyphenol content during sprouting was recommended as a very good option for addressing diabetes problem.

The oil content in mung bean seeds is relatively low (2.1–2.7 %). Oil of mung bean contains linoleic acid, palmitic acid and oleic acid as the dominant fatty acids.

Mung bean oil also contains vitamin E, including tocopherols and tocotrienol. The vitamin A content of 70 and 100 μg RAE (Retinol Activity Equivalent) has been reported in mung bean grains and sprouts, respectively.

Mung bean has low content of phytic acid as compared to other leguminous crops and cereals. Phytic acid is commonly found in cereals and other legume crops which have a negative impact on Fe and Zn availability in plant–based diets. Therefore, due to its palatable taste, other nutritional qualities and low cost, mung bean has been used as good source of iron especially in baby foods. Additionally, Vitamin C the most potent enhancer of iron absorption though low in mung bean grains (0.05 g $kg{-}1$ dry weight basis) is significantly high in its sprouts (1.38 g kg -1 dry weight basis).

12.4 Major Areas that Need Improvement in Vigna

Mung bean and urdbean are planted in kharif season and have to cope up with the problems of water management, nonsynchronous late maturity, diseases and pest. Therefore, attempts should be made to develop varieties with early and synchronous maturity, resistant to major diseases and pest which could be successfully cultivated in spring/summer and/or in kharif season (Khanal et al. 2005). There is an urgent need to developed varieties that are widely adaptable, tolerant to both low and high moisture stress, resistant to biotic stress such as cercospora leaf spot, web blight, thrips sucking bugs, etc. Also there is need to explore and decipher possibilities of growing mung bean and urdbean as inter or mixed crop with vegetables, fruits and maize. In addition to this there is requirement of developing suitable weed management technologies. Finally, the need to explore wild genepool for identification of desirable traits cannot be overemphasized.

12.5 Major Diseases in Mung bean

Mung bean is susceptible to diseases caused by fungi, bacteria and viruses. Yellow Mosaic is one of the most devastating diseases of mung bean. It is caused by bipartite begomo viruses namely mung bean yellow mosaic virus (MYMV), mung bean yellow mosaic India virus (MYMIV) and HgYMV in different mung bean growing areas of the world. The virus is transmitted by whitefly (*Bemisia tabaci*). The diseased plants have characteristic yellow coloration on the leaves and results in poorly developed pods thus affecting the yield significantly. First reported by Capoor and Verma (1948) in case of lima bean (*Phaseolus lunatus*) and dolichos bean, it was soon reported to affect mung bean with its to its first detail description from the farms of IARI (Nariani 1960). In the first two decades after its detection in mung bean, the efforts were made to screen germplasm for the resistant lines. These attempts have been successful in identifying some resistant lines in the cultivated genepool aand some wild species. Genetic studies have led to the identification of genetic loci controlling the resistance.

Tobacco streak virus (TSV) causes a minor to severe impact on yields. Thrips are the only known vector of TSV. In Australia, several common broadleaf weed species are hosts of TSV with parthenium weed being a widespread and key host of the virus in central Queensland. The physical damage caused by thrips feeding on the plant allows infected pollen to enter the mung bean plant where the virus multiplies. TSV-infected mung bean plants are usually stunted and wilted with dead (necrotic) stems and tips.

Various leaf and stem pathogens, such as powdery mildew and bacterial blight, are frequently seen, especially in growing crops. Powdery mildew caused by the fungus, *Podosphaera fusca*, is favoured by cooler growing conditions and is often widespread in late crops. Charcoal rot caused by *Macrophomina phaseolina* results in infected plants to die prematurely and may reduce yield with its biggest impact upon the marketability of sprouting grade beans. Seed borne infection creates a soft, wet rot of the sprouts during the germination process. Legume little leaf affected plants often fail to produce pods, and if they do, they are generally distorted with the beans either failing to develop inside the pod or turning brown in colour.

Other diseases affecting mung bean include Sclerotinia white mould, Cercospora leaf spot, bacterial blight, etc.

12.6 Abiotic Stresses Affecting Mung bean

Abiotic stresses affect plant metabolism, disrupt cellular homeostasis and uncouple major physiological and biochemical processes (Arora et al. 2002; Srivalli et al. 2003). There are various abiotic stress factors that act as inhibitors to reduce vegetative growth, pod set, flower initiation in mung bean (Morton et al. 1982). The major environmental factors which affect *Vigna* production are water and salt stress. Mung bean is highly sensitive to these stresses. Initially, it is the root system which is affected more than the shoot and subsequently these stresses act on the growth rate by affecting the major physical and biochemical parameters and changes the growth pattern. Abiotic stresses can also lead to number of insect pests and several disease conditions which ultimately result in further loss of productivity and quality of mung bean (Singh and Singh 2011).

12.7 Drought Stress

Basic metabolism of the plant is mainly affected by water stress which controls the plant growth, plant yield and quality of a crop (Hsiao and Acevedo 1974; Begg and Tuener 1976). Mung bean is reported to be more susceptible to drought than many other grain legumes (Pandey et al. 1984). This affects the productivity of mung bean mainly during spring and summer season. Thomas et al. (2004) reported that mung bean plants under water stress attained maturity earlier than the well-watered treatment. Mung bean crop is more sensitive to water deficit during the flowering period. Severe drought reduces vegetative growth, flower initiation and pod set (Morton et al. 1982). Drought induces early flowering in mung bean which results in poor pod setting. Increased drought during the development of reproductive organs also has a pronounced effect on fruit development and yield mainly due to an increased rate of floral and pod abortion (Liu et al. 2003). Water stress also decreases the number and shape of root hairs (Worral and Roughley 1976) and when nodules are formed, drought also alters the nodule structure and weight (Ramos et al. 2003).

12.7.1 Water Logging Stress

Excess of water can also affect the growth of mung bean. Heavy rain coupled with strong winds damages the mature crop resulting in severe yield losses. Death of mung bean plants occurs due to waterlogging and surviving plants, if any, become very sensitive to various fungal diseases and insect pests. During early stages of growth, mung bean cannot withstand waterlogging (Tickoo et al. 2006). Main effect of waterlogging occurs on the dry weight of roots. Oxygen concentration reduces around the roots of submerged plants and restricts nodule activity and nitrogen fixation during the waterlogging. Infact, it is because of this reason that mung bean is not suited to the wet tropics where the annual precipitation is >1000 mm (Fernandez and Shanmugasundaram 1988). However, water logging has no effect on flower colour parameters (Musgrave and Vanhoy 1989). Some other effects of water logging are decrease in chlorophyll content, flower partitioning, pod setting, photosynthesis rate, number of nodules, membrane stability index, crop growth rate, root dry weight, quality and crop yield. Varma and Rao (1975) observed that the number of pods and seed yield/pod in mung bean reduces sharply with excess moisture level. Duong et al. (1988) reported that 48 h of waterlogging reduced plant height up to 76 %, leaf area up to 46 % and dry matter production up to 57 %. Interrelationship between root and shoot carbon budgets was shown by Musgrave and Vanhoy (1989) in mung bean as a response to waterlogging stress and damage. Mung bean genotypes exhibiting differential sensitivity during recovery from water logging have been identified and characterized. This selection is required to identify the stress tolerance in germinated seed for the advancement of the crop and to increase the yield of the legume grain. (Haqqani and Pandey 1994; Tickoo et al. 2006; Kumar et al. 2012a, b). At AVRDC also efforts have been made to identify waterlogging tolerant cultivars and several tolerant selections such as 'V 1968', 'V 2984', 'V 3092' and 'V 3372' have been made (Tickoo et al. 2006).

12.7.2 Salinity Stress

Salinity stress is one of the most important abiotic factors as it leads to significant loss in mung bean production (Abd Alla et al. 1998; Saha et al. 2010). It is estimated that 6.5 % of the total land of the world is affected by salinity stress which is around 831 M ha of land (Hasanuzzaman et al. 2013). Salt stress along with other pests (stem and pod borer) and yellow mosaic disease was found to cause 80–100 % yield loss in mung bean particularly during rainy season. Paliwal and Maliwal (1980) had reported that germinating mung bean can resist 6 m mhos/cm of salinity, whereas, a mung bean plant can tolerate 9–18 m mhos/cm salinity. To enhance the yield and growth of mung bean under saline conditions, seeds are inoculated with L-TRP and *Rhizobium* and used as complement (Zahir et al. 2010). With increasing salt concentration, delayed and decreased germination is observed in mung bean. However, no effect was observed in some of the concurrences of *Vigna radiata* when they are cultivated on extremely alkaline ($pH > 8.5$) calcareous soils (Lawn et al. 1988).

On treating the germinating seeds of mung bean it has been found that high salt concentration has a drastic effect on roots as compared to shoots. Length of the roots, branches and the number of root hairs decreases during salinity. Salt stress causes chlorosis, necrosis and reduced the chlorophyll content (Wahid 2004). Polyamine synthesis is also affected by an increased amount of salt concentration in mung bean grain (Friedman et al. 2006). Increase in proline oxidases (which convert proline to glutamate) leading to decrease in proline content has been reported.

In case of mung bean, salinity also induces desiccation which results in more flower shedding and pod shattering during the summer season. Therefore, reduced yield in mung bean under salt stress may be due to more flowers shedding, reduced photosynthetic efficiency per day of plant to fill the developing seeds and shattering of the pods (Wahid et al. 2004; Ahmed 2009). In another study, different doses of salt (NaCl) concentration given to three varieties of *V. radiata* (T 44, SML 66, Sarif) showed a reduction in seedling growth, germination percentage, relative growth rate and photosynthetic pigments after 15 days only (Arulbalanchandran et al. 2009).

Gulati and Jaiwal (1994b) examined accumulation of ions in cellular and whole plant response against different salt concentration and demonstrated that salt stress is regulated at cellular level. NaCl salt stress was overwhelmed by pretreatment with salt concentration which modified antioxidant enzyme activities; reduced malondialdehyde and H_2O_2 content so that accumulation of osmolytes (proline) increases (Saha et al. 2010). Thus, in this way, plants of mung bean could adapt to survive even in salt stress.

12.7.3 Temperature Stress

Like other crops, mung bean is also sensitive to change in temperature and photoperiods. Most of the genotype of mung bean has a photoperiod of 12–13 h and beyond this photoperiod a progressive decrease in flowering is observed. According to Poehlman (1978) if the photoperiod is extended up to 16 h then in some short seasons early strain flowering was delayed for only a few days, however, in photosensitive strains this delay was of 30–40 days and some strains even fail to flower. Limited information is available about flower shedding in mung bean and scanty work has been carried out in this area. Kumari and Verma (1983) reported that high temperature stress has a negative impact on the retention of flower and consequently on the pod formation also. They also observed that due to high temperature extent of flower shedding has been extended up to 79 %. A Genotype with maximum flower retention and productive pods have been screened and selected during temperature >40 °C.

12.8 Resources Available and Their Utilization

12.8.1 Status of Global Collections and National Germplasm

A total of over 10,550 accessions of various *Vigna* species comprised of mung bean (3704), urdbean (3131), mothbean (1486), ricebean (2045) and azuki bean (185) have been stored at −18 °C in a long-term repository of National Gene Bank at NBPGR, New Delhi. Globally, mung bean germplasm accessions are maintained by more than 35 institutions, which hold more than 25,000 accessions. AVRDC at Taiwan maintains 5510 accessions of mung bean, and over 12000 accessions belonging to various *Vigna* species are held in the conservation unit in Georgia, USA.

12.8.2 Mung bean Varieties—the Cultivated Genepool

Though, a large number of varieties have been released and recommended for cultivation in different agro-climatic zones of India, a satisfactory increase in the yield has not been achieved so far. In India, out of a total of 50 varieties released during 1985–2010, 37 were cross-bred, seven were selections from local landraces and five varieties were developed by mutation breeding. In addition, one variety was a selection from an exotic accession. A critical analysis of the pedigree data of the varieties released reveals limited use of vast germplasm and frequent use of previously released varieties as the background material for developing novel varieties. Lakhanpaul et al. (2000) had pointed out that source material for 13 distinct varieties released by different agricultural institutes/universities could be traced back to only four local collections originating from a single state of India.

Thus, a narrow genetic base owing to high degree of commonness in the pedigrees prevails in the cultivated germplasm of mung bean though large germplasm having significant variation in important morpho-agronomic traits is available.

12.8.3 Wild Germplasm and Its Utilization

Crop Wild Relatives (CWR)—the putative progenitors including wild plant species that are related to crop plants are potential gene donors for the desirable traits (Ford-Lloyd et al. 2011). Thus, wild relatives, the important constituent of the total genetic diversity of the cultivated species are extremely valuable for the improvement of a crop. Genetic resources thus are important source of useful genes for resistance to diseases or for adaptability to changing climatic conditions. Likewise, some of the wild *Vigna* species have many useful genes (Table 12.1), which are transferable to cultivated crops by direct crosses (Tomooka et al. 2008; Pandiyan et al. 2008). Fuzi and Miyazaki (1987) reported an accession (TC1966) of *V. radiata* var. *sublobata* that showed perfect resistance against azukibean weevil (*Collasobruchus chinensis*). The resistance was found to be controlled by a single dominant gene (Kitamura et al. 1988). Fuzii et al. (1989) further found that TC1966 is completely resistant against *C. maculatus, C. phaseoli* and *Zabrotes subfasciatus*. Tomooka et al. (1992) developed a bruchid resistant mung bean line in Thailand by using TC1966 as a gene source. In addition to bruchid resistance, high resistance to yellow mosaic virus, high methionine content in seeds (Babu et al. 1988), higher photosynthetic activity and tolerance to drought (Ignacimuthu and Babu 1987),

Table 12.1 Wild/related species of *Vigna* as sources of useful genes

Species	Trait of interest
V. radiata var. *sublobata*	-Resistance to Mung bean Yellow Mosaic Virus (MYMV) -High number of seeds per pod and pods per plant -High methionine content
V. trilobata	Resistance to Mung bean Yellow Mosaic virus (YMV)
V. mungo var. *sylvestris*	Resistance to YMV
V. mungo var. *mungo*	Synchronous maturity and tolerance to Cercospora leaf spot
V. umbellata	High yield, tolerance to bruchid, responsive to inputs

higher tolerance to saline and alkaline soils (Lawn et al. 1988) have been reported for *V. radiata* var. *sublobata*. In comparison to *V. radiata* var. *sublobata,* there are very few studies on evaluation of *V. mungo* var. *silvestris* as genetic resource. However, this variety is cross-compatible with *V. mungo* (Miyazaki 1982).

Nevertheless, the varying extent of crossability barriers of the crop taxa with the wild and weedy relatives are the important determinants of the success of such conventional breeding programs. Thus, an insight into the results obtained so far while making interspecific crosses involving mung bean is important.

12.8.4 Hybridization

Interspecific crosses within the genus *Vigna* have been attempted by various workers. Barriers while crossing mung bean with other species range from failure of pollen tube to penetrate the style or stigma (Chowdhary and chowdhary 1977) to embryo abortion after fertilization and poor pod or seed set. Bharathi et al. (2006) while crossing *V. radiata* as female parent with four other species obtained highest crossability with *V. umbellata* (29.63) followed by *V. trilobata* (8.48 %) and *V. aconitifolia* (7.69 %). Interestingly, a reciprocal cross using *V. umbellata* as the female parent showed least number of pods (0.005 %) with shrunked pods and shrivelled seeds.

Sehrawat et al. (2015) while evaluating mung bean germplasm in saline environment identified the most salt tolerant wild relative of mung bean. Two genotypes namely EC528960 and TCR86 belonging to *V. luteola* and *V. trilobata,* respectively, were identified and are the potential source of genes to be introgressed in the salt sensitive mung bean genotypes through breeding.

12.8.5 Phylogenetic Relationships: Mungo-radiata *and Their Allies*

The taxonomic and nomenclatural uncertainty in *Vigna* occurred due to wrong nomenclature by Linneaeus, who named urdbean as *Phaseolus mungo* and mung bean as *P. radiatus*. Verdcourt (1970) reviewed the genus and on the basis of morphological and biochemical evidences transferred Asian species of *Phaseolus* to genus *Vigna*. Later, *Phaseolus mungo* was named as *V. mungo* and *P. radiatus* as *V. radiata*. Also, it was considered that both these species evolved from a common ancestor *P. sublobatus*, which afterwards was named as *V. sublobata*. Further misunderstanding occurred due to wrong cytogenetical relationships based on cytological study and scanning electron microscopy of seed coat pattern (Jain and Mehra 1980). The authors contended that *V. sublobata* was not likely to be the progenitor of mung bean (*V. radiata*) but instead the progenitor of urdbean (*V. mungo*). Morphological and biochemical studies by Arora et al. (1973) on wild populations of *Vigna* provide the first constructive evidence about evolutionary

Fig. 12.2 Mung bean and some of its wild relatives occurring in India: **a**. *V. radiata*; **b**. *V. radiata* var. *sublobata*; **c**. *V. hosei*; **d**. *V. trinervia var. trinervia*; **e**. *V. khandalensis*; **f**. *V. trinervia var. bourneae*

relationships among *V. radiata-mungo-sublobata-silvestris* group. They identified two botanical varieties, which, respectively, gave rise to *V. radiata* and *V. mungo.*

Of the seven subgenera under the genus *Vigna*, only subgenus *Ceratotropis* has its centre of species diversity in Asia. In the subgenus *Ceratotropis*, there are crop complexes associated with each of the domesticated species. The Asian *Vigna* genepool includes 21 species out of which eight are used for human and animal food. The subgenus includes wild relatives and their wild forms.

The species of *mungo-radiata* group has wide distribution throughout Indian subcontinent. The group includes two cultivated species *V. mungo* and *V. radiata* and their respective wild forms namely, *V. mungo* var. *silvestris* and *V. radiata* var. *sublobata* and *V. radiata* var. *silvestris*. Group also includes another wild species, i.e. *V. hainiana* which can be regarded as most ancestral type. All the species included in the group are variable in morphological characters (Fig. 12.2).

Lukoki et al. (1980) accepted specific distinction between two forms and their relation as wild ancestor to the cultivated species. They noted that Plx 416 was similar to *V. mungo* biochemically in possessing glutamyl-methionine and its sulphoxide; and morphologically possesses narrower stipules, brighter yellow flowers, 6–8 ovules per ovary, erect pods with long white hairs (or glabrescent) and seeds with a raised arillate hilum. Therefore, they recognized these plants as the wild ancestor of black gram and described it as a new taxon, namely *V. mungo* var. *silvestris* Lukoki, Maréchal and Otoul. Lukoki et al. (1980) also described Plx 274 similar to *V. radiata* var. *radiata* in possessing α-glutamyl-s-methylcysteine and its sulphoxide in its seeds and have broader stipules, paler yellow flowers, 10–14 ovules per ovary, spreading pods with short brown hairs and seeds with a flat non-arillate hilum. Chandel et al. (1984) and Miyazaki (1982) also supported the view that mung bean and urdbean were domesticated from *V. radiata* var. *sublobata* and *V. mungo* var. *silvestris*, respectively, on the basis of their biochemical and morphological studies.

Babu et al. (1988) assessed the seed protein and amino acid composition of *V. radiata* var. *sublobata* and two cultivated forms *V. mungo* and *V. radiata*. They reported high range of variation for all the amino acids indicating broad genetic base and suggested the usefulness of wild populations of *V. radiata* var. *sublobata* in the nutritional upgrading of mung- and urdbean. The amino acid profiles were population-specific and wild populations of *V. radiata* var. *sublobata* were shown to differ contrastingly in cysteine content with *V. radiata* as well as *V. mungo*. *V. mungo* contained high methionine content, which could be due to Y-glutamylmethonine and its sulphoxide while *V. radiata* possessed cysteine and its sulphoxide. These results were similar to those reported by Lukoki et al. (1980). However, Babu et al. (1985) continued to stress upon that *V. mungo* and *V. radiata* var. *sublobata* are closely related. Some indirect evidences from nodulation studies also indicate the origin of *V. mungo* and *V. radiata* from *V. mungo* var. *silvestris* and *V. radiata* var. *sublobata,* respectively (Kavimandan and Chandel 1988).

Various workers have analysed phylogenetic relations between mung bean and closely related taxa using molecular markers as well as sequence level variation in the phylogenetically useful loci. Ajibade et al. (2000) used ISSR amplification to study genetic relationships among 18 species of the genus *Vigna*. Tomooka et al. (2002) performed AFLP analysis of the diploid species in the genus *Vigna* subgenus *Ceratotropis*. The phenograms generated revealed that the species belonging to sections *Ceratotropis* and *Aconitifoliae* are well separated. However, most of the species in section *Angulares* showed a high level of similarity suggesting a low level of genetic variation.

Goel et al. (2002) reconstructed molecular phylogeny based on internal transcribed spacer (ITS) sequences to resolve the taxonomic contradictions in *Vigna* and its relation to the genus *Phaseolus*. ITS phylogeny was congruent with classification based on morphological, biochemical and cytogenetical and palynological features. However, subgenus *Plectotropis* of Neotropical origin was revealed to be very closely related to subgenus *Vigna* instead of forming a link between African and Asiatic *Vigna*. Sequence-based analysis of ITS and atpB-rbcL intergenic spacer split the subgenus *Ceratotropis* into three distinct sections namely Aconitifoliae, Angulares and *Ceratotropis* (Doi et al. 2002). Earlier, Vaillancourt and Weeden (1993) reconstructed chloroplast DNA phylogeny of old world *Vigna* which supported monophyly of old world *Vigna* with the only exception *V. frutscens* of subgenus *Vigna* that fell within the new world. Kaga et al. (1996) studied species relationships in subgenus *Ceratotropis* of genus *Vigna* through RAPD analysis and observed that the wild forms were always grouped with their most closely related cultivated forms. The largest intraspecific variation was found in mung bean and least variation was found in adzuki bean.

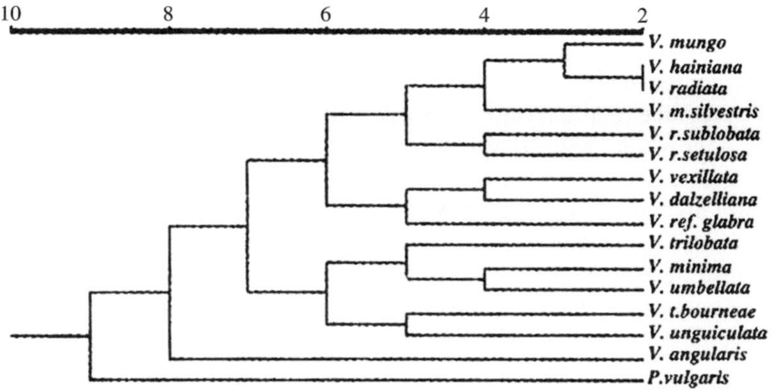

Fig. 12.3 Neighbour joining tree showing the relationship among 15 *Vigna* species using *Phaseolus vulgaris* as the outgroup (adapted from Vir et al. 2008)

Later, Vir et al. (2009) analysed population substructure, genetic differentiation and phylogenetic relationships among selected Asiatic *Vigna* species. Phylogenetic tree indicated that *V. mungo, V. radiata, V.mungo* var *silvestries, V. radiata* var *sublobata* and *V. radiata* var *setulosa* are distinct taxonomic groups. Also wild relatives were placed along with their cultivated forms reflecting their origin from a common ancestors (Fig. 12.3). Vir et al. (2010) also employed ISSR markers for diversity analysis, genotyping and assessment of species relationships in Asiatic *Vigna* from Indian subcontinent. The wild forms were grouped with their cultivated forms as expected whereas another wild forms *V. hainiana* emerged as a distinct taxon and seemed to be more primitive in comparison to the other wild relatives of green gram and black gram. Javadi et al. (2011) investigated phylogeny and biogeography of the subgenus *Ceratotropis* using chloroplast DNA sequence data. The study revealed three eco-geographical groups and late Pilocene-Pleistocene diversification.

12.8.6 Mung bean Germplasm and Core Collections

The effective utilization of huge germplasm collections maintained in the genebanks world over has necessitated the development of core set for each crop plant that has minimum similarities between the entries and yet represents the total collection in terms of genetic diversity (Frankel and Brown 1984; Van Hintum et al. 2000) With this realization, Bisht et al. (1998) developed a representative core set for Indian mung bean collections based upon their morpho-agronomic characteristics along with the passport data. The accessions from six major mung bean growing zones and one group of exotic collections subjected to principle component score strategy resulted in 152 accessions that could represent as a core of Indian mung bean germplasm collections. A comparison of the genetic diversity of this core collection with the total collection was also made to test the validity of the representative collection. Additionally, the variation pattern of the core set was also used to discriminate among accessions in order to enhance the accessibility and utilization of the germplasm. Recently, Schafleitner et al. (2015) established AVRDC—the world vegetable centre mung bean core collection by geographical stratification of the whole collection comprising more than 5000 accessions and subsequent diversity analysis based on eight phenotype descriptors. This core along and mini core collection together with the evaluation and genotypic data, is available for distribution to breeders. The core collection of 1481 entries has been under continuous evaluation for traits of interest for breeding biotic and abiotic stress tolerance and thus will add information of significant value to these collections.

12.8.7 Diversity Analysis

Analysis of genetic diversity in mung bean is important since diversity present within the species is paramount for maintaining its generic potential. Assessment of genetic diversity provides an opportunity to identify desirable traits such as higher yield, pest and disease resistant that are useful in the development of improved varieties. Also, genetic diversity within the species reflects its ability to adapt in changing environment. Genetic diversity within mung bean has been assessed by using different molecular markers such as RAPD (Lakhanpaul et al. 2000; Saini et al. 2008; Lavanya et al. 2008; Sony et al. 2012; Datta et al. 2012; Bhuyan et al. 2014) ISSR, AFLP (Chattopadhyay et al. 2005), SSR (Gwag et al. 2010; Chen et al. 2015) suggesting their utility in the analysis.

In addition, Vir et al. (2009) demonstrated the transferability of mung bean STMS primer to urdbean, wild relatives and other pulse taxa. Narsimhan et al. (2010) evaluated six mung bean Yellow Mosaic Virus resistant and susceptible genotypes of mung bean and urdbean using RGA primers from cowpea to assess the molecular diversity to develop suitable mapping population that can be used to identify and validate markers related to resistant loci.

Das and Singh (2014) conducted comparative analysis of genetic diversity across certain mung bean and urdbean cultivars of west Bengal using ISSR. Information generated was helpful in identification of the germplasm and diversity assessment. ISSR markers proved useful in assessment of genetic diversity, through detection of duplicate samples in germplasm collection and the selection of a core collection to enhance the efficacy of germplasm management for use in breeding and conservation programs. Recently, Chen et al. (2015) assessed genetic diversity and population structure of mung bean germplasm employing EST-based genomic SSR markers for germplasm evaluation.

12.9 Genomic Resources

12.9.1 Molecular Maps and Gene Tagging

In the pioneering study by Young et al. (1992), a major bruchid resistant gene was mapped in mung bean using RFLP markers. Bruchids are the most destructive pests of mung beans and other members of genus *Vigna*. Earlier bruchid resistance had been identified in wild mung bean accession TC1966. Fifty-eight F_2 progeny from a cross between TC1966 and a susceptible mung bean cultivar were analysed for 153 markers. Resistance mapped to a single locus on linkage group VIII, approximately 3.6 centimorgans from the nearest RFLP marker.

Menancio-Hautea et al. (1992) investigated genome relationships between mung bean and cowpea based on the linkage arrangement of random genomic restriction fragment length polymorphism (RFLP) markers. A common set of probes derived from cowpea, common bean, mung bean and soybean PstI genomic libraries were used to construct the genetic linkage maps. The mung bean and cowpea genomes were compared on the basis of the copy number and linkage arrangement of the 53 markers mapped in common between the two species. Results also suggested that the nucleotide sequences are conserved; however, there is variation in copy number and rearrangements in linkage order which happened after the divergence of species. Also, entire linkage groups were not conserved, but several large linkage blocks were maintained in genomes.

Humphry et al. (2002) constructed a genetic linkage map of mung bean consisting of 255 RFLP loci using a recombinant inbred population of 80 individuals. The population derived from an inter-subspecific cross between the cultivated mung bean variety "Berken" and a wild mung bean genotype "ACC 41" (*V. radiata* subsp. *sublobata*). The total length of the map, which comprised 13 linkage groups, spanned 737.9 cM with an average distance between markers of 3.0 cM and a maximum distance between linked markers of 15.4 cM. This map was compared with the previous released map of lablab using a common set of 65 RFLP probes. The chromosomal arrangement between these two species was found to be highly conserved. The two genomes, however, accumulated large number of duplications/deletions after they have diverged.

Gene tagging involves identification of molecular markers tightly linked to the trait of interest, an essential requirement of marker aided selection (MAS) in plant breeding programs for screening large populations in an unambiguous manner. Such markers have been developed in nearly all important crops. However, limited efforts have been made in case of mung bean. Nevertheless, Souframanien and Gopalakrishna (2005) identified a molecular marker closely linked to YMV resistance gene in case of black gram, a closely related species of green gram. A SCAR marker namely, $ISSR81_{1357}$ was specifically present in the resistant genotypes and consistently absent in susceptible ones.

12.9.2 *Whole Genome Sequence of Mung bean and Its Selected Allies*

Genome size of mung bean is 579 Mb with 2n = 2x = 22 chromosomes. Kang et al. (2014) constructed a draft genome of cultivated mung bean (*V. radiata* var. *radiata vc*1973a) on a chromosomal scale. This is the first draft genome sequence within the genus *Vigna.* For understanding domestication, polyploidization and speciation in the genus *Vigna*, whole genome sequence of the wild mung bean relative (*V. radiata var. sublobata*) and a tetraploid relative of mung bean (*V. reflex-pilosa* var. *glabra*), as well as transcriptome sequences of 22 *Vigna* accessions of 18 species were produced. In the study, 80 % of the *V. radiata* var. *radiata* genome was constructed with identification of 22,427 high-confidence protein coding genes and 160 *Vigna* gene clusters. Genomic sequencing provided insights into the polyploidy in legumes. It has been suggested that the domestication and cultivation of mung bean was initiated in the northwest and far south in India 4000–6000 years ago, based on the geographical distribution of the wild mung bean and archaeological records from India (Fuller 2007). As only one accession of *V. radiata var. sublobata* was included in the study, they could not observe any population substructure in *V. radiata var. sublobata* and thus were not able to determine whether there are V. radiata var. sublobata lineages more closely related to cultivated mung bean or they obtained evidences of multiple origin of the crop variety.

12.10 Biotechnological Interventions

Though dramatic results have been obtained in cereal crops by conventional breeding, the success in improving the crop yield has remained largely elusive in case of pulse crops in general and mung bean in particular. The narrow genetic base of the germplasm and crossability barriers between cultivated genepool and its allies has often been cited as the primary causes of failure of traditional methods for improvement. Nevertheless, the biotechnological interventions have now enabled the utilization of secondary and even tertiary genepool as gene transfers can be made across the wide taxa using modern techniques of genetic transformation with or without the help of in vitro regeneration protocols. Consequently, transgenic plants have been produced in most of the major crops.

However, legumes are generally not amenable to such biotechnological interventions easily due to the recalcitrance exhibited by these taxa to efficient tissue culture as well as transformation protocols. Likewise, success in developing transgenic mung bean has been rather limited using both direct and indirect method of organogenesis. The efforts by various workers towards optimizing usable protocols has been extensively reviewed and summarized by Sahoo et al. (2003).

Mung bean though a recalcitrant grain legume, *Agrobacterium*-mediated transformation has been achieved by Jaiwal et al. (2001). Hypocotyl and primary leaves excised from 2-day-old in vitro grown seedlings produced transgenic calli on B_5 basal medium supplemented with 5×10^{-6} M BAP, 2.5×10^{-6} M each of 2,4-D and NAA and 50 mg l^{-1} kanamycin after co-cultivation with *Agrobacterium tumefaciens* strains, LBA4404 (pTOK233), EHA105 (pBin9GusInt) and C58C1 (pIG121Hm), all containing β-glucuronidase (*gusA*) and neomycin phosphotransferase II (*npt*II) marker genes.

Sonia et al. (2007) developed fertile transgenic plants of mung bean with two transgenes namely bar and ά amylase inhibitors. Recently, Yadav et al. (2012) reported a reproducible and highly efficient protocol for *Agrobacterium*-mediated transformation of mung bean using double cotyledonary nodes (DCN). Transient and constitutive gene expression was observed in DCN explants and different tissues of T0 and T1 plants. Integration of annexin gene was further confirmed by southern blotting. However, undoubtedly, extensive and concerted research efforts are further required to develop efficient protocols for mung bean regeneration and transformation. Later, Mirza and Tazeen (2004) also optimized the *Agrobacterium tumefacians* mediated transformation protocol for mung bean by studying parameters like sensitivity of explants to kanamycin, pH of co-culture media, age of explants, types of explants, co-cultivation time and optical density of *Agrobacterium* culture medium. Transformed shoots were produced on shoot regeneration medium containing 50 mg/l kenamycin and 500 mg/l cefotaxine.

In conclusion, though germplasm holdings can be further enriched by conducting targeted explorations and through exchange programs, sufficient collections do exist in the national gene banks and international institutes such as AVRDC. Despite maintaining these large collections comprising of wild species, landraces, obsolete cultivars, etc., that are immensely endowed with a number of useful genes, efforts to utilize them in the ongoing improvement programs are less than satisfactory. Some of the major efforts for preventing the underutilization of mung bean germplasm include extensive screening, characterization and evaluation of the available holdings for identifying the elite genotypes and development of multiple core sets for the desirable traits. Intensive prebreeding efforts through strong interactions between germplasm curators and mung bean breeders are also required. On the other hand, developing novel varieties exploiting distant genepools necessitates the optimization of efficient protocols for regeneration and transformation for successful intervention of biotechnogical tools.

Efforts using a combination of traditional breeding and recent molecular approaches are the need of the hour to break the yield barriers in pulse crops such as mung bean that are the major source of nutrition for large populations and should be considered as the immediate and the urgent challenge for scientific community engaged in crop improvement programs.

References

Abd-Alla MH, Vuong TD, Harper JE (1998) Genotypic differences in nitrogen fixation response to NaCl stress in intact and grafted soybean. Crop Sci 38:72

Ahmed S (2009) effect of soil salinity on the yield and yield components of mungbean. Pak J Bot 41:263–268

Ajibade SR, Weeden NF, Chite SM (2000) Inter simple sequence repeat analysis of genetic relationships in the genus *Vigna*. Euphytica 111:47–55

Arora RK, Chandel KPS, Joshi BS (1973) Morphological diversity in *Phaseolus sublobatus* Roxb. Curr Sci 42:359–361

Arora A, Sairam RK, Srivastava GC (2002) Oxidative stress and antioxidative systems in plants. Curr Sci 82:1227–1238

Arulbalachandran D, Ganesh KS, Subramani A (2009) Changes in metabolites and antioxidant enzyme activity of three *Vigna* species induced by NaCl stress. Am Eurasian J Agron 2:109–116

Babu CR, Sharma SK, Johri BM (1985) Taxonomic revision of Indian *Phaseolus, Vigna, Macroptilium* and *Dysolobium*. Bull Bot Surv India 27:1–28

Babu CR, Sharma SK, Chatterjee SR, Arbol YP (1988) Seed protein and amino acid composition of *V. radiata* var. *sublobata* (Fabaceae) and two cultigens, *V. mungo* and *V. radiata*. Eco Bot 42(1):54–61

Baudin JP, Marechal R (1988) Taxonomy and evolution of genus *Vigna*. In: Mungbean, proceedings 2nd international symposium Asian vegetable research and development center. Shanhua, Taiwan pp 2–12

Begg JE, Turner NC (1976) Crop water deficits. Adv Agron 28:191–217

Bekheta MA, Talaat IM (2009) Physiological response of mungbean "*Vigna radiata*" plants to some bioregulators. J Appl Bot Food Qual 83:76–84

Bharathi A, Vijay KS, Veerabadhiran P, Subba BL (2006) Crossability barriers in mungbean (*Vigna radiata* L. Wilczek) with its wild relatives. Indian. J Crop Sci 1(1–2):120–124

Bhuyan SI, Hossain S, Islam MM, Begum SN, Urbi Z, Hossain S (2014) Molecular assessment of genetic diversity and relationship in selected mungbean germplasms. Biotechnology 13:126–134

Bisht IS, Mahajan RK, Kawalkar TG (1998) Diversity in mungbean (*Vigna radiata* L. Wilczek) germplasm collection and its potential use in crop improvement. Ann Appl Biol 132:301–312

Capoor SP, Varma PM (1948) Yellow mosaic of *Phaseolus lunatus* L. Current Science 17:152–53

Chandel KPS, Lester RN, Starling RJ (1984) The wild ancestor of urd and mungbeans (*V. mungo* (L.) Hepper and *V. radiata* (L.) Wilczek). Bot J Linn Soc 89:85–96

Chattopadhyay K, Ali N, Sarkar HK, Mandai N, Bhattacharyya S (2005) Diversity analysis by RAPD and ISSR markers among the selected mungbean (*Vigna radiata* L. Wilczek) genotypes. Indian J Genet Plant Breed 65(3):173–175

Chen H, Wang L, Wang S, Liu C, Blair MW, Cheng X (2015) Transcriptome sequencing of mung bean (*Vigna radiate* L.) genes and the identification of EST-SSR markers. PLoS ONE 10(4): e0120273

Chowdhury RK, Chowdhury JB (1977) Intergeneric hybridization between *Vigna mungo* L. Hepper and *Phaseolus calcaratus* RoxB. Indian J Agric Sci 47:117–121

Das I, Singh AP (2014) Effect of PGPR and organic manures on soil properties of organically cultivated mungbean. Bioscan 9(1):27–29

Datta S, Gangwar S, Kumar S, Gupta S, Rai R, Kaashyap M, Singh P, Nagaswamy N (2012) Genetic diversity in selected Indian mungbean [*Vigna radiata* (L.) Wilczek] cultivars using RAPD markers. Am J Plant Sci 3:10851091

De DN, Krishnan R (1968b) Cytological studies in *Phaseolus*, I Autotetroploid *Phaseolus aureus* x tetraploid species of *Phaseolus* and the backcross. Indian J Genet 28:12–22

Doi K, Kaga A, Tomooka N, Vaughan DA (2002) Molecular phylogeny of genus *Vigna* subgenus *Ceratotropis* based on rDNA ITS and atpB-rbcL intergenic spacer for cpDNA sequences. Genetica 114:129–145

Duong M, Nauyen PB, Vo MO, Nguyen HT, Duong NV, Nguyen CH (1988) Mungbean varietal development for problem soils in the Mekong Delta of Vietnam. TVIS Newsl 3:4–9

Fernandez GCJ, Shanmugasundaram S (1988) The AVRDC mungbean improvement program: the past, present and future. In: Shanmugasundaram S, McLean BT (eds) Procedings 2nd international symposium mungbean. AVRDC, Shanhua, Taiwan pp 58–70

Ford-Lloyd BV, Schmidt M, Armstrong SJ, Barazani O, Engels J, Hadas R, Hammer K, Kel SP, Kang D, Khoshbakht K, Li Y, Long C, Lu BR, Ma K, Viet Tung Nguyen KVT, Qiu L, Ge S, Wei W, Zhang Z, Maxted N (2011) Crop wild relatives—undervalued, underutilized and under threat? Bioscience 61(7):559–565

Frankel OH, Brown AHD (1984) Plant genetic resources today: a critical appraisal. In: Holden JHW, Williams JT (eds) Crop genetic resources: conservation and evaluation. Allen and Unwin, Winchester

Friedman R, Altman A, Levin N (2006) The effect of salt stress on polyamine biosynthesis and content in mungbean plants and in halophytes. Physiol Plant 76:295–302

Fuller DQ (2007) Contrasting patterns in crop domestication and domestication rates: recent archaeobotanical insights from the old world. Ann Bot 100:903–924

Fuzi K, Miyazaki S (1987) Infestation resistance of wild legumes *Vigna sublobata* to azukibean weevil, *Callosobruchus chinensis* (L.) (Coleoptera: Bruchidaceae) and its relationship with cytogenetic classification. Appl Entomol Zool 22:229–230

Fuzii K, Ishimoto M, Kitamura K (1989) Patterns of resistance to bean weevils (Bruchidae) in *Vigna radiata-mungo-sublobata* complex inform the breeding of new resistant varieties. Appl Entomol Zool 24(1):126–132

Goel S, Raina SN, Ogihara Y (2002) Molecular evolution and phylogenetic implications of internal transcribed spacer sequences of nuclear ribosomal DNA in the *Phaseolus-Vigna* group. Mol Phylogenet Evol 22(1):1–19

Gulati A, Jaiwal PK (1990) Culture conditions effecting plant regeneration from cotyledon of *Vigna radiata* (L.) Wilczek. Plant Cell Tiss Org Cult 23:1–7

Gulati A, Jaiwal PK (1994b) Plant regeneration from cotyledonary node explants of mungbean (*Vigna radiata* (L.) Wilczek). Plant Cell Rep 13:523–527

Gwag JG, Dixit A, Park YJ, Ma KH, Kwon SJ, Cho GT, Lee GA, Lee SY, Kang HK, Lee SH (2010) Assessment of genetic diversity and population structure in mungbean. Genes Genomics 32(4):299–308

Haqqani AM, Pandey RK (1994) Response of mungbean to water stress and irrigation at various growth stages and plant densities. II. Yield and yield components. Trop Agric (Trinidad) 71:289–294

Hasanuzzaman M, Nahar K, Fujita M (2013) Plant response to salt stress and role of exogenous protectants to mitigate saltinduced damages. In: Ahmad P, Azooz MM, Prasad MNV (eds) Ecophysiology and responses of plants under salt stress. Springer, New York, pp 25–87

Hsiao TC, Acevedo F (1974) Plant response to water deficits, water use efficiency and drought resistance. Agric Meteorol 14:69–84

Humphry ME, Konduri V, Lambrides CJ, Magner T, McIntyre CL, Aitken EAB, Liu CJ (2002) Development of a mungbean (*Vigna radiata*) RFLP linkage map and its comparison with lablab (*Lablab purpureus*) reveals a high level of co linearity between the two genomes. Theor Appl Genet 105:160–166

Ignacimuthu S, Babu CR (1987) Mutagenesis for resistance to storage pest in the wild and cultivated urd and mungbean. J Nucl Agric Biol 16:169–176

Jain HK, KL Mehra (1980) Evolution, adaptation, relationships and uses of the species of *Vigna* cultivated in India. In: Summerfield RJ, Bunting AH (eds) Advances in legume science. Royal Botanic Gardens, Kew, pp 459–468

Jaiwal PK, Kumari R, Ignacimuthu S, Potrykus I, Sautter C (2001) *Agrobacterium tumefaciens*-mediated genetic transformation of mungbean (*Vigna radiata* L. Wilczek)–a recalcitrant grain legume. Plant Sci 161:239–247

Javadi F et al (2011) Molecular phylogeny of the subgenus *Ceratotropis* (genus *Vigna*, Leguminosae) reveals three eco-geographical groups and Late Pliocene-Pleisticene diversification: evidence from four plastid DNA region sequences. Ann Bot (Oxford) 108:367–380

Kaga A, Tomooka N, Egawa Y, Hosaka K, Kamijima O (1996) Species relationships in the subgenus *Ceratotropis* (genus *Vigna*) as revealed by RAPD analysis. Euphytica 88:17–24

Kang YJ, Kim SK, Kim MY, Lestari P, Kim KH, Ha BK, Jun TH, Hwang WJ, Lee T, Lee J, Shim S, Yoon MY, Jang YE, Han KS, Taeprayoon P, Yoon N, Somta P, Tanya P, Kim KS, Gwag JG, Moon JK, Lee YH, Park BS, Bombarely A, Doyle JJ, Jackson SA, Schafleitner R, Srinives P, Varshney RK, Lee SH (2014) Genome sequence of mungbean and insights into evolution within *Vigna* species. Nat Commun. doi:10.1038/ncomms6443

Kavimandan SK, Chandel KPS (1988) Nodulation in some wild species of *Vigna* and symbiotic behavior of their root nodule bacteria. New Bot 15(4):253–258

Khanal NP, Khanal NN, Gurung GB, Shepra LT, Gupta KP, Thapa S, Joshi KD, Harris D, Kumar Rao JVDK, Darai R (2005) Mungbean in cereal fallows: Experience of farmers' participatory research and development activities in terai of Nepal. In: Kharkwal MC (ed) Proceedings of the fouth international legumes research conference (IFLRC-IV), October 18–22, 2005, New Delhi, India. Food Legumes for Nutritional Security and Sustainable Agriculture

Kumar BS, Prakash M, Sathiya N, Gokulakrishnan J (2012a) Breeding for salinity tolerance in mungbean. In: 2nd international conference on Asia agriculture and Animal (ICAAA 2012), vol 4. APCBEE Procedia, pp 30–35

Kumar RR, Goswami S, Sharma, Singh K, Gadpayle KA, Kumar N, Rai GK, Singh M, Rai RD (2012b) Protection against heat stress in wheat involves change in cell membrane stability, antioxidant enzymes, osmolyte, H2O2 and transcript of heat shock protein. Int J Plant Physiol Biochem 4(4):83–91

Kumari P, Varma SK (1983) Genotypic differences in flower production/shedding and yeild in mungbean *(Vigna radiata*) Indian J Plant Physi 26:402-405

Lakhanpaul S, Chadha S, Bhat KV (2000) Random amplified polymorphic DNA (RAPD) analysis in Indian mungbean (*V. radiata* (L.) Wilczek) cultivars. Genetica 109:227–234

Lavanya GR, Srivastava J, Ranade SA (2008) Molecular assessment of genetic diversity in mungbean germplasm. J. Genet 87:65–74

Lawn R, Williams W, Imrie BC (1988) Potential of wild germplasm as a source of tolerance to environmental stresses in mungbean. In: Mungbean Proceedings 2nd international symposium. AVRDC, Taiwan, pp 136–145

Liu F, Andersen MN, Jensen CR (2003) Loss of pod set caused by drought stress is associated with water status and ABA content of reproductive structures in soybean. Funct Plant Biol 30:271–280

Lukoki L, Maréchal R, Otoul E (1980) The wild ancestors of cultivated beans *V. radiata* and *V. mungo*. Bull Jard Bot Nat Belg 28:23–30

Marechal R, Mascherpa JM, Stainier F (1978) Etude taxonomique d' un group groupe d' especes des generes *Phaseolus* et *Vigna* (Papillionaceae) sur la base de donnees morphologiques, et polliniques traitees par l' analyse informatique. Boissera 28:1–273

Menancio-Hautea DM, Kumar L, Danesh D, Young ND (1992) A genome map for mungbean [*Vigna radiata* (L.) Wilczek] based on DNA genetic markers (2 N = 2X = 22). In: O'Brien SJ (ed) Genome maps, vol 6. Cold Spring Harbor Laboratory, Cold Spring Harbor, NY, pp 259–261

Misra N, Gupta AK (2006) Interactive effects of sodium and calcium on proline metabolism in salt tolerant green gram cultivar. Am J Plant Physiol 1:1–12

Miyazaki H (1982) Classification and phylogenetic relationship of the *Vigna radiata—silvestris—sublobata* group. Bull Natl Ints Sci D 33:1–61

Morton F, Smith RE, Poehlman JM (1982) The mungbean. College of Agricultural Sciences, Department of Agronomy and Soils, Mayaguez, Puerto Rico, Spec. Publ

Musgrave ME, Vanhoy MA (1989) A growth analysis of waterlogging damage in mungbean (*Phaseolus aureus*). Can J Bot 67:2391–2395

Narasimhan A, Patil BR, Datta S, Kaashyap M (2010) Genetic diversity assessment across different genotypes of mungbean and urdbean using molecular markers. Electron J Plant Breed 1(4):379–383

Nariani TK (1960) Yellow mosaic of mung (ZPhaseolus aureus L.). Indian Phytopathol 13:24–29

Paliwal KV, Maliwal GL (1980) Growth and nutrient uptake relationship of some crops in saline substrate. Ann Arid Zone 19:251–253

Pandey RK, Herrera WAT, Pendleton JW (1984) Drought response of grain legume under irrigation gradient. I. Yield and yield components. Agron J 76:549–553

Pandiyan M, Ramamoorthi N, Ganesh SK, Jebaraj S, Nagarajan P, Balasubramanian P (2008) Broadening the genetic base and introgression of MYMV resistance and yield improvement through unexplored genes from wild relatives in mungbean. Plant Mutat Rep 2:3343

Poehlman JM (1978) What we have learnt from the international mungbean nurseries. In: Proceedings of the 1st international mungbean symposium. Asian Vegetable Research and Development Centre, Shanhua, Tainan pp 97–100

Ramos MLG, Parsons R, Sprent JI, James EK (2003) Effect of water stress on nitrogen fixation and nodule structure of common bean. Pesquisa Agropecuária Brasileira 38:339–347

Saha P, Chatterjee P, Biswas AK (2010) NaCl pre-treatment alleviates salt stress by enhancement of antioxidant defense and osmolyte accumulation in mungbean (*Vigna radiata* L. Wilczek). Indian J Exp Biol 48:593–600

Sahoo L, Sugla T, Jaiwal PK (2003) In vitro regenearion and transformation of cowpea, mungbean and adzuki bean. In: Jaiwal PK, Singh RP (eds) Applied genetics of leguminosae biotechnology, pp 89–120. Kluwer Academin Publishers

Saini A, Reddy SK, Jawali N (2008) Intra-individual and intra-species heterogeneity in nuclear rDNA ITS region of *Vigna* species from subgenus *Ceratotropis*. Genet Res 90:299–316

Savi G (1826) Troisieme memoire sur les genres Phaseolus et Dolichos par le Dr. Gaetano Savi. Linnean 1:331

Schafleitner R, Nair RM, Rathore A, Wang YW, Lin CY, Chu SH, Lin PY, Chang JC, Ebert AW (2015) The AVRDC—the world vegetable center mungbean (*Vigna radiata*) core and mini core collections Schafleitner et al. BMC Genomics 16:344

Sehrawat N, Yadav M, Bhat KV, Sairam RK, Jaiwal PK (2015) Effect of salinity stress on mungbean [*Vigna radiata* (L) Wilczek] during consecutive summer and spring seasons. J Agric Sci 60(1):23–32

Singh BV, Ahuja MR (1977) *Phaseolus sublobatus* Roxb. A source of resistance to yellow mosaic virus for cultivated mung. Indian J Genet 37:130–132

Singh NB, Kochhar S (2005) Challenges for increasing area, production and productivity of pulses in the Indian sub-continent. In: Proceedings of the fouth international legumes research conference (IFLRC-IV), October 18–22, 2005, New Delhi, India; Kharkwal MC (ed) (2005) Food legumes for nutritional security and sustainable agriculture. In: Proceedings of the fouth international legumes research conference (IFLRC-IV), October 18–22, New Delhi, India; Kharkwal MC (ed) Food legumes for nutritional security and sustainable agriculture

Singh DP, Singh BB (2011) Breeding for tolerance to abiotic stresses in mungbean. J Food Legum 24(2):83–90

Smykal PP, Coyne CJ, Ambrose MJ, Maxted N, Schaefer H, Blair MW, Berger J, Greene SL, Nelson MN, Besharat N, Vymyslick T, Toker C, Saxena RK, Roorkiwal M, Pandey MK, Hu J, Li YH, Wang LX, Guo Y, Qiu LJ, Redden RJ, Varshney RK (2015) Crit Rev Plant Sci 34:43–104

Sonia SR, Singh RP, Jaiwal PK (2007) *Agrobacterium tumifaciens*-mediated transfer of *Phaseolus vulgaris* [-amylase inhibitor-1 gene into mungbean (*Vigna radiata* L. Wilczek) using bar as selectable marker. Plant Cell Rep 26:187–198

Souframanien J, Gopalkrishna T (2005) Validation of molecular markers linked to mungbean yellow mosain virus in black gram (Vigna mungo (L.) Hepper). Pages 53. In: Proceedings of international food legume research conference, New Delhi, India. Indian soc of genet pl br

Sony SK, Ahashanhabib MD, Islam MN (2012) Genetic diversity analysis of thirteen mungbean [*Vigna radita* (L.) Wilczek] cultivars using RAPD markers. Bangladesh J Bot 41(2):169–175

Srivalli B, Chinnusamy V, Chopra RK (2003) Antioxidant defence in response to abiotic stresses in plants. J Plant Biol 30:121–139

Tazeen S, Mirza B (2004) Factors affecting *Agrobacterium tumefaciens* mediated genetic transformation of *Vigna radiata* (L.) Wilczek. Pak J Bot 36(4):887–896

Thirumaran AS, Seralathan MA (1988) Utilization of mungbean. In: Shanmugasundaram S (ed) Second International Mungbean Symposium Proceedings. AVRDC, Tainan, pp 470–485

Thomas R, Fukai MJS, Peoples MB (2004) The effect of timing and severity of water deficit on growth, development, yield accumulation and nitrogen fixation of mungbean. Field Crop Res 86:67–80

Tickoo SK, deParalta-Venturia MN, Harik LR, Worcester HD, Salama ME, Young AN, Moch H, Amin MB (2006) Am J Surg. Pathol. 30:141

Tomooka N, Egawa Y, Lairungreang C, Thavarasook C(1992) Collection of wild Ceratotropis species on the Nansei Archipelago, Japan and evaluation of bruchid resistance JARQ, 26:3 pp 222–230

Tomooka N, Yoon MS, Doi K, Kaga A, Vaughan D (2002) AFLP analysis of diploid species in the genus *Vigna* subgenus *Ceratotropis*. Genet Res Crop Evol 49:521–530

Tomooka N, Senthil N, Pandiyan M, Ramamoothi N, Kaga A, Vaughan DA (2008) Collection and conservation of leguminous crops and their wild relatives in Tamil Nadu, India, 2008. Annual report on exploration and introduction of plant genetic resources (NIAS, Tsukuba, Japan), vol 24, pp 113–125

Vaillancourt RE, Weeden NF (1993) Lack of isozyme similarity between *Vigna unguiculata* and other species of genus *Vigna* (Leguminosae). Can J Bot 71:586–591

Varma AK, Rao NSS (1975) Effect of different levels of soil moisture on growth, yield and some physiological aspects of nodulation in greengram.Indian J Agric Sci 45:11–16

Van Hintum TJL, Brown AHD, Spillane C, Hodgkin T (2000) Core collections of plant genetic resources. IPGRI Technical Bulletin No. 3. International Plant Genetic Resources Institute, Rome, Italy

Varma AK, Rao NSS (1975) Effect of different levels of soil moisture on growth, yield and some physiological aspects of nodulation in greengram. Indian J Agric Sci 45:11–16

Verdcourt B (1970) Studies in the Leguminosae-Papilionoidae the ‘Flora of Tropical East Africa’. Kew Bull 24:507–569

Vir R, Bhat KV, Lakhanpaul S (2008) Analysis of population substructure, genetic differentiation and phylogenetic relationships among selected Asiatic *Vigna* Savi species. Genet Resour Crop Evol 56:783–795

Vir R, Bhat KV, Lakhanpaul S (2009) Transferability of Sequence Tagged Microsatellite Sites (STMS) to the other pulse yielding taxa belonging to tribe Phaseolae. Int J Integr Biol 5:62–65

Vir R, Jehan T, Bhat KV, Lakhanpaul S (2010) Genetic characterization and species relationships among selected Asiatic Vigna savi. Genet Resour Crop Evol 57:1091–1107

Wahid A (2004) Analysis of toxic and osmotic effects of sodium chloride on leaf growth and economic yield of sugarcane. Bot Bull Acad Sin 45:133–141

Wahid A, Hameed M, Rasul E (2004) Salt-induced injury symptoms, changes in nutrient and pigment composition, and yield characteristics of mungbean. Int J Agric Biol 6:1143–1152

Worrall VS, Roughly RJ (1976) The effects of moisture stress on infection of *Trifolium subterraneum* L. by *Rhizobium trifoli* Dang. J Exp Bot 27:1233–1241

Yadav SK, Katikala S, Yellisetty V, Kannepalle A, Narayana JL, Maddi V, Mandapaka M, Shanker AK, Bandi V, Bharadwaja KP (2012) Optimization of *Agrobacterium* mediated genetic transformation of cotyledonary node explants of *Vigna radiata*. SpringerPlus 1:59

Yao Y, Chen F, Wang M, Wang J, Ren G (2008) Antidiabetic activity of mung bean extracts in diabetic KK-Ay mice. J Agric Food Chem 56(19):8869–8873

Young ND, Kumar L, Menancio-Hautea D, Danesh D, Talekar NS, Shanmugasundaram S, Kim DH (1992) RFLP mapping of a major bruchid resistance gene in mungbean (*Vigna radiata* L. Wilczek). Theor Appl Genet 84:839–844

Zahir ZA, Shah MK, Naveed M, Akhter MJ (2010) Substrate dependent auxin production by *Rhizobium phaseoli* improves the growth and yield of *Vigna radiata* L. under salt stress conditions. J Microbiol Biotechnol 20:1288–1294

Chapter 13
Genetic Improvement in the Genus *Eleusine*

Renuka Agrawal and Ankur Maheshwari

Abstract *Eleusine* (Poaceae) is a small genus (x = 8, 9, 10) consisting of six diploid and three tetraploid species. *Eleusine coracana* (finger millet; $2n = 4x = 36$) is the only cultivated species of the genus which holds a prominent place in the cropping sequences in semi-arid regions of India, Africa and other South Asian countries. The grain of finger millet is used to prepare diverse types of food items and also have medicinal value. Finger millet and its wild relatives are rich source of disease resistance and other agronomically important alleles. These alleles can be used successfully for crop improvement programs. Therefore, in recent years considerable interest has been centred towards the understanding of the relationships between the cultivated crop plant and its wild relatives. This chapter gives an overview on current understanding on the phylogenetic relationships within the genus *Eleusine* and other resources that can be utilized for successful genetic improvement programs.

Keywords *Eleusine* · Finger millet · Progenitor · Allotetraploid

13.1 Introduction

The genus *Eleusine* Gaertn., is a member of the tribe Eragrosteae, subfamily Chloridoideae and family Poaceae. It is a small genus consisting of nine species that includes six diploid ($2n = 2x = 16, 18, 20$) and three polyploid species ($2n = 4x = 36, 38$)

R. Agrawal (✉)
Department of Botany, Miranda House, University of Delhi,
Patel Chest Marg, Delhi 110007, India
e-mail: renuka.du@gmail.com

A. Maheshwari
Department of Zoology, Zakir Husian Delhi College, University of Delhi,
Jawahar Lal Nehru Marg, New Delhi 110002, India
e-mail: maheshwariduz@hotmail.com

V.R. Rajpal et al. (eds.), *Gene Pool Diversity and Crop Improvement*,
Sustainable Development and Biodiversity 10,
DOI 10.1007/978-3-319-27096-8_13

(Bisht and Mukai 2002; Liu et al. 2011; Agrawal et al. 2014), built on $x = 8, 9, 10$. These annual or perennial species are widely distributed in the tropical and subtropical regions of Africa, Asia and South America (Phillips 1972). East Africa is considered to be the centre of diversity for the genus (Bisht and Mukai 2002). Eight species, *E. africana*, *E. coracana*, *E. kigeziensis*, *E. indica*, *E. multiflora*, *E. floccifolia*, *E. intermedia* and *E. jaegeri*, occur in this region (Mehra 1963; Phillips 1972). *E. coracana* ($2n = 4x = 36$), commonly known as finger millet or ragi, is the only economically important species of the genus and holds a prominent place in cropping sequences in semi-arid regions of India, Nepal and Africa. After Sorghum and pearl millet, finger millet ranks third in cereal production in semi-arid regions of the world (Bisht and Mukai 2002). The grain is widely used for preparing bread, cakes, soup, puddings, porridge and fermented beverages (Hilu and De Wet 1976a; Chandrashekar 2010; Neves 2011; Agrawal et al. 2014). Finger millet is a rich source of essential amino acids, polyphenols, and it has comparatively higher levels of calcium and iron than other known cereals (Barbeau and Hilu 1993; Chandrashekar 2010). It has a number of medicinal properties; particularly it has been shown to be effective in controlling blood sugar levels in diabetic patients (Duke and Wain 1981; Chandrashekar 2010; Pradhan et al. 2010). The species is also reported to be diaphoretic, diuretic and is used as prophylaxis for dysentery (Bhatnagar 1952). Ragi is a folk remedy for leprosy, liver diseases, measles, pleurisy, pneumonia and small pox (Watt and Breyer-Brandwijk 1962; Duke and Wain 1981).

Wild relatives of crop species often carry beneficial alleles effective against various biotic and abiotic stresses and therefore hold the key for successful improvement programs through gene(s) introgression from wild to cultivated crop species (Dida and Devos 2006; Dida et al. 2007, 2008). These genes can be easily transferred to finger millet and also the other crops particularly to cereals, which have gene synteny with finger millet. Therefore, understanding the genetic relationship among the species of genus *Eleusine* through molecular genetic studies would be helpful in better understanding the crossing barriers and the capability of incorporating these traits from wild species to cultivated species. This requires detailed genetical and cytogenetical studies using different molecular and cytological markers. Therefore, considerable interest has centred upon understanding of genetic relationships between the cultivated species and its immediate wild ancestor(s), and other wild relatives constituting primary, secondary and tertiary gene pools. Primary gene pool of finger millet *Eleusine coracana* includes, land races/varieties or cultivars evolved in diverse geographic areas after its domestication. Secondary gene pool of this millet are its wild progenitor species *E. africana* or other wild taxa with which finger millet can exchange genes in nature or which are genomically homologous. Other genetic resources of finger millet are wild species separated by ploidy level. It is in this context that the assessment of genomic relationships between *E. coracana* and its allied species has been a subject of comprehensive investigations at morphological, cytogenetical, biochemical and DNA level. Chromosome research has demonstrated the prevalence of polyploidy and aneuploidy in the evolution of the genus *Eleusine* (Chennaveeraiah and Hiremath 1974a, b; Hiremath and Chennaveeraiah 1982; Hiremath and Salimath 1991; Bisht and Mukai 2000, 2001a, b, 2002). Biochemical, nuclear and chloroplast DNA markers have provided valuable insight into species relationships, and on the origin of the crop species (Hilu et al. 1978; Hilu 1988, 1995; Hilu and Johnson 1992, 1997; Werth et al. 1993, 1994; Salimath et al. 1995a; Neves et al. 2005; Dida et al. 2007, 2008; Liu et al. 2011; Agrawal et al. 2014). Despite the abundance of the published work, there still exists considerable disagreement not only with respect to the identification of the diploid ancestors of the three polyploid species but the level of speciation and evolutionary relationships between the nine *Eleusine* species also remain unclear and (or) contradictory (Phillips 1972, 1995; Hilu and Johnson 1997; Lye 1999; Bisht and Mukai 2000, 2001a, 2002; Neves et al. 2005; Devarumath et al. 2005; Liu et al. 2011). This chapter is an attempt to gather all the current knowledge and resources available for elucidating the phylogenetic relationship between the diploid and polyploid taxa of the genus *Eleusine* and the genetic improvement of finger millet.

13.2 Morphology, Taxonomy, Distribution and Economic Importance

Eleusine Gaertn. is a member of tribe Eragrosteae (subtribe Eleusininae) in the C_4 grass subfamily Chloridoidea and family Poaceae. The genus *Eleusine* consists of nine species which includes annual and perennial herbs. Six species are diploid ($2n = 2x = 16, 18, 20$) and three are tetraploid ($2n = 4x = 36, 38$). Most of the *Eleusine* species are of African origin with only one species (*E. tristachya*), which is native to South America. East Africa is considered to be the centre of diversity for the genus and eight (*E. africana*, *E. coracana*, *E. kigeziensis*, *E. indica*, *E. multiflora*, *E. floccifolia*, *E. intermedia* and *E. jaegerii*) out of nine species occur in this region (Mehra 1963; Phillips 1972). Most of the species of *Eleusine* are distributed in tropical and subtropical regions of Africa, Asia and South America. Some species of the *Eleusine* (*E. coracana*, *E. indica* and *E. tristachya*) grow in wider range of open habitats from sea level to highlands whereas other species (*E. kigeziensis*, *E. multiflora*, *E. floccifolia*, *E. intermedia* and *E. jaegeri*) are adapted to upland habitats, growing at altitudes above 1,000 m (Phillips 1972, 1995). *E. indica,* commonly known as goosegrass, is a most successful cosmopolitan weed. *Eleusine* species are herbaceous plants with flattened culms or stems. The genus is characterized by presence or digitate or subdigitate inflorescence (rarely shortly racemose) with sub-sessile spikelet arranged along the main axis and 1–16 primary long branches. Lemmas are three nerved and awn less. The fruit (grain) is a caryopsis which is ornamented. The pericarp can be easily removed by soaking the grain in water (Phillips 1972, 1995; Liu et al. 2011). Generally grasses have smooth grains with fused or free pericarp. The character combination of an ornamented grain with free pericarp is highly uncommon in grasses (Phillips 1972) and occurs only in *Eleusine* and four other chloridoid genera (*Acrachne* Chiov., *Dactyloctenium* Willd., *Coelachyrum* Hochst. and Nees, and a few species of genus

Table 13.1 List of *Eleusine* species

Species	$2n$	Genome[a] formula	Growth habit
Eleusine coracana (L.) Gaertn.	36	AABB	Annual
E. africana Kennedy-O'Byrne	36	AABB	Annual
E. tristachya (Lam.) Lam.	18	AA	Annual
E. indica (L.) Gaertn.	18	AA	Annual
E. floccifolia (Forssk.) Spreng.	18	BB	Perennial
E. multiflora Hochst. ex A. Rich	16	CC	Annual
E. jaegeri Pilger	20	DD	Perennial
E. kigeziensis S.M. Phillips	38	AADD	Perennial
E. intermedia (Chiov.) S.M. Phillips	18	AB	Perennial

[a]Bisht and Mukai (2002)

Sporobolus). Therefore, ornamented fruits of *Eleusine* can be a very useful character for identification of the genus. Phillips (1972, 1974, 1995) provided keys for identification of *Eleusine* species. Presently, genus *Eleusine* Gaertn. includes 9–10 (Table 13.1) species namely *E. coracana* (L) Gaertn, *E. africana* Kennedy-O'Byrne, *E. kigeziensis* Phillips, *E. semisterilis* Phillips, *E. jaegeri* Pilger, *E. floccifolia* (Forssk) Spreng, *E. intermedia* Phillips, *E. indica* (L) Gaetrn, *E. multifora* Hochst and *E. tristachya* (Lam.) Lam. *Eleusine semisterilis* Phillips is not included in this chapter as it is described only from a single herbarium specimen collected from the southern part of Kenya and Mombassa (Phillips 1972) and located at Kew herbarium, in England. General description of each species of *Eleusine* is given below:

13.2.1 Eleusine coracana

It is an annual allotetraploid ($2n = 4x = 36$; AABB) species commonly known as finger millet or ragi. It is the only cultivated species of genus *Eleusine*. *Eleusine coracana* is a valuable crop since centuries in arid and semi-arid regions of the world particularly in eastern and southern Africa and India (National Research Council 1996). It is also cultivated in Nepal, china and other south Asian countries. India ranks first in terms of land under finger millet cultivation (Naylor et al. 2004). Morphological characters include robust culms with soft, glabrous, herbaceous leaves sometimes with pilose margins. Inflorescence is typically digitate or subdigitate with thick and stout spikes, which are often incurved (Fig. 13.1a). Grains are globose and can be blackish, brown, reddish or even whitish. Other diagnostic characters include non-shattering (persistent) spikelets and the exposed grains (Phillips 1972, 1995; Neves 2011). There is considerable morphological variation in the inflorescence of finger millet, this variation forms the basis for recognition of several races described in the literature (Hilu and De Wet 1976b; De Wet et al. 1984; Dida and Devos 2006; Upadhyaya et al. 2007). There are five races which are

Fig. 13.1 **a** Whole plants of *Eleusine coracana* with spikes and **b** spike of *Eleusine indica*. (Courtesy: Dr. Anil Kumar)

most commonly recognized: “coracana”, “elongate”, “plana”, “compacta” and “vulgaris” (De Wet et al. 1984; Dida and Devos 2006). Economically it is not only cultivated as staple crop in many regions but it also has number of other uses such as in making alcoholic beverages, in traditional medicine, as forage crop, in papermaking and soil retention (Neves 2011). The grain is used for making bread, cakes, soup, porridge, puddings, malt beer (Iyengar et al. 1945–1946; Hilu and De

Wet 1976a; National Research Council 1996; Nout 2009). Finger millet is a rich source of all essential amino acids except lysine and also calcium and iron as compared to other cereals (Barbeau and Hilu 1993; Chandrashekar 2010). The crop has resistance to a number of diseases and tolerance to soil moisture stress; it also has excellent storage qualities (Hilu and De Wet 1976a; Bhandari 1974; Rao and Krishnamoorthy 1981). Finger millet has an average life cycle of 90 days and it is a day neutral plant.

13.2.2 Eleusine africana

It was first reported from Africa by Moffet and Hurcombe (1949) as a tetraploid ($2n = 4x = 36$) form of diploid *E. indica* ($2n = 2x = 18$) due to its morphological similarities with *E. indica*. Later, Kennedy-O'Byrne (1957) on the basis of difference in chromosome number and various other morphological differences such as difference in length of lemma, separated this tetraploid form from diploid form and raised it to species level. Phillips (1974) recognized two species, *E. coracana* and *E. indica*, with '*africana*' as subspecies of *E. indica* due to its long, easily shattering spiklets. Later on, Phillips (1995) treated three taxa (*E. coracana*, *E. africana* and *E. indica*) as separate species due to presumed genetic isolation beteween *E. indica* and two tetraploids. Recently, Neves et al. (2005) regarded '*africana*' as the subspecies of *E. coracana* as '*africana*'. It is presumed to be the wild progenitor of finger millet. Therefore, the taxonomic status is still controversial. In the present review, the taxa are treated as three separate species. This species status was also supported by Chennaveeraiah and Hiremath (1974b), Hiremath and Salimath (1991), Phillips (1995), Bisht and Mukai (2002), Liu et al. (2011). *E. africana* is also an annual allotetraploid ($2n = 4x = 36$; AABB) grass and is considered to be the wild progenitor of *E. coracana* as it is genetically very similar to *E. coracana* (Chennaveeraiah and Hiremath 1974a; Hilu 1988; Hilu and De Wet 1976a; Hilu et al. 1978; Hilu and Johnson 1997; Bisht and Mukai 2002; Agrawal et al. 2014). It is a wild native of East Africa with its distribution in tropical and southern Africa—mainly southern and eastern uplands (Neves 2011). Morphological characters include moderately robust culms with soft usually glabrous leaves. Digitate and subdigitate inflorescence. Grains ovate to oblong with black to brownish in color (Neves 2011). It is used as forage grass and also a weed when associated to cultivated finger millet fields.

13.2.3 Eleusine indica

It is an annual diploid ($2n = 2x = 18$; AA) grass of African origin commonly known as goosegrass. *E. indica* is a cosmopolitan weed which creates problem in many countries throughout the world. Earlier, it was a major problem in tropics only, but now it has invaded temperate regions also. It is very difficult to eradicate it as it is herbicide resistant. The culms are slender with soft glabrous to sparsely pilose leaves. Inflorescence digitate to subdigitate with straight and slender spikes (Fig. 13.1b). Grains are elliptic, trigonous and blackish in color (Neves 2011). Goosegrass is used as forage and is used as traditional medicine in parts of Africa and Asia (Neves 2011).

13.2.4 Eleusine tristachya

It is also an annual diploid ($2n = 2x = 18$; AA) species which is only the non African member of genus *Eleusine*. It is native to South America; however, now this species has its distribution in different parts of the world including parts of North America, Australia, Africa and Europe (Phillips 1972; Hilu 1980). *E. tristachya* can be easily recognized by its digitate inflorescence with oblong spikes, which are tightly clustered at the top of the axis, and the perpendicular arrangement of spikelets to spike axis (Hilu 1980). Other morphological features include the presence of slender culms with soft glabrous leaves. Grains are oblong to globose, trigonous and blackish in color Neves (2011). It is an important forage grass particularly in Argentina (Lovisolo and Galati 2007).

13.2.5 Eleusine floccifolia

It is a perennial diploid ($2n = 2x = 18$, AA or BB) grass with its distribution along the mid to high altitudes of North East Africa and Arabia such as Ethiopia, Eritrea, Somalia, Kenya and Yemen. *E. floccifolia* has some unique features which help in the identification of this species such as the presence of small tuffs of white hairs that are scattered along the margins of leaves (Phillips 1972, 1974, 1995). Morphological characters include moderately robust culms with leaves having small tufts of white hairs scattered along the smooth margins. Its inflorescence is subdigitate with elliptic to oblong, trigonous, blackish grains (Neves 2011). Due to the presence of hairs, it is unpalatable and thus not consumed by livestock. However, it is used by locals for making baskets and other handicraft items, particularly in Ethiopia.

13.2.6 Eleusine intermedia

It is also a perennial diploid ($2n = 2x = 18$; AB) grass with very limited distribution. It grows in the uplands of northern Kenya and Southern Ethiopia (Phillips 1972). *E. intermedia* was first identified as a variety of *E. indica* but later due to number of differences, it was given the status of species by Phillips (1972). Morphological characters include moderately robust culms with herbaceous glabrous to pilose leaves. Inflorescence, shortly racemose or subdigitate, with elliptic, trigonous and blackish grain (Neves 2011). It can be sometimes confused with another perennial grass E. jaegeri but E. jaegeri has tough, glabrous leaves with rough margins whereas *E. intermedia* has softer, slightly pilose leaves with smooth margins (Phillips 1972). *E. intermedia* may be of some use as forage grass.

13.2.7 Eleusine jaegeri

It is another perennial diploid species with a different basic chromosome number ($2n = 2x = 20$; DD). Not much information is available about this species. It has a very restricted distribution in the East African highlands (Phillips 1972). *E. jaegeri* is reported to be the most robust species of *Eleusine* forming dense tussocks of pale green saw-edged leaves. Morphological characters include the presence of robust culms with leathery glabrous leaves having rough margins. Inflorescence is shortly racemose or subdigitate with elliptic-oblong, trigonous, blackish grains (Neves 2011). Again, it is an unpalatable grass avoided by livestock. Due to toughness and robustness, it is used by locals for making baskets.

13.2.8 Eleusine kigeziensis

It is a perennial tetraploid ($2n = 2x = 38$; AADD) endemic to Africa with its presence in the mountainous regions extending from Kigezi Province, Uganda and adjacent parts of the Congo and Rwanda southwards into Burundi (Phillips 1972). *E. kigeziensis* has the characters of both annuals and perennials. Morphologically, it appears as the hybrid between *E. indica* and one of the perennials. Its inflorescence resembles with annuals *E. indica*, *E. africana* and *E. coracana* but it differs from annuals in the presence of short ascending rhizome and perennial habit. On the other hand, it differs from other diploid perennials by the presence of slender rhizome, soft, sometimes pilose leaves with no tufts of hairs in the leaf margins and lemmas with a central 3-nerved keel whereas other perennials have simple one nerved lemmas (Phillips 1972, 1974, 1995). Other morphological features involve presence of moderately robust culms with soft leaves which are sparsely pilose on upper surface and glabrous on lower surface. Digitate inflorescence and elliptic, trigonous and blackish grains. It may be of some value as forage grass (Neves 2011).

13.2.9 Eleusine multiflora

It is another annual diploid with a different basic number ($2n = 2x = 16$; CC) and unique features such as short, oblong to ovate spikes that alternate at the top of the inflorescence axis. *E. multiflora* is distributed along the mid to high altitudes of Ethiopia, Eritrea, Kenya and Tanzania. Morphological characters include presence of slender culms with soft, sparsely pilose leaves. Inflorescence racemose with oblong laterally compressed blackish grains. Taxonomically, *E. multiflora* differs a lot from other members of genus *Eleusine*, and has some similarities with another genus *Acrachne* (Phillips 1972). *Eleusine multiflora* has intermediate morphological characters between *Eleusine* and *Acrachne* and therefore represents a link between these two genera (Phillips 1972; Clayton and Renvoize 1986). *Eleusine* multiflora generally disarticulate in the manner which is typical of other *Eleusine* taxa, i.e. the spikelets disarticulate beneath each floret but on the other hand occasionally a lemma falls before its palea, which remains on rachilla as in genus *Acrachne* (Phillips 1972; Hilu and Johnson 1997). In *Acrachne*, the lemma keel is drawn out into a mucro or awn-point whereas in *E. multiflora* the lemma keel is produced into a mucro or cusp, and in other *Eleusine* species the keel is not extended at all, the tip being simply acute or obtuse (Phillips 1972). The shedding of gains in *E. multiflora* is also similar to that of *Acrachne racemosa* (Phillips 1972). Potentially *E. multiflora* may be used as forage grass and sometimes it may be problem as a weed.

13.2.10 Other Related Taxa

Eleusine semisterilis Phillips is one species which is described only from a single specimen collected from the southern part of Kenya and Mombassa (Phillips 1972). It differs from other species of *Eleusine* in presence of its abortive spikelets at each end of spikes and the laxly arranged spikelets, as compared to the highly overlapping arrangement as present in *Eleusine* (Phillips 1972). A number of attempts were made for collection of this species, however, no evidence exists to establish that this species is found in wild. A study by Phillips mentioned that presence of atypical inflorescence characteristic of this specimen may be due to anomalous development. Two more species, *E. reniformis* Divak. and *E. compressa* (Forssk) Ascher, were also included by Chennaveeraiah and Hiremath (1974a, b) in their study. Subsequently, after detailed cytological and morphological analysis, they considered *E. reniformis* to be only a variety of *E. coracana*. *Eleusine compressa*, which is a tetraploid species having $2n = 40$ (Krishnaswamy 1940; Hiremath and Chennaveeraiah 1982) was later excluded from the genus *Eleusine* and now incorporated in another genus *Octhochloa* as *O.* compressa (Forssk.) Hilu (Phillips 1972; Hilu 1981). *Eleusine racemosa* Heyne and *E. verticillata* Roxb. were mentioned in few studies, the detailed information isn't available for these species (Hiremath and Salimath 1991), later both the species have been merged and mentioned under monotypic genus *Acrachne* as *A. racemosa* (Heyne) Ohwi.

13.3 Phylogenetic Position of *Eleusine*

Eleusine is placed with other grasses in family Poaceae and C4 grass subfamily Chloridoideae (Hilu and Alice 2001), tribe Eragrostideae, subtribe Eleusininae (Peterson et al. 2010). Its placement among Chloridoid members is uncertain (Hilu and Alice 2001; Columbus et al. 2007). *Eleusine* was earlier considered to be closely related to genera *Acrachne* and *Dactyloctenium* due to similar morphological characters (Phillips 1972, 1982, 1995; Clayton and Renvoize 1986), However, this relationship found to be void based on molecular analysis (Hilu and Alice 2001; Neves et al. 2005; Columbus et al. 2007). On the basis of numerical analysis of morphological data identified *Eleusine*, *Acrachne*, *Dactyloctenium*, *Coelachyrum* and *Sclerodactylon* as one group (Phillips 1982), moreover, the authentication for this using molecular analysis is either not available fully or not corroborated. In the phylogenetic analysis of subfamily chloridoideae based on *matK* sequences placed *Eleusine* close to *Dinebra*, *Leptochloa*, *Coelachyrum*, *Diplachne* and *Astrebla* (Hilu and Alice 2001).

13.4 Cytogenetics, Crossing Studies and Genome Size in *Eleusine*

Origin, evolution and domestication of finger millet have been a subject of considerable controversy. De Candolle (1886), Burkill (1935), Cobley (1956) believed cosmopolitan weed *E. indica* ($2n = 2x = 18$) cannot be the direct progenitor of finger millet, *E. coracana* ($2n = 4x = 36$) as former taxon is a diploid and latter one is allotetraploid. *E. africana* ($2n = 4x = 36$) is a wild grass distributed in East Africa and often occur as a weed along with cultivated finger millet. Morphologically these two species are similar in many respects. Natural hybridization between these two species has been reported (Mehra 1963).

Chennaveeraiah and Hiremath (1974b) studied the morphology, cytology and fertility of *E. coracana*, *E. africana* and their interspecific hybrid. They showed that *E. coracana* and its putative parent *E. africana* are allotetraploids ($2n = 4x = 36$) with about 87 % of the PMCs showing regular 18 bivalents in *E. coracana* × *E. africana* hybrids. Preponderance of bivalent formation, good fertility of F_1 and F_2 segregants suggest that the genomes of these two species are basically similar. Further, they concluded that *E. coracana* has originated through selection and further cultivation of a large grain mutant of *E. africana*. A common genomic notation of AABB has been proposed for these two species. Thus, allopolyploidy has played a significant role in the origin and evolution of finger millet and its wild progenitor species (Hiremath and Chennaveeraiah 1982).

13.4.1 Basic Chromosome Number

Eleusine tristachya was the first *Eleusine* species for which chromosome number was reported by Avdulov in 1928. Subsequently, Hiremath and Chennaveeraiah (1982), Hiremath and Salimath (1991) showed for the first time that *Eleusine* contains three basic chromosome numbers: $x = 8$, 9 and 10. Among the diploid taxa, basic chromosome numbers are $x = 8$ in *E. multiflora* ($2n = 2x = 16$), $x = 9$ in *E. floccifolia*, *E. tristachya*, *E. intermedia* and *E. indica*, all with $2n = 2x = 18$, and $x = 10$ in *E. jaegeri* ($2n = 2x = 20$). Two tetraploid species, *E. africana* and *E. coracana* ($2n = 4x = 36$) are based on $x = 9$. Whereas *E. kigeziensis* ($2n = 4x = 38$) probably evolved from a cross between *E. jaegeri* ($x = 10$) and *E. indica* ($x = 9$) thus combines two basic chromosome numbers. On the basis of the fact that majority of the *Eleusine* species exhibits $x = 9$ as the basic chromosome number and also on the basis of other evidences, Salimath (1990) suggested that $x = 9$ is the original basic chromosome number in genus *Eleusine*. Species with $x = 8$ and $x = 10$ appears to be derived from dysploid reduction or gain of a chromosome to the basic set of nine chromosome.

13.4.2 Karyotype and Species Differentiation

Karyotypes of finger millet and its wild species were analyzed and most of the chromosomes were found to be medium sized with little size difference within a complement (Hiremath and Chennaveeraiah 1982; Salimath 1990). *Eleusine multiflora* ($2n = 16$) differs from the rest of the diploid species of *Eleusine*, having larger chromosomes than other species with a prominent secondary constriction on the longest pair of the chromosomes (Bisht and Mukai 2000); its genome is represented as 'C'. Another diploid species *E. jaegeri* ($2n = 20$) has a single pair of satellite median chromosome and eight pairs of median chromosomes (Hiremath and Chennaveeraiah 1982); its genome is represented as 'D'. Four diploid species, *E. indica*, *E. tristachya*, *E. intermedia* and *E. floccifolia* have $2n = 18$, the karyotype of *E. indica* and *E. tristachya* both have one pair of satellite sub median chromosome, five pairs of median chromosomes and 3–4 pairs of sub median chromosomes. In contrast, *E. floccifolia* has two pairs of satellite median chromosomes and seven pairs of median chromosomes (Hiremath and Chennaveeraiah 1982). Salimath et al. 1995b, suggested a common genome 'A' or a differentiated form of 'A' genome for *E. indica*, *E. tristachya* and *E. floccifolia* which is based on association of the chromosomes in the hybrids of these species at metaphase-I. Two tetraploid species *E. coracana* and *E. africana* both have $2n = 36$ with genomic notation AABB (Hiremath and Salimath 1992), the chromosomes in both species were small in size without much difference in size of the largest and smallest pair of chromosomes. The karyotype in both the species is similar in chromosome size, absolute chromosome length, type and number of SAT-chromosomes and satellite size (Hiremath and Chennaveeraiah 1982; Hiremath and Salimath 1991). However, another study reported that in *E. coracana* the secondary constrictions were clearly visible in one pair of chromosomes but in *E. africana* there was no clear secondary constriction (Bisht and Mukai 2000).

Finger millet karyotypes of 14 collections covering the major growing areas of the world showed differences in the number of metacentric and submetacentric and SAT-chromosomes. Thus, karyotypic heteromorphism exists in different collections of finger millet and is attributed to chromosomal repatterning. Karyotypes of several collections of *E. coracana* and *E. africana* exhibited similarity in chromosome size, absolute chromosome length, type and number of SAT-chromosomes, and satellite size. Introgression of genes also occurs between these two taxa. Based on these facts, Hiremath and Chennaveeraiah (1982) suggested that *E. africana* is a wild progenitor of finger millet.

13.4.3 Crossing Studies

E. coracana and *E. africana* are allotetraploid ($2n = 4x = 36$) as it forms regular 18 bivalent during meiosis. Genomes of both these species are similar and genomic notation of AABB is assigned to them as mentioned earlier. *Eleusine africana* being an allotetraploid, must have originated as a result of hybridization between two diploid species having AA and BB genomes followed by chromosomal doubling (Fig. 13.2).

In an attempt to discover A and B genome donor(s) to finger millet *Eleusine coracana* or its progenitor species *E. africana*, Hiremath and Salimath (1992) crossed all the five diploid *Eleusine* species to above two tetraploid species. Crosses were successful with only *E. coracana*. They obtained and analyzed three combinations of triploid hybrids. *E. coracana* × *E. indica*, *E. coracana* × *E. floccifolia* and *E. coracana* × *E. multiflora*. In *E. coracana* × *E. indica* hybrids mean chromosome pairing of 8.84I + 8.8 II + 0.03 III + 0.1 IV per cell was found. About 87 % of the PMCs showed typical 9I + 9 II configuration suggesting that *E. indica* (AA) is one of the diploid genome donors to cultivated species *E. coracana*. Morphologically *E. indica* is very similar to *E. africana* and *E. coracana*.

Mean chromosome pairing of 11.08 I + 7.63 II + 0.16 III + 0.04 IV per cell was found in *E. coracana* × *E. floccifolia* triploid hybrids. Typical 9I + 9II configuration was observed in 45 % of the PMCs and remaining 53 % of the PMCs revealed 2–8 bivalents and varying number of univalents. These results suggest that one of the genomes of *E. coracana* is partially homologous with diploid *E. floccifolia* genome. But the question arises whether the *E. floccifolia* genome represents A or B genome? Salimath (1990) observed that in *E. indica* × *E. floccifolia* diploid hybrids, 80 % of the PMCs showed regular nine bivalent formation thus, suggesting that the 'A' genome of *E. indica* has partial homology with the *E. floccifolia* genome. Apparently, *E. floccifolia* is a member of the 'A' genome group of diploid *Eleusine* and its genome can be designated as AfAf. Morphologically *E. floccifolia* is quite

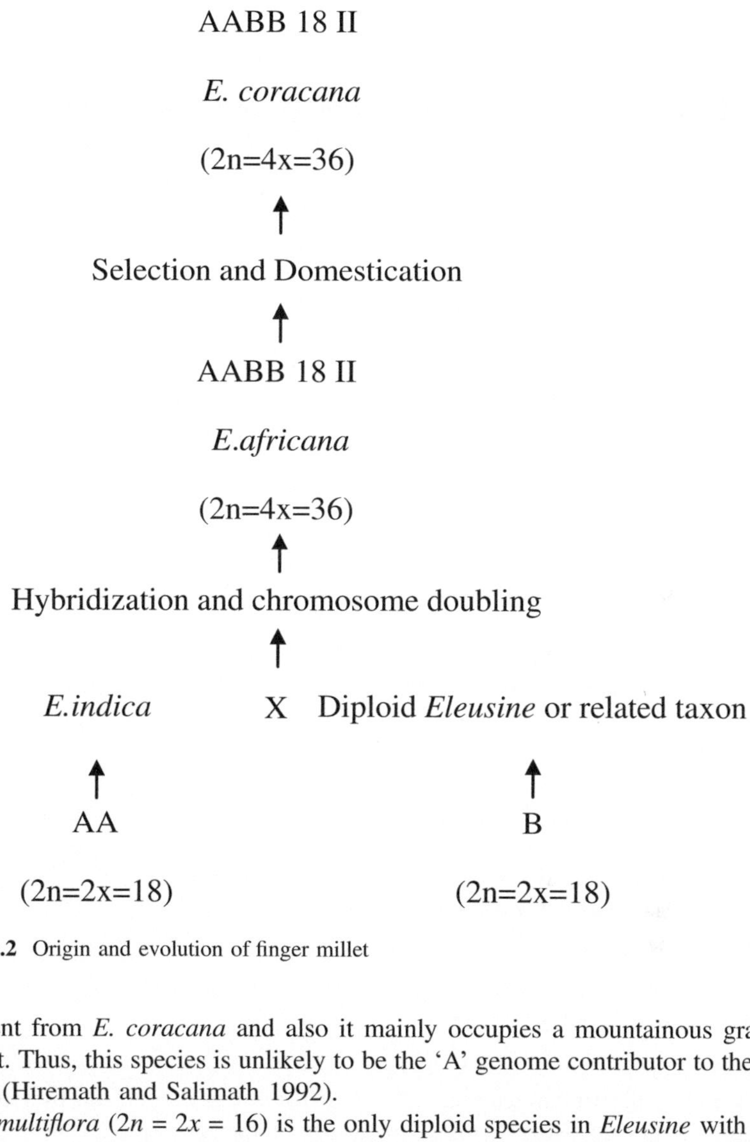

Fig. 13.2 Origin and evolution of finger millet

different from *E. coracana* and also it mainly occupies a mountainous grassland habitat. Thus, this species is unlikely to be the 'A' genome contributor to the finger millet (Hiremath and Salimath 1992).

E. multiflora ($2n = 2x = 16$) is the only diploid species in *Eleusine* with a base number $x = 8$. In *E. coracana* × *E. multiflora* triploid hybrids ($2n = 26$), mean chromosome pairing of 21.45 I + 1.97 II + 0.13 III + 0.04 IV per cell was found and only a very small percentage of the cells contained bivalents or multivalents. This observation suggests that *E. multiflora* lacks genomic homology with 'A' or 'B' genome of *E. coracana*. The formation of a few bivalents or multivalents, however, may be due to an ancestral genomic homology from which all the *Eleusine* taxa have evolved. Morphologically, *E. multiflora* is distinct from the rest of the diploid species and also it does not cross with any of the diploid *Eleusine* taxa. This indicates that *E. multiflora* is not a member of the 'A' or 'B' genome group of species. Genomically, it is a distinct species and genomic symbol CC is assigned to it (Hiremath and Salimath 1992).

Hiremath and Salimath (1992) also performed *E. coracana* X *E. tristachya* crosses, the resultant hybrids germinated and grew well but they failed to flower so chromosome pairing data could not be obtained. In another study by Devarumath et al. (2005), the *E. tristachya* x *E. coracana* hybrids showed mean chromosome pairing of 9.4 I + 8.9 II + 0 III + 0.1 IV per cell. In nearly 71 % of the PMCs typical 9 I + 9 II configurations were noticed. It is apparent from the data that the genome of *E. tristachya* is partially homologous with one genome of finger millet *E. coracana*. Weather *E. tristachya* genome belongs to 'A' or 'B' genome group can be tested by crossing it to confirm 'A' genome donor *E. indica*. In this diploid hybrid presence of 9 II would indicate that *E. tristachya* belongs to 'A' genomic group of *Eleusine*s. In contrast univalent configurations would suggest *E. tristachya* to be 'B' genome donor (Devarumath 1997). In *E. tristachya* x *E. indica* hybrids 89 % of PMCs revealed 9 II formation and thus suggesting that *E. tristachya* belong to 'A' genomic group of *Eleusine* taxa (Salimath et al. 1995a). E. tristachya is a distinct annual species with restricted distribution in South America and it grows no where sympatrically with *E. coracana*. Thus, this species could not be a direct 'A' genome donor to finger millet. Genomic symbol of At is assigned to this taxon (Devarumath et al. 2005).

Genome analytical studies (Hiremath and Salimath 1992) have shown that *E. indica* is the 'A' genome donor of finger millet and its 'A' genome is partially homologous to the genome of *E. floccifolia* and *E. tristachya* (Salimath 1990; Hiremath and Salimath 1992). Thus, *E. indica*, *E. floccifolia* and *E. tristachya* belong to AA genomic group of *Eleusine*s and form close genetic assemblage within the genus *Eleusine*. No conclusion can be drawn regarding genetic affinities of *E. intermedia* and *E. jaegeri* as crosses between them and *E. coracana* were unsuccessful. Thus, the identity of 'B' genome donor to the finger millet remains elusive (Hiremath and Salimath 1992).

13.4.4 Genome Size

First attempt to determine genomic size of *Eleusine* was carried out by Hiremath and Salimath (1991) using Feulgen micro-densitometry. The 2C DNA content ranged from 2.60 pg in *E. multiflora* to 5.78 pg in *E. coracana*. Later Mysore and Baird (1997) re-evaluated genomic size of *Eleusine* using laser flow-cytometry. Their results suggested that there was an overestimation of nuclear DNA content for most of the species calculated by Hiremath and Salimath (1991). The nuclear DNA content (genome size estimated by Mysore and Baird (1997) for *Eleusine* taxa were as follows (Table 13.2).

Table 13.2 Nuclear 2C DNA amounts in *Eleusine* species

Species	2*n*	Nuclear 2C value (pg[a])	
		Micro-densitometry	Flow-cytometry
E. coracana	36	5.13–5.78	3.36–3.87
E. africana	36	5.11	3.34
E. indica	18	2.73–2.95	1.61–1.76
E. tristachya	18	2.90	1.51
E. floccifolia	18	3.26	2.00
E. multiflora	16	2.60	2.65
E. jaegeri	20	3.33	1.90

[a]1 pg = 980 Mbp

13.5 Evolution of the Genus *Eleusine* and Origin of Finger Millet

The relationships in the genus *Eleusine* have been investigated using a diverse range of techniques, where cytogenetics has demonstrated the importance of polyploidy and aneuploidy in this genus, having basic chromosome numbers of $x = 8, 9, 10$, with $x = 9$ being the most common (Chennaveeraiah and Hiremath 1974a; Hiremath and Chennaveeraiah 1982; Hiremath and Salimath 1991; Bisht and Mukai, 2001a, 2002; Liu et al. 2014). The chemical and molecular analysis, including flavonoids (Hilu et al. 1978), isozymes (Werth et al. 1993, 1994), plastids (Hilu 1988; Hilu and Johnson 1997; Neves 2005; Liu et al. 2011; Agrawal et al. 2014), nuclear ribosomal (Hilu and Johnson 1992; Neves et al. 2005), multilocus and low copy DNA markers (Hilu 1995; Salimath et al. 1995a; Liu et al. 2011, 2014) have provided valuable insights into species relationships and on the origin of the crop. *Eleusine coracana* is an allotetraploid ($2n = 4x = 36$) (AABB) and *E. africana* ($2n = 4x = 36$) (AABB) is considered to be its wild progenitor, and both are genetically very similar (Chennaveeraiah and Hiremath 1974a; Hilu and De Wet 1976a; Hilu et al. 1978; Hilu 1988, 1995; Hiremath and Salimath 1992; Werth et al. 1994; Hilu and Johnson 1997; Bisht and Mukai 2000, 2001a, b; Neves et al. 2005; Devarumath et al. 2005; Dida et al. 2007, 2008; Liu et al. 2011; Agrawal et al. 2014). *E. indica* ($2n = 2x = 18$) (AA) found to be the maternal genome donor of *E. coracana* and *E. africana* (Hilu 1988; Hiremath and Salimath 1992; Hilu and Johnson 1997; Bisht and Mukai 2000, 2001a; Neves et al. 2005; Agrawal et al. 2014). *E. floccifolia* ($2n = 2x = 18$) could be the BB donor species of *E. coracana* was proposed by Bisht and Mukai (2000, 2001a, 2002). This was continuously refuted by others for a long time (Hiremath and Salimath 1992; Neves et al. 2005; Devarumath et al. 2010; Liu et al. 2011). As per these authors, the BB genome donor species remains unidentified and extinct possibly. *E. kigeziensis* ($2n = 4x = 36$ or 38) (AADD) is the third tetraploid species of the genus *Eleusine*. *E. indica* ($2n = 2x = 18$) (AA) and *E. jaegeri* ($2n = 2x = 20$) (DD) are thought to be genome donor of *E. kigeziensis* (Bisht and Mukai 2002; Devarumath et al. 2010). But this needs to be confirmed through mapping of rDNA and other repetitive DNA sequences on the chromosomes of *E. kigeziensis* or genomic in situ hybridization (GISH). Neves and colleagues in 2005 proposed that *E. kigeziensis* is autotetraploid with *E. indica* being closely related to *E. kigeziensis* but not the direct genome donor to *E. kigeziensis*. Liu et al. (2011) suggested that the three tetraploids (*E coracana, E. africana and E. kigeziensis*) are of allotetraploid origin and proposed independent origins of *E. kigeziensis* and *E. africana–E. coracana*. They also mentioned that both events may have involved the diploids *E. indica* and *E. tristachya* as maternal parents, but the paternal parents remain unidentified.

E. indica and *E. tristachya* are thought to be very similar and a number of studies based on chromosome research (Hiremath and Chennaveeraiah 1982; Hiremath and Salimath 1991), 2C DNA content (Hiremath and Salimath 1991), crossability data (Salimath et al. 1995b) and ribosomal DNA polymorphism (Hilu and Johnson 1992; Werth et al. 1994; Bisht and Mukai 2000), Internal transcribing spacer (ITS) sequence data (Neves et al. 2005) and cpDNA (Hilu and Johnson 1997; Neves et al. 2005; Liu et al. 2011, Agrawal et al. 2014) also support the close affinity between *E. indica* and *E. tristachya* but the degree of relationship between the two remains unclear. There is a close affinity between three diploid perennial species, *E. floccifolia* and *E. intermedia*. This is supported by 2C DNA values (Hiremath and Salimath 1991), ribosomal DNA variation (Hilu and Johnson 1992), and ITS sequence data (Neves et al. 2005) and cpDNA analyses (Liu et al. 2011; Agrawal et al. 2014). Furthermore, results indicate that *E. intermedia* may be a hybrid between *E. indica* and *E. floccifolia* (Bisht and Mukai 2001a, b). The hybrid is maintained at diploid level and produces fertile seeds. Hybridization studies were carried out to produce hybrids of *E. indica* and *E. floccifolia*. The hybrids failed to produce seeds but hybrid showed 81 % bivalent formation in PMC (Salimath et al. 1995b). The perennial nature of both E. floccifolia and *E. intermedia* and similar 2C DNA values suggest that these two species are closely associated (Hiremath and Salimath 1991).

The position of *E. multiflora* with chromosome number $2n = 2x = 16$ (CC genome) within the genus *Eleusine* is questionable mainly on the basis of its unusual inflorescence morphology which is considered to be intermediate between *Eleusine* and *Acrachne* (Phillips 1972; Clayton and Renvoize 1986), distinct chemical composition (Hilu et al. 1978), molecular data (Hilu and Johnson 1992; Hilu 1995), chromosome number ($2n = 16$), 2C DNA value and other cytogenetic features (Mysore and Baird 1997; Bisht and Mukai 2000) and cpDNA analyses (Agrawal et al. 2014).

13.6 Utilization of Molecular Techniques

13.6.1 Analysis of Genetic Diversity and Population Structure

A number of different molecular tools and techniques have been utilized for the analysis of species relationship, and to study genetic diversity and structure of population. Analysis of several molecular markers has demonstrated low levels of genetic diversity in accessions or lines of cultivated finger millet. The reason for low levels of genetic diversity is narrow genetic pool of cultivated finger millet, which in turn can be the consequence of domestication of crop. The crop evolved from a small subset of wild population with very limited genetic variability followed by low levels of introgression with wild relatives due to the highly inbred or self pollinating nature of finger millet. As a consequence the gene pool of finger millet remained restricted (Hilu and Johnson 1992; Werth et al. 1994; Salimath et al. 1995a; Dida et al. 2007). When genetic diversity was studied using RFLPs, it showed no or very low levels of diversity (Salimath et al. 1995a; Dida et al. 2007). Salimath et al. (1995a) used multiple DNA markers such as RFLPs, RAPDs and ISSRs to assess genetic variation in the crop. They found 14, 10 and 26 % polymorphism using RFLPs, RAPDs and ISSRs, respectively, in 17 accessions of finger millets from Asia and Africa. Among the three marker types they used, ISSRs were found to be most promising ones. Recently, AFLPs were also utilized as markers and have been utilized for construction of genetic map of *E. coracana* (Dida et al. 2007). Microsatellites or simple sequence repeats (SSRs) are currently the most useful markers for studying genetic diversity or population structure. Presently, large number of SSRs is available for analysis of genetic studies and population structure for cultivated *E. coaracana*. These have been either developed directly for finger millet to facilitate marker-assisted selection in finger millet (Dida et al. 2007) or have been developed in other major cereal crops and have been transferred successfully in minor grass species including *E. coracana* (Wang et al. 2005). Dida et al. (2008) utilized 45 SSRs genotyping analysis of 79 accessions of cultivated *E. coracana*, 14 accessions of *E. africana*, two accessions of *E. indica* and one accession of *E. kigeziensis*. They observed clear evidence of gene flow between cultivated and wild subpopulations. A large number of ESTs (expressed sequence tags) have been developed for *Eleusine*. Buell (2009) reported 1,749 ESTs for *E. coracana*. ESTs were also obtained and utilized by Dida et al. (2007) for the development of genetic map of finger millet.

13.6.2 Development of Genetic Map of Finger Millet and Comparative Genomics

The construction of genetic linkage map is prerequisite for marker-assisted selection for any crop improvement program. Dida et al. (2007) utilized restriction fragment length polymorphism (RFLP), amplified fragment length polymorphism (AFLP), expressed-sequenced tag (EST), and simple sequence repeat (SSR) to generate first genetic map of finger millet. Due to low levels of genetic diversity in cultivated *E. coracana*, the map was developed using a cross between *E. coracana* cultivar Okhale-1 and its wild progenitor *E. africana* accession MD-20.

Comparative genetic mapping have demonstrated the colinearity (conserved gene order) among many grass species (Gale and Devos 1998; Devos 2005). It has shown that species belonging to same subfamilies are characterized by a few common rearrangements. Comparative genomics can be of great help for finger millet improvement program. Therefore, to further supplement the genetic linkage map of finger millet and to allow finger millet breeders and researchers to efficiently use the information from other major cereals for which a lot of progress has been done for example rice, where the entire genome have been sequenced, the comparative analysis of colinearity between finger millet and rice genome has been done (Srinivasachary et al. 2007). Finger millet is the first member of sub family Chloridoideae for which a comprehensive comparative map has been developed that can be analyzed for structural similarity with other grass genomes. The comparative analysis between finger millet and rice genome demonstrate that both the genomes have remained relatively conserved since the time of divergence of these two lineages from a common ancestor around 60 million year ago. There are only 10 % of markers which were found at non-syntenic position mostly located at distal 14 % of chromosome arms, besides the other expected rearrangements to explain the difference in chromosome number between rice ($2n = 2x = 24$) and finger millet ($2n = 36$). This breakdown of synteny between rice and finger millet in the distal regions might be correlated with high recombination rates, as it was observed earlier in case of wheat (Srinivasachary et al. 2007).

13.7 Germplasm Resources

The germplasm in the form of seed samples are conserved at various locations in the world. A total of 6804 finger millet germplasm accessions from 25 countries are conserved for use in research and development at International Crops Research Institute for the Semi-Arid Tropics (ICRISAT) genebank at Patencheru, Hyderabad, India (http://www.icrisat.org/genebank-fingermillet.htm). Out of 6804 finger millet germplasm accessions, the maximum of 1487 accessions are from Uganda followed by 1400 accessions from India and 1148 accessions from Zimbabwe. ICRISAT also have a core subset of finger millet germplasm consisting of 622 accessions. Core collection is based on origin and data on 14 quantitative traits developed from the entire global collection of 5940 accessions available in the genebank at ICRISAT. The comparison of various statistical tests such as means, variances, frequency distribution, Shannon–Weaver diversity index (H) and phenotypic correlations indicated that sampling was optimal and the diversity has been captured very well in core subset. The National Plant Germplasm System, Beltsville, Maryland, USA (GRIN, USDA), has listed 741 accessions of *Eleusine* and out of these 721 are of finger millet (http://www.ars-grin.gov/npgs/searchgrin.html). Twenty three accessions of *Eleusine* incuding 14 accessions of *E. coracana*, are also available at IPK genebank, Gatersleben, Germany (http://www.ipk-gatersleben.de/en/genebank/).

References

Agrawal R, Agrawal N, Tandon R, Raina SN (2014) Chloroplast genes as genetic markers for inferring patterns of change, maternal ancestry and phylogenetic relationships among *Eleusine* species. AoB Plants 6:plt056

Avdulov NP (1928) Systematic karyology of family gramineae. Daily Paper All Union Botanists, London. 1:65–67

Barbeau WE, Hilu KW (1993) Protein, calcium, iron and amino acid content of selected wild and domesticated cultivars of finger millet. Plant Foods Hum Nutr 43:97–104

Bhandari MM (1974) Famine foods in the Rajasthan desert. Econ Bot 28(1):73–81

Bhatnagar SS (1952) The wealth of India, vol III. Council of Scientific and Industrial Research, New Delhi, pp 160–166

Bisht MS, Mukai Y (2000) Mapping of rDNA on the chromosomes of *Eleusine* species by fluorescence in situ hybridization. Genes Genet Syst 75:343–348

Bisht MS, Mukai Y (2001a) Genomic in situ hybridization identifies genome donor of finger millet (*Eleusine coracana*). Theor Appl Genet 102:825–832

Bisht MS, Mukai Y (2001b) Identification of genome donors to the wild species of finger millet, *Eleusine africana* by genomic in situ hybridization. Breed Sci 51:263–269

Bisht MS, Mukai Y (2002) Genome organization and polyploid evolution in the genus *Eleusine* (Poaceae). Plant Syst Evol 233:243–258

Buell CR (2009) Poaceae Genomes: going from unattainable to becoming a model clade for comparative plant genomics. Plant Physiol 149:111–116

Burkill IH (1935) A dictionary of the economic products of the Malay Peninsula, vols 2. Crown Agents for the Colonies, London, United Kingdom

Chandrashekar A (2010) Finger millet *Eleusine coracana*. Adv Food Nutr Res 59:215–262

Chennaveeraiah MS, Hiremath SC (1974a) Genome analysis of *Eleusine coracana* (L.) Gaertn. Euphytica 23:489–495

Chennaveeraiah MS, Hiremath SC (1974b) Character association pattern in *Eleusine* Gaertn. Karnataka Univ J Sci 19:257–275

Clayton WD, Renvoize SA (1986) Genera *Graminum:* grasses of the World. Kew Bull Addl Ser 13. Her majesty's Stationary Office, London

Cobley LS (1956) An introduction to the botany of tropical crops. Longmans, Green

Columbus JT, Cerros-Tlatilpa R, Kinny MS, Siqueiros-Del-gado ME, Bell HL, Griffith MP, Refulio-Rodriguez NF (2007) Phlylogenetics of chloridoideae (gramineae): a preliminary study based on nuclear ribosomal internal transcribed spacers and chloroplast *trnL-F* sequences. Aliso 23:565–579

De Candolle A (1886) Origin of cultivated plants. Hafner Publishing Co, New York

De Wet JMJ, Prasada Rao KE, Brink DE, Mengesha MH (1984) Systematics and Evolution of *Eleusine coracana* (Gramineae). Am J Bot 71(4):550–557

Devarumath RM (1997) Genome homology and differentiation in *Eleusine* coracana and its wild species. Ph.D. thesis, Karnataka University, Dharwad, India

Devarumath RM, Hiremath SC, Rao SR, Kumar A, Sheelavanthmath SS (2005) Genome interrelationship in the genus *Eleusine* (Poaceae) as revealed through heteroploid crosses. Caryologia 58:300–307

Devarumath RM, Sheelavanthmath SS, Hiremath SC (2010) Chromosome pairing analysis in interspecific hybrids among tetraploid species of *Eleusine* (Poaceae). Indian J Genet 70:299–303
Devos KM (2005) Updating the 'Crop Circle'. Curr Opin Plant Biol 8:155–162
Dida MM, Devos KM (2006) Finger millet. In: Kole C (ed) Genome mapping and molecular breeding in plants, vol 1., Cereals and milletsSpringer, Heidelberg, pp 333–343
Dida MM, Srinivasachary RS, Bennetzen JL, Gale MD, Devos KM (2007) The genetic map of finger millet, *Eleusine coracana*. Theor Appl Genet 114:321–332
Dida MM, Wanyera N, Dunn MLH, Bennetzen JL, Devos KM (2008) Population structure and diversity in finger millet (*Eleusine coracana*) germplasm. Trop Plant Biol 1:131–141
Duke JA, Wain KK (1981) Medicinal plants of the world. Computer index with more than 85:12–16
Gale MD, Devos KM (1998) Plant comparative genetics after 10 years. Science 282:656–659
Hilu KW (1980) Noteworthy collections: *Eleusine tristachya*. Madroño 27:177–178
Hilu KW (1981) Taxonomic status of the disputable *Eleusine compressa* (Gramineae). Kew Bull 36:559–563
Hilu KW (1988) Identification of the "A" genome of finger millet using chloroplast DNA. Genetics 118:163–167
Hilu KW (1995) Evolution of finger millet: evidence from random amplified polymorphic DNA. Genome 38:232–238
Hilu KW, Alice LA (2001) A phylogeny of the Chloridoideae (Poaceae) based on *matK* sequences. Syst Bot 26:386–405
Hilu KW, De Wet JMJ (1976) Domestication of *Eleusine coracana*. Econ Bot 30:199–208
Hilu KW, DeWet JMJ (1976) Racial evolution in *Eleusine coracana* ssp. *coracana* (finger millet). Am J Bot 63(10):1311–1318
Hilu KW, Johnson JL (1992) Ribosomal DNA variation in finger millet and wild species of *Eleusine* (Poaceae). Theor Appl Genet 83:895–902
Hilu KW, Johnson JL (1997) Systematics of *Eleusine* Gaertn. (Poaceae, Chloridoideae): chloroplast DNA and total evidence. Ann Missouri Bot Gard 84:841–847
Hilu KW, De Wet JMJ, Seigler D (1978) Flavonoids and systematics of *Eleusine*. Biochem Syst Ecol 6:247–249
Hiremath SC, Chennaveeraiah MS (1982) Cytogentical studies in wild and cultivated species of *Eleusine* (Gramineae). Caryologia 35:57–69
Hiremath SC, Salimath SS (1991) The quantitative nuclear DNA changes in *Eleusine* (Gramineae). Plant Syst Evol 178:225–233
Hiremath SC, Salimath SS (1992) The "A" genome donor of *Eleusine coracana* (L.) Gaertn. (Gramineae). Theor Appl Genet 84:747–754
Iyengar KG, Doraiswamy LS, Iyengar RS (1945–1946) Ragi (*Eleusine coracana* Gaertn). Mysore Agr J 24:33–49
Kennedy-O'byrne J (1957) Notes on African grasses. XXIX. A new species of *Eleusine* from tropical and South Africa. Kew Bull 11:65–72
Krishnaswamy N (1940) Investigation on the cytology and systematic of the Gramineae. Beih Bot Centralbl 60:1–56
Liu Q, Triplett JK, Wen J, Peterson M (2011) Allotetraploid origin and divergence in *Eleusine* (Chloridoideae, Poaceae): evidence from low-copy nuclear gene phylogenies and a plastid gene chronogram. Ann Bot 108:1287–1298
Liu Q, Jiang B, Wen J, Peterson PM (2014) Low-copy nuclear gene and McGISH resolves polyploid history of *Eleusine coracana* and morphological character evolution in *Eleusine*. Turk J Bot 38:1–12
Lovisolo MR, Galati BG (2007) Ultrastructure and development of the megagametophyte in *Eleusine tristachya* (Lam.) (Poaceae). Flora 202:293–301
Lye KA (1999) Nomenclature of finger millet (Poaceae). Lidia 4:x149–151
Mehra KL (1963) Differentiation of cultivated and wild *Eleusine* species. Phyton 20:189–198
Moffet AA, Hurcombe R (1949) Chromosome numbers of South African grasses. Heredity 3:69–373

Mysore KS, Baird V (1997) Nuclear DNA content in species of *Eleusine* (Gramineae): a critical re-evaluation using laser flow cytometry. Plant Syst Evol 207:1–11

National Research Council (1996) Lost Crops of Africa, vol I. Grains. National Academy Press, Washington, DC, USA

Naylor RL, Falcon WP, Goodman RM, Jahn MM, Sengooba T, Tefera H, Nelson RJ (2004) Biotechnology in the developing world: a case for increased investments in orphan crops. Food Policy 29:15–44

Neves SS (2011) *Eleusine*. In: Kole C (ed) Wild crop relatives: genomic and breeding resources, millets and grasses. Springer, Berlin, pp 113–133

Neves SS, Clark GS, Hilu KW, Baird WV (2005) Phylogeny of *Eleusine* (Poaceae: Chloridoideae) based on nuclear ITS and plastid *trnT–trnF* sequences. Mol Phylogenet Evol 35:395–419

Nout MJR (2009) Rich nutrition from the poorest-Cereal fermentations in Africa and Asia. Food Microbiol 26:685–692

Peterson PM, Romaschenko K, Johnson G (2010) A classification of the Chloridoideae (Poaceae) based on multi-gene phylogenetic trees. Mol Phylogenet Evol 55:580–598

Phillips SM (1972) A survey of the *Eleusine* Gaertn. (Gramineae) in Africa. Kew Bull 27:251–270

Phillips SM (1974) *Eleusine* Gaertn. In: Polhill RM (ed) Flora of Tropical East Africa. London, UK, Crown Agents for Overseas Governments and Administrations, pp 260–267

Phillips SM (1982) A numerical analysis of the Eragrostideae (Gramineae). Kew Bull 37:133–162

Phillips SM (1995) Poaceae (Gramineae). In: Hedberg I, Edwards S (eds) Flora of Ethiopia and Eritrea, vol 7. Addis Ababa University and Uppsala University, Addis

Pradhan A, Nag SK, Patil SK (2010) Dietary management of finger millet (*Eleusine coracana* L. Gaerth) controls diabetes. Curr Sci 98:763–765

Rao P, Krishnamoorthy P (1981) Studies on dry matter production and nutrient uptake by ragi (*Eleusine coracana*) under different moisture stresses. J Res Andhra Pradesh Agric Univ 3:57–63

Salimath SS (1990). Cytology and genome relations in some species of *Eleusine* and its allies. Ph. D. thesis, Karnatak University, Dharwad, India

Salimath SS, de Oliveira AC, Godwin ID, Bennetzen J (1995a) Assessment of genome origins and genetic diversity in the genus *Eleusine* with DNA markers. Genome 38:757–763

Salimath SS, Hiremath SC, Murthy HN (1995b) Genome differentiation patterns in diploid species of *Eleusine* (Poaceae). Hereditas 122:189–195

Srinivasachary DMM, Gale MD, Devos KM (2007) Comparative analyses reveal high levels of conserved colinearity between the finger millet and rice genomes. Theor Appl Genet 115:489–499

Upadhyaya HD, Gowda CLL, Reddy VG (2007) Morphological diversity in finger millet germplasm introduced from Southern and Eastern Africa. J SAT Agr Res 3(1):1–3

Wang ML, Barkley NA, Yu JK, Dean RE, Newman ML, Sorrells ME, Pederson GA (2005) Transfer of simple sequence repeat (SSR) markers from major cereal crops to minor grass species for germplasm characterization and evaluation. Plant Genet Resour 3:45–57

Watt JM, Breyer-Brandwijk MG (1962) The medicinal and poisonous plants of Southern and Eastern Africa, 2nd edn. E & S Livingstone, Edinburgh and London

Werth CR, Hilu K, Langner CA, Baird WV (1993) Duplicated gene expression for isocitrate dehydrogenase and 6-phosphogluconate dehydrogenase in diploid species of *Eleusine* (Gramineae). Am J Bot 80:705–710

Werth CR, Hilu KW, Langner CA (1994) Isozyme of *Eleusine* (Gramineae) and the origin of finger millet. Am J Bot 81:1186–1197

Chapter 14
An Ancient Medicinal Plant at the Crossroads of Modern Horticulture and Genetics: Genetic Resources and Biotechnology of Sea Buckthorn (*Hippophae* L., Elaeagnaceae)

Igor V. Bartish

Abstract Sea buckthorn (*Hippophae* L., Elaeagnaceae) has been exploited by humans for thousands of years on the Quinghai–Tibetan Plateau (QTP) and nearby areas. However, the considerable modern economic potential of this plant has started to receive full appreciation only recently. Expanding its traditional use in harsh climatic zones as important source of nutrients, vitamins, and as wood in treeless areas, today this plant is used also on large scales as landscape protection tools against corrosion of soil, and as a source of wide range of products in pharmaceutic, cosmetic, and nutritional supplement industries. This review aims to provide the latest insights from studies on the evolutionary history and biogeography of the genus, structure, and phylogeography of genetic diversity within its species. Understanding the genic and genomic interactions among populations and phylogenetically distant lineages within species of *Hippophae* should help to improve the efficiency of exploitation of genetic resources in this crop. Research efforts in the past century in breeding, systematics, cytogenetics, biochemistry, and genetics of *Hippophae* have created a solid background for advances in modern biotechnology of this crop. Recent studies reported application of next-generation sequencing (NGS) technologies and identification of thousands of genes in transcriptomes of sea buckthorn. Analyses of the transcriptomes provided better understanding of gene expression in biochemical pathways of unsaturated fatty acids, some other secondary metabolites, and regulation of gene complexes responsible for adaptation to different categories of abiotic stress. Further studies should focus on the creation of genetic maps of breeding populations; identification of quantitative trait loci, biochemical pathways of synthesis of bioactive secondary

I.V. Bartish (✉)
Department of Genetic Ecology, Institute of Botany, Academy of Sciences of Czech Republic, Zamek 1, 25243 Pruhonice, Czech Republic
e-mail: Igor.Bartish@ibot.cas.cz

V.R. Rajpal et al. (eds.), *Gene Pool Diversity and Crop Improvement*,
Sustainable Development and Biodiversity 10,
DOI 10.1007/978-3-319-27096-8_14

metabolites and correspondent genes, molecular mechanisms of tolerance and resistance to abiotic stress, diseases, and pests; and cloning of genes of agricultural importance. Advances in these research areas can lead to genetic engineering of plants with a combination of traits of high horticultural, medicinal, or nutrient value, adapted to specific environments of areas of their cultivation.

Keywords Biogeography · Genetic diversity · Genetic engineering · *Hippophae rhamnoides* · Homoploid hybridization · Marker-assisted selection · Molecular breeding · Molecular cloning · Phylogeography · Population structure

14.1 General Introduction

Several reviews on sea buckthorn have been published in the past two decades on horticulture (Li and Schroeder 1996; Li 2003), biochemistry (Kalia et al. 2011), medicinal and nutritional properties (Warnock and Miskin 2009; Singh and Ahmed 2010; Suryakumar and Gupta 2011; Wang et al. 2011a; Kanayama et al. 2012), causes and effects of dried-shrink disease and susceptibility to insect pests (Ruan et al. 2013); and biotechnological advances in the development of this crop (Kalia et al. 2011; Ruan et al. 2013). However, three important factors make it necessary to continue summarizing information on studies of this plant. First, all these reviews focused on only one species in the genus *Hippophae*, namely *H. rhamnoides*, leaving discussions of genetic diversity and evolution in the whole genus mostly untreated. Second, it is hard to overview all aspects of research on this wonderful plant in one paper. Directions of these studies cover a wide range of research areas: taxonomy and systematics, evolution and biogeography, population genetics, biochemistry, ecology; horticultural, ecological, and medicinal applications; ethnobotany, ethnopharmacology, and its impact on economy of indigenous people; conventional breeding, different aspects of biotechnology (in vitro cultivation, molecular breeding, cloning of genes, genetic transformation), etc. Inevitably each review can focus only on some of these areas, while treating others in a cursory way or leaving them out completely. Third, research work on *Hippophae* and *H. rhamnoides* has considerably intensified in the past two decades. This leads to a fast increase in the amount of information published each year, making each review partly outdated soon after publication. This study is aimed to provide an overview of results in those areas that have not been reflected in sufficient detail in earlier reviews, such as evolution, biogeography, phylogeography, and population genetics of *Hippophae* in different geographical regions. Besides, promising advances have been made in recent years in the biotechnology of this plant, and a brief summary of these results is also provided.

14.2 Characterization of the Genus

14.2.1 Systematics and Taxonomy

Sea buckthorn is a common English name for plants from the genus *Hippophae*. This genus belongs to Elaeagnaceae, a small family of three genera (two other genera are *Shepherdia* and *Elaeagnus*). Circumscription of species and subspecies of *Hippophae* has varied over the past century, mainly due to the homoplastic morphology and different opinions as to what constitute species, subspecies, or hybrids. In the first monograph of the genus, Servettaz (1908) recognized one species with three subspecies (Table 14.1). In the second major revision of the genus, Rousi (1971) raised these subspecies to species level and described seven new subspecies of *H.rhamnoides*. Lian et al. (1998) recognized six species in two sections with some additional new subspecies described by themselves, i.e., *H. goniocarpa* subsp. *goniocarpa*, *H. goniocarpa* subsp. *litangensis*, and *H. neurocarpa* subsp. *stellatopilosa* (Lian et al. 1995). In one of the latest treatments of the genus, Swenson and Bartish (2002) mostly accepted suggestions of Rousi (1971) and Lian et al. (1998) and recognized eight subspecies within *H. rhamnoides*, two subspecies within *H. neurocarpa*, and three other non-hybrid and currently monotypic species (*H. gyantsensis*, *H. salicifolia*, and *H. tibetana*). However, they also raised *H. litangensis* to the species level, which together with *H. goniocarpa* created a group of two species of hybridogeneous origin and therefore unclear taxonomic status (for a more detailed discussion on interspecific hybridization and hypotheses of hybrid origin for different taxa within the genus see chapter "Homoploid hybridization in *Hippophae*"). Swenson and Bartish (2002) did not recognize sections *Hippophae* and *Gyantsensis*, suggested by Lian and Chen (1993). Circumscription of the sections was based on differences in the morphology of seed coat among species, but molecular markers did not support these groups (Bartish et al. 2000a, 2002; Sun et al. 2002; Jia et al. 2012; Jia 2013). On the other hand, all five currently recognized non-hybrid species in the genus can be clearly defined by morphological characters (Rousi 1971; Lian et al. 1998; Swenson and Bartish 2002).

Several additional systematic modifications have been suggested for *Hippophae* soon after the publication of the taxonomic synopsis of the genus by Swenson and Bartish (2002). Lian et al. (2003a) described *H. rhamnoides* subsp. *wolongensis*. Analysis of random amplified polymorphic DNA (RAPD) markers suggested a close relationship between subsp. *wolongensis* and subsp. *sinensis* (Sheng et al. 2006). However, analyses of sequences of *trn*L-*trn*F and *trn*S-*trn*D genes of chloroplast DNA (cpDNA) and internal transcribed spacer (ITS) sequences of ribosomal DNA indicated instead that this taxon is a part of *H. rhamnoides* subsp. *yunnanensis* (Jia et al. 2012). The status of *H. rhamnoides* subsp. *wolongensis* is therefore not clear yet, and it has to be confirmed in further analyses of

Table 14.1 Overview of the main systematic treatments of *Hippophae* L

Servettaz (1908)	Rousi (1971)	Lian et al. (1998)	Swenson and Bartish (2002)	Tsvelev (2002)
Sect. Hippophae				
H. rhamnoides subsp. *rhamnoides*	*H. rhamnoides*	*H. rhamnoides*	*H. rhamnoides*	*H. rhamnoides*
	subsp. *carpatica*	subsp. *carpatica*	subsp. *carpatica*	
	subsp. *caucasica*	subsp. *caucasica*	subsp. *caucasica*	*H. caucasica*
	subsp. *fluviatilis*	subsp. *fluviatilis*	subsp. *fluviatilis*	*H. caucasica* subsp. *fluviatilis* (?)
	subsp. *gyantsensis*			
	subsp. *mongolica*	subsp. *mongolica*	subsp. *mongolica*	*H. mongolica*
	subsp. *rhamnoides*	subsp. *rhamnoides*	subsp. *rhamnoides*	
	subsp. *sinensis*	subsp. *sinensis*	subsp. *sinensis*	*H. sinensis*
	subsp. *turkestanica*	subsp. *turkestanica*	subsp. *turkestanica*	*H. turkestanica*
	subsp. *yunnanensis*	subsp. *yunnanensis*	subsp. *yunnanensis*	*H. yunnanensis*
subsp. *salicifolia*	*H. salicifolia*	*H. salicifolia*	*H. salicifolia*	*H. salicifolia*
Sect. Gyantsensis				
		H. gyantsensis	*H. gyantsensis*	*H. gyantsensis*
		H. goniocarpa	*H. goniocarpa*	
		subsp. *goniocarpa*		
		subsp. *litangensis*	*H. litangensis*	
		H. neurocarpa	*H. neurocarpa*	*H. neurocarpa*
		subsp. *neurocarpa*	subsp. *neurocarpa*	
		subsp. *stellatopilosa*	subsp. *stellatopilosa*	
subsp. *tibetana*	*H. tibetana*	*H. tibetana*	*H. tibetana*	*H. tibetana*

morphological, ecological, and genetic differentiations of a comprehensive sample of populations including *H. rhamnoides* subsp. *yunnanensis* and *sinensis*. Lian et al. (2003b) also did not recognize a species status of *H. litangensis* and suggested instead to treat it as a subspecies of *H. goniocarpa*, arguing that the putative parents of *H. litangensis* are subspecies of *H. rhamnoides* and *H. neurocarpa*. In another taxonomic study, Tsvelev (2002) analyzed three subspecies of *Hippophae rhamnoides* from Russia (subsp. *caucasica*, *mongolica*, and *turkestanica*) and suggested raising these and two Chinese taxa (subsp. *sinensis* and *yunnanensis*) to species level. He was not certain regarding the treatment of *H. rhamnoides* subsp. *fluviatilis*. Noting that it is morphologically distinct from the other European taxa, he suggested a possible position for this taxon as a subspecies of *H. caucasica*. Besides, Tsvelev (2002) considered subsp. *carpatica* to be indistinguishable from subsp. *caucasica*. These taxonomic suggestions are summarized in Table 14.1. However, molecular data does not support close relations between subsp. *caucasica* and *carpatica* (Bartish et al. 2000a; Bartish 2006), or treatment of subsp. *fluviatilis* as one of taxa within *H. caucasica* (Bartish et al. 2002; Bartish 2006; Jia et al. 2012; Jia 2013).

Several molecular analyses (Bartish et al. 2000a, 2006; Jia et al. 2012; Jia 2013) recovered almost all subspecies of *H. rhamnoides* as monophyletic with respect to each other (reciprocally monophyletic). These studies could thus provide support for the proposition of Tsvelev (2002) to raise several subspecies of *H. rhamnoides* to species level. However, a clearly identified set of optical morphological characters to differentiate these taxa from each other in the field has not yet been developed (Swenson and Bartish 2002). The ability to distinguish taxa in the field using morphological characters is required for robust and stable taxonomic decisions in addition to strongly supported monophyly of correspondent clades (Backlund and Bremer 1998). A comprehensive analysis of a representative set of morphological traits of all subspecies of *H. rhamnoides* in a large sample of populations (including numerous adult individuals within these populations) across the whole range of the species have not been published yet. Rousi (1965, 1971) reported a variation in leaf and seed traits across all subspecies of *H. rhamnoides* and the whole genus. Unfortunately, with the exception of subsp. *fluviatilis* and *rhamnoides*, his samples in these two studies were restricted to few individuals within most of the populations and most of the scored traits, thus preventing any robust comparisons of variability within and between populations from different taxa he recognized. A clear understanding of distribution of diversity within and among populations and taxa of *H. rhamnoides* is therefore still lacking for most morphological traits. Careful statistical analyses of morphological diversity in combination with correspondent analyses of molecular markers on the same sample of multiple populations across the whole range of the species are required to advance discussions on taxonomical status of subspecies from this species. The main task of these analyses would be identification of morphological traits, in which variation among subspecies is clearly higher, than variation within all of them.

14.2.2 Morphological Characters

In addition to taxonomical applications, carefully scored morphological characters can provide important insight into the species' biology and ecology, and into agricultural potential of different regional genetic resources. Differences in the morphology of shoots and thorns can potentially reflect variable intensities of browsing by mammals in different regions, and also variable potential of local populations in breeding programs for thornless cultivars. Variations in seed and fruit sizes and yield can characterize variable levels of investment of individual plants into sexual and clonal reproduction, and also variable potential of local populations in breeding programs for yield and quality of fruits. Numerous studies have focused on analyses of fruit and seed morphologies, physiology, and biochemical composition, mostly in cultivated plant varieties of sea buckthorn. This literature, however, needs a special review and is not discussed here. On the other hand, as already mentioned above, there is currently a lack of standardized and well-documented statistical analyses on variations within the natural populations of multiple morphological characters across the whole range of the genus. This situation probably reflects insufficient appreciation by taxonomists of variation within populations and taxa of *Hippophae* in most of the studied morphological characters.

Despite the lack of systematic analyses of morphological traits in representative samples of populations across the genus, several studies did provide some insights into morphological variations (or its lack) in *Hippophae*. An analysis of seeds and seedlings across 46 natural populations (between 2 and 43 seedlings per population) of six western subspecies of *H. rhamnoides* (Rousi 1965), and analysis of 2264 herbarium specimens across the whole genus (Rousi 1971), are still probably the most advanced attempts to study morphological variations in both the species and the genus. These analyses revealed significant differences among all taxa recognized by Rousi (1971) in at least one of several quantitative biometric characters: leaf length, breadth, and length/breadth ratio; seed length, width, thickness, length/width ratio, width/thickness ratio, and weight. The only exception was the absence of significant differences in any of these characters between *H. rhamnoides* subsp. *mongolica* and *rhamnoides* (Rousi 1971). Pollen morphology was analyzed by Sorsa (1971) across several taxa of the genus including those recognized by Rousi (1971), but no traits to differentiate these taxa could be identified.

Yao and Tigerstedt (1995) studied variations in hardiness and plant height among populations of three subspecies of *Hippophae rhamnoides* (subsp. *rhamnoides*, *sinensis,* and *turkestanica*). They found strong correlation between ecological (hardiness) and morphological (plant height) traits. This study supports the idea of Rousi (1971) that particular morphological characters in different species of the genus can reflect their adaptation to variable environments and ecological challenges. However, different morphological characters can be associated with environmental factors in different ways. Unlike tree height, fruit and seed characters of *Hippophae* can be unrelated to climatic conditions of localities of their sampling (Aras et al. 2007). It is also not clear how adaptations to different environments, possibly expressed in variation and differentiation of relevant morphological characters, are associated with genetic differentiation between populations within the species of this genus.

In conclusion, the lack of standardized and statistically sound data on variation within and between individual plants, populations, and taxa for multiple morphological characters across the whole range of the genus prevents (i) reliable differentiation between subspecies of *H. rhamnoides* and *H. neurocarpa*; (ii) clear identification of hybrids between different sympatric taxa in the field; (iii) analyses of potential association of morphology of particular traits with environments of specific localities; (iv) efficient identification of perspective regional resources of variable agricultural traits for breeding programs; and (v) analyses of genetic linkage groups of characters and search for molecular markers of quantitative trait loci (QTL), controlling morphological traits in natural populations of the genus.

14.2.3 Evolution and Biogeography

14.2.3.1 Molecular Phylogenies of Elaeagnaceae and Hippophae

According to APG III, Elaeagnaceae are included into Rosales, an order of nine families (APG III 2009). In the latest phylogenetic reconstructions Elaeagnaceae are sisters to a small clade of Barbeyaceae and Dirachmaceae, but this relationship received only moderate statistical support (Wang et al. 2009; Zhang et al. 2011). Phylogenetic relationships within Elaeagnaceae have been clearly resolved (Jia and Bartish, unpubished) and species from *Hippophae* formed a clade, which was strongly supported by molecular markers from chloroplast and nuclear genome by different statistical methods, further supporting the findings of Bartish et al. (2002). The genus was sister to *Shepherdia* and this relationship was strongly supported by molecular markers from both chloroplast and nuclear genomes.

Phylogenetic relationships within *Hippophae* have been subject to several molecular studies. In analyses of RAPD markers, Bartish et al. (2000a) obtained a clade of all subspecies of *H. rhamnoides*, in which, however, only six western subspecies received strong statistical support, while placement of subsp. *sinensis* and *yunnanensis* was not supported. Topology of relationships among other taxa did not receive any robust support in these analyses. Bartish et al. (2002) analyzed variations in restriction sites from several genes of cpDNA and a set of morphological characters, and found strong statistical support for monophyly of the genus. They reconstructed a (weakly supported) clade of eight subspecies of *H. rhamnoides*, and a clade of four other (non-hybrid) species, which also received only weak statistical support. Sun et al. (2002) added described by Lian et al. (2003a) a Chinese subspecies *H. rhamnoides* subsp. *wolongensis* to their sample. They used in their analyses ITS sequences and mostly confirmed results of the earlier RAPD- and cpDNA-based studies. The exceptions were placement of *H. tibetana* in strict consensus tree as sister to the clade of all other taxa, and high statistical support

(97 and 91 %, respectively) for sister clades of nine subspecies of *H. rhamnoides*, and three remaining non-hybrid species (*H. gyantsensis*, *H. neurocarpa*, and *H. salicifolia*).

In a recent analysis of all non-hybrid taxa in the genus (with the exception of *H. rhamnoides* subsp. *wolongensis*), Jia (2013) obtained two sets of sequences from chloroplast and nuclear genomes (five genes in each). All taxa in this study were represented by three populations across the whole range of each taxon. This study indicated that the current systematics of the genus by Swenson and Bartish (2002) is relatively robust. However, together with phylogenetic reconstructions by Jia et al. (2012), it also revealed the problems to be resolved in further studies. These include (i) the uncertain status of *H. rhamnoides* subsp. *yunnanensis* and *wolongensis*, (ii) possibility of raising the status of all subspecies to species level, provided these species can be identified and circumscribed by morphological characters, which can be easily identified in the field; (iii) uncertain status of hybrid taxa (see also chapter on hybridization in the genus).

14.2.3.2 *Biogeography of* Hippophae

Bobrov (1962) was probably the first to suggest east to west migrations along the mountain ranges of Eurasia currently occupied by *H. rhamnoides* as the main biogeographic hypothesis for this genus. Based only on paleobotanical data, he suggested the late Miocene as the most likely geological epoch of the putative migrations. Hyvönen (1996) conducted the first phylogenetic analysis in *Hippophae* using morphological characters. Contrary to Bobrov (1962), he suggested the western part of the genus' range as its most likely ancestral area, and an ancient range fragmentation in the genus. The first molecular phylogenies (Bartish et al. 2000a, 2002; Sun et al. 2002) mostly supported the hypothesis of East Asian origin of the genus. This hypothesis received further support in the explicit tests of probability of ancestral area for several taxa in *Hippophae* by Jia et al. (2012), who identified an area to the east of Qinghai–Tibetan Plateau (QTP) as the most likely ancestral area of the genus and suggested a likely route for migration of *H. rhamnoides* from East Asia to Europe. These authors also suggested that diversification in the genus started in the late Miocene to Pliocene. However, sampling of populations in this study was not well-balanced taxonomically and geographically. It focused mainly on two taxa from East Asia (*H. rhamnoides* subsp. *sinensis* and *yunnanensis*), which might bias the results. Moreover, molecular phylogenies in this study were dated using secondary calibrations, which can be misleading (Ho and Phillips 2009; Sauquet et al. 2012).

Therefore, Jia (2013) carried out biogeographic analyses of a well-balanced sample of populations across ranges of all five non-hybrid species and ten subspecies in the genus. He calibrated his tree using a set of several carefully selected fossil records across Rosales. He also tested his dating by a recently published record of *Hippophae* from the late Miocene (about seven million annuals [Ma] before present) of Anatolia and Greece (Biltekin 2010). Results of this study provide strong support for the hypothesis of Bobrov (1962) and suggest also a detailed set of biogeographic scenarios for evolutionary processes in the genus during a period of considerable orogeneses and climatic changes in Eurasia (the Miocene, the Pliocene, and the Quaternary).

14.2.3.3 Genetic Diversity of Populations and Phylogeography

Genetic Diversity of Natural Populations from Different Taxa of Hippophae

Molecular studies in the past two decades provide a relatively detailed picture of the population genetic structure within most taxa of *Hippophae* (Table 14.2). Yao and Tigerstedt (1993) used isozyme loci in the first application of molecular markers for analysis of genetic structure in three species of *Hippophae* (*H. neurocarpa*, *H. rhamnoides*, and *H. tibetana*) and three subspecies of *H. rhamnoides* (*rhamnoides*, *sinensis*, and *turkestanica*). They found that most of the genetic variations in *H. rhamnoides* were allocated within populations (53.9 %) and between subspecies (41.6 %), while differentiation between populations within subspecies can be very low (4.5 %; Dsp = 0.011). This pattern was further confirmed in general in the first study of DNA markers (RAPDs) in the genus, an analysis of population genetic structure of *H. rhamnoides* subsp. *rhamnoides* (Bartish et al. 1999), and also in analyses of a group of western subspecies of this species (Bartish et al. 2000a).

Fixation indices (Gst) varied considerably (0.068–0.470) in four studies of *H. rhamnoides* subsp. *sinensis*, even though the marker system used in these studies was the same (inter-simple sequence repeats, ISSRs), and sample sizes of populations were similar (Table 14.2). Sun et al. (2006) reported intermediate values of Gst (0.183) in analyses of RAPD markers in this subspecies. The unusually high differentiation among populations reported by Wang et al. (2011b) probably reflects the sampling strategy of these authors. They selected populations from three regions across the whole range of the taxon with contrasting vegetations and environments.

High differentiation among populations was reported for *Hippophae rhamnoides* subsp. *turkestanica* (Gst = 0.457) for amplified fragment length polymorphisms (AFLPs) by Raina et al. (2012) and for subsp. *yunnanensis* (Gst = 0.459) for RAPDs by Chen et al. (2010). The authors of both the studies attributed this unusually high for the genus and dominant markers differentiation among populations to climatic heterogeneity and complex landscape characteristics of localities of their sampling. Raina et al. (2012) found also relatively low differentiation among populations in two other species, *H. salicifolia* and *H. tibetana* (0.291 and 0.194, respectively). These estimates were similar to those observed in *H. rhamnoides* subsp. *rhamnoides* by Bartish et al. (1999). It should be noted, however, that analyses of *H. salicifolia* and *H. tibetana* were based on a relatively low number of populations (three in each species), so they should be considered as preliminary. In recently reported analyses of genetic variation in *H. rhamnoides* subsp. *turkestanica,* Srihari et al. (2013) applied several molecular marker systems (RAPD, ISSR, simple sequence repeats [SSR], two MAD-box loci and ITS) to

Table 14.2 Analyses of population genetic structure in different taxa of *Hippophae* using molecular markers

Taxon	Marker system	Region	No. of populations	No. of plants	No. of loci	Gst	Reference
Dominant markers							
H. rhamnoides ssp. *rhamnoides*	RAPD	N. Europe	10	106	174	0.151[a]	Bartish et al. (1999)
H. rhamnoides ssp. *sinensis*	RAPD	E. Asia	13	232	107	0.183	Sun et al. (2006)
H. rhamnoides ssp. *sinensis*	RAPD	E. Asia	5	–	151	0.406	Zhao et al. (2007)
H. rhamnoides ssp. *sinensis*	ISSR	E. Asia	10	145	326	0.182	Chen et al. (2008)
H. rhamnoides ssp. *sinensis*	ISSR	E. Asia	11	220	207	0.068	Tian et al. (2004a)
H. rhamnoides ssp. *sinensis*	ISSR	E. Asia	7	140	288	0.418	Tian et al. (2004b)
H. rhamnoides ssp. *sinensis*	ISSR	E. Asia	12	218	252	0.470	Wang et al. (2011b)
H. rhamnoides ssp. *turkestanica*	AFLP	C. Asia	32	348	163	0.457[b]	Raina et al. (2012)
H. rhamnoides ssp. *turkestanica*	SAMPL[c]	C. Asia	32	348	99	0.552[b]	Raina et al. (2012)
H. rhamnoides ssp. *yunnanensis*	ISSR	E. Asia	7	140	288	0.279	Tian et al. (2004b)
H. rhamnoides ssp. *yunnanensis*	RAPD	E. Asia	6	113	112	0.459	Chen et al. (2010)
Co-dominant markers							
H. rhamnoides ssp. *rhamnoides*	Isozymes	N. Europe	9	1825	6	0.045[d]	Yao and Tigerstedt (1993)
H. rhamnoides ssp. *sinensis*	cpSSR[e]	E. Asia	12	218	6	0.602	Wang et al. (2011b)
Sequences of chloroplast genes							
H. gyantsensis	trnL-trnF trnS-trnG	E. Asia	8	63	2	0.730	Cheng et al. (2009)
H. neurocarpa ssp. *neurocarpa*	trnL-trnF	E. Asia	7	35	1	0.535	Meng et al. (2008)
H. neurocarpa ssp. *stellatopilosa*	trnL-trnF	E. Asia	7	35	1	0.606	Meng et al. (2008)
H. rhamnoides ssp. *yunnanensis*	trnL-trnF trnS-trnG	E. Asia	6	46	2	0.627	Cheng et al. (2009)

(continued)

Table 14.2 (continued)

Taxon	Marker system	Region	No. of populations	No. of plants	No. of loci	Gst	Reference
H. tibetana	trnL-trnF trnS-trnG	E. Asia	21	183	2	0.798	Jia et al. (2011)
H. tibetana	trnT-trnF	E. Asia	37	891	1	0.611	Wang et al. (2010)

In most cases differentiation among populations within each taxon is reported as Gst (coefficient of genetic variation among all populations in the sample belonging to the same taxon). N. Europe—North Europe; E. Asia—East Asia; C. Asia—Central Asia

[a]The estimate is based on the proportion of total variance in the sample attributed to variation among populations

[b]These estimates are means of two groups of populations from Jammu and Kashmir and Himachal Pradesh. Ten populations from Uttaranchal were excluded from calculations because authors of this study were not certain about their taxonomic affiliation

[c]SAMPL—selective amplification of microsatellite polymorphic loci

[d]The estimate is the proportion of the total gene diversity in all isozymes attributed to variation between the geographical populations within subspecies

[e]cpSSR—chloroplast simple sequence repeats

evaluate genetic relationships among 15 populations of this taxon from the same regions and approximately the same localities, as in the study of Raina et al. (2012), and in one population of *H. salicifolia*. However, these authors used only three individual plants per each of the populations, combined populations of *H. rhamnoides* and *H. salicifolia*, and combined different marker systems in their analyses of molecular variance (AMOVA). It is therefore not possible to infer partitioning of genetic variation within *H. rhamnoides* subsp. *turkestanica* by different marker systems from these analyses.

Several reports of partitioning of genetic variations within the species from QTP were based on analyses of variation in haplotypes of chloroplast DNA (chlorotypes). Meng et al. (2008) sequenced *trn*L-F gene in 70 individuals from 14 populations of two subspecies of *Hippophae neurocarpa*. They found moderate relatively to other studies of chlorotypes levels of differentiation among populations (0.535 and 0.606 for *H. neurocarpa* subsp. *neurocarpa* and *stellatopilosa*, respectively). Cheng et al. (2009) studied diversity and differentiation of chlorotypes of *trn*L-*trn*F and *trn*S-*trn*D genes in 109 individual plants from 14 populations of *H. gyantsensis* and *H. rhamnoides* subsp *yunnanensis*. They found stronger differentiation among populations in these two taxa (0.730 and 0.627 for *H. gyantsensis* and *H. rhamnoides* subsp *yunnanensis*, respectively). However, it should be noted that reconstructed in this study lineages did not correlate with taxonomic circumscriptions of the sampled material. The incongruence between taxonomic identifications of populations and reported cladogenic reconstructions might result from (i) sampling of hybrid populations, (ii) incomplete lineage sorting among recently differentiated taxa, or (iii) incorrect taxonomic identifications (no morphological data were reported in this paper). In all of these cases estimates of differentiation among populations within species will be inflated, so these results should be treated with a certain degree of precaution. Wang et al. (2010) found similarly high levels of differentiation (0.611) in a sample of 37 populations of *H. tibetana*, representing most of the range of this species. Finally, Jia et al. (2011) analyzed genetic variation in chlorotypes from a sample of 21 populations of *H. tibetana* across the main areas of distribution of the species and reported even higher levels of differentiation among populations (0.798).

Results on partitioning of genetic variation within and among populations should not be directly compared and discussed, if genetic estimates were based on different molecular marker systems. Nuclear and chloroplast genomes are transmitted between generations in a different way: only nuclear genome information is carried by pollen, while both nuclear and chloroplast genomes can be dispersed via seeds in *Hippophae* (Bartish et al. 2002) as in most other angiosperms. The difference in transmission routes from parent to offspring generations for cpDNA and nuclear DNA (nDNA) results in differences in effective population sizes between exclusively maternally (via seeds) transmitted cpDNA and both maternally and paternally (via pollen) transmitted nDNA. In populations of strictly outcrossing dioecious *Hippophae,* this theoretically means that (all else being equal) effective population size of chloroplast genes is one-quarter of nuclear genes (Birky et al. 1983). Besides, anonymous dominant markers (AFLPs, ISSRs, RAPDs), which are amplified mostly from nuclear genomes, can represent different rates of mutation comparatively to sequences of individual genes of cpDNA (Wolfe et al. 1987). These differences are a likely reason for variable levels of resolution in estimates of genetic differentiation between the two types of molecular marker systems.

Although results of the sequencing analyses of particular cpDNA genes thus are not comparable directly with results based on dominant markers, they nevertheless indicate relatively low levels of gene flow and high isolation of maternally transmitted lineages in the samples of populations of all six studied taxa with ranges on QTP and around it (*Hippophae gyantsensis*, *H. neurocarpa* subsp. *neurocarpa* and *stellatopilosa*, *H. rhamnoides* subsp. *turkestanica* and *yunnanensis*, *H. tibetana*). When fixation indices are compared for similar types of markers between different taxa, a contrast between high levels of differentiation among populations from QTP and neighboring areas and relatively low levels of differentiation among populations from lowlands (*H. rhamnoides* subsp. *rhamnoides* and *sinensis*), is indicated. Further analyses should test explicitly if this contrast can be explained by the landscape complexity and climatic gradients of the QTP and surrounding areas, as has been suggested by some authors (Chen et al. 2010; Wang et al. 2010; Raina et al. 2012), by different ages and demography of taxa and populations, by different ecological adaptations, or by a combination of any of these factors. Comprehensive analyses of population genetic structure are so far lacking for *H. rhamnoides* subsp. *carpatica*, *caucasica*, *fluviatilis*, and *mongolica*, as well as for *H. salicifolia.*

Phylogeography of Different Species of Hippophae

Phylogeographic studies focused so far on two of several main regions from the range of *Hippophae*, namely Europe and QTP (Table 14.3). Long before the advances of molecular techniques and based exclusively on analyses of palinological data, Gams (1943) suggested recolonization of Central Europe and Scandinavia by *H. rhamnoides* following deglaciation of these regions after the Last Glacial Maximum (LGM). According to his inferences, recolonization likely originated from a glacial refugium in southeastern Europe. This was one of the first explicitly formulated hypotheses regarding the late Quaternary phylogeography in the genus. This hypothesis has been tested in a study of four taxa from Europe and Asia Minor (*H. rhamnoides* subsp. *carpatica*, *caucasica*, *fluviatilis*, and *rhamnoides*) by Bartish et al. (2006). These authors analyzed DNA sequences of chalcone synthase intron (*Chs*i) and carried out a restriction fragment length polymorphism (RFLP) analyses of variation in cpDNA within the sample of 26 populations. This study supported the hypothesis of Gams (1943) and revealed a detailed picture of demographic and evolutionary processes in *H. rhamnoides* from the region. In particular, Bartish et al. (2006) (i) identified southeastern Europe as the most likely source area of recolonization into central Europe and Scandinavia after LGM, (ii) revealed multiple lineages of *Chs*i and possibility of existence of several microrefugia in the area; (iii) identified northern Alps as a contact zone between populations from the Alps and the East/Central European–Scandinavian clade; (iv) detected at least four episodes of population growth, all within about the last 40 ky (thousand years); and (v) found nearly synchronized timing of population expansions in the area of sampling, most likely correlating with the Younger Dryas Stadial shortly after LGM. This study reported the highest levels of nucleotide diversity within European populations of *H. rhamnoides* in the area close to northern and northeastern Alps. The observed pattern of genetic diversity in this species likely reflects the admixture of haplotypes after secondary contact of lineages from different glacial refugia, following a general trend of widespread European trees and shrubs (Petit et al. 2003). Surprisingly high levels of genetic diversity in the two most northern populations from Sweden and Norway suggest the possibility of an additional hybrid zone in northern Scandinavia. Genetic structure of populations from Asia Minor indicated demographic trends, which were different from European, and possibility of variable evolutionary responses to global climatic processes in different regions.

The first phylogeographic study of *Hippophae* from QTP was reported by Meng et al. (2008). These authors analyzed phylogeographic patterns in *H. neurocarpa* subsp. *neurocarpa* and *stellatopilosa* and found weak genealogical concordance in the sequenced fragment of cpDNA (*trn*L-*trn*F) with morphological differentiation between these two taxa. Their nested cladogram was based on eight haplotypes and classified these haplotypes into three lineages: one consisted of endemic haplotypes of subsp. *neurocarpa* and the other two included haplotypes from both subspecies. Several unique haplotypes were recovered in the high altitudes, suggesting that *H. neurocarpa* might have survived in these arid habitats at the LGM or even earlier in

Table 14.3 Phylogeographic analyses of different taxa of *Hippophae*

Taxa	Marker system	Region	N of pops	N of plants	Main patterns revealed	Reference
H. gyantsensis	*trn*L-*trn*F *trn*S-*trn*G	S QTP	8	63	Allopatric divergence between lineages from different parts of QTP	Cheng et al. (2009)
H. neurocarpa ssp. *neurocarpa* and *stellatopilosa*	*trn*L-*trn*F	E and N QTP	14	70	Allopatric divergence and CRE in E and N QTP	Meng et al. (2008)
H. rhamnoides ssp. *carpatica, caucasica, fluvialtilis,* and *rhamnoides*	Chsi, *trn*C-*trn*D, *trn*D-*trn*T, *trn*S-*trn*fM	Europe; Asia Minor	27	128	Allopatric divergence between lineages from different mountain ranges in the Pleistocene; CRE from SE to NW Eur after LGM; recent hybridization in NA	Bartish et al. (2006)
H. rhamnoides ssp. *sinensis*	*trn*L-*trn*F; *trn*S-*trn*G; ITS	E QTP; NEC	26	295	HDE in NEC in the LIP or FG	Jia et al. (2012)
H. rhamnoides ssp. *yunnanensis*	*trn*L-*trn*F *trn*S-*trn*G	E QTP	6	46	Allopatric divergence between lineages from different parts of QTP	Cheng et al. (2009)
H. rhamnoides ssp. *yunnanensis*	*trn*L-*trn*F; *trn*S-*trn*G; ITS	E QTP	15	171	Allopatric divergence in the Pliocene/Pleistocene; no range expansion	Jia et al. (2012)
H. tibetana	*trn*T-*trn*F	E, S QTP	37	891	Allopatric divergence in the Pliocene/Pleistocene; range expansion in S QTP in the Pleistocene	Wang et al. (2010)
H. tibetana	*trn*L-*trn*F; *trn*S-*trn*G; ITS	E, S QTP	21	183	Allopatric divergence between lineages from different parts of QTP; range expansion in E QTP in FG; range expansion in E and S QTP after LGM	Jia et al. (2011)

Chsi—chalcon synthase intron (nuclear gene). E, N, S, QTP—east, north, and south Qinghai–Tibetan Plateau; SE, NW Eur—southeast, northwest Europe; NA—northern Alps; NEC—northeast China. CRE—contiguous range expansion; HDE—historical demographic expansion. EG—Early Glacial (114-74 ky); FG—Full Glacial (74-14.6 ky); LGM—Last Glacial Maximum (24-14.6 ky), LIP—Last Interglacial Period (125-114 ky). Dates of geological epochs are according to Tzedakis et al. (2013)

multiple refugia. The authors of this study also report contiguous range expansion from three main putative glacial refugia identified in their analyses.

Similar to the study of Meng et al. (2008), analyses of phylogeographic patterns in chlorotypes of *H. gyantsensis* and *H. rhamnoides* subsp. *yunnanensis* by Cheng et al. (2009) did not reconstruct clades of haplotypes, which would be concordant with morphological taxa. Their data could not therefore distinguish between two alterative hypotheses regarding origin of *H. gyantsensis*: through homoploid hybridization or allopatric speciation. These authors explained their results by complex maternal lineage sorting between the studied taxa. They also argued that the unique haplotypes recovered in separate populations from each of the taxa they sampled might indicate maintainance of both these taxa in multiple refugia across their ranges during the LGM.

Wang et al. (2010) studied phylogeography of *H. tibetana* across most of the range of the species and revealed three main clades of chlorotypes, reconstructed from sequences of cpDNA *trn*T-*trn*F region. These clades were geographically structured within eastern, southeastern, and western groups of populations along respective edges of QTP. Results of this study suggested existence of multiple microrefugia of the species across its geographical range during the LGM and even earlier glaciations. Besides, the putative LGM microrefugia of *H. tibetana* may have been maintained at 4000 m above sea level and higher, the highest of all known and reported so far refugia at the global scale. Results of this study further supported the theory of the recent and rapid uplift of the QTP. The support was based on estimates of divergence times among and within the three lineages (from the problematic secondary calibrations, however), distinct geographic structures of the lineages, and general co-occurrence of these lineages with landscape-determined dispersal barriers. The authors concluded that uplift of the plateau, which likely was especially rapid starting from the late Pliocene (about 3.5 Ma), and the associated with this process climatic changes may have affected the dispersal and differentiation of *H. tibetana* and shaped its phylogeographic structure.

Populations of *H. tibetana* from approximately the same areas as in the study of Wang et al. (2010) were sampled by Jia et al. (2011), but these authors sequenced different genes in their analyses (*trn*S-*trn*D and *trn*L-*trn*F from cpDNA; ITS from nDNA; Table 14.3). They found two main clades with strong geographical structure. For most populations in both eastern and western regions they found a single widespread chlorotype. This pattern indicated a recent postglacial expansion within each region, while mismatch analyses of all chlorotypes within the eastern group of populations suggested an earlier regional expansion before the LGM. Existence of different chlorotypes across populations indicated possibility of multiple refugia in both regions. Coalescent tests rejected the hypothesis of origin of extant populations from a single refugium during the LGM, but supported hypothesis of diversification of the two main lineages before the late Pleistocene. Similarly to Wang et al. (2010), Jia et al. (2011) concluded that the patterns they found indicated evolutionary response in the species to orogenic processes and the climate changes on QTP and around it in the Quaternary.

Finally, Jia et al. (2012) collected populations across most of the ranges of *H. rhamnoides* focusing their sampling on east QTP and two most eastern subspecies of this species, subsp. *sinensis* and *yunnanensis* (41 populations from these two taxa). Applying phylogeographic analyses of both cpDNA (*trn*L-F and *trn*S-D) and nDNA (ITS) genomic sequences simultaneously, they identified (i) the most likely ancestral areas for *H. rhamnoides* subsp. *sinensis* and *yunnanensis*; (ii) a range expansion in subsp. *sinensis* (dated by a secondary calibration for mean substitution rate, it was estimated to have occurred before the LGM); and (iii) possible hybrid origin of subsp. *mongolica* from a putative cross between subsp. *sinensis* (as a seed parent) and subsp. *turkestanica* (as a pollen parent). These authors further argued that the Quaternary climatic oscillations are likely to have fragmented the distribution of *H. rhamnoides* and triggered allopatric divergence and the formation of strongly differentiated clades within the species.

In conclusion, the published phylogeographic studies of different species of *Hippophae* at two extremes of the range of this genus (Europe and QTP with surrounding areas) revealed several main patterns, which likely reflect responses of these species to the Quaternary climatic fluctuations (Table 14.3). Both allopatric lineage divergence, likely during periods of climatic deterioration, and population range expansion, likely during periods of favorable for *Hippophae* climate, have been indicated by phylogenetic and demographic analyses. Throughout periods of unfavorable climate, populations of *Hippophae* could be maintained in multiple isolated microrefugia some of them at very high altitudes across QTP. Although nearly synchronized timing of population expansions has been identified in Europe (Bartish et al. 2006), it is not clear if population expansions were synchronized in the late Quaternary across different mountain ranges of Eurasia. Currently the data, which could allow comparisons between demographic and population genetic processes in taxa from different regions, are not available, are from different molecular markers, or lack resolution.

14.2.4 Homoploid Interspecific Hybridization in the Genus

Table 14.4 provides a summary of all reported so far putative hybridizations in the genus. As has been discussed above (see "Systematics and Taxonomy"), Lian et al. (1995) introduced new names: *Hippophae goniocarpa* subsp. *goniocarpa* and *H. goniocarpa* subsp. *litangensis*. They further suggested, based on morphological characters and unpublished isozyme data, that these two taxa had their origin through hybridization between *H. rhamnoides* subsp. *sinensis* /*H. neurocarpa*, and *H. rhamnoides* subsp. *yunnanensis*/*H. neurocarpa* subsp. *stellatopilosa*, respectively. This hypothesis was supported in analyses of RAPD markers by Bartish et al. (2000a) and ITS sequences by Sun et al. (2002). However, these studies were unable to specify direction of the crosses. Bartish et al. (2002) used a cladistic approach and cpDNA RFLPs in their phylogenetic analyses of all taxa in the genus, including two subspecies of *H. goniocarpa*, described by Lian et al. (1995).

Table 14.4 Taxa of putative hybrid origin and hybridization events, suggested by analyses of molecular markers

Taxon	Seed parents	Pollen parent	Source of data	References
H. goniocarpa	*H. rhamnoides* ssp. *sinensis* or *H. neurocarpa* ssp. *neurocarpa*	*H. rhamnoides* ssp. *sinensis* or *H. neurocarpa* ssp. *neurocarpa*	Morphology and isozymes	Lian et al. (1995)
	H. rhamnoides ssp. *sinensis* or *H. neurocarpa* ssp. *neurocarpa*	*H. neurocarpa* ssp. *neurocarpa* or *H. rhamnoides* ssp. *sinensis*	RAPD	Bartish et al. (2000a)
	H. rhamnoides ssp. *sinensis* or *H. neurocarpa* ssp. *neurocarpa*	*H. neurocarpa* ssp. *neurocarpa* or *H. rhamnoides* ssp. *sinensis*	ITS	Sun et al. (2002, 2003)
	H. rhamnoides ssp. *sinensis*	*H. neurocarpa* ssp. *neurocarpa*	cpDNA RFLP	Bartish et al. (2002)
	H. rhamnoides ssp. *sinensis* and *H. neurocarpa* ssp. *neurocarpa*		*trn*L-F	Wang et al. (2008a)
H. litangensis	*H. rhamnoides* ssp. *yunnanensis* or *H. neurocarpa* ssp. *stellatopilosa*	*H. rhamnoides* ssp. *sinensis* or *H. neurocarpa* ssp. *stellatopilosa*	Morphology and isozymes	Lian et al. (1995)
	H. rhamnoides ssp. *yunnanensis* or *H. neurocarpa* ssp. *stellatopilosa*	*H. neurocarpa* ssp. *stellatopilosa* or *H. rhamnoides* ssp. *yunnanensis*	RAPD	Bartish et al. (2000a)
	H. rhamnoides ssp. *yunnanensis* or *H. neurocarpa* ssp. *stellatopilosa*	*H. neurocarpa* ssp. *stellatopilosa* or *H. rhamnoides* ssp. *yunnanensis*	ITS	Sun et al. (2002)
	H. rhamnoides ssp. *yunnanensis*	*H. neurocarpa* ssp. *stellatopilosa*	cpDNA RFLP	Bartish et al. (2002)
H. rhamnoides ssp. *mongolica*	*H. rhamnoides* ssp. *sinensis*	*H. rhamnoides* ssp. *turkestanica*	*trn*L-F, *trn*S-D, ITS	Jia et al. (2012)
Undescribed	Undescribed "*Grand Cannyon*"[a]	*H. rhamnoides* ssp. *yunnanensis*	*trn*L-F, *trn*S-D, ITS, *Chs*i	Jia et al. (2012); Xu et al. (personal communication, 2013)
Undescribed	Undescribed "*Grand Cannyon*"[a]	*H. gyantsensis*	*trn*L-F, *trn*S-D, ITS, *Chs*i	Jia et al. (2012); Xu et al. (personal communication, 2013)

(continued)

Table 14.4 (continued)

Taxon	Seed parents	Pollen parent	Source of data	References
Undescribed	*H. gyantsensis*	*H. rhamnoides* ssp. *yunnanensis*	*trn*L-F, *trn*S-D, ITS, *Chs*i	Jia et al. (2012); Xu et al. (personal communication, 2013)
Undescribed	*H. gyantsensis*	*H. neurocarpa* ssp. *stellatopilosa*	*trn*L-F, *Chs*i	Xu et al. (personal communication, 2013)

[a]"*Grand Cannyon*" is an informal name introduced by Dr. Kun Sun for currently undescribed cryptic taxon of *Hippophae,* revealed by recent molecular analyses of Jia et al. (2012) as a part of *H. rhamnoides* subsp. *yunnanensis sensu lato*

This study confirmed hybrid origin of these taxa, identified their putative parents, and suggested direction of crosses. *H. rhamnoides* subsp. *sinensis* and *yunnanensis* were identified as seed, while *H. neurocarpa* subsp. *neurocarpa* and *stellatopilosa* as pollen parents of subsp. *goniocarpa* and *litangensis*, respectively. Based on these results, Bartish et al. (2002) suggested raising the hybrid taxa to species level under new combinations: *H. goniocarpa* Y.S. Lian et al. ex Swenson & Bartish and *H. litangensis* Y.S. Lian & X.L. Chen ex Swenson & Bartish. These species were accepted in general by Swenson and Bartish (2002), but with a precautionary note that several morphological characters were not available for morphological studies of these hybrid species. Sun et al. (2003) confirmed hybrid origin of *H. goniocarapa* in analyses of ITS sequences, and suggested that hybridization between *H. rhamnoides* subsp. *sinensis* and *H. neurocarpa* was recent. Besides, their results indicated multiple hybridization events in origin of *H. goniocarpa.* In analysis of *trn*L-F sequences in 75 individuals of *H. goniocarpa*, *H. rhamnoides* subsp. *sinensis* and *H. neurocarpa* subsp. *neurocarpa* from two cites with sympatrically grown populations of these taxa, Wang et al. (2008a) revealed individuals of *H. goniocarpa* with cpDNA from both putative parental species. Their results thus confirmed the idea of multiple hybridization events in origin of this species. Thorough analyses of multiple morphological, cpDNA and nuclear markers in a representative sample of individuals of *H. goniocarpa* and *H. litangensis* from several localities are clearly necessary to better understand evolution of these species. Such analyses could allow settling the argument of earlier studies regarding the status of the taxa, which originally were described under these names (Lian et al. 1995; Bartish et al. 2002, Swenson and Bartish 2002). According to results of Sun et al. (2003) and Wang et al. (2008a), it is unlikely that *H. goniocarpa* is a genetically stable taxon, as suggested by Lian et al. (1995). More likely, it can be an evolutionary recent hybrid swarm of ephemeral populations. These populations should be of very different genetic backgrounds and originate from zones of secondary contact of numerous populations of the putative parental species. In the latter case circumscription of all these hybrid populations and individual plants in two subspecies of the same species, as suggested by Lian et al. (2003b), will be problematic.

Molecular analyses revealed several additional putative hybrid taxa in *Hippophae*. All these taxa are cryptic hybrids, since no earlier taxonomic studies indicated their hybrid origin. Results reported by Jia et al. (2012) from analyses of sequences of two loci of cpDNA and ITS from a representative sample of populations of *H. rhamnoides* subsp. *mongolica*, *turkestanica*, and *sinensis* indicated possibility of hybrid origin of subsp. *mongolica*. A single chlorotype of this taxon was reconstructed as sister to a clade of multiple chlorotypes of subsp. *sinensis*. However, two of ITS haplotypes of *H. rhamnoides* subsp. *mongolica* created a clade sister to a clade of haplotypes of subsp. *turkestanica*, and another two ITS haplotypes were included within the clade of subsp. *turkestanica*. The incongruence in phylogenetic relationships between chloroplast and nuclear genomes of the three Asian subspecies of *H. rhamnoides* was further confirmed in analyses of a larger sample of five genes from each of the two genomes (cpDNA and nDNA) by Jia (2013). This study provides support to the hypothesis of an ancient and likely unique hybrid origin of *H. rhamnoides* subsp. *mongolica* and dismisses the hypothesis of incomplete lineage sorting as a possible explanation for the incongruence between phylogenetic reconstructions. Given the monophyly of both genomes of *H. rhamnoides* subsp. *mongolica* reconstructed in this study, currently wide but isolated range of the subspecies, and distinct climatic differentiation from each of the two putative parental taxa, this taxon should be considered as an ancient and stable hybrid with unique origin and ecological adaptations.

Several poorly characterized putative hybrid taxa were further indicated by analyses of Jia et al. (2012). One of them can be a cryptic taxon, currently identified as *Hippophae rhamnoides* subsp. *yunnanensis*. This taxon was represented by a group of populations from south QTP, where the flow of the Brahmaputra River turns from west–east to north–south direction. The provisional name of this taxon is "Grand Canyon," reflecting its currently identified range. The group of populations under this name was characterized by a clade of chlorotypes (clade C), which was sister to a clade of all subspecies of *H. rhamnoides* excluding subsp. *yunnanensis*. The clade of ITS haplotypes sampled from the same group of populations (clade II) was, however, sister to all subspecies of *H. rhamnoides* including subsp. *yunnanensis*. These results made taxonomic status of *H. rhamnoides* subsp. *yunnanensis* unclear and beg for thorough analyses of morphological characters in specimens from different populations, currently circumscribed within this taxon. Finally, results of Jia et al. (2012) revealed also putative hybridization between *H. gyantsensis* and *H. rhamnoides* subsp. *yunnanensis*. This hybridization was indicated by presence in individuals of population from Bomi, Xizang, of chlorotypes from clade A, which was later identified by Jia and Bartish (unpublished) as one of cpDNA clades of *H. gyantsensis*, and ITS haplotypes from clade III, representing *H. rhamnoides* subsp. *yunnanensis* sensu stricto. Interestingly, individuals with chlorotypes of clade A were also found in the population from Litang, which was defined as *H. rhamnoides* subsp. *yunnanensis* by Jia et al. (2012), although all ITS haplotypes from this population belonged to a clade of haplotypes of *H. gyantsensis* (clade I). Adding to this confusion, the Litang population is located in the county referred to as locality of origin of the type of *H. litangensis* (Lian et al. 1995), a putative hybrid species, for which *H. rhamnoides* subsp. *yunnanensis* and *H. neurocarpa* subsp. *stellatopilosa* were suggested as parental taxa (see above). Hybridization between *H. gyantsensis* and *H. rhamnoides* has not been suggested in earlier studies in the genus. Populations from Bomi and Litang obviously need further investigation as localities of putative interspecific hybridizations, in which several species of *Hippophae* can be involved.

14.3 Biotechnological Advancements in *Hippophae Rhamnoides*

14.3.1 Assessment of Gene Bank Collections

Breeding programs of different crops are often focused on several agricultural traits most important for current economically efficient production, such as high yield and resistance to variable diseases and pests. Gene bank collections of crops represent an important part of breeding programs and are developed to conserve a wide range of naturally available traits together with achievements of past generations of breeders (Jeppsson et al. 1999; Nybom et al. 2003). Unfortunately, many traits are not easily determined on living plants, pedigree information on origin of cultivars is often lacking, and accessions are sometimes mislabeled. Different kinds of relatively cheap, easy-to-generate, and highly polymorphic molecular markers have been developed (Weising et al. 1994) and can help to improve effectiveness of using gene bank accessions (i.e., Hokanson et al. 1998). Analyses of genetic relatedness among accessions of a germplasm collection can provide valuable information on genetic background of different cultivars, helping to develop conservation strategies for preservation of important agricultural traits. Effectiveness of breeding programs can be improved using parental varieties with clearly defined taxonomic and geographic origins and highly resolved genetic relationships to other cultivars. Assessment of collections of cultivated varieties by molecular markers can also help to keep track of certified cultivars and prevent their illegal distribution (Congiu et al. 2000).

Modern breeding programs in sea buckthorn are relatively young and even the earliest of them from the ex-Soviet Union started only in the second half of the past century (Kalinina and Panteleyeva 1987). Nevertheless, a relatively large number of cultivars have been developed by now (Trajkovsky and Jeppsson 1999); for a list of cultivars from Lisavenko Institute of Horticulture for Siberia, Barnaul, the pioneer and one of the leaders of sea buckthorn breeding, see Zubarev et al. (2014). One of the most diverse gene banks of sea buckthorn varieties in the world was established at Balsgård, Department of Horticultural Plant Breeding of Swedish Agricultural

University (Bartish et al. 2000b). Assessment of genetic diversity in a subset of this collection (55 varieties) by RAPD analysis revealed three main groups of genotypes (Bartish et al. 2000b). The first of these groups represented Siberian cultivars with genetic background almost exclusively in *Hippophae rhamnoides* subsp. *mongolica* from Altai region. The second group included hybrids between cultivars from breeding programs at the Moscow Lomonosov University. In these programs breeders used selections of wild specimens of *H. rhamnoides* subsp. *rhamnoides* (mostly from the eastern Baltic coast) as one of the parents, and cultivars from Siberia as the second parent. Finally, the third group represented a mix of wild accessions and cultivars from crosses involving *H. rhamnoides* subsp. *carpatica*, *caucasica*, *fluviatilis*, and *rhamnoides* from Central and Eastern Europe and from Caucasus, and hybrids of these varieties. Although the three groups were significantly differentiated from each other, the major part of molecular variance (approximately 75 %) in the analyzed set of accessions was found within taxonomic or geographic groups. This study demonstrated the utility of molecular markers for clarification of geographic and taxonomic origin of accessions and cultivars of sea buckthorn.

Analyses of 14 cultivars involved into Chinese breeding programs using RAPD (Ruan et al. 2004), or the same cultivars plus one additional using AFLP molecular markers (Ruan and Li 2005, Ruan 2006) revealed strong presence of genetic component of *Hippophae rhamnoides* subsp. *mongolica* in these breeding programs. Nine of the analyzed cultivars were obtained from Russia and Mongolia (all represented *H. rhamnoides* subsp. *mongolica*). Among the remaining cultivars three represented hybrids between *H. rhamnoides* subsp. *mongolica* and *sinensis*, and the other three–*H. rhamnoides* subsp. *sinensis* (Ruan et al. 2004, Ruan and Li 2005). All three cultivars representing *H. rhamnoides* subsp. *sinensis* and all three hybrids were placed by cluster analysis of RAPD profiles either between two clusters of cultivars each representing *H. rhamnoides* subsp. *mongolica*, or among cultivars from this taxon. Besides, at least in the case of cultivars Hongguo (*H. rhamnoides* subsp. *sinensis*) and Hongyun (*H. rhamnoides* subsp. *mongolica*), representatives of two subspecies were difficult to distinguish from each other by phenotypic characters, and authors doubt that cultivar Zhongguoyou really represents subsp. *sinensis* (Ruan and Li 2005). These results thus indicated that all analyzed cultivars were hardly differentiated genetically from *H. rhamnoides* subsp. *mongolica*. Authors explained the unexpected placement of Chinese cultivars by geographic origin of the putative representatives of *H. rhamnoides* subsp. *sinensis* from Inner Mongolia, where natural hybridization between two subspecies might occur (Ruan and Li 2005). However, the gap between ranges of these subspecies is currently wide (Rousi 1971; Jia 2013). Neither RAPD (Bartish et al. 2000a), nor sequencing analyses (Jia et al. 2012) of representative samples of natural populations of two subspecies indicate recent hybridization between these taxa. Instead, these analyses revealed clear genetic differentiation between natural populations of *H. rhamnoides* subsp. *mongolica* and *sinensis*.

In a follow-up study from the same laboratory, 52 accessions from a wider taxonomic range of cultivars representing *Hippophae rhamnoides* subsp. *mongolica*, *rhamnoides*, *sinensis*, and *H. salicifolia* were analyzed by ISSR markers (Ruan et al. 2009). These analyses demonstrated utility of ISSR markers for assessment of breeding collections of sea buckthorn and confirmed strong representation of *H. rhamnoides* subsp. *mongolica* in Chinese breeding programs. In addition, these markers support higher genetic polymorphism in Siberian cultivars relatively to Chinese selections (Liu et al. 2007).

Assessment of genetic relationships among selections from native stands in other subspecies of *Hippophae rhamnoides* has been so far scarce. Shah et al. (2009) used AFLP markers in analyses of 25 ecotypes from ten localities in northern Pakistan, representing *Hippophae rhamnoides* subsp. *turkestanica*. The ecotypes were clustered in three groups, which did not correspond to geographic or climatic distances between the localities. The authors explained this result by high levels of seed dispersal in the area by birds and river flows. In analyses of RAPD markers in ten selections of *Hippophae rhamnoides* subsp. *caucasica* from Erzurum province of Turkey Ercisli et al. (2008) reported high polymorphism in these accessions (92.2 %). All investigated genotypes were clearly differentiated by Jaccard's similarity indices. Authors of this study also reported lack of congruence between fatty acid composition and RAPD-based genetic relationships among the ten selections. Simon-Gruita et al. (2012) analyzed five accessions of natural populations of *H. rhamnoides* subsp. *carpatica* and five cultivars from the breeding programs in Romania. These authors used RAPD markers and found a relatively close relationship between natural populations grown at the same altitude. They also concluded that Romanian commercial varieties resulted mainly from recent selections from natural populations with little indication of coherent breeding programs in their genetic composition. Finally, a preliminary analysis of exotic naturalized populations of *H. rhamnoides* in Canada by Chowdhury et al. (2000) using RAPDs suggested relatively high levels of Shannon's phenotypic diversity in these populations. Estimates of the diversity index for this species were comparable with corresponding estimates of diversity in native populations of co-familial buffaloberry (*Shepherdia argentea*), and higher than diversity of another native representative of Elaegnaceae, silverberry (*Elaeagnus commutata*). Authors of this study do not specify, however, which subspecies of *H. rhamnoides* naturalized in the area of their collections (southwestern Saskatchewan).

In conclusion, assessment by molecular markers of numerous gene bank collections of cultivated sea buckthorn in different regions of the world reveals a relatively narrow taxonomic representation of *Hippophae rhamnoides* in many of these collections. The most productive so far Asian breeding programs in China (Huang 1995) and Russia (Zubarev 2014) have been focusing mostly on introgression into cultivars of genetic material from a relatively small part of the range of *H. rhamnoides* subsp. *mongolica* (it was mainly derived from Altai region). European cultivars were mostly represented by progenies of *H. rhamnoides* subsp. *rhamnoides* and by hybrids between these progenies and Siberian cultivars (Bartish et al. 2000b).

14.3.2 Molecular Breeding and Development of Genetic Markers of Important Traits

Marker-assisted selection (MAS) of superior varieties with combination of economically important traits can be an efficient way of introgression of useful genetic material from wild populations into breeding programs (Tanksley 1993; Xiao et al. 1998). Genetic maps of linkage groups of traits of interest with variable molecular markers can be a useful tool in successful molecular breeding. However, development of mapping populations for specific traits in trees and shrubs and in young crops such as sea buckthorn is time-consuming and laborious. Besides, success of MAS largely depends on the extent of genetic linkage between markers and relevant QTL loci (Virk et al. 1996). To increase the resolution of the approach and the likelihood of linkage between markers and traits of interest, saturation of linkage maps should be as high as possible. High levels of map resolution require large number of markers, which preferably should also be transferable between mapping populations for different traits. Co-dominant markers, such as SSRs and single nucleotide polymorphisms (SNPs) are relatively evenly distributed throughout genome (Akkaya et al. 1995), abundant, and highly polymorphic (Morgante and Olivieri 1993). Besides, these markers are also relatively stable and therefore express high levels of transferability between mapping progenies from the same species (Antanaviciute et al. 2012; Fernandez-Fernandez et al. 2012). Due to these characteristics SSRs and SNPs have been the main tools in many mapping projects. Unfortunately, development of these markers requires sophisticated personnel and expensive laboratory equipment, and has so far been both expensive and laborious (Gupta and Varshney 2000). This could be the main reason for the current lack of genetic maps in *Hippophae rhamnoides*. Fortunately, new technologies are being currently developed, which allow creation of highly saturated genetic maps even in tree species with relatively large genomes as in *Malus* from Rosaceae (Antanaviciute et al. 2012). These technologies hold great promise for considerable improvement in efficiency, and decreasing the costs of generation of linkage maps in young and minor crops without sequenced genomes. First primers for amplification of microsatellite markers have been already reported for *H. rhamnoides* subsp. *sinensis* (Wang et al. 2008b) and subsp. *turkestanica* (Jian et al. 2010; Jian et al. 2014), and development of thousands of SSRs using next-generation sequencing (NGS) is underway (Ghanghal et al. 2013). Until now, however, MAS in *Hippophae rhamnoides* has been focused on generation of molecular markers for a few important traits under simple genetic control, such as sex determination (Persson and Nybom 1998; Sharma et al. 2010; Korekar et al. 2012) and dried-shrink disease resistance (Ruan et al. 2009; Li et al. 2010).

Markers of genetic sex determination can be used for discrimination between male and female plants of *H. rhamnoides* at early stages of development. The first of attempts to obtain these markers used RAPD analyses and revealed a male-specific marker (OPD15-600) in one of two progenies of controlled crosses (Persson and Nybom 1998). Both parents of the first progeny were commercial varieties from wild material selected in eastern Germany, within the range of *H. rhamnoides* subsp. *rhamnoides*. The marker was amplified in all 17 male plants and was absent in all 17 female plants of this progeny, used in the analyses. However, the marker was present in only one male plant of eight tested in the second progeny. The female parent of this progeny was derived from an open pollinated population of a Russian cultivar Dar Katuni, which was selected from plant material from the Altai, within the range of *H. rhamnoides* subsp. *mongolica*. The male parent was collected from a wild population of *H. rhamnoides* subsp. *rhamnoides* in Kilskiir, Uppland, Sweden. The different geographic and taxonomic origins of parents of these two crosses and high levels of mutations in RAPD loci (which decreases transferability of these markers between phylogenetically distant genomes) can be the main reason for the failure to amplify marker OPD15-600 in most males of the second progeny.

Sharma et al. (2010) obtained two gender-specific markers in isozyme and RAPD analyses of five female and five male plants of *Hippophae rhamnoides* subsp. *turkestanica* from Sunnam and Kinnaur districts of Himachal Pradesh, at the southeastern edge of the range of this taxon (I. V. Bartish, personal observation, 2013). One of these markers was a female-specific allele of peroxidase, and the other–a male-specific amplification of OPD20-911. However, the very low numbers of the tested in these analyses plants make utility of the developed markers uncertain.

In a more recent study, Korekar et al. (2012) reported development of sequence-characterized amplified region (SCAR) markers of genetic sex determination in *Hippophae rhamnoides* subsp. *turkestanica*. The analyzed plant material was collected from natural populations in Ladakh region. Authors of this study found consistent amplification of female-specific (FS) polymorphic fragments of 1,164, and 868 bp, named as OPA-04(FS) and OPT-06(FS), respectively. The fragments were cloned, sequenced, and specific to these fragments primers were developed. One of the fragments, OPA- 04(FS), was found to be 54 % similar with a hypothetical oxidoreductase and the other, OPT-06(FS), with BNR/Asp-box repeat domain containing protein of *Pyrenophora tritici-repentis* Pt-1C-BFP. These markers were tested on 120 female and 100 male plants (each from different locality) and showed consistent gender specificity. These results suggest that the markers developed by Korekar et al. (2012) can be a promising tool in detection of gender in plants of *H. rhamnoides* subsp. *turkestanica* at early stages of their development. However, it remains to be seen if these markers can be applied successfully across the whole taxonomic and geographic ranges of the species.

The second agronomical trait of sea buckthorn, which received considerable interest in molecular breeding studies, was dried-shrink disease (DSD) resistance. The disease is caused by several species of *Fusarium* and *Phomopsis* (Ruan et al. 2010), and work on development of molecular markers of resistance and susceptibility to this disease has started recently (Ruan et al. 2009; Li et al. 2010). Ruan et al. (2009) reported a significant correlation with resistance to DSD ($P < 0.001$) for four ISSR markers (809_{290}, and 811_{280}, 835_{700}, and 887_{190}) among analyzed accessions of cultivars and varieties from Chinese breeding programs. They suggested using these markers in a selection of lineages resistant to DSD, when no other genetic information is available. Li et al. (2010) estimated variability of sequence-related amplified polymorphism (SRAP) markers in 77 accessions of 22 cultivars of *H. rhamnoides*, most of which were used in an earlier study by Ruan et al. (2009). They searched for association of these markers with DSD resistance using multiple regression analysis and revealed 11 SRAPs significantly ($P < 0.001$) correlated with this trait. These authors suggest application of the obtained markers for screening of resistant and susceptible to DSD genotypes at early stages of plant development. The proportion of known genes and co-dominant loci is relatively high in the total pool of SRAPs (Li and Quiros 2001). These markers should therefore express high levels of transferability between closely related taxa (such as subspecies of *H. rhamnoides*), and could be a more efficient alternative to taxon-specific AFLPs, ISSRs, and RAPDs.

14.3.3 Gene Cloning

Cloning of genes controlling expression of agronomical traits can considerably advance molecular breeding (Glick et al. 2010). For example, genes of traits under simple genetic control, as resistance to some diseases and pests (Gusberti et al. 2013), control of ripening (Barry et al. 2006), or tree architecture (Dardick et al. 2013) can be transferred to, or specifically blocked in selected genomes, triggering expression of the desired traits in superior cultivars. However, the task can be complicated, if genes controlling the traits of interest are not known or have not been sequenced. Identification of these genes usually requires availability of mapping populations with segregation of the trait of interest, and highly resolved genetic maps (Collard et al. 2005). As already mentioned, these maps are not available for sea buckthorn yet. Nevertheless, the first steps in direction for better understanding of expression of specific traits at the level of individual genes have been made also in *Hippophae*. Xu et al. (2009) studied adaptation to drought stress in seedlings of *H. rhamnoides* from Jiuzhai population. This locality is in the range of *H. rhamnoides* subsp. *sinensis*, according to the data from Jia et al. (2012). Xu et al. (2009) used two-dimensional electrophoresis and mass spectrometry for identification of drought-associated proteins in this population. They reported 13 drought stress-responsive proteins, four of which have not been found in higher plants. Functions of J-type co-chaperone (Hsc20), a putative ABC transporter ATP-binging protein, a probable nitrogen regulation protein (NtrX), and heat shock protein (HslU) were predicted either from their conserved domains or homologies to other organisms. Results of this study elucidate mechanisms of responses to drought stress in plants and could potentially open a way to cloning of genes involved in stress tolerance in sea buckthorn.

Ghangal et al. (2012) constructed a cDNA library of clones of *Hippophae rhamnoides*. Authors of this study do not indicate which subspecies was used in their analyses, but they state that leaf tissue for this work was sampled from plants growing at Defense Institute of High Altitude Research, Leh, Ladakh. Because most populations from this region belong to *H. rhamnoides* subsp. *turkestanica* (Raina et al. 2012; I. V. Bartish, personal observations, 2013), the reported cDNA library was most likely created from genetic material of this taxon. The library comprises 3412 Expressed Sequence Tags (ESTs) and 1665 unigenes, of which 1278 were annotated by similarity search. Ghangal et al. (2012) reported identification of 43 unigenes responsive to biotic and abiotic stresses. The change in expression pattern under cold/freeze stresses was examined by real-time PCR in 13 of these genes, which earlier were shown to be associated with cold stress tolerance in *Arabidopsis*, and in three novel transcripts. These transcriptome analyses revealed dramatic changes in expression patterns in *H. rhamnoides* (3–12 fold increases) under freezing temperatures for *Pathogenesis Related Thaumatin*, *Chitinase*, *Rare Cold-Inducible* 2A, and *High Mobility Group* B2 genes. Authors suggest that further investigations might help in establishing some of the identified cold and freeze responsive genes as valuable resources for developing cold-resistant crop plants. In particular, breeding of better adapted to environmental stress cultivars of sea buckthorn might benefit from involvement of these and other associated with tolerance or resistance to cold and freeze genes into breeding programs. For example, Gupta et al. (2009) cloned partial cDNA (689 bp) of glycerol-3-phosphate acyltransferase (GPAT) and suggest that full-length cloning and overexpression of this gene could help cold-sussceptible plants to protect their photosynthetic machinery from photoinhibition under cold conditions and better resist freezing temperatures.

NGS techniques and methods have been used intensively in the past decade to generate a large amount of genetic data (Mardis 2008). Although these approaches are especially efficient in analyses of annotated genomes, they also hold great promise of considerable increase in efficiency of genetic analyses in non-model species. Fatima et al. (2012) applied high-throughput 454 sequencing methodology to uncover genes related to oil biosynthesis, other important metabolic pathways, and stress-response pathways in *Hippophae rhamnoides* subsp. *mongolica* (seeds of RC-4 cultivar from a Canadian breeding program were used in this study for RNA extractions). These authors reported identification of 89,141 putative unigenes in their dataset and most of the genes involved in fatty acid biosynthesis. In particular, Fatima et al. (2012) identified sequences for most enzymes involved in the biosynthesis and elongation of fatty acids (with the exception of ACP-S-malonyl transferase). They also identified genes involved in isoprenoid biosynthesis (113 sequences) and abiotic stress-related genes (gene ontology category "response to stress" resulted in 1525 sequences). The carotenoid biosynthesis genes were abundantly represented in the seed transcriptome, as would be expected. The largest number of sequences of stress-related genes fell within the heat, oxidative, osmotic, and cold stress categories, followed by wounding, DNA damage and water deprivation categories. Heat shock proteins were the most highly represented set in the mature seed transcriptome of the species. Fatima et al. (2012) concluded that their dataset forms a comprehensive genomic resource for sea buckthorn and establishes a basis for dissecting metabolic pathways related to the formation of oil and bioactive components. They also suggested that by using knowledge of the metabolic pathways, new strategies can be developed to enhance the agricultural production of variable desirable compounds. These compounds could potentially be utilized in pharmacology, cosmetic, and nutritional industries.

The promising perspectives of NGS technologies in analyses of transcriptome composition of sea buckthorn were further confirmed by a recent study of Ghanghal et al. (2013). Authors of this work used for their sequencing analyses RNA extracts from leaf and root tissues of plants from the same region as in the study of Ghangal et al. (2012). Applying Illumina HiSeq 2000 platform, six frequently used short read assemblers, and following two different strategies to select the best transcriptome assembly, Ghanghal et al. (2013) reported de novo short read assembly of 88,297 transcripts (>100 bp), representing about 53 Mb of sea buckthorn transcriptome and their functional annotation of this transcriptome revealed conservation of genes involved in various biological processes.

The studies by Fatima et al. (2012) and Ghangal et al. (2013) demonstrated the power of NGS for gene function discovery in *Hippophae*. The transcriptome data generated in these studies could provide a valuable resource for gene discovery and development of functional molecular markers in sea buckthorn breeding programs. By widening the taxonomic range of transcriptomes analyzed by NGS methods, further studies should advance our understanding of how genetic variability within *Hippophae* is translated into the incredibly high adaptability of its different species to abiotic stress and biotic pressure. This knowledge can be especially useful for efficient engineering of highly productive cultivars with wide range of adaptations to different environments of cultivation.

Acknowledgements Financial support for Ph.D. studies by Dongrui Jia and Alexey A. Borisyuk, field trips in 2012–2014 by Igor V. Bartish and Alexey A. Borisyuk, and research of Igor V. Bartish on *Hippophae* from Alexey L. Kudrin foundation "Strategy" is greatly appreciated.

References

Akkaya MS, Shoemaker RC, Specht JE, Bhagwat AA, Cregan PB (1995) Integration of simple sequence repeat DNA markers into a soybean linkage map. Crop Sci 35:1439–1445

Antanaviciute L, Fernandez-Fernandez F, Jansen J, Banchi E, Evans K, Viola R, Velasco R, Dunwell J, Troggio M, Sargent D (2012) Development of a dense SNP-based linkage map of an apple rootstock progeny using the Malus Infinium whole genome genotyping array. BMC Genom 13:203

APG III (2009) An update of the Angiosperm Phylogeny Group classification for the orders and families of flowering plants: APG III. Bot J Linn Soc 161:105–121

Aras A, Akkemik U, Kaya Z (2007) *Hippophae rhamnoides* L.: fruit and seed morphology and its taxonomic problems in Turkey. Pak J Bot 39:1907–1916

Backlund A, Bremer K (1998) To be or not to be – principles of classification and monotypic plant families. Taxon 47:391–400

Barry CS, Giovannoni JJ (2006) Ripening in the tomato green-ripe mutant is inhibited by ectopic expression of a protein that disrupts ethylene signaling. Proc Natl Acad Sci U.S.A. 103:923–7928

Bartish GI, Jeppsson N, Bartish IV, Nybom H (2000b) Assessment of genetic diversity using RAPD analysis in a germplasm collection of sea buckthorn. Agric Food Sci Finland 9:279–289
Bartish IV, Jeppsson N, Nybom H (1999) Population genetic structure in the dioecious pioneer plant species *Hippophae rhamnoides* investigated by random amplified polymorphic DNA (RAPD) markers. Mol Ecol 8:791–802
Bartish IV, Jeppsson N, Bartish GI, Lu R, Nybom H (2000a) Inter- and intraspecific genetic variation in *Hippophae* (Elaeagnaceae) investigated by RAPD markers. Plant Syst Evol 225:85–101
Bartish IV, Jeppsson N, Nybom H, Swenson U (2002) Phylogeny of *Hippophae* (Elaeagnaceae) inferred from parsimony analysis of chloroplast DNA and morphology. Syst Bot 27:41–54
Bartish IV, Kadereit JW, Comes P (2006) Late Quaternary history of *Hippophae rhamnoides* L. (Elaeagnaceae) inferred from chalcone synthase intron (Chsi) sequences and chloroplast DNA variation. Mol Ecol 15:4065–4083
Biltekin D (2010) Vegetation and climate of North Anatolian and North Aegean region since 7 Ma according to pollen analysis. Ph.D. thesis, Université Claude Bernard-Lyon I and Istanbul Technical University. http://tel.archives-ouvertes.fr/docs/00/72/08/92/PDF/TH2010_Biltekin_Demet.pdf
Birky CW, Maruyama T, Fuerst P (1983) An approach to population and evolutionary genetic theory for genes in mitochondria and chloroplasts, and some results. Genetics 103:513–527
Bobrov EG (1962) Review on genus *Myricaria* Desv. and its history. Botanicheskiy Zhurnal 52:924–936 (in Russian with English abstract)
Chen G, Wang Y, Zhao C, Korpelainen H, Li C (2008) Genetic diversity of *Hippophae rhamnoides* populations at varying altitudes in the Wolong natural reserve of China as revealed by ISSR markers. Silvae Genet 57:29–36
Chen W, Su X, Zhang H, Sun K, Ma R, Chen X (2010) High genetic differentiation of *Hippophae rhamnoides* ssp. *yunnanensis* (Elaeagnaceae), a plant endemic to the Qinghai-Tibet plateau. Biochem Genet 48:565–576
Cheng K, Sun K, Wen HY, Jia DR, Liu JQ (2009) Maternal divergence and phylogeographical relationships between *Hippophae gyantsensis* and *H. rhamnoides* subsp. *yunnanensis*. Zhiwu Shengtai Xuebao 33:1–11 (in Chinese with English abstract)
Chowdhury MA, Jana S, Schroeder WR (2000) Phenotypic diversity in four woody species on the Canadian prairies. Can J Plant Sci 80:137–142
Collard BCY, Jahufer MZZ, Brouwer JB, Pang ECK (2005) An introduction to markers, quantitative trait loci (QTL) mapping and marker-assisted selection for crop improvement: the basic concepts. Euphytica 142:169–196
Congiu L, Chicca M, Cella R, Rossi R, Bernacchia G (2000) The use of random amplified polymorphic DNA (RAPD) markers to identify strawberry varieties: a forensic application. Mol Ecol 9:229–232
Dardick C, Callahan A, Horn R, Ruiz KB, Zhebentyayeva T, Hollender C, Whitaker M, Abbott A, Scorza R (2013) PpeTAC1 promotes the horizontal growth of branches in peach trees and is a member of a functionally conserved gene family found in diverse plants species. Plant J 75:618–630
Ercisli S, Orhan E, Yildirim N, Agar G (2008) Comparasion of sea buckthorn genotypes (*Hippophaë rhamnoides* L.) based on RAPD and FAME data. Turkish J Agric For 32:363–368
Fatima T, Snyder CL, Schroeder WR, Cram D, Datla R, Wishart D, Weselake RJ, Krishna P (2012) Fatty acid composition of developing sea buckthorn (*Hippophae rhamnoides* L.) berry and the transcriptome of the mature seed. PLoS ONE 7:e34099
Fernandez-Fernandez F, Antanaviciute L, van Dyk MM, Tobutt KR, Evans KM, Rees DJG, Dunwell JM, Sargent DJ (2012) A genetic linkage map of an apple rootstock progeny anchored to the *Malus* genome sequence. Tree Genet Genomes 8:991–1002
Gams H (1943) Der Sanddorn (*Hippophae rhamnoides* L.) im Alpengebiet. Beihefte zum Botanischen Centralblatt, Abteilung B 2:68–96

Ghangal R, Raghuvanshi S, Sharma PC (2012) Expressed sequence tag based identification and expression analysis of some cold inducible elements in sea buckthorn (*Hippophae rhamnoides* L.). Plant Physiol Biochem 51:123–128

Ghangal R, Chaudhary S, Jain M, Purty RS, Sharma PC (2013) Optimization of *de novo* short read assembly of sea buckthorn (*Hippophae rhamnoides* L.) transcriptome. PLoS ONE 8:7

Glick BR, Pasternak J, Patten CL (2010) Molecular Biotechnology: Principles and Applications of Recombinant DNA. ASM Press, Washington, D.C.

Gupta PK, Varshney RK (2000) The development and use of microsatellite markers for genetic analysis and plant breeding with emphasis on bread wheat. Euphytica 113:163–185

Gupta SM, Ahmed Z, Kumar N (2009) Isolation of cDNA fragment of glycerol-3-phosphate acyltransferase gene from sea buckthorn. Def Sci J 59:147–151

Gusberti M, Gessler C, Broggini GAL (2013) RNA-seq analysis reveals candidate genes for ontogenic resistance in Malus-Venturia pathosystem. PLoS ONE 8:e78457

Jain A, Ghangal R, Grover A, Raghuvanshi S, Sharma PC (2010) Development of EST-based new SSR markers in sea buckthorn. Physiol Mol Biol Plants 16:375–378

Jain A, Chaudhary S, Sharma PC (2014) Mining of microsatellites using next generation sequencing of sea buckthorn (*Hippophae rhamnoides* L.) transcriptome. Physiol Mol Biol Plants 20:115–123

Jia DR, Liu TL, Wang LY, Zhou DW, Liu JQ (2011) Evolutionary history of an alpine shrub *Hippophae tibetana* (Elaeagnaceae): allopatric divergence and regional expansion. Biol J Linn Soc 102:37–50

Jia DR, Abbott RJ, Liu TL, Mao KS, Bartish IV, Liu JQ (2012) Out of the Qinghai-Tibet Plateau: evidence for the origin and dispersal of Eurasian temperate plants from a phylogeographic study of *Hippophaë rhamnoides* (Elaeagnaceae). New Phytol 194:1123–1133

Jia DR (2013) Influence of climatic fluctuations in Neogene on evolution of ecologically diverse plant genus: an example of *Hippophae* L. (Elaeagnaceae). Ph.D. thesis, Charles University in Prague, Czech Republic. https://is.cuni.cz/webapps/zzp/detail/99009/?lang=en

Jeppsson N, Bartish IV, Persson HA (1999) DNA analysis as a tool in sea buckthorn breeding. In: Janick J (ed) Perspectives on new crops and new uses. ASHS Press, Alexandria, VA, pp 338–341

Ho SYW, Phillips MJ (2009) Accounting for calibration uncertainty in phylogenetic estimation of evolutionary divergence times. Syst Biol 58:367–380

Hokanson SC, Szewc-McFadden AK, Lamboy WF, McFerson JR (1998) Microsatellite (SSR) markers reveal genetic identities, genetic diversity and relationships in a Malus x domestica Borkh. core subset collection. Theor Appl Genet 97:671–683

Huang Q (1995) A review of sea buckthorn breeding in China. In: Proceedings of international workshop on Seabuckthorn. Beijing, China, pp 111–117

Hyvönen J (1996) On phylogeny of *Hippophae* (Elaeagnaceae). Nordic J Bot 16:51–62

Kalia RK, Singh R, Rai MK, Mishra GP, Singh SR, Dhawan AK (2011) Biotechnological interventions in sea buckthorn (*Hippophae* L.): current status and future prospects. Trees Struct Funct 25:559–575

Kalinina IP, Panteleyeva YI (1987) Breeding of sea buckthorn in the Altai. In: Advances in agricultural science. Moscow, Russia (in Russian)

Kanayama Y, Kato K, Stobdan T, Galitsyn GG, Kochetov AV, Kanahama K (2012) Research progress on the medicinal and nutritional properties of sea buckthorn (*Hippophae rhamnoides*) —a review. J Hortic Sci Biotech 87:203–210

Korekar G, Sharma RK, Kumar R, Meenu RK, Bisht NC, Srivastava RB, Ahuja PS, Stobdan T (2012) Identification and validation of sex-linked SCAR markers in dioecious *Hippophae rhamnoides* L. (Elaeagnaceae). Biotechnol Lett 34:973–978

Li G, Quiros CF (2001) Sequence-related amplified polymorphism (SRAP), a new marker system based on a simple PCR reaction: its application to mapping and gene tagging in *Brassica*. Theor Appl Genet 103:455–461

Li H, Ruan CJ, Teixeira da Silva JA, Liu BQ (2010) Associations of SRAP markers with dried-shrink disease resistance in a germplasm collection of sea buckthorn (*Hippophae* L.). Genome 53:447–457

Li TSC, Schroeder WR (1996) Sea buckthorn (*Hippophae rhamnoides* L.): a multipurpose plant. HortTechnology 6:370–380

Li TSC (2003) See buckthorn (*Hippophae rhamnoides* L.): production and utilisation. National Research Council of Canada, Ottawa

Lian YS, Chen XL (1993) Study on the germplasm resource of the genus *Hippophae* L. In: International symposium on sea buckthorn (*Hippophae rhamnoides* L.). Novosibirsk, Russia, pp 157–161

Lian YS, Chen XL, Sun K (1995) New discoveries of the genus *Hippophae* L. In: Proceedings of international workshop on seabuckthorn. China Science and Technology Press, Beijing, pp 60–66

Lian YS, Chen XL, Lian H (1998) Systematic classification of the genus *Hippophae* L. Seabuckthorn Res 1:13–23

Lian YS, Chen XL, Sun K, Ma R (2003a) A new subspecies of *Hippophae* (Elaeagnaceae) from China. Novon 13:200–202

Lian YS, Chen XL, Sun K, Ma R (2003b) Clarification of the systematic position of *Hippophae goniocarpa* (Elaeagnaceae). Bot J Linn Soc 142:425–430

Liu R, Yang J, Gao L (2007) ISSR analysis of Chinese sea buckthorn and Russian sea buckthorn. Acta Bot Boreal Occident Sin 27:671–677

Mardis ER (2008) Next-generation DNA sequencing methods. Annu Rev Genomics Hum Genet 9:387–402

Meng LH, Yang HL, Wu GL, Wang YJ (2008) Phylogeography of *Hippophae neurocarpa* (Elaeagnaceae) inferred from the chloroplast DNA trnL-F sequence variation. J Syst Evol 46:32–40

Morgante M, Olivieri AM (1993) PCR-amplified microsatellite markers in plant genetics. Plant J 3:393–427

Nybom H, Bartish IV, Garkava-Gustavsson L, Persson H, Werlemark G, Esselink D (2003) Evaluating genetic resources in minor fruits. In: Janick J (ed) Genetics and breeding of tree fruits and nuts. International Society of Horticultural Science, Leuven, pp 81–94

Persson HA, Nybom H (1998) Genetic sex determination and RAPD marker segregation in the dioecious species sea buckthorn (*Hippophae rhamnoides* L.). Hereditas 129:45–51

Petit RJ, Aguinagalde I, de Beaulieu JL, Bittkau C, Brewer S, Cheddadi R, Ennos R, Fineschi S, Grivet D, Lascoux M, Mohanty A, Muller-Starck GM, Demesure-Musch B, Palme A, Martin JP, Rendell S, Vendramin GG (2003) Glacial refugia: hotspots but not melting pots of genetic diversity. Science 300:1563–1565

Raina SN, Jain S, Sehgal D, Kumar A, Dar TH, Bhat V, Pandey V, Vaishnavi S, Bhargav A, Singh V, Rani V, Tandon R, Tewari M, Mahmoudi A (2012) Diversity and relationships of multipurpose sea buckthorn (*Hippophae* L.) germplasm from the Indian Himalayas as assessed by AFLP and SAMPL markers. Genet Resour Crop Evol 59:1033–1053

Rousi A (1965) Observations on the cytology and variation of European and Asiatic populations of *Hippophaë rhamnoides*. Ann Bot Fenn 2:1–18

Rousi A (1971) The genus *Hippophaë* L: a taxonomic study. Ann Bot Fenn 8:177–227

Ruan CJ, Qin P, Zheng JW, He ZX (2004) Genetic relationships among some cultivars of sea buckthorn from China, Russia and Mongolia based on RAPD analysis. Sci Hortic 101:417–426

Ruan CJ, Li DQ (2005) AFLP fingerprinting analysis of some cultivated varieties of sea buckthorn (*Hippophae rhamnoides*). J Genet 84:311–316

Ruan CJ (2006) Genetic relationships among sea buckthorn varieties from China, Russia and Mongolia using AFLP markers. J Hortic Sci Biotechnol 81:409–414

Ruan CJ, Li H, Mopper S (2009) Characterization and identification of ISSR markers associated with resistance to dried-shrink disease in sea buckthorn. Mol Breeding 24:255–268

Ruan C-J, Teixeira da Silva JA, Li Q, Li H, Zhang J (2010) Pathogenicity of dried-shrink disease and evaluation of resistance in a germplasm collection of sea buckthorn (*Hippophae* L.) from China and other countries. Sci Hortic 127:70–78

Ruan CJ, Rumpunen K, Nybom H (2013) Advances in improvement of quality and resistance in a multipurpose crop: sea buckthorn. Crit Rev Biotechnol 33:126–144

Sauquet H, Ho SY, Gandolfo MA, Jordan GJ, Wilf P, Cantrill DJ, Bayly MJ, Bromham L, Brown GK, Carpenter RJ (2012) Testing the impact of calibration on molecular divergence times using a fossil-rich group: the case of *Nothofagus* (Fagales). Syst Biol 61:289–313

Servettaz C (1908) Monographie der Elaeagnaceae. Beihefte zum Botanischen Centralblatt 25:1–420

Shah AH, Ahmad SD, Khaliq I, Batool F, Hassan L, Pearce RS (2009) Evaluation of phylogenetic relationship among sea buckthorn (*Hippophae rhamnoides* L. ssp. *turkestanica*) wild ecotypes from Pakistan using amplified fragment length polymorphism (AFLP). Pak J Bot 41:2419–2426

Sharma A, Zinta G, Rana S, Shirko P (2010) Molecular identification of sex in *Hippophae rhamnoides* L. using isozyme and RAPD markers. For Stud China 12:62–66

Sheng H, An L, Chen T, Xu S, Liu G, Zheng X, Pu L, Liu Y, Lian Y (2006) Analysis of the genetic diversity and relationships among and within species of *Hippophae* (Elaeagnaceae) based on RAPD markers. Plant Syst Evol 260:25–37

Simon-Gruita A, Tataru E, Constantin N, Duta Cornescu G, Pavlusenco Camelia E, Rati V, Rati L, Stoian V (2012) The assessment of the genetic diversity of sea buckthorn populations from Romania using RAPD markers. Rom Biotechnol Lett 17:7749–7756

Singh R, Ahmed Z (2010) Sea buckthorn: a multipurpose medicinal plant. In: Singh VK, Govil JN (eds) Drug plants IV. Studium Press Ltd, Houston, pp 227–239

Sorsa P (1971) Pollen morphological study of the genus *Hippophaë* L., including new taxa recognized by A Rousi. Ann Bot Fenn 8:228–236

Srihari JM, Verma B, Kumar N, Chahota RK, Singh V, Rathour R, Singh SK, Sharma SK, Sharma TR (2013) Analysis of molecular genetic diversity and population structure in sea buckthorn (*Hippophae* spp L.) from north-western Himalayan region of India. J Med Plants Res 7:3183–3196

Sun K, Chen X, Ma R, Li C, Wang Q, Ge S (2002) Molecular phylogenetics of *Hippophae* L. (Elaeagnaceae) based on the internal transcribed spacer (ITS) sequences of nrDNA. Plant Syst Evol 235:121–134

Sun K, Ma R, Chen X, Li C, Ge S (2003) Hybrid origin of the diploid species *Hippophae goniocarpa* evidenced by the internal transcribed spacers (ITS) of nuclear rDNA. Belg J Bot 136:91–96

Sun K, Chen W, Ma R, Chen X, Li A, Ge S (2006) Genetic variation in *Hippophae rhamnoides* ssp. *sinensis* (Elaeagnaceae) revealed by RAPD markers. Biochem Genet 44:186–197

Suryakumar G, Gupta A (2011) Medicinal and therapeutic potential of sea buckthorn (*Hippophae rhamnoides* L.). J Ethnopharmacol 138:268–278

Swenson U, Bartish IV (2002) Taxonomic synopsis of *Hippophae* (Elaeagnaceae). Nordic J Bot 22:369–374

Tanksley SD (1993) Mapping polygenes. Annu Rev Genet 27:205–233

Tian CJ, Lei YD, Shi H, Nan P, Chen JK, Zhong Y (2004a) Genetic diversity of sea buckthorn (*Hippophae rhamnoides*) populations in northeastern and northwestern China as revealed by ISSR markers. New Forest 27:229–237

Tian CJ, Nan P, Shi SH, Chen JK, Zhong Y (2004b) Molecular genetic variation in Chinese populations of three subspecies of *Hippophae rhamnoides*. Biochem Genet 42:259–267

Trajkovski V, Jeppsson N (1999) Domestication of sea buckthorn. Bot Lithuanica suppl. 2:37–46

Tsvelev NN (2002) On the genera *Elaeagnus* and *Hippophae* (Elaeagnaceae) in Russia and adjacent states. Botanicheskiy Zhurnal 87:74–86 (in Russian with English abstract)

Tzedakis PC, Emerson BC, Hewitt GM (2013) Cryptic or mystic? Glacial tree refugia in northern Europe. Trends Ecol Evol 28:696–704

Virk PS, Ford-Lloyd BV, Jackson MT, Pooni HS, Clemeno TP, Newbury HJ (1996) Predicting quantitative variation within rice germplasm using molecular markers. Heredity 76:296–304

Wang AL, Schluetz F, Liu JQ (2008a) Molecular evidence for double maternal origins of the diploid hybrid *Hippophae goniocarpa* (Elaeagnaceae). Bot J Linn Soc 156:111–118

Wang AL, Zhang Q, Wan DS, Yang YZ, Liu JQ (2008b) Nine microsatellite DNA primers for *Hippophae rhamnoides* ssp *sinensis* (Elaeagnaceae). Conserv Genet 9:969–971

Wang B, Lin L, Ni Q, Su CL (2011a) *Hippophae rhamnoides* Linn. for treatment of diabetes mellitus: a review. J Med Plants Res 5:2599–2607

Wang H, Moore MJ, Soltis PS, Bell CD, Brockington SF, Alexandre R, Davis CC, Latvis M, Manchester SR, Soltis DE (2009) Rosid radiation and the rapid rise of angiosperm-dominated forests. Proc Natl Acad Sci U.S.A. 106:3853–3858

Wang H, Qiong LA, Sun K, Lu F, Wang Y, Song Z, Wu Q, Chen J, Zhang W (2010) Phylogeographic structure of *Hippophae tibetana* (Elaeagnaceae) highlights the highest microrefugia and the rapid uplift of the Qinghai-Tibetan Plateau. Mol Ecol 19:2964–2979

Wang Y, Jiang H, Peng S, Korpelainen H (2011b) Genetic structure in fragmented populations of *Hippophae rhamnoides* ssp. *sinensis* in China investigated by ISSR and cpSSR markers. Plant Syst Evol 295:97–107

Warnock M, Miskin D (2009) Sea buckthorn (*Hippophae rhamnoides L*): a review and its potential as a crop in Scotland. In: Singh VK, Govil JN (eds) Standardization of herbal/ayurvedic formations. Studium Press Llc, Houston, pp 257–272

Weising K, Nybom H, Wolff K, Meyer W (1994) DNA Fingerprinting in Plants and Fungi. CRC Press, Boca Raton

Wolfe KH, Li WH, Sharp PM (1987) Rates of nucleotide substitution vary greatly among plant mitochondrial, chloroplast, and nuclear DNAs. Proc Natl Acad Sci U.S.A. 84:9054–9058

Xiao J, Li J, Grandillo S, Ahn SN, Yuan L, Tanksley SD, McCouch SR (1998) Identification of trait-improving Quantitative Trait Loci alleles from a wild rice relative, *Oryza rufipogon*. Genetics 150:899–909

Xu G, Li CY, Yao YN (2009) Proteomics analysis of drought stress-responsive proteins in *Hippophae rhamnoides* L. Plant Mol Biol Report 27:153–161

Yao YM, Tigerstedt PMA (1993) Isozyme studies of genetic diversity and evolution in *Hippophae*. Genet Resour Crop Evol 40:153–164

Yao YM, Tigerstedt PMA (1995) Geographical variation of growth rhytm, height, and hardiness, and their relations in *Hippophaë rhamnoides*. J Am Soc Hortic Sci 120:691–698

Zhang SD, Soltis DE, Yang Y, Li DZ, Yi TS (2011) Multi-gene analysis provides a well-supported phylogeny of Rosales. Mol Phylogenet Evol 60:21–28

Zhao C, Chen G, Wang Y, Korpelainen H, Li C (2007) Genetic variation of Hippophae rhamnoides populations at different altitudes in the Wolong Nature Reserve based on RAPDs. Chin J Appl Environ Biol 13:753–758

Zubarev YA, Gunin A, Oderova EV (2014) Characteristics of Russian sea bukthorn (*Hippophaë rhamnoides* subsp. *mongolica*) varieties. In: Singh V (ed) Seabuckthorn (*Hippophaë* L.) a multipurpose wonder plant, vol IV: emerging trends in research and technologies. Daya Publishing House, Astral International Pvt. Ltd., New Dehli, pp 89–98

Chapter 15
Unlocking the Potential of Genetic Resources for Improvement of Sesame (*Sesamum indicum* L.): The Current Scenario

Vibhuti Singh, Sachin Kumar, Amrita Singh, Niti Pathak Bhaduri, Kangila Venkataramana Bhat and Suman Lakhanpaul

Abstract Sesame (*Sesamum indicum* L.) economically valued worldwide for its seeds and seed oil has been designated as 'queen of oilseeds'. Antioxidants such as lignans and their derivatives prevent oxidation of the oil and provide longer shelf life making sesame oil one of the most stable oils. Due to the presence of several bioactive compounds it has been often listed among the world's healthiest foods. However, the attempts to improve sesame crop remain scanty resulting in lack of superior genotypes having high yield potential and resistance to biotic and abiotic stresses. Further, traits such as indeterminate growth habit and capsule shattering are also responsible for its reduced yields making it less favorable for large-scale farming. Wild relatives of sesame are important reservoir of useful genes and need to be exploited for sesame improvement. These wild species exhibit crossability with the cultivated gene pool to varying extents and can be utilized for transferring the desirable traits using conventional breeding approaches assisted with modern techniques. Extensive research efforts are therefore desirable in several aspects such as identification of different gene pools of sesame genetic resources, phylogenetic relationships, assessment of genetic diversity in the cultivated gene pool, etc. Molecular approaches to develop genetic map, hybrid testing, identification of core collections, DNA fingerprinting are already underway. Biotechnological interventions required for the successful production of transgenic plants have also been initiated. Recently, most of the genes and biosynthetic pathways involved in oil and other useful components of sesame seeds have been unraveled. An integrated approach based on conventional and modern tools for identification and utilization of useful genes followed by their successful incorporation in the cultivated gene

V. Singh · S. Kumar · A. Singh · N.P. Bhaduri · S. Lakhanpaul (✉)
Department of Botany, University of Delhi, Delhi 110007, India
e-mail: sumanlp2001@yahoo.com

K.V. Bhat
National Bureau of Plant Genetic Resources, Indian Agricultural Research Institute, Pusa Campus, New Delhi 110012, India
e-mail: kvbhat2001@yahoo.com

V.R. Rajpal et al. (eds.), *Gene Pool Diversity and Crop Improvement*, Sustainable Development and Biodiversity 10,
DOI 10.1007/978-3-319-27096-8_15

pool is desirable for large-scale cultivation of sesame, to meet the increasing demands of healthy food crops due to the ever increasing heath awareness world over.

Keywords Sesame · Germplasm · Wild relatives · Crop improvement · Genomic resources

15.1 Sesame—Introduction

Sesamum indicum L. (Family Pedaliaceae) commonly known as sesame is an important oilseed crop. It is referred as 'queen of oilseeds' due to its regard by the users and owing to its oil quality (Bedigian and Harlan 1986). It is one of the most ancient crops in the world known to mankind, with archeological evidences dating back to 2250 and 1750 BC at Harappa in the Indus valley (Najeeb et al. 2012). Ironically, it is considered as an 'orphan crop' due to meager research efforts attributed to the fact that it is not a mandate crop for any international crop research institute (Bhat et al. 1999).

Sesame is mainly a crop of warmer areas including Asia and Africa (Ashri 1988), but newer cultivars have extended their range to temperate regions. It is an excellent rotation crop of cotton, maize, groundnut, wheat, and sorghum. It reduces nematode populations that attack cotton and groundnut (Elbadri and Yassin 2010). Its deep and extensive root system makes it an excellent soil builder. It also improves soil texture, retains moisture and reduces soil erosion. The left over composted sesame leaves also help in moisture retention of the soil making favorable conditions for planting the next crop. Sesame is resistant to drought, it is a low cost crop and is therefore known as one of the best alternative specialty crops.

15.1.1 Economic Importance

Sesame is cultivated on a worldwide basis for its seeds, oil and protein. Seeds are used either as a whole or decorticated in sweets or confectioneries, in bakery products and milled to get oily paste (tahini) (Abou-Gharbia et al. 1997). Sesame seeds also find place in several rituals in countries such as India, Nepal etc. Interestingly, sesame seed is a reservoir of nutritional components with numerous beneficial effects and promotes human health. The bioactive components present in the seed include vital minerals, vitamins, phytosterols, polyunsaturated fatty acids, tocopherols and a unique class of antioxidant compounds, lignans that impart its antioxidative potential. The seed meal is also high in protein (35–50 %) and contains significant amount of amino acids such as tryptophan and methionine.

Sesame oil is used in cooking, in preparation of salads and finds use in the production of margarine, soaps, pharmaceuticals, paints and lubricants. The seed oil content ranges from 32.5 to 58.8 %, which is generally greater in white than black seeds (Moazzami and Kamal-Eldin 2006). In majority, sesame oil consists of triacylglycerols (95 %), diacylglycerols (2.6 %), and unsaponifiables (2.3 %) (Kamal-Eldin 2010). A high proportion of polyunsaturated fats in sesame oil mainly consist of linoleic (35–50 %), oleic (35–50 %), stearic (3.5–16 %) and palmitic (7–12 %) acids (Kamal-Eldin 2010). Linolenic acid is also present but only up to 1 % of the total fatty acids. Sesame oil used in combination with soybean oil enhances the nutritive value of the lipids and increases vitamin E activity (Namiki 1995) and in combination with canola or mustard oil has been recommended for healthy diets (Ghafoorunissa 1996). Sesame oil has several health benefits like growth arresting and apoptosis prevention properties in cancer cells (Yokota et al. 2007); useful in the treatment of several chronic diseases, including hepatitis, diabetes and migraines (Kita et al. 1995); alleviating depression and fatigue; and has antibacterial and anitviral properties for common skin pathogens. In addition, sesame oil maintains good cholesterol (HDL) and lowers bad cholesterol (LDL) (Sugano et al. 1990) and acts as a UV-protectant (Elleuch et al. 2007).

The beneficial properties of sesame oil are owing to three major constituents namely lignans, tocopherols and phytosterols. Lignans are a group of phenylpropanoid compounds, known for their innate nonenzymatic antioxidant defense mechanism against reactive oxygen species. These important plant phenolics are characterized by the coupling of two phenylpropanoid (C6–C3) units by a bond between β-positions in the propane side chains. Unlike other vegetable oils, where unsaponifiable fraction consists of phytosterols as a major component, sesame oil is an exception which is dominated by the presence of lignans. Two major groups of lignans exist in sesame seeds, namely oil soluble lignans (such as sesamin, sesamolin, pinoresinol) and glycosylated water soluble lignans (such as sesaminol triglucoside, pinoresinol triglucoside) (Katsuzaki et al. 1994; Moazzami and Kamal-Eldin 2006). In addition, sesamol, a free 3, 4-methylenediphenoxy phenol is usually present in traces (Salunkhe et al. 1991).

Sesame seeds have been found to possess high amounts of γ-tocopherol ranging from 468.5 to 517.9 mg/kg lipid along with low quantities of α and δ-tocopherols (Yoshida et al. 2007; Williamson et al. 2008). However, the overall level of tocopherol in sesame is low in comparison to other vegetable oils (Kamal-Eldin 2010). Further, in the unsaponifiable fraction of sesame oil, phytosterols form the second component after lignans in terms of quantity. Phytosterols are triterpenes structurally similar to cholesterol. Predominant phytosterol present in sesame oil is β-sitosterol (231.7–305.2 mg/100 g sesame seed), followed by campesterol and stigmasterol (Shahidi and Tan 2010). In comparison to other seeds and nuts, sesame seeds along with wheat germ have been reported to contain the highest (400–413 mg 100 g^{-1}) phytosterol content (Phillips et al. 2005).

Other nonfood applications of sesame are use of sesame flowers to prepare perfume in Africa (Morris 2002), use of sesamin as a synergist for pyrethrum or rotenone insecticides (Haller et al. 1942) and use of chlorosesamone extracted from roots of sesame as a fungicide (Hasan et al. 2001). In addition, investigations on use of sesame oil as a source for biodiesel have resulted in a product with fuel properties in parity with mineral diesel but with superior environmental performance (Ahmad et al. 2010).

15.1.2 Sesame Production

The largest producers of sesame in the world are from Asiatic region: India, China, and Myanmar. India ranks first in the world in terms of sesame-growing area (24 %) (Raikwar and Srivastava 2013) however, area harvested and production has shown a significant decrease in past years (Fig. 15.1). The trend is similar in case of sesame production worldwide (Fig. 15.1). Average productivity of sesame has lowered ranging from 144 to 234 kg/ha compared to past 20 years (Raikwar and Srivastava 2013) which has led to a gap in the demand and the supply. In this constrained scenario, increasing the production of sesame by overcoming the hurdles that limit the crop yield are the need of the hour.

15.1.3 Taxonomy and Distribution

Sesame is a member of family Pedaliaceae (Order Lamiales), which is phylogenetically close to families, such as Acanthaceae and Scrophulariaceae (Olmstead et al. 1993). Family Pedaliaceae consists of 65 species belonging to 15 genera distributed largely on sea shores and desert areas of three regions viz. tropical

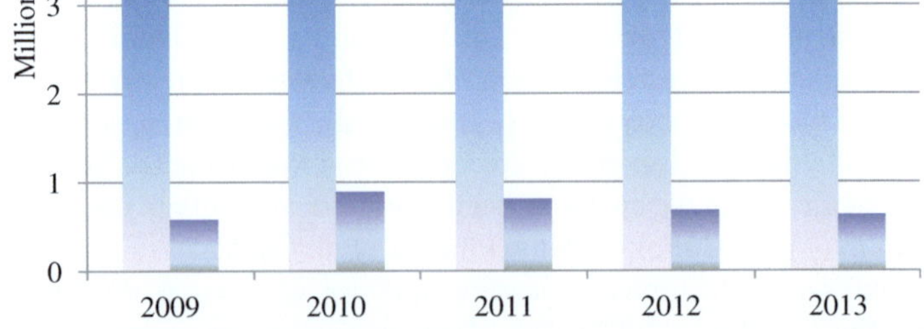

Fig. 15.1 Status of sesame seed production during past five years (FAOSTAT 2015)

Africa, Indian subcontinent (including Sri Lanka) and Far East (Duhoon et al. 2002). The genus *Sesamum* has about 23 species (Bedigian 2004a), of which eight species occur in India. *S. indicum* has its origin in India (Bedigian 2003; Bhat et al. 1999) and is the only species recognized in cultivated form in the family comprising diverse array of cultivars.

With respect to plant habit *S. indicum* is an erect, pubescent, or puberulous herb with usually alternate oblong upper leaves; lower opposite and lobed leaves. Flowers are axillary, pure white or usually with purple, yellow marks; solitary or few, and fascicled on a short-pedicel. Two extrafloral nectaries yellow in color, are usually present at the base of the flower pedicel. Calyx is small, 5-partite. Corolla is tubular with 2-lipped limb and five-rounded lobes. Stamens are four in number, didynamous, included, epipetalous; anthers are sagittate with two subparallel cells. Ovary is 2-celled, early falsely 4-celled bearing many ovules with axile placentation. A nectar disk is present at the base of ovary. Capsule is erect, oblong, loculicidally 2-valved (splitting occurs in the middle of the locule); unarmed and appears to be 4-celled. Seeds are numerous, smooth and obliquely oblong (Hooker 1885). Sesame is normally self-pollinated, although insect cross-pollination has also been reported (Mahfouz et al. 2012).

15.1.4 Wild Relatives of Sesame

Wild species of sesame vary in their habitat, morphological features and ploidy level, latter of which is represented by three chromosome groups 26, 32, 64 (Joshi 1961). Wild relatives have also been delineated into gene pools based on available hybridization studies for limited number of taxa. As per Harlan (1992) the gene pool-1 consists of the crop plant *S. indicum* and its progenitor *S. indicum* subsp. *malabarcium*. Two species namely, *S. alatum* and *S. prostratum* have been placed under gene pool-2 due to barriers in hybridization with *S. indicum* (Raghavan and Krishnamurty 1947; Ramanathan 1950; Tarihal 2003; Rajeswari and Ramaswamy 2004). Although, in few reports no seed set has been observed for *S. alatum* during hybridization studies (Lee et al. 1991). *S. radiatum* is placed in gene pool-3 due to lack of capsule formation (Ram et al. 2006), no seed set (Lopez and Mazzani 1964; Prabhakaran et al. 1992) and use of embryo rescue methods (Dasharath et al. 2007). A recent phylogenetic reconstruction based on two chloroplast loci also supports this categorization as *S. indicum* and *S. indicum* subsp. *malabaricum* were seen to have a closer affinity. *S. prostratum* appeared to be a recent divergence in comparison to *S. alatum* (Gormley et al. 2015).

Interestingly, the wild species of sesame are known to exhibit tolerance and resistance to different pests and diseases and some abiotic stresses (Fig. 15.2). They can thrive on nutritionally poor soils and sites with low moisture availability. They can tolerate heat and drought and have been referred as weedy due to their negligible demanding nature evident by their successful colonization in swarms (Bedigian 2010). Some of the adaptive features in wild species include fleshy roots,

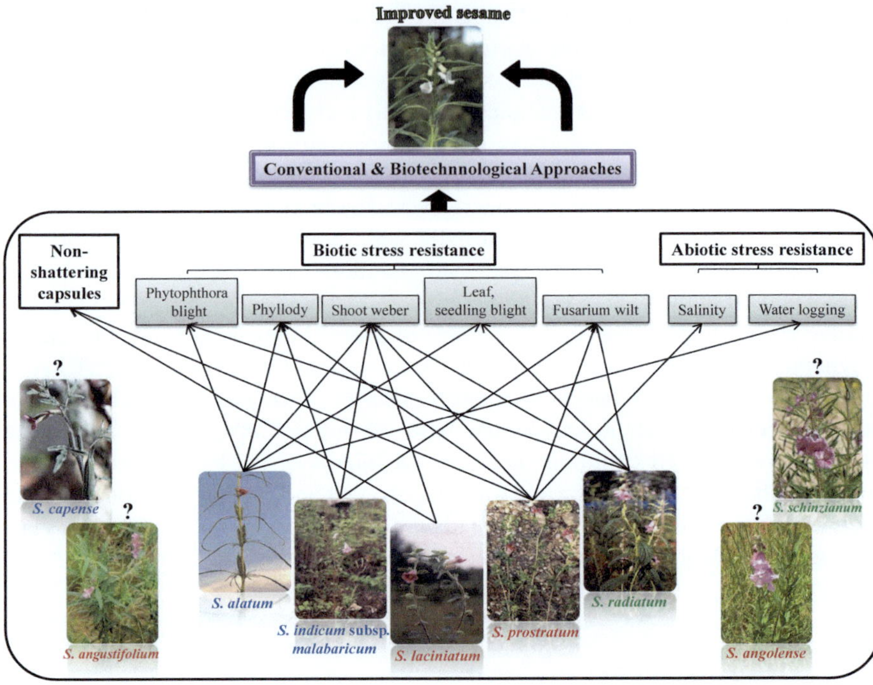

Fig. 15.2 Summary of agronomically useful attributes present in wild relatives of sesame. Question mark (?) indicates major wild species still not screened for useful traits. Font color for species name refers to chromosome number: *Blue* n = 16, *Red* n = 32, *Green* n = 64

small linear leaves, large number of stomata on the adaxial surface, hairiness and increased fruit set in the dry season (Nimmakayala et al. 2011). Seeds of wild relatives show the property of dormancy due to the presence of chemical inhibitors (Bedigian 1984). Germination is brought about by washing away of the inhibitors by natural rainfall in turn protecting the species from unfavorable conditions. A brief description of major wild relatives of sesame is given below. Morphological description is as per ‘plants.jstor.org’ unless mentioned.

Sesamum indicum subsp. *malabaricum* (Burm.) Bedigian: Recently, a subspecies status for erstwhile taxon *S. malabaricum* has been proposed by Bedigian (2014) which is known as the progenitor of sesame. Plant is an erect annual herb, with unpleasant odor. Stem often divaricately branched, pubescent to glabrescent. Leaves heteromorphic, lower palmately 3-foliolate, tripartite, margin coarsely dentate to serrate, acute apex, obtuse, or ovate base, upper leaves oblong to lanceolate, entire with acute base and apex. Lower surface is densely glandular. Flower pink or white, usually with intense deep purple pigmentation at lower lip. Capsule long, quadrangular, round at base, acuminate into beak at apex. Seeds broadly ovoid, conspicuously reticulate and rugose (Bedigian 2014).

S. alatum Thonn.: An erect annual herb with simple branched, glaberescent stem. Leaves heteromorphic, lower leaves palmately 3–5 foliate or partite, lobes lanceolate or linearly lanceolate, central lobe longest, cuneate at base, acute or rounded at apex, margins entire or undulate; upper leaves simple, linear-lanceolate. Flowers pink or purple sometimes spotted red within. Capsule narrowly obconical, long beaked, broader in upper part and gradually narrowed at base, slightly compressed. Seeds obliquely overlapping, foveolate with suborbicular long wing at base and apex. It is widely distributed in tropical Africa as a flourishing monoculture plant and is an introduced weed in India. Leaves, flowers, and young shoots of the species are used as a vegetable when cooked (Bedigian 2010). Seed oil is an aphrodisiac, cures diarrhoea and intestinal disorders.

S. capense Burm.f.: An erect, tall conspicuous, sparsely branched herb. Stem is angular and sulcate or subterete. Leaves digitately 3–5 foliolate, leaflets obovate–oblong to linear–lanceolate, subobtuse–obtuse, narrowed at base into petiolule, intermediate the longest, glaucous above. Corolla violet outside, violet–purple inside, obliquely campanulate. Capsule broad, strongly three-nerved, short horns at base, beak long acuminate. Seed faces muriculate–foveolate with wing running all around. It is widely distributed throughout Southern Africa. Seeds of *S. capense* are edible and the plant is used in traditional medicine.

S. latifolium Gillett: An erect herb with densely pubescent quadrangular stem. Leaves generally heteromorphic, lower ones long petioled, large ovate-cordate or 3-lobed, upper leaves smaller ovate-lanceolate, inconspicuous to coarsely serrate, acuminate at apex, densely pubescent below. Flower pale pink or pinkish mauve, bracteoles conspicuous. Capsule shortly pubescent, oblong-quadrangular abruptly narrowed into a short beak. Seed faces foveolate, sides deeply pitted, margins sharply angled. The species has been found growing on granitic outcrops in Eastern Sudan, in Nuba mountains in south of Talodi, west Zalengei, and Agadi in central Sudan (Bedigian 1981).

S. angolense Welw.: An erect annual or perennial herb with simple quadrangular stem. Leaves subsessile or shortly petiolate, narrowly oblong/oblanceolate/elliptic, lower surface white tomentose, cuneate at base, apiculate at apex. Flowers reddish pink or pale mauve with deeper markings. Capsules narrowly oblong, subquadrangular, 4-sulcate, densely pubescent, gradually narrowed into a flattened rather broad and short beak. Seeds with double fringe, faintly rugose on the sides and faces. It shows widespread distribution in the Democratic Republic of Congo, Rwanda, Burundi, Uganda, Tanzania, etc. and grows in dense monoculture. The seeds are flattened, obconical in shape and faintly rugose (Bedigian 2010). Leaves of *S. angolense* are mixed with legumes and made into sauce which is served with cereal such as maize or millet. It is used as a medicinal plant for treatment of skin diseases such as measles and sores (Bedigian 2003) and is also utilized as ornamental and for soap making or as green manure.

S. angustifolium (Oliv.) Engl.: An erect annual or perennial herb. Leaves sessile or subsessile, variable in shape, linear to cultrate, margins entire inrolled or narrowly lanceolate irregularly toothed, cuneate at base, acute at apex, densely glandular below. Flowers purple, pink, or mauve often spotted within. Capsule cultrate in lateral view, appressed to stem, parchment-like when mature, beak straight or bent toward stem. Seeds with transversal ribs on both faces. It shows widespread distribution in Kenya, Sudan, Tanzania, and Uganda in zones with slightly higher rainfall. The seeds are small, black with rugose surface pattern. The mucilaginous leaves and flowers are cooked with other ingredients as leaf vegetables. The seeds are used in sauce or soup after crushing. The leaf mucilage is used to treat eye troubles, burns, wounds, and diarrhea in children (Bedigian 2003). Leaves are also utilized as ornamental and for soap making or as green manure.

S. prostratum Retz.: Herb with prostrate habit, the leaves are villous, orbicular, or obovate crenate or obtusely lobed with white indumentum beneath. Capsule is ovoid and compressed. Seeds are black in color with reticulate texture and foveolate testa.

S. laciniatum Klein. ex Willd.: Herb with prostrate habit, with hispid, ovate and lobed leaves which are deeply subpedately pinnatified. Flower purple. The capsule is ovoid and compressed. The seeds are black in color with reticulate texture and foveolate testa.

S. radiatum Schumach. & Thonn.: An erect annual woody herb with simple or branched, pubescent stem. Leaves scarcely heteromorphic, petiolate; lower ones ovate, coarsely toothed, acute at both ends; upper leaves lanceolate, entire. Leaves sparingly and persistently hairy and mealy glandular below. Flower purple or purplish. Capsule narrowly oblong in lateral view, with short broad beak. Seeds are rugose or pitted, with sculptural lines radiating perpendicular to the edge all around the margin of the seed faces. Origin of crop is Africa but now found throughout tropics in India, Sri Lanka, etc. The fresh leaves are used in soup and in sauces eaten with porridge (Bedigian 2004b). Mucilage of the plant is used for medical purposes. A cold infusion of macerated fresh young leaves facilitate childbirth (Bedigian 2003).

S. schinzianum Aschers. ex Schinz.: An erect branched annual, softly glandular-hairy, leaves lanceolate to narrowly elliptic, subobtuse, or apiculate, cuneate at base, entire or obscurely repand, pedicels short. Flower pale rose villous. Capsule rounded-quandrangular, deeply 4-sulcate, beak acuminate. Nectary stipitate. Seeds are strongly compressed, faces finely granular. Plants occur in bushes. Its distribution is restricted to Namibia.

15.2 Genetic Improvement of Sesame

Inadequate research efforts due to either lack of funding or their insufficient continuity and limited international cooperation for germplasm exchanges in the past have been responsible for the unsatisfactory improvement work in sesame. Also, genetic and breeding improvement efforts in sesame have been limited and slow as it is a crop of developing countries and within these countries it is a small holder's crop. For improvement of yield potential and traits of economic importance of crop plants it is essential to identify the superior genotypes among cultivars, introgressed linesand within wild relatives. The presence of untapped genetic variation existing in wild relatives and landraces of crop plants should be exploited gainfully for development of agronomically superior cultivars.

15.2.1 Areas for Improvement

Sesame suffers from both biotic and abiotic stresses leading to drastic reduction in crop yields. This along with the plant architecture which includes indeterminate growth/nonuniform capsule ripening and shattering of seeds at capsule maturity is a cause for its poor adaptability to mechanical harvesting. Thus, major areas for sesame improvement that need urgent attention are as follows.

15.2.1.1 Higher Yielding Genotypes

High and stable yield with applied inputs such as irrigation and farming is an important aspect to consider. Potential area of heterosis during hybrid combination also needs attention. The importance for this area is evident by the fact that India has the largest area under sesame cultivation while China is the largest producer, mainly due to low yielding cultivars in India. Further, there has been a significant and persistent decline in sesame production during the last 2 decades. Average productivity of sesame cultivation in India is approximately 422 kg/ha in comparison to the potential yield of 2000 kg/ha (Mkamilo and Bedigian 2007).

15.2.1.2 Improved Seed and Oil Quality

Expanding use of sesame seeds has resulted in reports on immunoglobulin E (IgE)-mediated food allergies (Agne et al. 2003). Oleosins are major allergens of sesame seeds. Altering of allergenic epitopes in sesame seed storage proteins and oleosins using protein engineering could be beneficial in decreasing sesame seed allergenicity. In addition, high contents of phytic and oxalic acid hinder the use of sesame protein as food (Johnson et al. 1979). Suppression or reduction of such constituents is another challenging area that needs to be taken care of.

15.2.1.3 Modification of Growth Habit and Seed Retention

An important consideration in case of sesame involves the continued and recalcitrant persistence of traits that are generally present in wild plants and have been negatively selected during the process of domestication in most of the other crop plants. Indeterminate growth habit in sesame is one of these highly undesirable traits as it leads to asynchronous capsule ripening, that is mainly responsible for lack of its amenability to modern farming technologies. Determinate sesame mutants obtained from mutation breeding have low yields compared to indeterminate lines (Ashri 1995), however, high yield in such mutants were reported by altering the production practices (Uzun and Cağirgan 2006). Advancement in such areas is mandatory to increase mechanization of sesame harvesting.

Similarly, dehiscence of the capsule at maturity is yet another trait that results in significant loss of yields and all attempts to develop nonshattering capsules have been unsuccessful so far. Shattering capsules of sesame makes the crop unsuitable for mechanical harvest and thus restricts its commercial potential. Mutants showing closed capsule character have been obtained using γ-ray radiation experiments (Ashri 1987). However, mutant genotypes show deleterious characters, such as semi-sterility, twisted stems, cupped leaves, short capsules and low yield caused either due to pleiotropy or multiple mutations. Therefore, neither the hundreds of years of selection during the process of domestication nor the current traditional as well as recent methods of improvement have resulted in shatter proof and determinate sesame cultivars.

In fact, due to these reasons, sesame is rightly grouped under the category of semi-domesticated crops at times. Therefore, any attempts to incorporate these highly desirable traits require research efforts to delineate these traits in sesame.

15.2.1.4 Resistance to Biotic Stresses

A large group of crop pathogens including fungi, bacteria and viruses are known to affect sesame and are responsible for major biotic stresses such as phyllody disease, *Fusarium* wilt, *Phytophthora* blight, seedling blight etc.

Several fungi reduce sesame seed germination by 30–61 %. *Fusarium oxysporum* is the major one responsible for 34 % of seedling rot (Khati and Pandey 2004) followed by *Alternaria sesami* with losses ranging from 4 to 25 % (Narayanaswamy et al. 2012). Severe leaf infection by *A. sesami* also affects the seed weight component of yield significantly (Narayanaswamy et al. 2012) while *A. alternata* induces blight symptoms of spots in stems and on leaves (Prasad and Reddy 1997). In addition, leaf spot disease caused by *Cercospora sesami* occurs in severe forms in India resulting in an estimated loss of about 30 % (Chowdhury 1945; Prasad and Reddy 1997). In fact, wilt and root rot diseases caused by the soilborne pathogens *F. oxysporum* f. sp. *sesamicola* and *Macrophomina phaseolina*, respectively, are responsible for sesame yield losses in all areas of its cultivation (Elewa et al. 2011). Screening studies have shown that sesame genotypes highly resistant to *Fusarium* wilt result in very low yields, thus demanding attention for improvement (El-Bramawy and Abd Al-Wahid 2009). *Phytophthora parasitica* var. *sesami* causes blight of sesame, observed characteristically as water-soaked spots ensuing in blackening of affected tissues in stems especially near the soil level. Incidence of 78.33 % has been reported during the month of August and losses range from 66 to 100 % especially when infection occurs at seedling stage

(Verma et al. 2005). Powdery mildew in sesame causes yield loss of 42 % and with every 1 % increase in disease intensity yield loss of 5.63 kg/ha has been recorded in India (Adiver and Kumari 2010). The casual fungi include *Luveilluta taurica, Erysiphe cichoracearum* and *Sphaerotheca fulginea* (Anyanga and Obongo 2001). The disease is air-borne and affects all aerial parts, leaves, flowers and pods (Egonyu et al. 2005).

Among bacterial pathogens, phytoplasma (Order Mollicutes) causes one of the most destructive diseases called phyllody in sesame which is responsible for losses of up to 99 % in certain tracts in India. Yields are dramatically decreased especially in warm climates (Salehi and Izadpanah 1992). Major symptom is conversion of flowers into leaf-like structures leading up to 80–100 % loss in capsule formation. Affected plants can easily be identified from the bushy appearance of the apical region at maturity. In addition, *Pseudomonas syringae pv. sesami* cause leaf spot disease and is responsible for substantial yield reduction (Sutica and Dowson 1962) along with *Xanthomonas campestris* pv. *sesami* (Ciferri 1955; Schumutterer and Kranz 1965). Sesame is also susceptible to turnip mosaic virus, watermelon mosaic virus, and tobacco ringspot virus under experimental inoculation conditions.

Nearly, thirty-eight insect pest species have been recorded to infest sesame in Uganda (Egonyu et al. 2005) and twenty nine in India (Baskaran et al. 1997). Of these, sesame webworm (*Antigastra catalaunalis* Dup.) and the gall midge (*Asphondylia sesami* Felt.) are considered most important pests worldwide with elevated occurrences of almost 62 and 98.8 %, respectively (Egonyu et al. 2005). Furthermore, *A. catalaunalis* is found to have the highest relative incidence of 0.63 and mean percentage capsule damage due to *Asphondylia sesami* occurs in a range of 29–34.3 % among the various sesame varieties (Egonyu et al. 2005). Avoidable yield losses of *A. catalaunalis* are known to vary from 6.2 to 43.1 % in different genotypes of *Sesamum* (Gupta et al. 2000) and 18.4–25.7 % in multilocation trials across India (Singh et al. 2014). Such yield losses attain a higher magnitude in rainy season (Baskaran et al. 1997).

15.2.1.5 Tolerance/Resistance to Abiotic Stresses

Abiotic stress is the negative impact of nonliving factors on the living organisms in a specific environment. Sesame suffers from three major abiotic stresses namely water logging, chilling and salinity.

Sesame is very sensitive to waterlogging. Seasonal rainfall often causes water-logging damage to the sesame production in south of China, India and Burma. The stress induces symptoms of wilting and leaf chlorosis, with susceptible accessions showing early onset. Formation of adventitious roots above the flooding level and development of abundant aerenchyma in root and stem have been observed in tolerant accessions (Wei et al. 2013). Moreover, stage dependent yield losses due to water logging have been observed in sesame. Sesame mutant SM7 shows 15 %

yield reduction when water logged for 48 h, at 25 days after sowing at seedling stage, with respect to the control plants. However, similar water logging treatments in vegetative, flowering, and seed filling stage shows yield losses ranging from 12 to 14 % (IAEA 2001). Thus, intermittent heavy rainfall is detrimental to the sesame crop.

Susceptibility to chilling is evident in sesame due to early photoinhibition in comparison to the plants resistant to chilling stress (Hetherington et al. 1989). This is supported by the fact that sesame shows significant reduction in growth below 20 °C, and growth and germination are totally inhibited below 10 °C. Sesame seed shows marked reduction in content of lignans (an antioxidant) viz. sesamin and sesamolin in the oil (Beroza and Kinman 1955) during frost damage. Additionally, sesame lies under the category of salt sensitive plants as indicated by downregulation of *SeMIPS* (Myo-inositol 1-phosphate synthase) gene (Chun et al. 2003) as in case of other susceptible plants like *Arabidopsis*, sunflower (Ishitani et al. 1996; Fernandez et al. 2008).

Further, sesame cultivation is generally confined to semiarid regions where soil salts are a limiting factor. Previous reports indicate that sesame is sensitive to soil salinity (Yousif et al. 1972; Cerda et al. 1977) mostly due to the presence of calcium chloride (Nassery et al. 1979). Salinity stress is shown to affect all traits significantly, causing reduction in germination percentage, germination, normal seedling percentage, seedling length, and dry weight (Tabatabaei and Naghibalghora 2014).

Studies in sesame suffering from different abiotic stresses have shown activation of antioxidative enzymes, proline, plant defense systems, etc. (Gehlot et al. 2005; Koca et al. 2007). Therefore, there is a need to develop or identify stress combating sesame cultivars.

15.2.2 Resources Available

Due to the efforts of scientists involved in germplasm collection, management, and conservation, various resources are now available for sesame improvement which includes wild germplasm, core collections and genomic resources.

15.2.3 Wild Germplasm

A total of 466 accessions belonging to *Sesamum* species that are wild relatives of *S. indicum* are conserved at NBPGR National Gene Bank, India (PGR Portal 2015; Pathak et al. 2014). Other species of genus *Sesamum* that are distantly related to sesame occur in parts of Africa and efforts are required to collect, evaluate, and conserve them.

15.2.3.1 Sesame Germplasm and Core Collections

National Gene Bank at NBPGR, New Delhi, India maintains over 9598 accessions of sesame (Pathak et al. 2014; PGR Portal 2015). Indigenous sesame core comprising of 343 accessions was identified (Bisht et al. 1998), while another set of sesame core comprising of 172 accessions has been developed from recent characterization of world sesame collections (Mahajan et al. 2007).

Further, 7698 sesame accessions, consisting of 3538 exotic collections, 2660 indigenous collections, 1072 improved genetic stocks and 428 others have been conserved by the Gene Bank of Rural Development Administration (RDA) in Suwon, Korea (Kang et al. 2006). Other organization and gene banks in the world undertaking conservation efforts of sesame include the United States Department of Agriculture (USDA), Agricultural Research Service (ARS), Plant Resources Conservation Unit (PGRCU) which have conserved 1226 sesame accessions belonging to Europe, Africa, Asia, North America, and South America (Morris 2009).

15.2.3.2 *Genomic Resources*

Research carried out in past years have led to the availability of a huge repertoire of genomic resources for sesame in the form of Expressed Sequence Tags (ESTs), transcriptomes and the whole genome sequence data.

Suh et al. (2003) obtained 3328 ESTs from a cDNA library of 5–25 days-old immature sesame seeds. Successful identification of genes involved in biosynthesis of sesame lignans, sesamin and sesamolin was carried out. A possible metabolic pathway for the generation of cofactors required for synthesis of storage lipid in non-green oilseeds was elucidated. Therefore, ESTs generated by large-scale single pass cDNA sequencing is a valuable approach for the identification of novel genes involved in specific metabolic pathways.

Wei et al. (2011) reported transcriptome from five different tissues of sesame plants using the approach of next generation sequencing. More than 22,000 unigenes could be successfully mapped onto 119 pathways using KEGG database. Further, in order to exploit the natural variation in the genes involved in lignan and lipid biosynthetic pathways, SSR markers from over 7700 unigenes were developed. The usefulness of these EST-based SSR markers was validated by their ability to detect polymorphism in diverse sesame accessions with varying oil and lignan contents.

Later, Zhang et al. (2013a–c) of Henan Sesame Research Centre, China sequenced the whole genome of *S. indicum* under the aegis of Sesame Genome Working Group (SGWG). A total of 27,148 genes were estimated to be present on the de novo assembled genome having a contig N50 of 52.2 kb and a high scaffold N50 of 2.1 Mb. The comparative genomic and transcriptomic analyses revealed candidate genes and oil biosynthetic pathway contributing to the high oil content in sesame. The expansion of Type 1 lipid transfer genes by tandem duplications and contraction of lipid degradations genes are some of the most significant findings in this study. Additionally, presence of high genetic diversity in lipid-related genes among twenty nine accessions from 12 different countries is also highly relevant. Such genomic delineation of synthesis and variation in lipid-related genes can be used as an important platform for qualitative and quantitative improvement of oil yield in sesame.

15.3 Mining of Sesame Wild Germplasm

15.3.1 Hybridization

Several studies are available pertaining to interspecific and intergeneric crosses of *Sesamum* sp. This is due to the advantage of epipetalous flowers and thus, emasculation which results in easy pollination in sesame making it amenable to controlled crossing. Moreover, low seed rate (2.0–2.5 kg/ha) and high seed multiplication ratio (1:300) facilitate ease of hybridization and recombination breeding programs (Nimmakayala et al. 2011). The interspecific crosses of sesame with wild relatives have led to the establishment of data depicting either close affinity of compatibility or incompatibility among the species.

Intergeneric crosses of *Sesamum* sp. with other genera of family Pedaliaceae have suggested its relatedness to *Martynia* and *Ceratotheca*. Nonviable seeds were obtained in intergeneric crosses between *S. indicum* and *Martynia* (Srinivasan 1942). However, crosses between *M. annua* and *S. laciniatum*, *S. radiatum* independently showed uninhibited pollen germination and growth of pollen tube in the styles and entrance in the ovule have been observed but the pollinated flowers dropped after 4 or 5 days of hybridization in all the crosses (Subramanian 1995). On the other hand, *Ceratotheca* showed fertile seeds in crosses with *S. indicum* (Falusi et al. 2002).

Cytoplasmic male sterile (CMS) lines in sesame were developed by hybridizing *S. indicum* with its wild relative *S. indicum* subsp. *malabaricum* (Bhuyan et al. 1997). CMS lines provide an opportunity to facilitate cross-pollination process for the production of hybrid seeds. Bhuyan and Sarma (2003) obtained 36 hybrid combinations of diverse origin using CMS system.

Somatic hybridization has also been used to develop sesame hybrid plants. Distantly related species can be fused using somatic hybridization. Dasharath et al. (2007) used ovary and ovule culture to develop interspecific hybrids between cultivated *S. indicum* and its wild relatives, *S. radiatum* and *S. occidentale*.

Parani et al. (1997) carried out hybridization studies between *S. alatum* and *S. indicum* with the objective of transferring phyllody resistance. Identical chromosome number of the parental species used, ambiguous morphological characters, and lack of segregation in F_2 led them to the use of protein and isozyme-based markers for the confirmation of the putative hybrid plants. Sodium dodecyl sulfate-polyacrylamide gel electrophoresis (SDS-PAGE)-based studies aimed at protein polymorphism revealed a transfer of five male specific protein bands in the hybrid. Out of four loci of esterases namely, *EST* A, B, C, and D, *EST* D was found in the heterozygous condition in the hybrid and homozygous in the parents thus indicating their hybridity. Similarly, out of five loci of peroxidases (*PRX* A, *PRX* B, *PRX* C, *PRX* D, and *PRX* E), two loci, namely, *PRX* A and *PRX* E were found useful.

Further, Parani et al. (1997) also carried out RAPD analysis of the two parental species and the putative hybrid *S. indicum* × *S. alatum* using 20 decamer primers. A total of 56 fragments out of 127 showed polymorphism between the parents. Almost all the fragments observed in the hybrid were shared either by *S. indicum* or by *S. alatum* thereby establishing their hybridity. Twenty-five male specific fragments identified were found to be present in the hybrid progeny and could be used as potential markers to differentiate the hybrids from the selfed progeny. Of the three markers employed, RAPD analysis was advocated as easy, economical, and reliable method for the large-scale screening of the hybrids.

15.3.2 Molecular Analysis of Species Diversity

RAPD analysis was carried out in 42 accessions of cultivated sesame and one accession of wild taxa, namely, *S.indicum* subsp. *malabaricum* using 10 primers (Bhat et al. unpublished). A total of 101 amplicons were scored at an average of 10.1 bands per primer with an overall high polymorphism of 61.4 %. The analysis of genetic diversity among cultivated forms revealed a low level of within accession variation (H_s = 0.011), while coefficient of gene differentiation (G_{st}) was 0.549, indicating high genetic differentiation among accessions which were landraces cultivated by farmers in different agroecological regions. The accessions from Eastern and Northeast regions of India were highly diverse indicating that these regions contain higher diversity requiring more intensive germplasm collections. The results indicated that although wild species, *S. indicum* subsp. *malabaricum* was slightly more genetically diverse (H_t = 0.067) than cultivated *S. indicum* (H_t = 0.024); it did not exhibit any substantial divergence from the cultivated types. Reports on low genetic diversity are also available in *Ceratotheca sesamoides* and *S. radiatum* based on AFLP markers (Adéoti et al. 2011).

15.3.3 Phylogenetic Relationships

Phylogenetic relationship using PCR-RFLP of unilocus markers, such as ribosomal DNA, chloroplast DNA and multilocus multiallellic marker systems such as amplified fragment length polymorphism (AFLP) exhibited high genetic similarity between *S. indicum* and *S. indicum* subsp. *malabaricum* (Fig. 15.3). Similarly,

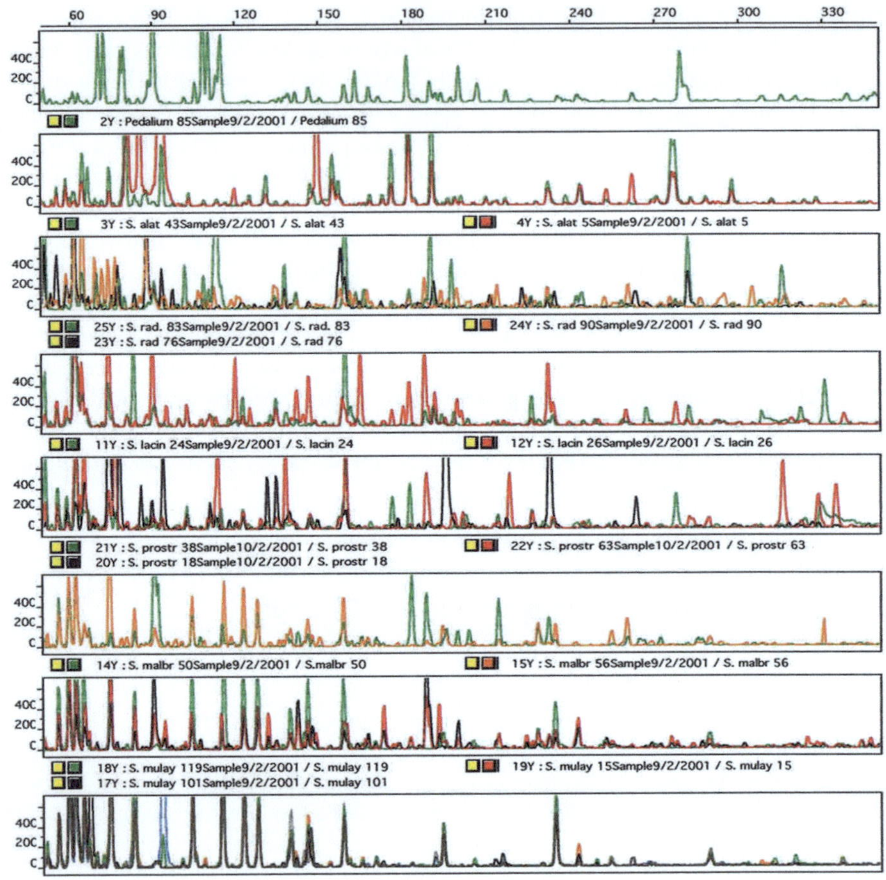

Fig. 15.3 Representative AFLP chromatogram showing profiles for wild relatives of sesame

S. laciniatum and *S. prostratum* appear to be closely related to each other and significantly diverged from other species. *S. radiatum* ($2n = 64$) appear to be distinct from all other species under the genus *Sesamum* occurring in India (Bhat et al. unpublished).

Akhila and Beevy (2011) characterized sesame morphologically and analyzed its seed protein in *S. indicum* L. and *S. occidentale*. Data on 13 quantitative and three qualitative characters of the cultivated species and seven accessions of the wild taxa were analyzed. A dendrogram based on UPGMA analysis of seed protein suggested intraspecific relationships of the wild taxa as evidenced from the morphological characterization.

15.3.4 Seed Oil and Antioxidants

Breeding for enhancing oil quality and content is very important objective in any oilseed crop. Genetic diversity in seed oil content and fatty acid composition in six wild species of genus *Sesamum* viz., *S. mulayanum, S. capense, S. laciniatum, S. latifolium, S. occidentale, and S. schinzianum* was studied and compared with the cultivated species, *S. indicum* (Hiremath et al. 2007). Seed oil content was low in wild species ranging from 20.3 to 33.9 % in comparison to cultivated sesame which was from 46.13 to 53.8 %. In contrast, wild species of sesame are known to contain higher percentages of unsaponifiable fraction (Fig. 15.4; Kamal-Eldin 2010). Analysis of sterols composition in *S. indicum* and three wild species (Fig. 15.5) depicts comparable proportion of sterols in *S. latifolium* while high proportion of campesterol and δ5-avenasterol in *S. alatum* (Kamal-Eldin and Appelqvist 1994). *S. alatum* also stands distinct from all other species with respect to methylated sterol composition under the category of desmethyl sterols (Kamal-Eldin 2010).

With respect to fatty acid composition of oil, wild species exhibit wide range of variation in palmitic and stearic acid contents. Stearic acid and linoleic acid content in all the wild species is significantly higher than the cultivated sesame. On the other hand, wild species have lower oleic acid (Kamal-Eldin 1993). Studies by Mondal et al. (2010) have also revealed high diversity in fatty acid content within Indian sesame germplasm, with oleic and linoleic acid forming the major proportions, i.e., 45.9 and 45 %. Very high linoleic acid content was observed in accessions of three wild species, *S. mulayanum* (49.3 %), *S. malabaricum* (48.2 %), and *S. radiatum* (51.6 %).

As per reports, all wild *Sesamum* spp. have black or brown thick, rough seed coat whereas domesticated sesame, *S. indicum* has thin and smooth seed coat with various shades of white, brown or black color (Kamal-Eldin 1993). Thus, it has

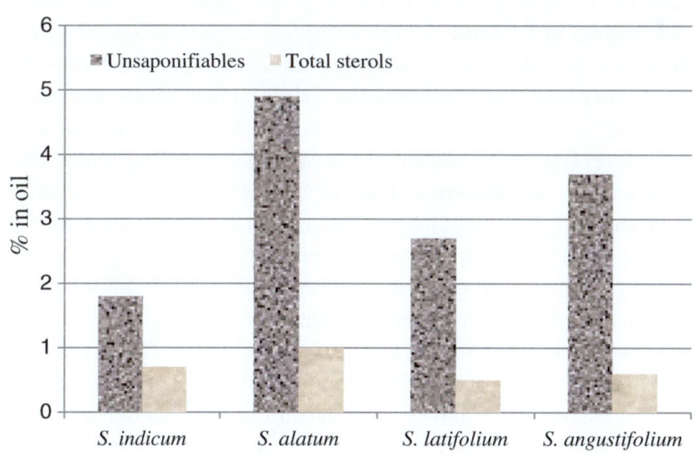

Fig. 15.4 Maximum percentage of unsaponifiables and total sterol fraction in seed oil of sesame and its wild relatives (Adapted from Kamal-Eldin 2010)

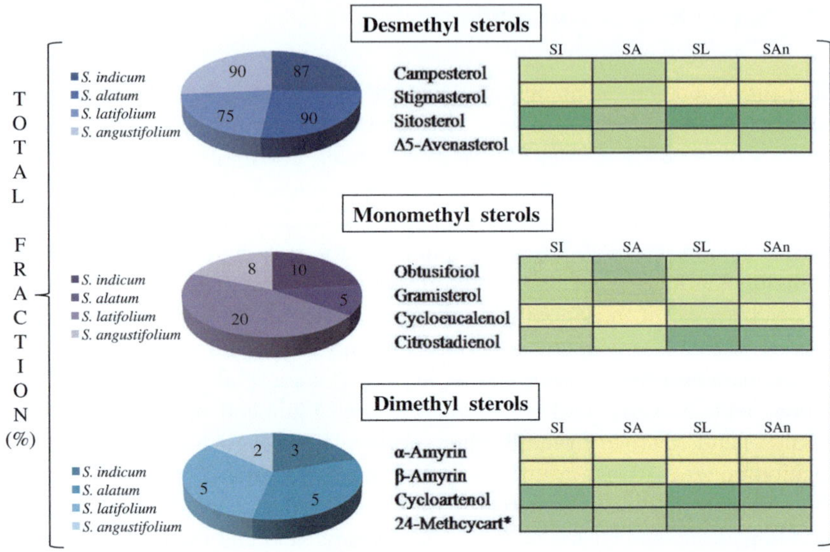

Fig. 15.5 Relative proportions of sterol fractions alongwith their composition in sesame (SI) and three wild species *S. alatum* (SA), *S. latifolium* (SL), *S. angustifolium* (SAn) (Adapted from Kamal-Eldin 2010). Percentage variation: 3 — 60 %

been proposed that seed oil content can possibly be increased in wild species if selections are made for thinner seed coat (Kamal-Eldin 1993). The content of lignan is variable within sesame ranging from 1550 to 18,600 mg/kg for sesamin and 1230–10,600 mg/kg for sesamolin (Kamal-Eldin 2010). Highest sesamin and sesamolin in capsules is observed at 30 days after flowering (Yasumoto et al. 2005). White seeded sesame is shown to contain high amount of lignans in comparison to the black seeded (Kang et al. 2003). Similarly low lignan content was observed in black seeded sesame by Namiki (1995). Cultivars from north-eastern states in India were found to be rich in lignans (Hemlatha and Ghafoorunissa, 2004). Wild species vary in the amount of lignans present in the seed oil (Fig. 15.6). Few species such as *S. triphyllum* and *S. rigidum* lack the presence of lignans, sesamin and sesamolin while traces are present in *S. alatum*, *S. pedalioides* and *S. capense* (Ashri 2006). Other lignans such as sesangolin is present in *S. angolense* and *S. angustifolium*, while high amount of 2-*epi*-sesalatin is present in *S. alatum* (Kamal-Eldin 2010). Recent studies revealed existence of exploitable levels of lignan (sesamin and sesamolin) and tocopherol contents in sesame germplasm (Pathak et al. 2014). The study carried out on 143 sesame samples showed the average content of sesamin and sesamolin as 0.86 and 0.50 mg g^{-1} seed, respectively. The average tocopherol content (292 μg g^{-1} seed) found in this study indicated presence of very high amount of γ-tocopherol in Indian sesame germplasm. *Sesamum* species namely, *S. indicum* subsp. *malabaricum* and *S. mulayanum* showed high lignan and tocopherol contents (1.31 mg/g seed and 239 μg/g seed; 1.15 mg/g and 281 μg/g seed) and thus could be utilized in sesame breeding programs for nutritional enhancement.

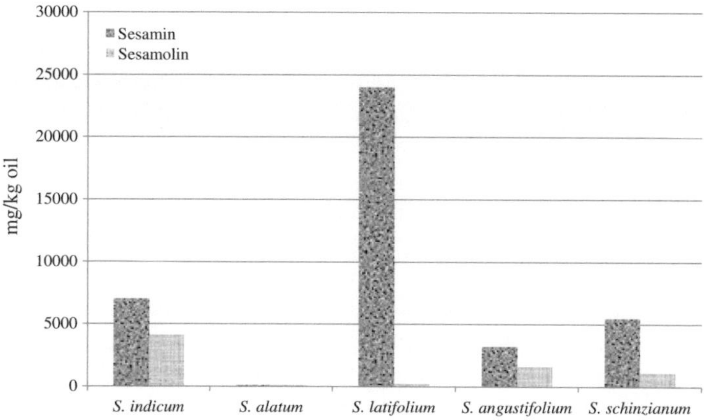

Fig. 15.6 Levels of oil soluble lignans in sesame and its wild relatives. Maximum values of sesamin have been depicted and the respective value for sesamolin has been retained (Adapted from Kamal-Eldin 2010)

Ono et al. (2006) identified a cytochrome P450 gene, CYP8101 responsible for catalysis of sesamin biosynthesis from pinoresinol. The protein is unique due to its dual catalytic ability that results in the formation of two methylenedioxy bridges, which is limited to a single bridge in all P450 proteins known to date. Transient expression system has revealed the localization of sesamin biosynthesis to the cytoplasmic surface of endoplasmic reticulum. Presence of a functional homolog from *S. radiatum* (contains sesamin) and a nonfunctional homolog from *S. alatum* (nearly lacks sesamin) provided functional validation for the role of CYP8101. Recent study by Pathak et al. (2015) examined the functional expression of *sesamin synthase* gene (CYP81Q1) during capsule maturation (0–40 days after flowering) in three wild *Sesamum* species and four sesame cultivars. Among the cultivated accessions of *S. indicum*, only one (CO-1) exhibited transcript abundance of sesamin synthase along with high sesamin content similar to *S. indicum* subsp. *malabaricum* suggesting a close relationship between the two. Additionally, the study demonstrated that the stage at 25 DAF in sesame capsule maturation should be selected for oil collection. Through this work, it has been proposed that interspecific crosses between *S. indicum* subsp. *malabaricum* and elite cultivars of *S. indicum* will prove advantageous in sesame breeding in order to obtain superior genotypes with high-quality oil.

15.3.5 Disease Resistance

Resistance to phyllody disease was reported in the wild species *S. alatum* through artificial screening (Srinivasulu and Narayanaswamy 1992; Singh et al. 2007). Rajeshwari et al. (2010) optimized a protocol for the production of hybrids of a cross between *S. alatum* and *S. indicum* through ovule culture. The developed hybrids exhibited moderate phyllody resistance. In another study, phyllody resistant sesame lines were developed through intra and interspecific crosses among different cultivated and wild species of sesame. It was reported that disease resistance is governed by one dominant (wild species) and one recessive (cultivated species) gene (Singh et al. 2007).

Further, treatment of *S. prostratum* callus with *F. oxysporum f. sesame* showed accumulation of hydrogen peroxide concentration and lipid peroxidation after 6 h. In addition, total phenolics content, activity of antioxidative enzymes and phenylalanine ammonia-lyase (*PAL*) showed an increase. Results indicate that *S. prostratum* which is resistant against *F. oxysporum* produces reactive oxygen species as defense barriers against the invading pathogen (Rajab et al. 2009).

Possibility of transforming of plant cells with appropriate gene/s involved in signal transduction of phytoalexin induction, that would activate Ca^{2+}-cascade and enhance the biosynthetic activity of secondary metabolites relating plant defense responses was assessed by transforming *S. schinzianum* plants, which is known to show high transformation and re-differentiation efficiency (Mitsuma et al. 2004). Calmodulin gene, *cam-4*, a CAM gene specifically expressed in oligogalacturonide-treated carrot which showed elicitation of a phytoalexin was expressed in the wild *Sesamum* sp. using *Agrobacterium*-mediated transfection method (Mitsuma et al. 2004). Transgenic plants for CAM gene showed enhanced production of phenylpropane derivatives in sesame suggesting that the engineering of signal transduction processes by the transfection of appropriate genes can prove beneficial in molecular breeding of sesame.

15.4 Mining of Sesame Cultivars

Primary gene pool of sesame comprising currently used and obsolete cultivars form an important genetic resource due to their high genetic diversity revealed and absence of crossability barrier among them. Sarwar and Haq (2006) evaluated 106 sesame genotypes from different parts of the world and documented heritability for yield related parameters, such as seed yield, capsule number, and branches per plant. It was concluded that selection of sesame elite genotypes for seed yield is possible on the basis of these characters. Various high yielding sesame varieties have been selected on the basis of these phenotypic and genotypic marker traits.

15.4.1 Molecular Diversity

Isshiki and Umezaki (1997) studied patterns of variations for seven enzyme systems in 68 accessions of sesame from Japan (12), Korea (15), and Thailand (41). Only one out of seven enzyme systems, namely, isocitrate dehydrogenase exhibited variations and was shown to be controlled by a single locus (*Idh*) with two alleles. The two alleles were found to be very widely distributed in the accessions. The locus *Idh* was proposed as an important genetic marker as few gene markers with simple genetic control are available in this crop. In addition, little variation in the germplasm analyzed suggested a narrow genetic base of the sesame germplasm grown in these countries.

Diaz et al. (1999) studied the isozyme variability in sesame accessions from six centers of diversity namely, India, Korea, Western Asia, Africa, China–Japan, and Central Asia. An analysis of the five putative loci belonging to five isozymes systems (acid phosphatase, isocitrate dehydrogenase, 6-phosphogluconate dehydrogenase, and shikimate dehydrogenase) showed similarity in the extent of polymorphic loci in sesame with the other cultivated plants particularly self-pollinated crop taxa. However, the extent of total diversity was reportedly lower than the other cultivated self-pollinated species. Differences between the values of the total diversity were not found to be statistically significant. Interestingly, 80 % of the species variability and center-wise total variability was found due to differences among populations.

Kim et al. (2002) reported genetic relationships among 75 accessions of Korean and exotic sesame germplasm using ISSR markers. The material analyzed comprised Korean cultivars (26), breeding lines and land races (17), including 32 introductions from 13 countries. UPGMA analysis divided these 75 accessions into seven groups where the largest group consisted of 25 Korean cultivars, 8 Korean breeding lines, and 17 worldwide accessions. No clear division was indicated on the basis of their geographic origin. The grouping of all Korean cultivars in the same cluster indicated their narrow genetic base. Nevertheless, Korean breeding lines differed markedly from the cultivars and thus can prove to be sources of useful traits in the improvement programs. In addition, grouping of exotic accessions with the Korean germplasm indicated that geographical separation did not generally result in the greater genetic distance and was ascribed to the exchange of materials from widely separate locations.

Bhat et al. (1999) analyzed genetic diversity in sesame germplasm of-Indian subcontinent and compared it with germplasm of 21 other sesame growing countries using RAPD analysis. The extent of genetic diversity was greater in the germplasm from Indian subcontinent in comparison to other countries. Among the Indian accessions, the collections from the states of Rajasthan and northeast were highly diverse. Such high diversity in the Indian sesame germplasm indicates nativity of the sesame crop. Relatively lower level of genetic diversity in the exotic germplasm has been ascribed to the comparatively recent introduction of a limited germplasm into these areas.

RAPD analysis of 60 sesame cultivars and improved lines released for general cultivation in India revealed presence of low to moderate genetic diversity. The Jaccard's similarity coefficient values for the 285 amplicons compared ranged from 0.48 to 0.94. This was in contrast to the high genetic diversity reported earlier in sesame germplasm (Bhat et al. 1999). Among the 58 sesame germplasm accessions the similarity coefficients ranged from 0.19 to 0.89. This clearly shows that only a small fraction of the total genetic diversity present in sesame germplasm has been represented in the improved cultivars of the crop. Results of the RAPD analysis was in agreement with the pedigrees of the cross-bred cultivars. The relative distance of varieties in the dendrogram was comparable to the pedigree data that was available.

An analysis of the Turkish sesame populations (Ercan et al. 2004) has also been carried out using RAPD markers. A total of 38 accessions of sesame germplasm from different regions of Turkey were sampled and were subjected to RAPD analysis using 12 decamer primers. Five out of 12 primers showed monomorphism. Low genetic variation among four main geographic regions was reported. However, analysis of molecular variance (AMOVA) revealed highest genetic variation among the populations within the regions. Of the total genetic diversity, 8.1 % was attributable to differences among regions and 91.8 % was due to population differences. Low level of differentiation among regions was ascribed to high rates of gene flow among populations within the region as a result of both human migration and agricultural trade. Performing AMOVA analysis separately for each region also identified the areas with highest genetic variation.

Nimmakayla et al. (2005) analyzed a set of 124 sesame genotypes collected from different parts of the world and tested for the SSR polymorphism using 14 microsatellites primers. A total of 144 SSR alleles were scored where the largest number of alleles (18) were identified by SSR primer PRU1 that had a repeat motif of $(GA)_7$ $(Y)_{99}GT)_7(GA)$ and the least by PRU 4 with a repeat motif of $(CA)_6CT$ $(CA)_5$ that amplified only 3 alleles. An AFLP analysis using nine primer pairs resulted in high polymorphisms resolving genome wide diversity among 124 genotypes.

Fifty microsatellite sequences have been isolated from an enriched library of sesame (Dixit 2005). Usefulness of 10 polymorphism microsatellites was tested for the diversity analysis using a total of 16 sesame accessions. Three to six alleles per locus at an average of 4.6 alleles having a fragment size of 150 to 307 bp was reported. Microsatellites used were found to be highly informative as the expected heterozygosity (He) and polymorphism information content (PIC) ranged from 0.437 to 0.858 and 0.34 to 0.80, respectively.

15.4.2 Gene Tagging

Identifying a molecular marker closely linked to the closed capsule character can enhance efficiency of the breeding programs aimed at eliminating such negative features. Uzun et al. (2003) identified an AFLP marker linked to closed capsule mutant trait using bulk segregant approach and later confirmed it by analyzing AFLP profile from single plants. The closed capsule mutant *cc3* was cross with a Turkish cultivar and the segregating population for the recessive mutant trait closed capsule was obtained. An F_2 population of 150 individuals was screened for the mutant trait phenotypically. Two bulks contrasting for the trait of interest were prepared and were subjected to bulk segregant analysis using AFLP markers.

A total of 72 primer combinations were screened for linkage to the desirable trait. Only one AFLP marker was found to be polymorphic chiefly due to the almost isogenic nature of the two parents used. Therefore, a 258 nucleotide AFLP marker tightly linked to the closed mutant trait was identified. A large deletion in the closed capsule mutant lines has been suggested resulting in the loss of function mutation though the hypothesis needs further confirmation.

15.4.3 Construction of High-Density Genetic Map and QTL Studies

The first QTL mapping in sesame was done by Zhang et al. (2013a). A high-density linkage map of sesame with 653 marker loci in 14 LGs was assembled. Using this linkage map, it was shown that seed coat color is controlled by two major genes with additive-dominant-epistatic effects plus polygenes with additive-dominant-epistatic effects. Four QTLs, namely, QTL1-1, 11-1, 11-2, and 11-3 were detected and found distributed in three linkage groups. This study provides a platform for further genetics and molecular marker-assisted selection (MAS) breeding research in sesame. Interestingly, seed coat color in sesame is an important agronomic trait. It is associated with biochemical functions involved in protein and oil metabolism, antioxidant activity and disease resistance. It has been suggested that seed coat color trait is a more suitable trait for estimating sesame evolution than geographic origin, since the direction of evolution in sesame has been suggested from wild species to black cultivars and then white cultivars (Zhang et al. 2013a).

A high-density genetic map for sesame has been constructed using Specific Length Amplified Fragment sequencing (SLAF-seq) (Zhang et al. 2013b). It is a recently developed high-resolution strategy for large-scale de novo SNP discovery and genotyping. In total, 28.21 Gb of data containing 201,488,285 pair-end reads were obtained after sequencing. From this data 71,793 high-quality SLAFs were detected of which 3673 SLAFs were polymorphic and 1272 of the polymorphic markers met the requirements for use in the construction of a genetic map. The final map included 1233 markers on the 15 linkage groups (LGs) and was 1,474.87 cm in length with an average distance of 1.20 cm between adjacent markers. A large number of polymorphic markers were developed in a short time using the SLAF-seq approach. The resultant high-density genetic map would be useful in gene/QTL fine mapping, map-based gene isolation, and molecular breeding for sesame. It will also act as a reference for positioning sequence scaffolds on a physical map, thereby assisting in the assembling of sesame genome sequence.

A molecular map of the important agro-botanic traits in sesame has been developed by Rao et al. (2014). Two sesame genotypes differing in important agro-botanic traits were crossed to study the inheritance pattern of nine traits. Seventeen QTLs were identified for these traits by single marker analysis. Out of the total QTLs detected, five explaining high phenotypic variation are promising which include one QTL for corolla color, two for capsule shape, and one each for capsule hair density and number of nodes.

QTL mapping and generation of high-density genetic maps paves way for better understanding of the genome structure, location of genes of interest on chromosomes, and their linkage with DNA markers. This would also help in further producing a more saturated high-density molecular linkage map, thus assisting in marker-assisted breeding for economically valuable traits such as yield, biotic, and abiotic resistance.

15.4.4 Heterosis to Increase Crop Yield

Commercial exploitation of heterosis is a fast and simple conventional breeding approach to increase crop yield. Though hybrid development is a costly and laborious on-field process, the high outcrossing rate in sesame favors the exploitation of heterosis.

Assessment of the extent of heterosis in sesame for 15 quantitative traits including seed yield per plant was done by crossing twelve lines and three testers in a line x tester fashion to develop forty eight F_1 hybrids (Vavdiya et al. 2013). The analysis of variance indicated highly significant differences among the parents and hybrids for all the characters. This indicates the presence of sufficient amount of genetic diversity for all traits studied. Heterosis was worked-out over better parent and standard variety, G.Til-4. The standard heterosis for seed yield per plant ranged from −12.32 to 137.39 %. The crosses NIC-75 × G.Til-10, IC-81564 × G.Til-10, NIC-75 × G.Til-4, AT-238 × G.Til-10, and Borda-1 × G.Til-10 were good heterotic combinations for seed yield per plant, which recorded 137.39, 128.74, 111.34, 100.42, and 90.84 % standard heterosis, respectively. The heterosis for seed yield per plant was associated with the heterosis expressed by its component characters.

Genetic diversity in parents is considered desirable to exploit heterosis in any breeding program. An investigation was carried out to search whether any relationship existed between heterosis of cross combinations with phenetic divergence, combining ability, and genetic divergence of parents in sesame. For this, seven sesame genotypes and their 21 cross-combinations developed through half-diallel mating were assessed for morphological markers, microsatellite markers, and seed storage protein polymorphism along with estimation of different parameters. The clustering pattern of parents varied for morphological, protein, and simple sequence repeats (SSRs), though some concordance was observed between phenetic and genetic divergence of parents (Das et al. 2013). Mid-parent heterosis % and better parent heterosis % was found to be positively and significantly correlated with specific combining ability and hybrids per se, but no specific trend was seen between morphological, protein and SSR marker data. However, heterotic crosses were more reliably predicted by SSR based genetic diversity value of above 0.5 between parents (Das et al. 2013). Therefore, it was inferred that parental diversity, based on morphological and seed storage protein polymorphism did not corroborate well with heterotic expression of characters in hybrids. However, the study based on microsatellite markers suggested that heterosis could be explained by parental diversity to some extent (Das et al. 2013).

15.4.5 Biotechnological Interventions in Sesame

Conventional breeding techniques have limited scope in improvement of resistance to biotic stresses and in quality improvement owing to low genetic variability for these traits and crossability barriers. Biotechnology is a viable option for developing sesame genotypes that can perform better under biotic and abiotic stresses. Furthermore, post fertilization barriers restrict the transfer of resistance genes from wild species to cultivated crops. This barrier cannot be overcome through conventional breeding. Therefore, the only option left for improvement of sesame is to transfer genes from other sources through genetic transformation techniques.

Plant tissue culture technology has been extensively used by plant breeders for crop improvement in several oil seed crops. The first report on tissue culture in sesame was that of Lee et al. (1985) on shoot tip culture followed by George et al. (1987) using different explants of sesame. Herbicide tolerant lines of sesame were obtained using in vitro selection by Chae et al. (1987). However, the recalcitrant nature of sesame provides hindrance in genetic transformation. Shoot regeneration with low frequencies from cotyledon and/or hypocotyl explants have been reported (Taskin and Turgut 1997; Were et al. 2006; Seo et al. 2007). Induction of somatic embryos has also been reported from hypocotyl-derived calluses but no plant regeneration was achieved (Mary and Jayabalan 1997). Susceptibility of sesame to *A. tumefaciens* has been shown by Taskin et al. (1999) but no transformed shoot/plant was recovered. However, hairy root cultures using *Agrobacterium rhizogenes* have been established (Ogasawara et al. 1993; Jin et al. 2005).

However, advancement in plant tissue culture approaches has led to increased opportunities for sesame improvement. For instance, regenerated plants can be grown to maturity in less than 4 months from shoot apical meristems and hypocotyls segments (Ram et al. 1990). Somatic embryos have been successfully induced directly from the surface of the zygotic embryos of sesame. As somatic embryogenesis produces stable variants, in vitro mutagenesis has been done on embryogenic sesame cultures to increase variability. Tissue culture methods have also been used to facilitate wide crosses using embryo culture techniques (Ram et al. 1990). Apart from these, nodal explants (Gangopadhyay et al. 1998) and leaf (Sharma and Pareek 1998) have also been used in micropropagating sesame. High frequency plant regeneration through direct adventitious shoot formation from deembryonated cotyledon segments of sesame have now been achieved (Seo et al. 2007). Enhanced shoot regeneration frequency was obtained by pre-culturing cotyledon explants in a high sucrose concentration (6 or 9 %) for 2 weeks.

Protocol optimization for genetic transformation and plant regeneration of sesame is reported (Were et al. 2006). *Agrobacterium*-mediated transformation protocol for the first time has been established for generation of fertile transgenic sesame plants (Yadav et al. 2010). The method is efficient for plant regeneration with direct multiple shoot organogenesis from cotyledon explants including establishment of an optimal selection medium.

Studies targeting specific areas of improvement in sesame are incredibly meager. Nevertheless, steps are being taken to generate transgenic sesame seeds with lower phytate expression using an antisense expression cassette to enhance the use of sesame protein as food (Suh et al. 2010). An initial step toward developing improved sesame by the production of GFP expressing transgenic sesame plants using an *A. tumefaciens* mediated transformation system has been attempted. The promoter of the seed-specific microsomal δ-12 desaturase gene (*SeFAD2*) has been isolated from sesame. Transient expression of the GUS gene under the seed-specific *SeFAD2* promoter in developing sesame has been successfully achieved after microprojectile bombardment. Further investigations into seed-specific promoter in developing sesame seeds (*SeFAD2*) would help in generating seeds with improved qualities (Kim et al. 2008). Out of eight toco-isomers, sesame seeds are richest in γ-tocopherol. Apart from enhancing the nutritional functionality in humans, the production of other seven toco-isomers in sesame can also help the sesame plant in protection against oxidative stress. Studies pertaining to genetic improvement of nutritional components such as tocochromanols biosynthetic enzymes and the characterization of their recombinant products are underway. These studies can be extended to obtain transgenic sesame producing high tocochromanols thereby enhancing its amelioration ability.

15.5 Future Perspectives

A thorough molecular diversity analysis using robust techniques such as AFLPs, STMS, and SNPs can greatly help in detailed characterization of wild germplasm of sesame. This would also help in designation of diversity rich areas and thereby in the initiation of on-farm conservation activities.

The wild species of *Sesamum* are rich source of genes for resistance to biotic and abiotic stresses. There is a need to undertake detailed molecular phylogeny studies preferably under international network mode so that nomenclature of various species and their relationships with the cultigens are unambiguous.

Molecular tagging and mapping studies have enhanced the use of molecular markers to bring about greater specificity in crop improvement programs for incorporation of desired traits. Such studies are urgently required if sesame as a crop has to survive the future day competition from other more ‘efficient’ and remunerative oil seed crops such as soybean and brassicae. Finally, optimization of efficient protocols for in vitro regeneration and development of transgenic plants are the most important areas for sesame improvement that can bring a paradigm shift in enhancing the production and utilization of this unique oil yielding crop.

References

Abou-Gharbia HA, Shahidi F, Shehata AAY, Youssef MM (1997) Effects of processing on oxidative stability of sesame oil extracted from intact and dehulled seed. J Am Oil Chem Soc 74:215–221

Adéoti K, Rival A, Dansi A, Santoni S, Brown S et al (2011) Genetic characterization of two traditional leafy vegetables (*Sesamum radiatum* Thonn. ex Hornem and *Ceratotheca sesamoides* Endl.) of Benin, using flow cytometry and amplified fragment length polymorphism (AFLP) markers. Afr J Biotechnol 10(65):14264–14275

Adiver SS, Kumari (2010) Fungal diseases of oilseed crops and their management. In: Arya A, Perelló AE (eds) Management of fungal plant pathogens. CABI, pp 263–274

Agne PSE, Rance F, Bidat E (2003) Sesame seed allergy. Rev Franc 43(8):507–516(10)

Ahmad M, Khan MA, Zafar M, Sultana S (2010) Environment friendly renewable energy from sesame biodiesel energy sources. Energy Sources Part A 32:189–196

Akhila H, Beevy SS (2011) Morphological and seed protein characterization of the cultivated and the wild taxa of *Sesamum* L. (Pedaliaceae). Plant Syst Evol 293:65–70

Anyanga WO, Obongo YY (2001) Sesame (*Sesamum indicum* L.). In: Mukiibi JK (ed) Agriculture in Uganda, vol II. Crops, pp 97–102

Ashri A (1987) Report on FAO/IAEA expert consultation on breeding improved sesame cultivars. FAO/AGP, Rome

Ashri A (1988) Sesame breeding-objectives and approaches. In: Oil crops-sunflower, linseed and sesame. In: Proceedings of 4th oil crop network workshop, Njoro

Ashri A (1995) Sesame research overview: current status, perspectives and priorities. In: Bennett MR, Wood IM (eds) Proceedings of the 1st Australian sesame workshop. NT Dept. Primary Industry and Fisheries, Darwin, pp 1–17

Ashri A (2006) Sesame (*Sesamum indicum* L.) In: Singh RJ (ed) Genetic resources, chromosome engineering, and crop improvement oilseed crops, vol 4. CRC Press, pp 231–289

Baskaran RKM, Mahadevan NR, Sathiyananddam VKR, Thangavelu S (1997) Approaches for prevention and control sesame pests. In: Sinha MP (ed) Recent advances in ecobiological research, vol 2. APH Publishing, New Delhi, pp 133–143

Bedigian D (1981) Origin, diversity, exploration and collection of sesame. In: Sesame-status and improvement FAO plant production and protection, Paper 29, pp 164–169

Bedigian D (1984) *Sesamum indicum* L. Crop origin, diversity, chemistry and ethnobotany, PhD, University of Illinois, Champaign-Urbana University Microfilms DA8502071 dissertation. Abs Int 45:3410-B

Bedigian D, Harlan JR (1986) Evidence for cultivation of sesame in the ancient world. Econ Bot 40(2):137–154

Bedigian D (2003) Sesame in Africa: origin and dispersals. In: Neumann K, Butler A, Kahlheber S (eds) Food, fuel and fields—progress in African archaeobotany. Africa, pp 17–36

Bedigian D (2004a) History and lore of sesame in southwest Asia. Econ Bot 58(3):329–353

Bedigian D (2004b) Slimy leaves and oily seeds: distribution and use of wild relatives of sesame in Africa. Econ Bot 58(Suppl):S3–S33

Bedigian D (2010) Cultivated sesame and wild relatives in the genus *Sesamum* L. In: Sesame the genus *Sesamum*, CRC Press, Boca Raton, pp 53–61

Bedigian D (2014) A new combination for the Indian progenitor of sesame, *Sesamum indicum* (Pedaliaceae). Novon 23(1):5–13

Beroza M, Kinman ML (1955) Sesamin, sesamolin, and sesamol content of the oil of sesame seed as affected by strain, location grown, ageing, and frost damage. J Am Oil Chem Soc 32:348–350

Bhat KV, Babrekar PP, Lakhanpaul S (1999) Study of genetic diversity in Indian and exotic sesame (*Sesamum indicum* L.) germplasm using random amplified polymorphic DNA (RAPD) markers. Euphytica 110:21–33

Bhuyan J, Ramalingm RS, Sree-Rangaswamy SR (1997) Development of cytoplasmic-genic male sterile lines in sesame (*Sesamum indicum* L.) through genome substitution. Bull Pure Appl Sci B16:17–20

Bhuyan J, Sarma MK (2003) Identification of heterotic crosses involving cytoplasmic-genic male sterile lines in sesame (*Sesamum indicum* L.). Sesame Safflower Newsl 18:7–11

Bisht IS, Mahajan RK, Loknathan TR, Agrawal RC (1998) Diversity in Indian sesame collection and stratification of germplasm accessions in different diversity groups. Genet Res Crop Evol 45(4):325–335

Chae YA, Park SK, Anand IJ (1987) Selection in vitro for herbicide tolerant cell lines of *Sesamum indicum* 2: Selection of herbicide tolerant calli and plant regeneration. Korean J Plant Breed 19:75–80

Chowdhury S (1945) Control of *Cercospora* blight of til. Indian J Agric Sci 15:110–142

Chun JA, Jin UH, Lee JW, Yi YB, Hyung NI et al (2003) Isolation and characterization of a myo-inositol 1-phosphate synthase cDNA from developing sesame (*Sesamum indicum* L.) seeds: functional and differential expression, and salt-induced transcription during germination. Planta 216(5):874–880

Cerda A, Bingham FT, Hoffman GJ (1977) Interactive effect of salinity and phosphorus on sesame. Soil Sci Soc Am J 41:915–918

Ciferri R (1955) Preliminary list of noteworthy diseases of cultivated plants in continental eastern China. Plant Dis Report 39(10):785–792

Das A, Pandey S, Dasgupta T (2013) Association of heterosis with combining ability and genetic divergence in sesame (*Sesamum Indicum* L.). Intern J Sci Tech Res 2(12):307–314

Dasharath K, Sridevi O, Salimath PM, Ramesh T (2007) Production of interspecific hybrids in sesame through embryo rescue. Ind J Crop Sci 2(1):193–196

Diaz AJP, Layrisse A, Pugh TM (1999) Analisis de la diversidad genetic en el ajan joli mediante electrophoresis de isoenzimes (abstract in english). Trop Agron 49:169–186

Dixit A, Jin MH, Chung JW, Yu JW, Chung HK (2005) Development of polymorphic microsatellite markers in sesame (*Sesamum indicum* L.). Mol Ecol Notes 5(4):736–738

Duhoon SS, Jyotishi A, Deshmukh MR, Singh NB (2002) Optimization of sesame (*Sesamum indicum* L.) production through bio/natural inputs. J Oilseeds Res 19(1):73–78

Egonyu JP, Kyamanywa S, Anyanga W, Ssekabembe CK (2005) Review of pests and diseases of sesame in Uganda. Afr Crop Sci Conf Proc 7:1411–1416

Elbadri GA, Yassin AM (2010) Sesame's protective role in crop nematode control. In: Bedigian D (ed) Sesame: the genus *Sesamum*, CRC Press, Boca Raton, pp 211–217

El-Bramawy MASA, Abd Al-Wahid OA (2009) Evaluation of resistance of selected sesame (*Sesamum indicum*) genotypes to *Fusarium* wilt disease caused by *Fusarium oxysporum* f. sp. sesami. Tunis J Plant Prot 4:29–39

Elewa IS, Mostafa MH, Sahab AF, Ziedan EH (2011) Direct effect of biocontrol agents on wilt and root-rot diseases of sesame. Arch Phytopathol Plant Prot 44(5):493–504

Elleuch M, Besbes S, Roiseux O, Blecker C, Attia H (2007) Quality characteristics of sesame seeds and by-products. Food Chem 103:641–650

Ercan AG, Taskin M, Turgut K (2004) Analysis of genetic diversity in Turkish sesame (*Sesamum indicum* L.) populations using RAPD markers. Genet Res Crop Evol 51(6):599–607

Falusi OA, Salako EA, Funmi FM (2002) Inheritance of hairiness of stem and petiole in a selection from local (Nigeria) germplasm of sesame. Tropicultura 20(3):156–158

FAOSTAT (2015) http://faostat3.fao.org/browse/Q/QC/E (Accessed 20 June 2015)

Fernandez P, Rienzo JD, Fernandez L, Hopp HE, Paniego N et al (2008) Transcriptomic identification of candidate genes involved in sunflower responses to chilling and salt stresses based on cDNA microarray analysis. BMC Plant Biol 8:11

Gangopadhyay G, Poddar R, Gupta S (1998) Micropropagation of sesame (*Sesamum indicum* L.) by in vitro multiple shoot production from nodal explants. Phytomorphology 48(1):83–90

Gehlot HS, Purohit A, Shekhawat NS (2005) Metabolic changes and protein patterns associated with adaptation to salinity in *Sesamum indicum* cultivars. J Cell Mol Biol 4:31–39

George L, Bapat VA, Rao PS (1987) *In vitro* multiplication of sesame (*Sesamum indicum*) through tissue culture. Ann Bot 60:17–21

Ghafoorunissa (1996) Fats in Indian diets and their nutritional and health implications. Lipids 31 (Suppl): S287–S291

Gupta MP, Rai HS, Jakhmola SS (2000) Assessment of avoidable losses in grain yield due to the incidence of leaf roller and capsule borer *Antigastra catalaunalis* Dup. in sesame varieties. Extended summaries. National seminar on oilseed and oils research and development needs in the millennium directorate of oilseeds research, Hyderabad, India, pp 226

Gormley C, Bedigian D, Richard G (2015) Phylogeny of Pedaliaceae and Martyniaceae and the placement of *Trapella* in Plantaginaceae. Syst Bot 40(1):259–268

Haller HL, LaForge FB, Sullivan WN (1942) Some compounds related to sesamin: their structures and their synergistic effect with pyrethrum insecticides. J Org Chem 07(2):185–188

Harlan JR (1992) Crops and Ancient Man, 2nd edn. American Society for Agronomy, Madison

Hasan AF, Furumoto T, Begum S, Fukui H (2001) Hydroxysesamone and 2,3-epoxysesamone from roots of *Sesamum indicum*. Phytochemistry 58:1225–1228

Hetherington SE, He J, Smillie RM (1989) Photoinhibition at low temperature in chilling-sensitive and—resistant plants. Plant Physiol 90:1609–1615

Hiremath SC, Patil CG, Patil KB, Nagasampige MH (2007) Genetic diversity of seed lipid content and fatty acid composition in some species of *Sesamum* L. (Pedaliaceae). Afr J Biotechnol 6 (5):539–543

Hooker JD (1885) Pedaliaceae. In: Flora of British India, London 4:387

IAEA (2001) Sesame improvement by induced mutations. Final reports of an FAO/IAEA co-ordinated research project organized by the Joint FAO/IAEA division of nuclear techniques in food and agriculture 1993–1998:1–180

Ishitani M, Majumder AL, Bornhouser A, Michalowski CB, Jensen RG et al (1996) Co-ordinate transcriptional induction of myo-inositol metabolism during environmental stress. Plant J 9:537–548

Isshiki S, Umezaki T (1997) Genetic variations of isozymes in cultivated sesame (*Sesamum indicum* L.). Euphytica 93(3):375–377

Jin UH, Chun JA, Han MO, Lee JW, Yi YB et al (2005) Sesame hairy root cultures for extra-cellular production of a recombinant fungal phytase. Process Biochem 40(12):3754–3762

Johnson LA, Suleiman TM, Lusas EW (1979) Sesame protein: a review and prospectus. J Am Oil Chem Soc 56:463–468

Joshi AB, Narayanan ES, Vasudeva RS (1961) Sesamum. Indian Oilseeds Central Committee, Hyderabad, p 109

Kamal-Eldin (2010) Chemical studies on the lignans and other minor constituents of sesame seed oil. In: Bedigian D (ed) Sesame: the genus *Sesamum*, CRC Press, Boca Raton, pp 79–91

Kamal-Eldin A (1993) Seed oils of *Sesamum indicum* L. and some wild relatives: a compositional study of the fatty acids, acyl lipids, sterols, tocopherols and lignans, Ph.D. thesis submitted to Swedish University Agricultural Science

Kamal-Eldin A, Appelqvist LA (1994) Variations in the composition of sterols, tocopherols and lignans in seed oils from 4 *sesamum* species. J Am Oil Chem Soc 71:149–156

Kang CW, Kim SY, Lee SW, Mathur PN, Hodgkin T et al (2006) Selection of a core collection of korean sesame germplasm by a stepwise clustering method. Breed Sci 56:85–91

Kang MH, Choi JS, Ha TY (2003) Chemical properties of sesames cultivated in Korea and China. Food Sci Biotechnol 12(6):621–624

Katsuzaki H, Kawakishi S, Osawa T (1994) Sesaminol glucosides in sesame seeds. Phytochemistry 35(3):773–776

Khati P, Pandey KN (2004) Seed mycoflora of sesame and their role in plant health. J Mycol Pathol 34:378–380

Kim DH, Zur G, Danin-Poleg Y, Lee SW, Shim KB et al (2002) Genetic relationships of sesame germplasm collection as revealed by inter-simple sequence repeats. Plant Breed 121(3):259–262

Kim MJ, Go YS, Ahn SJ, Chung CH, Suh MC (2008) Functional complementation of a periila ω3 fatty acid desaturase under the seed-specific *SeFAD2* promoter. J Plant Biol 51(3):174–179

Kita S, Matsumura Y, Morimoto S, Akimoto K, Furuya M et al (1995) Antihypertensive effect of sesamin. II. Protection against two-kidney, one-clip renal hypertension and cardiovascular hypertrophy. Biol Pharm Bull 18:1283–1285

Koca H, Bor M, Ozdemir F, Turkan I (2007) The effect of salt stress on lipid peroxidation, antioxidative enzymes and proline content of sesame cultivars. Environ Exp Bot 60:344–351

Lee JI, Park YH, Park YS, Im BG (1985) Propagation of sesame (*Sesamum indicum* L.) through shoot tip culture. Korean J Breed 17(4):367–372

Lee JI, Lee BH, Seong NS, Kang CW (1991) Studies on interspecific hybridization in sesame. I. Characteristics and cross affinity of wild sesame. Korean J Breed 22:356–360

Lopez AM, Mazzani B (1964) Fruit length, number of seeds per fruit and seed size in seven sesame cultivars. Agronomia Trop Venezuela 14:133–135

Mahajan RK, Bisht IS, Dhillon BS (2007) Establishment of a core collection of world sesame (*Sesamum indicum* L.) germplasm accessions. Sabrao J Breed Genet 39:53–64

Mahfouz H, Kamel S, Belal A, Said M (2012) Pollinators visiting sesame (*Sesamum indicum* L.) seed crop with reference to foraging activity of some bee species. Cercetări Agronomice în Moldova 46(2):150

Mary RJ, Jayabalan N (1997) Influence of growth regulators on somatic embryogenesis in sesame. Plant Cell Tiss Org 49(1):67–70

Mkamilo GS, Bedigian D (2007) Sesamum indicum L. In: van der Vossen HAM, Mkamilo GS (eds) Vegetable Oils 14. Wageningen, Plant Resources of Tropical Africa (PROTA), pp 153–158

Mitsuma S, Ishigaki E, Sugiyama R, Asamizu T, Yamada K et al (2004) Activation of 000phenylpropanoid metabolism in sesame by over-expression of carrot calmodulin gene. Biol Pharm Bull 27(10):1621–1625

Moazzami AA, Kamal-Eldin A (2006) Sesame seed is a rich source of dietary lignans. J Am Oil Chem Soc 83:719–723

Mondal N, Bhat KV, Srivastava PS (2010) Variation in fatty acid composition in Indian germplasm of sesame. J Am Oil Chem Soc 87:1263–1269

Morris JB (2002) Food, industrial, nutraceutical and pharmaceutical uses of sesame genetic resources. In: Janick J, Whipkey A (eds) Trends in new crops and new uses. ASHS Press, Alexandria, pp 153–156

Morris JB (2009) Characterization of Sesame (*Sesamum indicum* L.) Germplasm Regenerated in Georgia, USA. Genetic Res Crop Evol 56:925–936

Najeeb U, Mirza MY, Jilani G, Mubashir AK, Zhou WJ (2012) Sesame. In: Gupta SK (ed), Technological innovations in major world oil crops, vol 1. Springer, New York, pp 131–145

Namiki M (1995) The chemistry and physiological functions of sesame. Food Rev Int 11:281–329

Narayanaswamy T, Surendra HS, Rathod S (2012) Fitting of statistical models for growth pattern of root and shoot morphological traits in sesame (*Sesamum indicum* L.). Environ Ecol 30 (4):1362–1365

Nassery H, Ogata G, Maas EV (1979) Sensitivity of sesame to various salts. Agron J 71(4):595–597

Nimmakayala P, Kaur P, Bashet AZ, Bates GT, Langham R, Reddy OUK (2005) Molecular characterization of *Sesamum* using SSRs and AFLPs. In: Plant and animal genomes XIII conference, San Diego, pp 15–19

Nimmakayala P, Perumal R, Mulpuri S, Reddy UK (2011) Sesamum. In: Kole C (ed) Wild corp relatives: genomic and breeding resources oilseeds. Springer, Berlin, pp 261–273

Ogasawara T, Chiba K, Tada M (1993) Production in high-yield of a naphthoquinone by a hairy root culture of *Sesamum indicum*. Phytochemistry 33(5):1095–1098

Olmstead RG, Bremer B, Scott KM, Palmer JD (1993) A parsimony analysis of the Asteridae *sensu lato* based on rbcL sequences. Ann Mo Bot Gard 80:700–722

Ono E, Nakai M, Fukui Y, Tomimori N, Fukuchi-Mizutani M et al (2006) Formation of two methylenedioxy bridges by a *Sesamum* CYP81Q protein yielding a furofuran lignan, (+)-sesamin. Proc Natl Acad Sci USA 103:10116–10121

Parani M, Singh KN, Rangasamy S, Ramalingam RS (1997) Identification of *Sesamum alatum* × *Sesamum indicum* hybrid using protein, isozyme and RAPD markers. Indian J Genet 57:381–388

Pathak N, Rai AK, Saha S, Walia SK, Sen SK, Bhat KV (2014) Quantitative dissection of antioxidative bioactive components in cultivated and wild sesame germplasm reveals potentially exploitable wide genetic variability. J Crop Sci Biotechnol 17(3):127–139

Pathak N, Bhaduri A, Bhat KV, Rai AK (2015) Tracking sesamin synthase gene expression through seed maturity in wild and cultivated sesame species—a domestication footprint. Plant Biol pp 1–8 (in press)

PGR Portal (2015) www.nbpgr.ernet.in/pgrportal (Accessed 09 December 2015)

Phillips KM, Ruggio DM, Ashraf-Khorassani M (2005) Phytosterol composition of nuts and seeds commonly consumed in the United States. J Agric Food Chem 30:53(24):9436–45

Prabakaran AJ (1992) Identification of male sterile sources through wide hybridization and induced mutagenesis in sesame (*Sesamum indicum* L.). Ph. D. Thesis, TNAU, Coimbatore

Prasad PR, Reddy SR (1997) Diseases of sesame in two districts (Warangal and Karimnagar) of Andhra Pradesh. In: Reddy MS, Srivastava H, Purohit DK, Reddy SR (eds) Microbial biotechnology, Scientific Publishers, Jodhpur, pp 169–174

Raghavan TS, Krishnamurthy KV (1947) Cytogenetical studies in *Sesamum*. Part 1.Cytology of the parents, *Sesamum orientale* Linn., and *Sesamum prostratum* Retz. and the cytology of the sterile hybrid between them and of the fertile amphidiploid. Proc Indian Acad Sci B 26(6):236–275

Raikwar RS, Srivastva P (2013) Productivity enhancement of sesame (*Sesamum indicum* L.) through improved production technologies. Afr J Agric Res 8(47):6073–6078

Rajab R, Rajan SS, Satheesh LS, Harish SR, Sunukumar SS et al (2009) Hypersensitive response of *Sesamum prostratum* Retz. elicitated by *Fusarium oxysporum* f. *sesami* (Schelt) Jacz Butler. Ind J Exp Biol 47(10):834–838

Rajeswari S, Ramaswamy NM (2004) Effect of growth regulators on retention of floral pedicels and capsules in sesame. Sesame Safflower Newsl 19:1137–1617

Rajeswari S, Thiruvengadam V, Ramaswamy NM (2010) Production of interspecific hybrids between *Sesamum alatum* Thonn and *Sesamum indicum* L. through ovule culture and screening for phyllody disease resistance. S Afr J Bot 76(2):252–258

Ram R, Catlin D, Romero J, Cowley C (1990) Sesame: new approaches for crop improvement. In: Janick J, Simon JE (eds) Advances in new crops. Timber Press, Portland, pp 225–228

Ram SG, Sundaravelpandian K, Kumar M, Vinod KK, Bapu JK et al (2006) Pollen–pistil interaction in the inter-specific crosses of *Sesamum* sp. Euphytica 152:379–385

Ramanathan K (1950) A note on the interspecific hybridization in *Sesamum*. Madras Agric J 37:397–400

Rao P, Prasuna K, Anuradha G, Srividya A, Vemireddy LR et al (2014) Molecular mapping of important agro-botanic traits in sesame. Electronic J Plant Breeding 5(3):475–488

Salehi M, Izadpanah K (1992) Etiology and transmission of sesame phyllody in Iran. J Phytopathol 135:37–47

Salunkhe DK, Chavan JK, Adsule RN, Kadam SS (1991) Sesame. In: Röbbelen G, Downey RK, Ashri A (eds) World oilseeds-chemistry, technology, and utilization. Van Nostrand Reinhold, New York, pp 371–402

Sarwar G, Haq MA (2006) Evaluation of sesame germplasm for genetic parameters and disease resistance. J Agric Res 44(2):89–96

Schumutterer H, Kranz J (1965) On *Cylindrosporium sesame* (Huss.) causing brown spot disease of sesame. Phytopathologische Zeitshrift 254:193–201

Seo HY, Kim YJ, Park TI, Kim HS, Yun SJ et al (2007) High-frequency plant regeneration via adventitious shoot formation from deembryonated cotyledon explants of *Sesamum indicum* L. Vitro Cell Dev Biol Plant 43(3):209–214

Shahidi F, Tan Z (2010) Physiological effects of sesame bioactive and antioxidant compounds. In: Bedigian D (ed), Sesame-The genus *Sesamum*, CRC Press, Boca Raton, pp 139–153

Singh PK, Akram M, Vajpeyi M, Srivastava RL, Kumar K et al (2007) Screening and development of resistant sesame varieties against phytoplasma. Bull Insectol 60:303–304

Singh TVK, Satyanarayana J, Peshin R (2014) Crop loss assessment in India-past experiences and future strategies. In: Peshin R, Pimental D (eds) Integrated Pest Management. Springer, New York, pp 227–243

Srinivasan AR (1942) Contribution to the morphology *Pedalium murex* Linn. and *Sesamum indicum* D.C. In: Proceedings of Indian acadamic sciences 16(5):155–164

Srinivasulu B, Narayanaswamy P (1992) Four-tier technique for screening *Sesamum* entries against phyllody disease. Oil Crops Newsl 9:38–40

Subramanian M (1995) Pollen germination studies in Sesame. Ann Agric Res 16:225–226

Sugano M, Inoue T, Koba K, Yoshida K, Hirose N et al (1990) Influence of sesame lignans on various lipid parameters in rats. Agric Biol Chem 54:2669–2673

Suh MC, Hyung N, Chung C (2010) Molecular biotechnology of sesame. In: Bedigian D (ed) Sesame-The genus *Sesamum*, CRC press, Boca Raton, pp 219–243

Suh MC, Kim MJ, Hur CG, Bae JM, Park YI et al (2003) Comparative analysis of expressed sequence tags from *Sesamum indicum* and *Arabidopsis thaliana* developing seeds. Plant Mol Biol 52:1107–1123

Sutica D, Dowson WJ (1962) Bacterial Leaf Spot of *Sesamum* in Yugoslavia. J Phytopathol 45:57–65

Tabatabaei SA, Naghibalghora SM (2014) The effect of salinity stress on germination characteristics and changes of biochemically of sesame seeds. Cercetări Agronomice în Moldova 47(158):61–68

Tarihal R (2003) Study of fertilization barriers in crosses between *Sesamum indicum* and its wild relatives. Ind J Genet 63(2):132–136

Taşkin KM, Ercan AG, Turgut K (1999) *Agrobacterium tumefaciens* mediated transformation of sesame (*Sesamum indicum* L.). Turk J Bot 23:291–295

Taskin KM, Turgut K (1997) In vitro regeneration of sesame (*S. indicum* L.). Turk J Bot 21:15–18

Uzun B, Lee D, Donini P, Çagirgan ML (2003) Identification of a molecular marker linked to the closed capsule mutant trait in sesame using AFLP. Plant Breed 122(1):95–97

Vavdiya PA, Dobariya KL, Babariya CA, Sapovadiya MV (2013) Heterosis for seed yield and its components in sesame (*Sesamum indicum* L.). Electron J Plant Breed 4(3):1246–1250

Verma ML, Mehta N, Sangwan MS (2005) Fungal and bacterial diseases of sesame. In: Saharan GS, Mehta N, Sangwan MS (eds) Diseases of oilseed crops, Indus Publishing, New Delhi, pp 269–300

Wei W, Li D, Wang L, Ding X, Zhang Y et al (2013) Morpho-anatomical and physiological responses to waterlogging of sesame (*Sesamum indicum* L.). Plant Sci 208:102–111

Wei W, Qi X, Wang L, Zhang Y, Hua W et al (2011) Characterization of the sesame (*Sesamum indicum* L.) global transcriptome using Illumina paired-end sequencing and development of EST-SSR markers. BMC Genom 12:451

Were BA, Gudu S, Onkware AO, Carlsson AS, Welander M (2006) *In vitro* regeneration of sesame (*Sesamum indicum* L.) from seedling cotyledon and hypocotyl explants. Plant Cell Tiss Org 85(2):235–239

Williamson KS, Morris JB, Pye QN, Kamat CD, Hensley K et al (2008) A survey of sesamin and composition of tocopherol variability from seeds of eleven diverse sesame (*Sesamum indicum* L.) genotypes using HPLC-PAD-ECD. Phytochem Anal 19(4):311–322

Yadav M, Chaudhary D, Sainger M, Jaiwal PK (2010) *Agrobacterium tumefaciens*-mediated genetic transformation of sesame (*Sesamum indicum* L.) Plant Cell Tiss Org 103(3):377–386

Yasumoto S, Sugiura M, Komaki K, Katsuta M (2005) Change in sesamin and sesamolin contents of sesame (*Sesamum indicum* L.) seeds during maturation and their accurate evaluation of the contents. Japanese. J Crop Sci 74(2):165–171

Yokota T (2007) Sesamin, a lignan of sesame, down-regulates cyclin D1 protein expression in human tumor cells. Cancer Sci 98(9):1447–1453

Yoshida H (2007) Antioxidant distributions and triacylglycerol molecular species of sesame seeds (*Sesamum indicum*). J Am Oil Chem Soc 84:165–172

Yousif YH, Bingham FT, Yermanos DM (1972) Growth, mineral composition, and seed oil of sesame (*Sesamum Indicum* L.) as affected by NaCl. Soil Sci Soc Am J 36(3):450–453

Uzun B, Cağirgan M (2006) Comparison of determinate and indeterminate lines of sesame for agronomic traits. Field Crops Res 96:13–18

Zhang H, Miao H, Wei L, Li C, Zhao R et al (2013a) Genetic analysis and QTL mapping of seed coat color in sesame (*Sesamum indicum* L.). PLoS One 8(5):e63898

Zhang Y, Wang L, Xin H, Li D, Ma C et al (2013b) Construction of a high-density genetic map for sesame based on large scale marker development by specific length amplified fragment (SLAF) sequencing. BMC Plant Biol 13:141

Zhang Y, Wang L, Xin H, Li D, Ma C et al (2013c) Construction of a high-density genetic map for sesame based on large scale marker development by specific length amplified fragment (SLAF) sequencing. BMC Plant Biol 13:141

Index

V.R. Rajpal et al. (eds.), *Gene Pool Diversity and Crop Improvement*,
Sustainable Development and Biodiversity 10,
DOI 10.1007/978-3-319-27096-8

MIX
Papier aus verantwortungsvollen Quellen
Paper from responsible sources
FSC® C105338

If you have any concerns about our products,
you can contact us on
ProductSafety@springernature.com

In case Publisher is established outside the EU,
the EU authorized representative is:
Springer Nature Customer Service Center GmbH
Europaplatz 3, 69115 Heidelberg, Germany

Printed by Libri Plureos GmbH
in Hamburg, Germany